COMMUNICATING PROCESS ARCHITECTURES 2008

Concurrent Systems Engineering Series

Series Editors: M.R. Jane, J. Hulskamp, P.H. Welch, D. Stiles and T.L. Kunii

Volume 66

Previously published in this series:

Volume 65, Communicating Process Architectures 2007 (WoTUG-30), A.A. McEwan,
   S. Schneider, W. Ifill and P.H. Welch
Volume 64, Communicating Process Architectures 2006 (WoTUG-29), P.H. Welch, J. Kerridge
   and F.R.M. Barnes
Volume 63, Communicating Process Architectures 2005 (WoTUG-28), J.F. Broenink,
   H.W. Roebbers, J.P.E. Sunter, P.H. Welch and D.C. Wood
Volume 62, Communicating Process Architectures 2004 (WoTUG-27), I.R. East, J. Martin,
   P.H. Welch, D. Duce and M. Green
Volume 61, Communicating Process Architectures 2003 (WoTUG-26), J.F. Broenink and
   G.H. Hilderink
Volume 60, Communicating Process Architectures 2002 (WoTUG-25), J.S. Pascoe, P.H. Welch,
   R.J. Loader and V.S. Sunderam
Volume 59, Communicating Process Architectures 2001 (WoTUG-24), A. Chalmers, M. Mirmehdi
   and H. Muller
Volume 58, Communicating Process Architectures 2000 (WoTUG-23), P.H. Welch and
   A.W.P. Bakkers
Volume 57, Architectures, Languages and Techniques for Concurrent Systems (WoTUG-22),
   B.M. Cook
Volumes 54–56, Computational Intelligence for Modelling, Control & Automation,
   M. Mohammadian
Volume 53, Advances in Computer and Information Sciences '98, U. Güdükbay, T. Dayar,
   A. Gürsoy and E. Gelenbe
Volume 52, Architectures, Languages and Patterns for Parallel and Distributed Applications
   (WoTUG-21), P.H. Welch and A.W.P. Bakkers
Volume 51, The Network Designer's Handbook, A.M. Jones, N.J. Davies, M.A. Firth and
   C.J. Wright
Volume 50, Parallel Programming and JAVA (WoTUG-20), A. Bakkers
Volume 49, Correct Models of Parallel Computing, S. Noguchi and M. Ota

Transputer and OCCAM Engineering Series

Volume 45, Parallel Programming and Applications, P. Fritzson and L. Finmo
Volume 44, Transputer and Occam Developments (WoTUG-18), P. Nixon
Volume 43, Parallel Computing: Technology and Practice (PCAT-94), J.P. Gray and F. Naghdy
Volume 42, Transputer Research and Applications 7 (NATUG-7), H. Arabnia

ISSN 1383-7575

Communicating Process Architectures 2008

WoTUG-31

Edited by

Peter H. Welch
Computing Laboratory, University of Kent, UK

Susan Stepney
Department of Computer Science, University of York, UK

Fiona A.C. Polack
Department of Computer Science, University of York, UK

Frederick R.M. Barnes
Computing Laboratory, University of Kent, UK

Alistair A. McEwan
Department of Engineering, University of Leicester, UK

Gardiner S. Stiles
Electrical and Computer Engineering, Utah State University, USA

Jan F. Broenink
Department of Electrical Engineering, University of Twente, the Netherlands

and

Adam T. Sampson
Computing Laboratory, University of Kent, UK

Proceedings of the 31st WoTUG Technical Meeting,
7–10 September 2008, University of York,
York, UK

IOS Press

Amsterdam • Berlin • Oxford • Tokyo • Washington, DC

© 2008 The authors and IOS Press.

All rights reserved. No part of this book may be reproduced, stored in a retrieval system, or transmitted, in any form or by any means, without prior written permission from the publisher.

ISBN 978-1-58603-907-3
Library of Congress Control Number: 2008907202

Publisher
IOS Press
Nieuwe Hemweg 6B
1013 BG Amsterdam
Netherlands
fax: +31 20 687 0019
e-mail: order@iospress.nl

Distributor in the UK and Ireland
Gazelle Books Services Ltd.
White Cross Mills
Hightown
Lancaster LA1 4XS
United Kingdom
fax: +44 1524 63232
e-mail: sales@gazellebooks.co.uk

Distributor in the USA and Canada
IOS Press, Inc.
4502 Rachael Manor Drive
Fairfax, VA 22032
USA
fax: +1 703 323 3668
e-mail: iosbooks@iospress.com

LEGAL NOTICE

The publisher is not responsible for the use which might be made of the following information.

PRINTED IN THE NETHERLANDS

Preface

The thirty-first *Communicating Process Architectures Conference*, CPA 2008, organised under the auspices of WoTUG and the Department of Computer Science of the University of York, is being held in York, UK, 7–10 September 2008.

The York Department of Computer Science is delighted to host this conference again. The thirteenth *Occam User Group Technical Meeting* (from which these CPA conferences are descended) on *"Real-Time Systems with Transputers"* was held at this University in September, 1990. York has a long history of close contacts with this conference, from the early heady days when occam was new, to current collaborations with the University of Kent using occam-π for simulating complex systems. The satellite workshop *"Complex Systems Modelling and Simulation (CoSMoS)"*, arising from this collaboration, is part of this year's conference.

The city of York, one of Europe's most beautiful cities, combines evidence of a history going back to Roman times with a bustling modern city centre. York Minster, built on the foundations of the Roman city and an earlier Norman cathedral, is among the finest Gothic cathedrals, and dominates the city. Romans, Vikings, and more recent history are commemorated in a number of top-class museums, as well as being apparent in the architecture of the city.

We are delighted to have two excellent invited speakers, covering both theoretical aspects and industrial applications of Communicating Processes. Professor Samson Abramsky, FRS, is the *Christopher Strachey Professor of Computing* at the University of Oxford, where he leads the Theory and Automated Verification group. He has worked in the areas of semantics and logic of computation, and concurrency. His work on game semantics considers interaction and information flow between multiple agents and their environment. This has yielded new approaches to compositional model-checking and to analysis for programs with state, concurrency, probability and other features. Professor Colin O'Halloran is the head of the *Systems Assurance Group* at QinetiQ. He has been instrumental in the uptake of formal methods in the development and verification of high assurance systems on an industrial scale. His research interests are in automating the use of formal methods, and using these techniques at reasonable cost and on an industrial scale.

This conference and workshop were partially supported by AWE, EPSRC, Microsoft Research, and WoTUG. The editors would like to thank all the paper reviewers for the detailed feedback they provided the authors, the authors for their diligence in responding well to (sometimes harsh) criticism, the staff at the University of York – especially Bob French and Jenny Baldry – for the website and local arrangements – and, finally, to those individuals at the Universities of Kent and Tromsø – especially Carl Ritson, Neil Brown, John Markus Bjørndalen and Jon Simpson – without whom these Proceedings would not have made their press deadline!

Peter Welch (*University of Kent*), Susan Stepney (*University of York*),
Fiona Polack (*University of York*), Frederick Barnes (*University of Kent*),
Alistair McEwan (*University of Leicester*), Dyke Stiles (*Utah State University*),
Jan Broenink (*University of Twente*), Adam Sampson (*University of Kent*).

Editorial Board

Prof. Peter H. Welch, *University of Kent, UK (Chair)*
Prof. Susan Stepney, *University of York, UK*
Dr. Fiona A.C. Polack, *University of York, UK*
Dr. Frederick R.M. Barnes, *University of Kent, UK*
Dr. Alistair A. McEwan, *University of Leicester, UK*
Prof. Gardiner (Dyke) Stiles, *Utah State University, USA*
Dr. Jan F. Broenink, *University of Twente, The Netherlands*
Mr. Adam T. Sampson, *University of Kent, UK*

Reviewing Committee

Dr. Alastair R. Allen, *Aberdeen University, UK*
Mr. Paul S. Andrews, *University of York, UK*
Dr. Iain Bate, *University of York, UK*
Dr. John Markus Bjørndalen, *University of Tromsø, Norway*
Dr. Phil Brooke, *University of Teesside, UK*
Mr. Neil C.C. Brown, *University of Kent, UK*
Dr. Radu Calinescu, *University of Oxford, UK*
Mr. Kevin Chalmers, *Napier University, UK*
Dr. Barry Cook, *4Links Ltd., UK*
Dr. Ian East, *Oxford Brookes University, UK*
Dr. José L. Fiadeiro, *University of Leicester, UK*
Dr. Bill Gardner, *University of Guelph, Canada*
Dr. Michael Goldsmith, *Formal Systems (Europe) Ltd., UK*
Mr. Marcel Groothuis, *University of Twente, The Netherlands*
Mr. Roberto Guanciale, *University of Leicester, UK*
Dr. Matthew Huntbach, *Queen Mary, University of London, UK*
Mr. Jason Hurt, *University of Nevada, USA*
Dr. Wilson Ifill, *AWE, UK*
Ms. Ruth Ivimey-Cook, *Creative Business Systems Ltd., UK*
Dr. Jeremy Jacob, *University of York, UK*
Prof. Jon Kerridge, *Napier University, UK*
Dr. Albert M. Koelmans, *Newcastle University, UK*
Dr. Adrian E. Lawrence, *University of Loughborough, UK*
Dr. Jeremy M.R. Martin, *GlaxoSmithKline, UK*
Dr. Richard Paige, *University of York, UK*
Dr. Jan B. Pedersen, *University of Nevada, USA*
Dr. Mike Poppleton, *University of Southampton, UK*
Mr. Carl G. Ritson, *University of Kent, UK*
Ir. Herman Roebbers, *TASS Software Professionals, The Netherlands*
Prof. Steve Schneider, *University of Surrey, UK*
Mr. Zheng Shi, *University of York, UK*
Mr. Jon Simpson, *University of Kent, UK*
Dr. Marc L. Smith, *Vassar College, USA*
Mr. Bernard Sufrin, *University of Oxford*
Peter Vidler, *TTE Systems Ltd., UK*
Dr. Brian Vinter, *University of Copenhagen, Denmark*
Mr. Paul Walker, *4Links Ltd., UK*
Dr. Russel Winder, *Concertant LLP, UK*
Prof. Jim Woodcock, *University of York, UK*
Mr. Fengxiang Zhang, *University of York, UK*

Contents

Preface   v
    *Peter Welch, Susan Stepney, Fiona Polack, Frederick Barnes,
Alistair McEwan, Dyke Stiles, Jan Broenink and Adam Sampson*

Editorial Board   vi

Reviewing Committee   vii

Part A. Invited Speakers

Types, Orthogonality and Genericity: Some Tools for Communicating Process Architectures   1
    Samson Abramsky

How to Soar with CSP   15
    Colin O'Halloran

Part B. Conference Papers

A CSP Model for Mobile Channels   17
    Peter H. Welch and Frederick R.M. Barnes

Communicating Scala Objects   35
    Bernard Sufrin

Combining EDF Scheduling with occam Using the Toc Programming Language   55
    Martin Korsgaard and Sverre Hendseth

Communicating Haskell Processes: Composable Explicit Concurrency Using Monads   67
    Neil C.C. Brown

Two-Way Protocols for occam-π 85
    Adam T. Sampson

Prioritized Service Architecture: Refinement and Visual Design   99
    Ian R. East

Experiments in Translating CSP||B to Handel-C   115
    Steve Schneider, Helen Treharne, Alistair McEwan and Wilson Ifill

FPGA Based Control of a Production Cell System   135
    Marcel A. Groothuis, Jasper J.P. van Zuijlen and Jan F. Broenink

Shared-Clock Methodology for Time-Triggered Multi-Cores   149
    Keith F. Athaide, Michael J. Pont and Devaraj Ayavoo

Transfer Request Broker: Resolving Input-Output Choice   163
    Oliver Faust, Bernhard H.C. Sputh and Alastair R. Allen

Mechanical Verification of a Two-Way Sliding Window Protocol   179
Bahareh Badban, Wan Fokkink and Jaco van de Pol

RRABP: Point-to-Point Communication over Unreliable Components   203
Bernhard H.C. Sputh, Oliver Faust and Alastair R. Allen

IC2IC: a Lightweight Serial Interconnect Channel for Multiprocessor Networks   219
Oliver Faust, Bernhard H.C. Sputh and Alastair R. Allen

Asynchronous Active Objects in Java   237
George Oprean and Jan B. Pedersen

JCSPre: the Robot Edition to Control LEGO NXT Robots   255
Jon Kerridge, Alex Panayotopoulos and Patrick Lismore

A Critique of JCSP Networking   271
Kevin Chalmers, Jon Kerridge and Imed Romdhani

Virtual Machine Based Debugging for occam-π 293
Carl G. Ritson and Jonathan Simpson

Process-Oriented Collective Operations   309
John Markus Bjørndalen and Adam T. Sampson

Representation and Implementation of CSP and VCR Traces   329
Neil C.C. Brown and Marc L. Smith

CSPBuilder – CSP Based Scientific Workflow Modelling   347
Rune Møllegård Friborg and Brian Vinter

Visual Process-Oriented Programming for Robotics   365
Jonathan Simpson and Christian L. Jacobsen

Solving the Santa Claus Problem: a Comparison of Various Concurrent Programming Techniques   381
Jason Hurt and Jan B. Pedersen

Mobile Agents and Processes Using Communicating Process Architectures   397
Jon Kerridge, Jens-Oliver Haschke and Kevin Chalmers

YASS: a Scaleable Sensornet Simulator for Large Scale Experimentation   411
Jonathan Tate and Iain Bate

Modelling a Multi-Core Media Processor Using JCSP   431
Anna Kosek, Jon Kerridge and Aly Syed

Part C. Fringe Presentation Abstracts

How to Make a Process Invisible   445
Neil C.C. Brown

Designing Animation Facilities for gCSP   447
Hans T.J. van der Steen, Marcel A. Groothuis and Jan F. Broenink

Tock: One Year On   449
Adam T. Sampson and Neil C.C. Brown

Introducing JCSP Networking 2.0   451
Kevin Chalmers

Mobile Processes in an Ant Simulation   453
Eric Bonnici

Santa Claus – with Mobile Reindeer and Elves   455
Peter H. Welch and Jan B. Pedersen

Subject Index   457

Author Index   459

Types, Orthogonality and Genericity: Some Tools for Communicating Process Architectures

Samson ABRAMSKY

Oxford University Computing Laboratory,
Wolfson Building, Parks Road, Oxford OX1 3QD, U.K.

samson@comlab.ox.ac.uk

Abstract. We shall develop a simple and natural formalization of the idea of *client-server architectures*, and, based on this, define a notion of *orthogonality* between clients and servers, which embodies strong correctness properties, and exposes the rich logical structure inherent in such systems. Then we generalize from pure clients and servers to *components*, which provide some services to the environment, and require others from it. We identify the key notion of *composition* of such components, in which some of the services required by one component are supplied by another. This allows complex systems to be built from ultimately simple components. We show that this has the logical form of the *Cut rule*, a fundamental principle of logic, and that it can be enriched with a suitable notion of *behavioural types* based on orthogonality, in such a way that correctness properties are preserved by composition. We also develop the basic ideas of how logical constructions can be used to develop *structured interfaces* for systems, with operations corresponding to logical rules. Finally, we show how the setting can be enhanced, and made more robust and expressive, by using *names* (as in the π-calculus) to allow clients to bind dynamically to generic instances of services.

Keywords. client, server, type, orthogonality, genericity, interaction.

Introduction

Concurrent programming has been a major topic of study in Computer Science for the past four decades. It is coming into renewed prominence currently, for several reasons:

- As the remit of Moore's law finally runs out, further performance increases must be sought from multi-core architectures.
- The spread of web-based applications, and of mobile, global and ubiquitous computing, is making programming with multiple threads, across machine boundaries, and interacting with other such programs, the rule rather than the exception.

These trends raise major questions of *programmability*: how can high-quality software be produced reliably in these demanding circumstances? Communicating process architectures, which offer general forms, templates and structuring devices for building complex systems out of concurrent, communicating components, have an important part to play in addressing this issue.

Hot Topics and Timeless Truths

Our aim in this paper is to explore some general mechanisms and structures which can be used for building communicating process architectures. We shall show how ideas from logic and semantics can play a very natural rôle here.

The paper is aimed at the broad CPA community with some prior exposure to formal methods, so we shall try to avoid making the technical development too heavy, and emphasize the underlying intuitions.

Our plan is as follows:

- We shall develop a simple and natural formalization of the idea of *client-server architectures*, and, based on this, define a notion of *orthogonality* between clients and servers, which embodies strong correctness properties, and exposes the rich logical structure inherent in such systems.
- Then we generalize from pure clients and servers to *components*, which provide some services to the environment, and require others from it. We identify the key notion of *composition* of such components, in which some of the services required by one component are supplied by another. This allows complex systems to be built from ultimately simple components. We show that this has the logical form of the *Cut rule*, a fundamental principle of logic, and that it can be enriched with a suitable notion of *behavioural types* based on orthogonality, in such a way that correctness properties are preserved by composition.
- We also develop the basic ideas of how logical constructions can be used to develop *structured interfaces* for systems, with operations corresponding to logical rules.
- Finally, we show how the setting can be enhanced, and made more robust and expressive, by using *names* (as in the π-calculus) to allow clients to bind dynamically to generic instances of services.

We shall build extensively on previous work by the author and his colleagues, and others. Some references are provided in the final section of the paper. Our main aim in the present paper is to find a simple and intuitive level of presentation, in which the formal structures flow in a natural and convincing fashion from the intuitions and the concrete setting of communicating process architectures.

1. Background: Clients and Servers

Our basic structuring paradigm is quite natural and familiar: we shall view interactions as occurring between *clients* and *servers*. A server is a process which offers services to its environment; a client makes uses of services.[1] Moreover, we will assume that services are stylized into a *procedural interface*: a service interface is structured into a number of *methods*, and each use of a method is initiated by a *call*, and terminated by a *return*. Thus service methods are very much like methods in object-oriented programming; and although we shall not develop an extensive object-oriented setting for the ideas we shall explore here, they would certainly sit comfortably in such a setting. However, our basic premise is simply that client-server interactions are structured into procedure-like call-return interfaces.

---

[1] At a later stage, we shall pursue the obvious thought that in a sufficiently global perspective, a given process may be seen as both a client and a server in various contexts; but it will be convenient to start with a simple-minded but clear distinction between the two, which roughly matches the standard network architecture concept.

1.1. Datatypes and Signatures

Each service method call and return will have some associated parameters. We assume a standard collection of basic data types, *e.g.* **int**, **bool**. The type of a service method m will then have the form

$$m : A_1 \times \cdots \times A_n \longrightarrow B_1 \times \cdots \times B_m$$

where each A_i and B_j is a basic type. We allow $n = 0$ or $m = 0$, and write **unit** for the empty product, which we can think of as a standard one-element type.

Examples

- A `integer counter` service has a method

    $$\text{inc} : \textbf{unit} \longrightarrow \textbf{nat}$$

    It maintains a local counter; each time the method is invoked, the current value of the counter is returned, and the counter is incremented.
- A `stack` service has methods

    $$\begin{aligned}\text{push} &: D \longrightarrow \textbf{unit} \\ \text{pop} &: \textbf{unit} \longrightarrow \textbf{unit} \\ \text{top} &: \textbf{unit} \longrightarrow D\end{aligned}$$

A *signature* Σ is a collection of such named, typed methods:

$$\Sigma = \{m_1 : T_1, \ldots, m_k : T_k\}.$$

A signature defines the interface presented by a service or collection of services to an environment of potential clients.

A more refined view of such an interface would distinguish a number of service types, each with its own collection of methods. This would be analogous to introducing *classes* as in object-oriented programming. We shall refrain from introducing this refined structure, since the notions we wish to explore can already be defined at the level of "flat" collections of methods.

1.2. Client-Server Interactions

Although we have written service method types in a convenient functional form

$$m : A_1 \times \cdots \times A_n \longrightarrow B_1 \times \cdots \times B_m$$

a method interaction really consists of two separate *events*, each of which consists of a *synchronized interaction* between a client and server. From the point of view of a server offering a method m, it will offer a call $m^{\downarrow}(x_1, \ldots, x_n)$ to the environment. Here the variables $x_1, \ldots, x_n$ will be bound by the call, and used in the subsequent computation to perform the service method. A client performing a method call on m is trying to perform an action $m^{\uparrow}(v_1, \ldots, v_n)$ with actual values for the parameters. The method is actually invoked when such a call by the client synchronizes with a matching call offer by the server.

Dually, a service return involves the synchronization of an action $m_{\downarrow}(w_1, \ldots, w_m)$ by the server with an action $m_{\uparrow}(y_1, \ldots, y_m)$ by the client, which binds the values returned by the server for future use in the client.

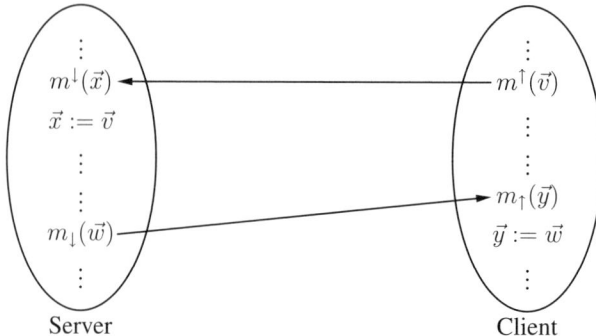

Whereas in a sequential program, the caller simply waits after the call for the return, and dually the subroutine simply waits to be invoked, here we are assuming a concurrent world where multiple threads are running in both the server and the client. We should really think of the 'server' and 'client' as parallel compositions

$$s = \|_{i \in I} s_i, \qquad c = \|_{j \in J} c_j$$

where each s_i is offering one or more service methods. Thus multiple methods are being offered concurrently by the servers, and are being called concurrently by the clients.

To proceed further, we need some formal representation of processes. We shall not need to delve into the details of process algebra, nor do we need to commit ourselves to a particular choice from the wide range of process models and equivalences which have been proposed. Nevertheless, for definiteness we shall assume that the behaviour of processes is specified by *labelled transition systems* [21], which can fairly be described as the workhorses of concurrency theory.

Recall that a labelled transition system on a set of labels $\mathcal{L}$ is a structure (Q, R) where Q is a set of states, and $R \subseteq Q \times \mathcal{L} \times Q$ is the transition relation. We write $p \xrightarrow{a} q$ for $(p, a, q) \in R$. We assume, in a standard fashion, that $\mathcal{L}$ contains a *silent* or *unobservable* action τ [21]; transitions $p \xrightarrow{\tau} q$ represent internal computational steps of a system, without reference to its environment.

Now fix a server signature Σ. We define

$$\Sigma_s = \{m^\downarrow(\vec{x}), m_\downarrow(\vec{w}) \mid m \in \Sigma\}$$

the set of server-side actions over Σ. Dually, we define

$$\Sigma_c = \{m^\uparrow(\vec{v}), m_\uparrow(\vec{y}) \mid m \in \Sigma\}$$

the set of client-side actions over Σ.

A Σ-*server* is a process defined by a labelled transition system over a set of labels $\mathcal{L}$ such that $\Sigma_s \subseteq \mathcal{L}$ and $\Sigma_c \cap \mathcal{L} = \varnothing$. A Σ-*client* is a process defined by a labelled transition system over a set of labels $\mathcal{M}$ such that $\Sigma_c \subseteq \mathcal{M}$ and $\Sigma_s \cap \mathcal{L} = \varnothing$.

We shall now define an operation which expresses how a client interacts with a server. Let s be a Σ-server, and c a Σ-client. We define the behaviour of $s \triangleleft_\Sigma c$ by specifying its labelled transitions. Firstly, we specify that for actions outside the interface Σ, s and c proceed concurrently, without interacting with each other:

$$\frac{s \xrightarrow{\lambda} s'}{s \triangleleft_\Sigma c \xrightarrow{\lambda} s' \triangleleft_\Sigma c} \ (\lambda \notin \Sigma_s) \qquad \frac{c \xrightarrow{\lambda} c'}{s \triangleleft_\Sigma c \xrightarrow{\lambda} s \triangleleft_\Sigma c'} \ (\lambda \notin \Sigma_c)$$

Then we give the expected rules describing the interactions corresponding to method calls and returns:

$$\frac{s \xrightarrow{m^{\downarrow}(\vec{x})} s' \quad c \xrightarrow{m^{\uparrow}(\vec{v})} c'}{s \triangleleft_{\Sigma} c \xrightarrow{m^{\Downarrow}(\vec{x}:\vec{v})} s'[\vec{v}/\vec{x}] \triangleleft_{\Sigma} c'} \qquad \frac{s \xrightarrow{m_{\downarrow}(\vec{w})} s' \quad c \xrightarrow{m_{\uparrow}(\vec{y})} c'}{s \triangleleft_{\Sigma} c \xrightarrow{m_{\Uparrow}(\vec{w}:\vec{y})} s' \triangleleft_{\Sigma} c'[\vec{w}/\vec{y}]}$$

Here $s'[\vec{v}/\vec{x}]$ denotes the result of binding the values $\vec{v}$ to the variables $\vec{x}$ in the continuation process s', and similarly for $c'[\vec{w}/\vec{y}]$.

2. Orthogonality

Intuitively, there is an obvious *duality* between clients and servers; and where there is duality, there should be some *logical structure*. We now take a first step towards articulating this structure, which we shall use to encapsulate the key *correctness properties* of a client-server system.

We fix a server signature Σ. Let s be a Σ-server, and c a Σ-client. A *computation* of the server-client system $s \triangleleft_{\Sigma} c$ is a sequence of transitions

$$s \triangleleft_{\Sigma} c \xrightarrow{\lambda_1} s_1 \triangleleft_{\Sigma} c_1 \xrightarrow{\lambda_2} \cdots \xrightarrow{\lambda_k} s_k \triangleleft_{\Sigma} c_k.$$

We shall only consider finite computations.

For every such computation, we have an associated *observation*, which is obtained by concatenating the labels $\lambda_1 \cdots \lambda_k$ on the transitions. For each method $m \in \Sigma$, we obtain an *m-observation* by erasing all labels in this sequence other than those of the form $m^{\Downarrow}(\vec{x}:\vec{v})$ and $m_{\Uparrow}(\vec{w}:\vec{y})$, and replacing these by $m^{\Downarrow}$ and $m_{\Uparrow}$ respectively.

We say that $s \triangleleft_{\Sigma} c$ is Σ-*active* if for some computation starting from $s \triangleleft_{\Sigma} c$, and some $m \in \Sigma$, the corresponding m-observation is non-empty. A similar notion applies to clients c considered in isolation.

We say that *s is orthogonal to c*, written $s \perp c$, if for every computation

$$s \triangleleft_{\Sigma} c \xrightarrow{\lambda_1} \cdots \xrightarrow{\lambda_k} s_k \triangleleft_{\Sigma} c_k$$

the following conditions hold:

Σ-**receptiveness.** If c_k is Σ-active, then so is $s_k \triangleleft_{\Sigma} c_k$.

Σ-**completion.** The computation can be extended to

$$s \triangleleft_{\Sigma} c \xrightarrow{\lambda_1} \cdots \xrightarrow{\lambda_k} s_k \triangleleft_{\Sigma} c_k \cdots \xrightarrow{\lambda_{k+l}} s_{k+l} \triangleleft_{\Sigma} c_{k+l}$$

with $l \geq 0$, whose m-observation o_m, for each $m \in \Sigma$, is a sequence of alternating calls and returns of m:

$$o_m \in (m^{\Downarrow} m_{\Uparrow})^*.$$

This notion of orthogonality combines several important correctness notions, "localized" to the client-server interface:

- It incorporates a notion of *deadlock-freedom*. Σ-receptiveness ensures that whenever the client wishes to proceed by calling a service method, a matching offer of such a service must eventually be available from the server. Σ-completion ensures that every method execution which is initiated by a call must be completed by the corresponding return. Note that, since this condition is applied along all computations, it takes account of non-deterministic branching by the server and the client; no matter how

the non-deterministic choices are resolved, we cannot get into a state where either the client or the server get blocked and are unable to participate in the method return.
- There is also an element of *livelock-freedom*. It ensures that we cannot get indefinitely delayed by other concurrent activities from completing a method return. Note that our definition in terms of extensions of finite computations in effect incorporates a *strong fairness assumption*; that transitions which are enabled infinitely often must eventually be performed. An alternative treatment could be given in terms of maximal (finite or infinite) computations, subject to explicit fairness constraints.
- Finally, notice that this notion of orthogonality also incorporates a *serialization requirement*: once a given method has been invoked, no further offers of that method will be made until the invocation has been completed by the method return. The obvious way of achieving this is to have each method implementation running as a single sequential thread within the server. This may seem overly restrictive, but we shall see later how with a suitable use of *names* as unique identifiers for multiple instances of generic service methods, this need not compromise expressiveness.

2.1. The Galois Connection

We shall now see how the orthogonality relation allows further structure to be expressed. Let $\mathcal{S}$ be the set of Σ-servers over a fixed ambient set of labels $\mathcal{L}$, and $\mathcal{C}$ the set of Σ-clients over a set of labels $\mathcal{M}$. We have defined a relation $\perp \subseteq \mathcal{S} \times \mathcal{C}$. This relation can be used in a standard way to define a *Galois connection* [13] between $\mathbb{P}(\mathcal{S})$ and $\mathbb{P}(\mathcal{C})$.

Given $S \subseteq \mathcal{S}$, define

$$S^\perp = \{c \in \mathcal{C} \mid \forall s \in S.\, s \perp c\}$$

Similarly, given $C \subseteq \mathcal{C}$, define

$$C^\perp = \{s \in \mathcal{S} \mid \forall c \in C.\, s \perp c\}$$

The following is standard [13]:

Proposition 2.1
1. For all $S, T \subseteq \mathcal{S}$: $S \subseteq T \Rightarrow T^\perp \subseteq S^\perp$.
2. For all $C, D \subseteq \mathcal{C}$: $C \subseteq D \Rightarrow D^\perp \subseteq C^\perp$.
3. For all $C \subseteq \mathcal{C}$ and $S \subseteq \mathcal{S}$: $S^\perp \supseteq C \iff S \subseteq C^\perp$.
4. For all $C \subseteq \mathcal{C}$ and $S \subseteq \mathcal{S}$: $S^\perp = S^{\perp\perp\perp}$ and $C^\perp = C^{\perp\perp\perp}$.

We define $\bar{S} = S^{\perp\perp}$, $\bar{C} = C^{\perp\perp}$. The following is also standard:

Proposition 2.2 *The mappings $S \mapsto \bar{S}$ and $C \mapsto \bar{C}$ are closure operators, on $\mathbb{P}(\mathcal{S})$ and $\mathbb{P}(\mathcal{C})$ respectively. That is:*

$$S \subseteq \bar{S}, \quad \bar{\bar{S}} = \bar{S}, \quad S \subseteq T \Rightarrow \bar{S} \subseteq \bar{T}$$

and similarly

$$C \subseteq \bar{C}, \quad \bar{\bar{C}} = \bar{C}, \quad C \subseteq D \Rightarrow \bar{C} \subseteq \bar{D}.$$

We can think of closed sets of servers as *saturated* under "tests" by clients: $\bar{S}$ is the largest set of servers which passes all the tests of orthogonality under interaction with clients that the servers in S do. Similarly for closed sets of clients. We shall eventually view such closed sets as *behavioural types*.

2.2. Examples

We shall give some simple examples to illustrate the definitions. We shall use the notation of CCS [21], but we could just as well have used CSP [16], or other process calculi.

Firstly, we define a signature with two operations:

$$\text{inc} : \text{unit} \longrightarrow \text{unit}$$
$$\text{dec} : \text{unit} \longrightarrow \text{unit}$$

This specifies an interface for a binary semaphore. We define a server $s \equiv \texttt{binsem}(0)$, where:

$$\texttt{binsem}(0) = \text{inc}^{\downarrow}.\text{inc}_{\downarrow}.\texttt{binsem}(1)$$
$$\texttt{binsem}(1) = \text{dec}^{\downarrow}.\text{dec}_{\downarrow}.\texttt{binsem}(0)$$

We define a client c by

$$c \equiv \text{inc}^{\uparrow} \,|\, \text{inc}_{\uparrow} \,|\, \text{dec}^{\uparrow} \,|\, \text{dec}_{\uparrow}.$$

Then s is orthogonal to c. However, $\text{dec}^{\uparrow}.c$ is not orthogonal to s, since Receptivity fails; while $\text{inc}_{\downarrow}.s$ is not orthogonal to c, since Completion fails.

3. The Open Universe

Thus far, we have assumed a simple-minded but convenient dichotomy between clients and servers. There are many reasons to move to a more general setting. The most compelling is that the general condition of systems is to be *open*, interacting with an environment which is not completely specified, and themselves changing over time. Rather than thinking of complete systems, we should think in terms of *system components*.

Relating this to our service-based view of distributed computation, it is natural to take the view that the general case of a system component is a process which:

1. *requires* certain services to be provided to it by its environment
2. *provides* other services to its environment.

Such a component is a *client* with respect to the services in (1), and a *server* with respect to the services in (2). The interface of such a component has the form

$$\Sigma \Longrightarrow \Delta$$

where Σ and Δ are server signatures as already defined; we assume that $\Sigma \cap \Delta = \varnothing$. The idea is that Σ describes the interface of the component to its environment *qua* client, and Δ the interface *qua* server.

A process p has this interface type, written $p :: \Sigma \Longrightarrow \Delta$, if it is described by a labelled transition system on the set of labels

$$\Sigma_c \cup \Delta_s \cup \{\tau\}.$$

Here Σ_c is the client-side set of actions of Σ, and Δ_s the server-side set of actions of Δ, as previously defined, while τ is the usual silent action for internal computational steps. Note that this interface type is now regarded as exhaustive of the possible actions of p. Note also that a *pure Δ-server* is a process $s :: \Longrightarrow \Delta$ with empty client-side signature, while a *pure Σ-client* is a process $c :: \Sigma \Longrightarrow$ with empty server-side signature. Thus servers and clients as discussed previously are special cases of the notion of component.

3.1. Composition

We now look at a fundamental operation for such systems: composition, *i.e.* plugging one system into another across its specified interface. It is composition which enables the construction of complex systems from ultimately simple components.

We shall write interface types in a partitioned form:

$$p :: \Sigma_1, \ldots \Sigma_k \Longrightarrow \Delta.$$

This is equivalent to $p :: \Sigma \Longrightarrow \Delta$, where $\Sigma = \bigcup_{i=1}^{k} \Sigma_i$. Writing the signature in this form implicitly assumes that the Σ_i are *pairwise disjoint*. More generally, we assume from now on that all signatures named by distinct symbols are pairwise disjoint.

Suppose that we have components

$$p :: \Delta' \Longrightarrow \Sigma, \qquad q :: \Sigma, \Delta'' \Longrightarrow \Theta.$$

Then we can form the system $p \triangleleft_\Sigma q$. Note that our definition of $s \triangleleft_\Sigma c$ in Section 1 was sufficiently general to cover this situation, taking $s = p$ and $c = q$, with $\mathcal{L} = \Delta'_c \cup \Sigma_s \cup \{\tau\}$, and $\mathcal{M} = \Sigma_c \cup \Delta''_c \cup \Theta_s \cup \{\tau\}$.

We now define

$$p \odot_\Sigma q = (p \triangleleft_\Sigma q) \backslash \Sigma.$$

The is the process obtained from $p \triangleleft_\Sigma q$ by replacing all labels $m^{\Uparrow}(\vec{x} : \vec{v})$ and $m_{\Uparrow}(\vec{w} : \vec{y})$, with $m \in \Sigma$, by τ. This internalizes the interaction between p and q, in which p provides the services in Σ required by q.

Proposition 3.1 *The process $p \odot_\Sigma q$ satisfies the following interface specification:*

$$p \odot_\Sigma q :: \Delta', \Delta'' \Longrightarrow \Theta.$$

The reader familiar with the sequent calculus formulation of logic will recognize the analogy with the *Cut rule*:

$$\frac{\Gamma \Longrightarrow A \quad A, \Gamma' \Longrightarrow B}{\Gamma, \Gamma' \Longrightarrow B}$$

This correspondence is no accident, and we will develop a little more of the connection to logic in the next section, although we will not be able to explore it fully here.

3.2. Behavioural Types

We now return to the idea of behavioural types which we discussed briefly in the previous Section. Given a signature Σ, we define

$$\mathcal{C}(\Sigma) = \{c \mid c :: \Sigma \Longrightarrow\}, \qquad \mathcal{S}(\Sigma) = \{s \mid s ::\Longrightarrow \Sigma\}$$

the sets of pure Σ-clients and servers. We define an orthogonality relation

$$\perp_\Sigma \subseteq \mathcal{C}(\Sigma) \times \mathcal{S}(\Sigma)$$

just as in Section 1, and operations $(\cdot)^\perp$ on sets of servers and clients. We use the notation $A[\Sigma]$ for a *behavioural type of servers on* Σ, i.e. a set $A \subseteq \mathcal{S}(\Sigma)$ such that $A = A^{\perp\perp}$.

Given a component $p :: \Sigma \Longrightarrow \Delta$ and behavioural types $A[\Sigma]$ and $B[\Delta]$, we define $p :: A[\Sigma] \Longrightarrow B[\Delta]$ if the following condition holds:

$$\forall s \in A.\, s \odot_\Sigma p \in B.$$

This is the condition for a component to satisfy a behavioural type specification, guaranteeing important correctness properties across its interface.

Proposition 3.2 *The following are equivalent:*

1. $p :: A[\Sigma] \Longrightarrow B[\Delta]$
2. $\forall s \in A, c \in B^\perp.\, (s \odot_\Sigma p) \perp_\Delta c$

3. $\forall s \in A, c \in B^\perp.\ s \perp_\Sigma (p \odot_\Delta c)$.

These properties suggest how $(\cdot)^\perp$ plays the rôle of a logical negation. For example, one can think of property (3) as a form of *contraposition*: it says that if c is in $B^\perp$, then $p \odot_\Delta c$ is in $A^\perp$. Note that $s \odot_\Sigma p$ is a Δ-server, while $p \odot_\Delta c$ is a Σ-client, so these expressions are well-typed.

3.3. Soundness of Composition

We now come to a key point: *the soundness of composition with respect to behavioural type specifications*. This means that the key correctness properties across interfaces are preserved in a compositional fashion as we build up a complex system by plugging components together.

Firstly, we extend the definition of behavioural type specification for components to cover the case of partitioned inputs. We define $p :: A_1[\Sigma_1], \ldots, A_k[\Sigma_k] \Longrightarrow B[\Delta]$ if:

$$\forall s_1 \in A_1, \ldots, s_k \in A_k.\ (\|_{i=1}^{k} s_i) \odot_\Sigma p \in B.$$

Here $\|_{i=1}^{k} s_i$ is the parallel composition of the s_i; note that, since the Σ_i are disjoint, these processes cannot communicate with each other, and simply interleave their actions freely.

Now we can state our key soundness property for composition:

Proposition 3.3 *Suppose we have* $p :: A'[\Delta'] \Longrightarrow B[\Sigma]$ *and* $q :: B[\Sigma], A''[\Delta''] \Longrightarrow C[\Theta]$. *Then*

$$p \odot_\Sigma q :: A'[\Delta'], A''[\Delta''] \Longrightarrow C[\Theta].$$

This can be formulated as an inference rule:

$$\frac{p :: A'[\Delta'] \Longrightarrow B[\Sigma] \qquad q :: B[\Sigma], A''[\Delta''] \Longrightarrow C[\Theta]}{p \odot_\Sigma q :: A'[\Delta'], A''[\Delta''] \Longrightarrow C[\Theta]}$$

4. Structured Types and Behavioural Logic

We shall now develop the logical structure inherent in this perspective on communicating process architectures a little further.

We shall consider some ways of combining types. The first is very straightforward. Suppose we have behavioural types $A[\Sigma]$ and $B[\Delta]$ (where, as always, we are assuming that Σ and Δ are disjoint). We define the new type

$$A[\Sigma] \otimes B[\Delta] = \{s \parallel t \mid s \in A \text{ and } t \in B\}^{\perp\perp}$$

over the signature $\Sigma \cup \Delta$.

Now we can see the partitioned-input specification

$$p :: A_1[\Sigma_1], \ldots, A_k[\Sigma_k] \longrightarrow \Delta$$

as equivalent to

$$p :: A_1[\Sigma_1] \otimes \cdots \otimes A_k[\Sigma_k] \Longrightarrow \Delta.$$

Moreover, we have the "introduction rule":

$$\frac{p :: A[\Sigma] \Longrightarrow B[\Delta], \qquad q :: A'[\Sigma'] \Longrightarrow B'[\Delta']}{p \parallel q :: A[\Sigma], A'[\Sigma'] \Longrightarrow B[\Delta] \otimes B'[\Delta']}$$

Thus $\otimes$ allows us to combine disjoint collections of services, so as to merge independent components into a single system.

More interestingly, we can form an implication type

$$A[\Sigma] \multimap B[\Delta].$$

This describes the type of components which need a server of type A in order to produce a server of type B:

$$A[\Sigma] \multimap B[\Delta] = \{p :: \Sigma \Longrightarrow \Delta \mid \forall s \in A.\, s \odot_\Sigma p \in B\}.$$

We can define an introduction rule:

$$\frac{p :: A[\Sigma], B[\Delta] \Longrightarrow C[\Theta]}{p :: A[\Sigma] \Longrightarrow B[\Delta] \multimap C[\Theta]}$$

and an elimination rule

$$\frac{C[\Theta] \Longrightarrow A[\Sigma] \multimap B[\Delta], \quad q :: C'[\Theta'] \Longrightarrow A[\Sigma]}{q \odot_\Sigma p :: C[\Theta], C'[\Theta'] \Longrightarrow B[\Delta]}$$

These correspond to abstraction and application in the (linear) λ-calculus [15].

This gives us a first glimpse of how logical and 'higher-order' structure can be found in a natural way in process architectures built from client-server interfaces. We refer the interested reader to [5,6,3] for further details, in a somewhat different setting.

4.1. Logical Wiring and Copycat Processes

Given a signature $\Sigma = \{m_1 : T_1, \ldots, m_k : T_k\}$, it is useful to create renamed variants of it:

$$\Sigma^{(i)} = \{m_1^{(i)} : T_1, \ldots, m_k^{(i)} : T_k\}.$$

We can define an "identity component" which behaves like a copycat process [9,2], relaying information between input and output:

$$\mathrm{id}_\Sigma :: \Sigma^{(1)} \Longrightarrow \Sigma^{(2)}$$

$$\mathrm{id}_\Sigma = \sum_{m \in \Sigma} m^{(2)\downarrow}(\vec{x}).m^{(1)\uparrow}(\vec{x}).m_\uparrow^{(1)}(\vec{y}).m_\downarrow^{(2)}(\vec{y}).\mathrm{id}_\Sigma$$

In a similar fashion, we can define a component for function application (or Modus Ponens):

$$\mathrm{app}_{\Sigma,\Delta} :: (\Sigma^{(2)} \multimap \Delta^{(1)}) \otimes \Sigma^{(1)} \Longrightarrow \Delta^{(2)}$$

$$\mathrm{app}_{\Sigma,\Delta} = \mathrm{id}_\Sigma \parallel \mathrm{id}_\Delta.$$

To express the key property of this process, we need a suitable notion of equivalence of components. Given $p, q :: \Sigma \Longrightarrow \Delta$, we define $p \approx_{\Sigma,\Delta} q$ iff:

$$\forall s \in \mathcal{S}(\Sigma), c \in \mathcal{C}(\Delta).\ (s \odot_\Sigma p) \perp_\Delta c \iff (s \odot_\Sigma q) \perp_\Delta c.$$

Now suppose we have processes

$$p :: \Theta_1 \Longrightarrow \Sigma^{(2)} \multimap \Delta^{(1)}, \qquad q :: \Theta_2 \Longrightarrow \Sigma^{(1)}.$$

Then, allowing ourselves a little latitude with renaming of signatures, we can express the key equivalence as follows:

$$(p \parallel q) \odot_{(\Sigma^{(2)} \multimap \Delta^{(1)}) \otimes \Sigma^{(1)}} \mathrm{app}_{\Sigma,\Delta} \approx_{\Theta_1 \otimes \Theta_2, \Delta} q \odot_\Sigma p.$$

5. Genericity and Names

The view of service interfaces presented by signatures as developed thus far has some limitations:

- It is rather rigid, depending on methods having unique globally assigned names.
- It is behaviourally restrictive, because of the serialization requirement on method calls.

Note that allowing concurrent method activations would cause a problem as things stand: there would not be enough information available to associate method returns with the corresponding calls. The natural way to address this is to assign a unique name to each concurrent method invocation; this can then be used to associate a return with the appropriate call. The use of distinct names for the various threads which are offering the same generic service method also allows for more flexible and robust naming.

The notion of names we are using is that of the π-calculus and other 'nominal calculi' [22]. Names can be compared for equality and passed around in messages, and new names can be generated in a given scope, and these are the only available operations. We find it notationally clearer to distinguish between name constants $\alpha, \beta, \gamma, \ldots$ and name variables $a, b, c, \ldots$ but this is not an essential point.

These considerations lead to the following revised view of the actions associated with a method m:

- On the server side, a server offers a *located instance* of a call of m, as an action $\langle \alpha \rangle m^\downarrow(\vec{x})$. Several such calls with distinct locations (*i.e.* names α) may be offered concurrently, thus alleviating the serialization requirement. The corresponding return action will be $\langle \alpha \rangle m_\downarrow(\vec{w})$. The name can be used to match this action up with the corresponding call.
- On the client side, there are two options for a call: the client may either invoke a located instance it already knows:

$$\langle \alpha \rangle m^\uparrow(\vec{v})$$

or it may invoke a generic instance with a name variable:

$$(a) m^\uparrow(\vec{v}).$$

This can synchronize with any located instance which is offered, with a being bound to the name of that instance. A return will be on a specific instance:

$$\langle \alpha \rangle m_\uparrow(\vec{y}).$$

The definition of the operation $s \triangleleft_\Sigma c$ is correspondingly refined in the rules specifying the client-server interactions, as follows:

$$\frac{s \xrightarrow{\langle \alpha \rangle m^\downarrow(\vec{x})} s' \quad c \xrightarrow{(a) m^\uparrow(\vec{v})} c'}{s \triangleleft_\Sigma c \xrightarrow{\langle \alpha : a \rangle m^{\downarrow\uparrow}(\vec{x}:\vec{v})} s'[\vec{v}/\vec{x}] \triangleleft_\Sigma c'[\alpha/a]}$$

$$\frac{s \xrightarrow{\langle \alpha \rangle m^\downarrow(\vec{x})} s' \quad c \xrightarrow{\langle \alpha \rangle m^\uparrow(\vec{v})} c'}{s \triangleleft_\Sigma c \xrightarrow{\langle \alpha \rangle m^{\downarrow\uparrow}(\vec{x}:\vec{v})} s'[\vec{v}/\vec{x}] \triangleleft_\Sigma c'} \qquad \frac{s \xrightarrow{\langle \alpha \rangle m_\downarrow(\vec{w})} s' \quad c \xrightarrow{\langle \alpha \rangle m_\uparrow(\vec{y})} c'}{s \triangleleft_\Sigma c \xrightarrow{\langle \alpha \rangle m_{\downarrow\uparrow}(\vec{w}:\vec{y})} s' \triangleleft_\Sigma c'[\vec{w}/\vec{y}]}$$

It is easy to encode these operations in the π-calculus [22], or other nominal calculi. They represent a simple design pattern for such calculi, which fits very naturally with the server-client setting we have been considering.

Note that in generic method calls, the client is *receiving* the name while sending the parameters to the call, while conversely the server is sending the name and receiving the parameters. This "exchange of values" [20] can easily be encoded, but is a natural feature of these interactions.

5.1. Orthogonality and Types Revisited

It is straightforward to recast the entire previous development in the refined setting. The main point is that the orthogonality relation is now defined in terms of the symbols $\langle\alpha\rangle m^{\Uparrow}$ and $\langle\alpha\rangle m_{\Downarrow\uparrow}$. Thus serialization is only required on instances.

6. Further Directions

The ideas we have presented can be developed much further, and to some extent have been; see the references in the next section.

One aspect which we would have liked to include in the present paper concerns *data assertions*. The notion of orthogonality we have studied in this paper deals with control flow and synchronization, but ignores the flow of data between clients and servers through the parameters to method calls and returns. In fact, the correctness properties of this data-flow can also be encapsulated elegantly in an extended notion of orthogonality. We plan to describe this in a future publication.

Acknowledgements

This paper builds on much previous work, mainly by the author and his colleagues, although the specific form of the presentation, and the emphasis on procedural interfaces and client-server interactions, has not appeared previously in published form.

We mention some of the main precursors.

- Particularly relevant is the work on *Interaction categories* and *Specification structures* — although no categories appear in the present paper! Much of this was done jointly with the author's former students Simon Gay and Raja Nagarajan [1,9,5,6,7,8]. See also [12] with Dusko Pavlovic.
- Ideas from Game semantics [2,10,9,11], many developed in collaboration with Radha Jagadeesan and Pasquale Malacaria, have also been influential; for example, the four types of actions associated with method calls correspond to the four-fold classification of moves in games as Player/Opponent and Question/Answer.
- Various notions of orthogonality in the same general spirit as that used here, but mainly applied in logical and type-theoretic settings, have appeared over the past 15–20 years. Some examples include [15,19,17]. The application to notions in concurrency originated with the present author [1,5,6,3].
- The closest precursor of the present paper is an unpublished lecture given by the author in 2002 [4].
- General background is of course provided by work in concurrency theory, especially process algebra [16,21,22].

Recent work in a related spirit includes work on session types, see e.g. [23]. Another possibly relevant connection, of which the author has only recently become aware, is with the Handshake algebra developed for asynchronous circuits [18], and a game semantics for this algebra which has appeared recently [14].

References

[1] S. Abramsky. Interaction categories (extended abstract). In G. L. Burn, S. J. Gay, and M. D. Ryan, editors, *Theory and Formal Methods 1993*, pages 57–69. Springer-Verlag, 1993.
[2] S. Abramsky. Semantics of Interaction: an introduction to Game Semantics. In P. Dybjer and A. Pitts, editors, *Proceedings of the 1996 CLiCS Summer School, Isaac Newton Institute*, pages 1–31, Cambridge University Press, 1997.
[3] S. Abramsky. Process realizability. In F. L. Bauer and R. Steinbrüggen, editors, *Foundations of Secure Computation: Proceedings of the 1999 Marktoberdorf Summer School*, pages 167–180. IOS Press, 2000.
[4] S. Abramsky. Reactive refinement. Oxford University Computing Laboratory seminar, 2002.
[5] S. Abramsky, S. Gay, and R. Nagarajan. Interaction categories and the foundations of typed concurrent programming. In M. Broy, editor, *Proceedings of the 1994 Marktoberdorf Summer Sxhool on Deductive Program Design*, pages 35–113. Springer-Verlag, 1996.
[6] S. Abramsky, S. Gay, and R. Nagarajan. Specification structures and propositions-as-types for concurrency. In G. Birtwistle and F. Moller, editors, *Logics for Concurrency: Structure vs. Automata—Proceedings of the VI I Ith Banff Higher Order Workshop*, pages 5–40. Springer-Verlag, 1996.
[7] S. Abramsky, S. Gay, and R. Nagarajan. A type-theoretic approach to deadlock-freedom of asynchronous systems. In M. Abadi and T. Ito, editors, *Theoretical Aspects of Computer Software*, volume 1281 of *Springer Lecture Notes in Computer Science*, pages 295–320. Springer-Verlag, 1997.
[8] S. Abramsky, S. J. Gay, and R. Nagarajan. A specification structure for deadlock-freedom of synchronous processes. In *Theoretical Computer Science*, volume 222, pages 1–53, 1999.
[9] S. Abramsky and R. Jagadeesan. Games and full completeness for multiplicative linear logic. *Journal of Symbolic Logic*, 59(2):543–574, 1994.
[10] S. Abramsky and R. Jagadeesan. New foundations for the geometry of interaction. *Information and Computation, 111(1)*, pages 53–119, 1994.
[11] S. Abramsky, R. Jagadeesan, and P. Malacaria. Full abstraction for PCF. In *Information and Computation*, volume 163, pages 409–470, 2000.
[12] S. Abramsky and D. Pavlovic. Specifying processes. In E. Moggi and G. Rosolini, editors, *Proceedings of the International Symposium on Category Theory In Computer Science*, volume 1290 of *Springer Lecture Notes in Computer Science*, pages 147–158. Springer-Verlag, 1997.
[13] B. A. Davey and H. A. Priestley. *Introduction to Lattices and Order*. Cambridge University Press, second edition, 2002.
[14] L. Fossati. Handshake games. *Electronic Notes in Theoretical Computer Science*, 171(3):21–41, 2007.
[15] J.-Y. Girard. Linear logic. *Theoretical Computer Science*, 1987.
[16] C. A. R. Hoare. *Communicating Sequential Processes*. Prentice Hall, 1985.
[17] M. Hyland and A. Schalk. Glueing and orthogonality for models of linear logic. *Theoretical Computer Science*, 294:183–231, 2003.
[18] M. B. Josephs, J. T. Udding, and J. T. Yantchev. Handshake algebra. Technical Report SBU-CISM-93-1, South Bank University, 1993.
[19] R. Loader. Linear Logic, Totality and Full Completeness. LICS 1994: 292-298, 2004.
[20] G. Milne and R. Milner. Concurrent processes and their syntax. *Journal of the ACM*, 26(2):302–321, 1979.
[21] R. Milner. *Communication and Concurrency*. Prentice Hall International, 1989.
[22] R. Milner. *Communication and Mobile Systems: the π-calculus*. Cambridge University Press, 1999.
[23] V. Vasconcelos, S. J. Gay, and A. Ravara. Type checking a multithreaded functional language with session types. *Theoretical Computer Science*, 368((1-2)):64–87, 2006.

How to Soar with CSP

Colin O'HALLORAN

Director, Systems Assurance Programme, QinetiQ

Abstract. In this talk, I shall discuss work on the necessary technology required for flight clearance of Autonomous Aircraft employing Agents by reducing the certification problem to small verifiable steps that can be carried out by a machine. The certification of such Agents falls into two parts: the validation of the safety of the Agent; and the verification of the implementation of the agent. The work focuses on the Soar agent language and the main results are:

- language subset for Soar, designed for formal analysis;
- a formal model of the Soar subset written in CSP;
- a prototype translator "Soar2Csp" from Soar to the CSP model;
- a framework for static analysis of Soar agents through model checking using FDR2;
- the identification of "healthiness conditions" required of any Soar Agent;
- a verifiable implementation of the CSP based Soar agents on an FPGA.

A CSP Model for Mobile Channels

Peter H. WELCH and Frederick R.M. BARNES

Computing Laboratory, University of Kent, Canterbury, Kent, CT2 7NF, England

Abstract. CSP processes have a static view of their environment – a *fixed* set of events through which they synchronise with each other. In contrast, the π-calculus is based on the dynamic construction of events (channels) and their distribution over pre-existing channels. In this way, process networks can be constructed dynamically with processes acquiring new connectivity. For the construction of complex systems, such as Internet trading and the modeling of living organisms, such capabilities have an obvious attraction. The occam-π multiprocessing language is built upon classical occam, whose design and semantics are founded on CSP. To address the dynamics of complex systems, occam-π extensions enable the movement of channels (and multiway synchronisation barriers) through channels, with constraints in line with previous occam discipline for safe and efficient programming. This paper reconciles these extensions by building a formal (operational) semantics for mobile channels entirely within CSP. These semantics provide two benefits: formal analysis of occam-π systems using mobile channels and formal specification of implementation mechanisms for mobiles used by the occam-π compiler and run-time kernel.

Keywords. channels, processes, mobility, modeling, occam-π, CSP, π-calculus.

Introduction

The dynamic creation of channels and processes, together with their communication through channels, enables network topology to evolve in response to run-time events. Systems requiring this capability abound – for example, the modelling of complex biological phenomena and commercial Internet applications. Formal specification and analysis has been pioneered through Milner's π-calculus. Here, we present a model of channel mobility using Hoare's CSP and explain our motivation and the benefits obtained.

Mobile channels have been introduced into the occam-π multiprocessing language [1,2,3], whose classical design and semantics are founded on CSP [4,5,6]. CSP processes synchronise on fixed sets of events (so cannot dynamically acquire new connections) and that static nature cannot easily be relaxed. However, CSP allows infinite event sets and recursion, which gives us a lot of freedom when modeling.

This paper presents a CSP model for channel mobility that yields semantics that are both *operational* and *denotational*. The operational aspect provides a formal specification for all data structures and algorithms for a supporting run-time kernel (from which the occam-π kernel is derived). The denotational side preserves the *compositional* nature of occam-π components (no surprises when processes are networked in parallel, *what-you-see-is-what-you-get*). It also allows formal specification and analysis of occam-π systems and, so long as the number of mobile channels can be bounded and that bound is not too large, the application of automated model checkers (such as FDR [7]).

Section 1 reviews the mobile channel mechanisms of occam-π. Section 2 introduces the technique of modeling channels with processes, essential for the formal semantics of mobility presented here. Section 3 builds the kernel processes underlying the semantics. Section 4 maps occam-π code to the relevant synchronisations with the kernel. Finally, Section 5 draws conclusions and directions for future work.

1. Mobile Channels in occam-π

Mobile *channels*, along with mobile *data* and mobile *processes*, have been introduced into the occam-π multiprocessing language, a careful blend of classical (CSP-based) occam2.1 with the network dynamics of the π-calculus [8]. The *mobility* concept supported reflects the idea of movement: something *moves* from source to target, with the source losing it (unless explicitly marked for *sharing*). Mobile objects may also be *cloned* for distribution.

occam-π introduces channel *bundles*: a record structure of individual channels (fields) carrying different protocols (message structures) and operating in different directions. occam-π also separates the concept of channel (or channel bundle) *ends* from the channels (or bundles) themselves: processes see only one *end* of the external channels with which they interact with their environment. For mobile channels, it is the *channel-ends* that can be moved (by communication or assignment) – not the channels themselves. With the current compiler, channel mobility is implemented only for the new channel bundle types.

1.1 Mobile Channel Bundles

Channel types declare a *bundle* of channels that will always be kept together. They are similar to the idea proposed for occam3 [9], except that the ends of our bundles are (currently always declared to be) mobile, directions are specified for the individual channels, and the bundle has distinct *ends*.

Figure 1: a channel bundle.

Figure 1 shows a typical bundle of channels supporting *client-server* communications. By convention, a *server* process takes the *negative* end of the bundle, receiving and answering questions from *client* processes sharing the *positive* end. The type is declared as follows:

```
CHAN TYPE RESOURCE.LINK
  MOBILE RECORD
    CHAN RESOURCE.ASK ask!:
    CHAN RESOURCE.ANS ans?:
:
```

Note that this declaration specifies field channel directions from the point of view of the positive end of the bundle. So, clients operate these channels in those declared directions (they *ask* questions and *receive* answers), whereas a server operates them the other way around (it *receives* questions and *delivers* answers).

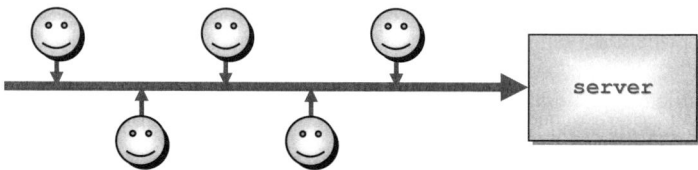

Figure 2: a client-server network.

1.2 Declaring, Allocating and Placing Ends of Mobile Channel Bundles

Variables are declared only to hold the ends of channel bundles – not the bundle as a whole. These ends are independently mobile. Bundle ends are allocated dynamically and in pairs. By default, a bundle end is *unshared*: it may only be connected to one process at a time. A bundle end may be declared as being *shared*: it may be connected to any number of parallel processes (which compete with each other to use). Resources for the bundles are automatically recovered when all references to them are lost.

Here is code that sets up an initial system (Figure 2) of many clients and one server:

```
RESOURCE.LINK- resource.server.end:
SHARED RESOURCE.LINK+ resource.client.end:
SEQ
  resource.server.end, resource.client.end := MOBILE RESOURCE.LINK
  PAR
    resource.server (resource.server.end, ...)
    PAR i = 0 FOR n.clients
      client (resource.client.end, ...)
:
```

where the server and client processes may have other connections (not shown in Figure 2). Note the polarity of the channel bundle types in the above declarations, indicating which end of the bundle is held by each variable.

1.3 Using and Moving Ends of Channel Bundles

1.3.1 Clients Holding Shared Positive Ends

This client process is aware that its link to the resource server is shared:

```
PROC client (SHARED RESOURCE.LINK+ resource.link,
             CHAN SHARED RESOURCE.LINK+ forward!, update?,
             ...)
  ...
:
```

In the above, `resource.link` is the (shared client) bundle end and the (classical) channels `forward!` and `update?` are for sending and receiving, respectively, new bundle ends.

To demonstrate use of this channel bundle, let us make the protocols used by its sub-channels more concrete:

```
PROTOCOL RESOURCE.ASK
  CASE
    size; INT
    deposit; RESOURCE
:

PROTOCOL RESOURCE.ANS IS RESOURCE:
```

where **RESOURCE** is some (expensive to allocate) mobile data structure. A client asks for a **RESOURCE** of a certain size (on its `ask!` sub-channel end) and duly receives one (on `ans?`):

```
CLAIM resource.link
  SEQ
    resource.link[ask] ! size; 42
    resource.link[ans] ? my.resource    -- RESOURCE
```

When the client has finished with the resource, it returns it back to the server:

```
CLAIM resource.link
  resource.link[ask] ! deposit; my.resource
```

Outside a CLAIM, a client may forward its end of the link to its server to another process:

```
forward ! resource.link
```

This dynamically introduces another client to the server. Because the bundle end is *shared*, the original client retains its link to the server.

When not in mid-transaction with its server (again, outside a CLAIM block), this client may choose to update its link:

```
update ? resource.link
```

It loses its original connection and is now the client of a (presumably) different server.

1.3.2 A Server Holding a Negative End

This server *pools* RESOURCE instances, retrieving suitable ones from its pool where possible when new ones are requested:

```
PROC resource.server (RESOURCE.LINK- resource.link,
                      CHAN RESOURCE.LINK- lend!, return?,
                      ...)
  ... declare dynamic structures for managing the RESOURCE pool
  SEQ
    ... initialise RESOURCE pool
    INITIAL BOOL running IS TRUE:
    WHILE running
      resource.link[ask] ? CASE
        INT n:
        size; n
          RESOURCE resource:
          SEQ
            IF
              ... a suitably sized RESOURCE is in the pool
                ... move it into the resource variable
              TRUE
                ... dynamically allocate a new resource (of size 'n')
            resource.link[ans] ! resource
        RESOURCE resource:
        deposit; resource
          ... deposit resource into the pool
:
```

At any time, this server may relinquish servicing its clients by forwarding its (exclusive) end of the link to another server:

```
lend ! resource.link
```

Because this link is *unshared*, the `resource.link` variable becomes *undefined* and this server can no longer service it – any attempt to do so will be trapped by the compiler. This server may do this if, for some reason, it cannot satisfy a client's request but the `forward` channel connects to a reserve server that can. To continue providing service, the forwarding had better be a loan – i.e. the other server returns it after satisfying the difficult request:

```
return ? resource.link
```

Our server may now resume service to all its clients. The above server coding may need slight adjustment (e.g. if the reserve server has already supplied the resource on its behalf).

2. Modeling Channels with Processes

To provide a formal semantics of channel mobility in occam-π, we cannot model its channels with CSP *channels*. There are three challenges: the dynamics of construction, the dynamics of mobility (which demands varying synchronisation alphabets) and the concept of channel *ends* (which enforces correct sequences of direction – also unprotected within the π-calculus).

Instead, following the techniques developed for the formal modeling of mobile barriers (multiway synchronisations) in occam-π [10], we model mobile channels as *processes*. Each mobile channel process is constructed on demand and given a unique *index* number. Conventional channels, through which such a process is driven, model the two different *ends* of the mobile channel. Application processes *interleave* in their use of these *ends*, with that interleaving governed by possession of the relevant index. *Mobility* derives simply from communicating (and, then, forgetting) that index.

Let P be a process and c be an external (i.e. non-hidden) channel that P uses only for output. Further, assume P never uses c as the first action in a branch of an external choice (a constraint satisfied by all occam-π processes). Using the notion of *parallel introduction* (section 3.1.2 of [11]), all communications on channel c may be devolved to a *buddy* process, ChanC, with no change in semantics – i.e. that P is failures-divergences equivalent to the system (expressed in CSP_M, the machine-readable CSP syntax defined for FDR [7]):

```
( (P' ; killC -> SKIP) [| {| writeC, ackC, killC |} |] ChanC )
\ {| writeC, ackC, killC |}
```

where `writeC`, `ackC`, and `killC` are events chosen outside the alphabet of P, and where:

```
ChanC = (writeC?x -> c!x -> ackC -> ChanC [] killC -> SKIP)
```

and where P' is obtained from P by delegating all communications (`c!a -> Q`) to its buddy process (`writeC!a -> ackC -> Q`).

Process P' completes a `writeC/ackC` sequence if, and only if, the communication `c!x` happens – and the `writeC/ackC` events are hidden (i.e. undetectable by an observer). Process P does not engage on c as the first event of an external choice; so neither does P' on `writeC`. This is a necessary constraint since, otherwise, the hiding of `writeC` would introduce non-determinism not previously present. Termination of the *buddy* process, ChanC, is handled with the addition of the (hidden) `killC` event – used only once, when/if the original process terminates.

Formally, the equivalence follows from the rich algebraic laws of CSP relating *event hiding*, *choice*, *sequential* and *parallel composition*, *prefix* and *SKIP* (outlined in section 4 of [11]).

Figure 3 applies this equivalence to transform a *channel*, c, connecting processes P and Q into a *process*, ChanC'. Process P' is defined above. For convenience, processes Q' and ChanC' are defined from Q and ChanC just by *renaming* their external (free) channel, c, as `readC` (where `readC` is an event chosen outside the alphabets of P and Q) We now have distinct terms to talk separately about the *writing-end* (`writeC/ackC`) and *reading-end* (`readC`) of our original channel c.

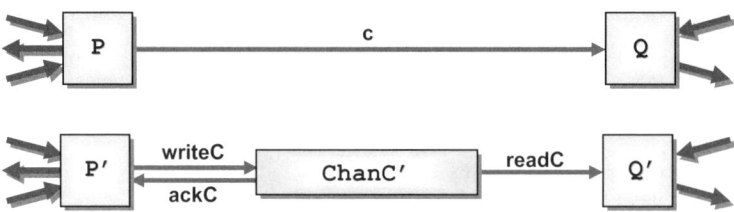

Figure 3: modeling a channel with a process.

Figure 3 applies this equivalence to transform a *channel*, c, connecting processes P and Q into a *process*, ChanC', connecting processes P' and Q'. The process P' is defined above. For convenience, processes Q' and ChanC' are defined from Q and ChanC just by *renaming* their external channel, c, as readC (where readC is an event chosen outside the alphabets of P and Q). This gives us distinct terms with which we can talk separately about the *writing-end* (writeC/ackC) and *reading-end* (readC) of our original channel c.

3. A Kernel for Mobile Channels

We present semantics for the mobile channels of occam-π, addressing channel bundles, dynamic allocation, the separate identities of channel bundle ends, the use of the channels within the bundles, sharing and mobility.

3.1 Processes within a Channel Bundle

A channel bundle is a parallel collection of processes: one holding a *reference count* (and responsible for termination), two *mutexes* (one for each possibly shared end) and one *channel process* for each field in the bundle:

```
Bundle (c, fields) =
  (Refs (c, 2) [| {kill} |]
    (Mutex (c, positive) [| {kill} |]
      (Mutex (c, negative) [| {kill} |]
        Channels (c, fields)
      )
    )
  )
) \ {kill}
```

where c is the unique index for the bundle, fields is the number of its channel fields and positive and negative are constants with values 0 and 1, respectively. This is visualised in Figure 4, showing the channels independently engaged by the sub-processes.

The reference counting process is initialised to 2, since one variable for each bundle end knows about it upon construction. This process engages in enrol and resign events for this bundle and is responsible for terminating its sibling processes should the count ever reach zero:

```
Refs (c, n) =
  enrol.c -> Refs (c, n + 1) []
  resign.c -> Refs (c, n - 1)                  , if n > 0

Refs (c, 0) = kill -> SKIP
```

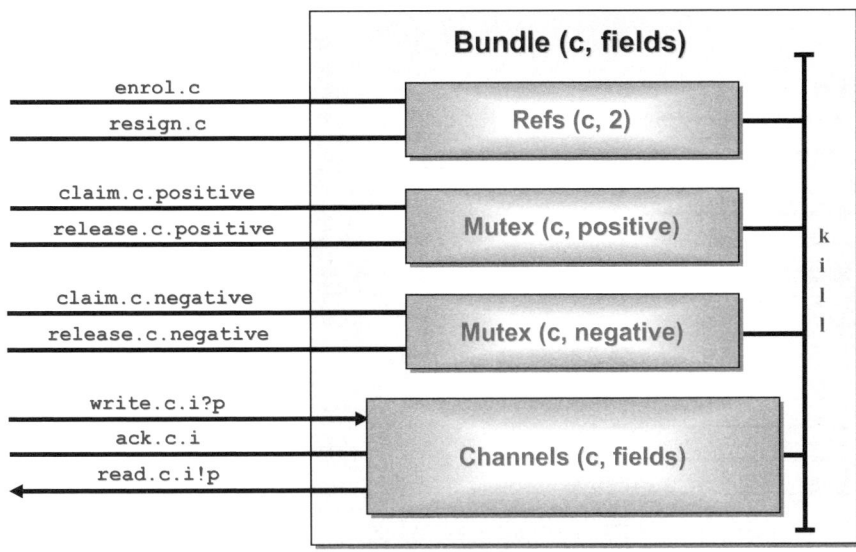

Figure 4: formal model of an occam-π channel bundle.

The *mutex* processes provide mutually exclusive locking for each end (`positive` or `negative`) of the bundle, engaging in `claim` and `release` events for this bundle. They can be terminated at any time by a `kill` signal (from the reference counting process). They enforce access to a bundle end by only one application process at a time. Strictly, they are only necessary for each end that is declared *shared*; their over-engineering here in this model simplifies it and is harmless. The coding is elementary and standard:

```
Mutex (c, x) =
  claim.c.x -> release.c.x -> Mutex (c, x)   []
  kill -> SKIP
```

where `x` : {`positive`, `negative`}.

The `channels` process (Figure 5) is a parallel collection of channel processes, one for each field, synchronising only on the termination signal:

```
Channels (c, fields) =
  [| {kill} |] i:{0..(fields - 1)} @ Chan (c, i)
```

where each channel process follows the pattern presented in Section 2:

```
Chan (c, i) =
  write.c.i?p -> read.c.i!p -> ack.c.i -> Chan (c, i)   []
  kill -> SKIP
```

3.2 Dynamic Generation of Channel Bundles

Mobile channel bundle processes are generated upon request by the following server:

```
MC (c) =
  setMC?f -> getMC!c -> (Bundle (c, f) ||| MC (c + 1))   []
  noMoreBundles -> SKIP
```

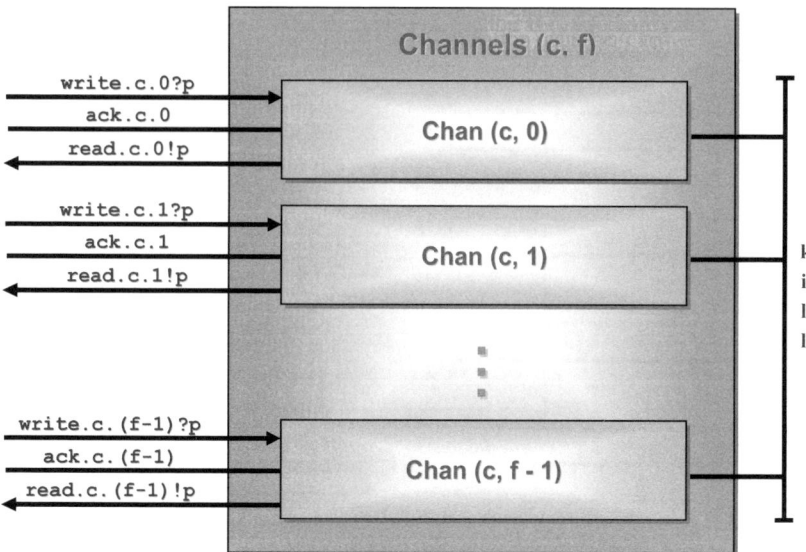

Figure 5: processes modelling the channels in a bundle.

Application processes will interleave on the `setMC` and `getMC` channels when constructing mobile channel bundles: the number of fields for the bundle is requested on `setMC` and a unique index number for the generated bundle is returned on `getMC`.

This channel bundle generator will be started with index `1`. We reserve index `0` for an *undefined* bundle:

```
undefined = 0

UndefinedBundle =
  resign.undefined -> UndefinedBundle []
  noMoreBundles -> SKIP
```

Note that both the channel bundle generator, `MC`, and `UndefinedBundle` terminate on the `noMoreBundles` signal. That signal is generated when, and only when, the application is about to terminate. The `resign` signal, accepted but ignored by `UndefinedBundle`, is there to simplify some technical details in Section 4.

3.3 Mobile Channel Kernel

The *mobile channel kernel* consists of the generator and undefined bundle processes:

```
MOBILE_CHANNEL_KERNEL = MC (1) [| {noMoreBundles} |] UndefinedBundle
```

In addition to `noMoreBundles`, it engages (but does not syncrhonise internally) upon the following set of channels:

```
kernel_chans =
  {| enrol, resign, claim, release, write, read, ack,
     setMC, getMC, noMoreBundles |}
```

```
                APPLICATION_SYSTEM'                    MOBILE_CHANNEL_KERNEL
```

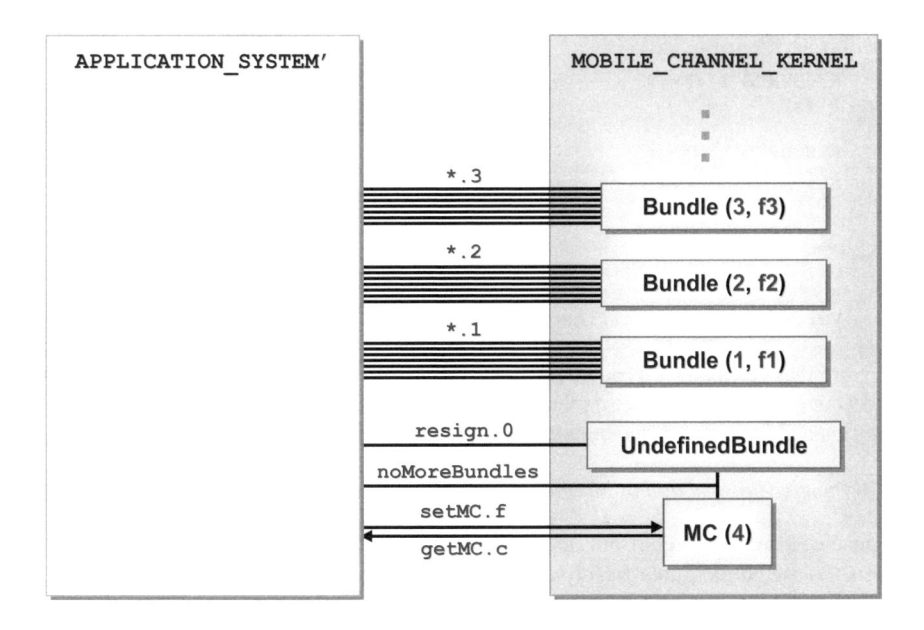

Figure 6: application system and support kernel
(after the construction of three mobile channel bundles).

If `APPLICATION_SYSTEM` is an occam-π application and `APPLICATION_SYSTEM'` is the CSP model of its use of mobile channel bundle primitives (Section 4), the full model is:

```
((APPLICATION_SYSTEM' ; noMoreBundles -> SKIP)
 [| kernel_chans |]
 MOBILE_CHANNEL_KERNEL) \ kernel_chans
```

Figure 6 gives a visualisation after three mobile channels have been constructed.

4. The Semantics of occam-π Mobile Channels

This section defines the mapping from all occam-π mobile channel mechanisms to CSP. The mapping is simplified using some syntactic extensions from Circus [12], a CSP algebra combined with elements of Z for the formal specification of rich state transformations. The extensions used here are restricted to the declaration of *state variables*, their assignment as new primitive CSP processes and their use in expressions. These extensions have a trivial mapping down to pure CSP (which is described in section 2.3 of [10]).

The mappings will be presented by induction and example, using the definitions from Section 1 for concreteness. In the following, if `P` (or `P(x)`) is an occam π process, then `P'` (or `P'(x)`) denotes its mapping into CSP.

4.1 Declaring Mobile Channel Bundle End Variables

All mobile channel bundle end variables – regardless of polarity, shared status, number of fields and channel protocols – are represented by variables holding a natural number index, unique for each *defined* bundle. Initially, they are all set to `undefined` (see Section 3.2).

Each of the following declarations:

```
RESOURCE.LINK+ x:              SHARED RESOURCE.LINK+ x:
P (x)                          P (x)

RESOURCE.LINK- x:              SHARED RESOURCE.LINK- x:
P (x)                          P (x)
```

↝ (maps to)

```
Var x:N • x := undefined; P'(x); resign.x -> SKIP
```

Note that the mapped process resigns from the mobile bundle just before termination. We do not have to check whether the bundle variable is defined at this point, because of the definition and inclusion of the `UndefinedBundle` process in the kernel (Section 3.2).

4.2 Dynamic Construction of Mobile Bundles

Suppose `client` and `server` are opposite polarity `RESOURCE.LINK` variables, shared or unshared. To assign them to a freshly constructed mobile bundle, send the kernel the number of fields required (`#RESOURCE.LINK`), receive the index of channels to the bundle process created and assign this to the variables (not forgetting to resign from any bundles they may previously have been referencing):

```
client, server := MOBILE RESOURCE.LINK
```

↝

```
setMC!#RESOURCE.LINK -> getMC?tmp ->
   ( resign.client -> (client := tmp) |||
     resign.server -> (server := tmp) )
```

Note that these processes interleave their use of the kernel channels. In the unlikely event that both variables were previously referencing opposite ends of the *same* bundle, the interleaved resignations are still handled correctly. As mentioned in Section 4.1, resigning from variables holding undefined references is also safe. We use *interleaving* in the above specification to give maximum freedom to the occam-π compiler/kernel in implementing these mechanisms.

4.3 Claiming a Shared Mobile Channel Bundle End

Suppose that `client` and `server` are `SHARED` bundle-ends. This time, we need to know their polarity. Suppose that `client` is positive and `server` is negative (which is the normal convention for a *client-server* relationship). An occam-π `CLAIM` obtains exclusive use of the bundle-end for the duration of the `CLAIM` block. So:

```
CLAIM client
  P
```

↝

```
claim.client.positive -> P'; release.client.positive -> SKIP
```

and:

```
CLAIM server
  P
```

⤳

```
claim.server.negative -> P'; release server.negative -> SKIP
```

where these `claim` and `release` events synchronise with the relevant *mutex* processes in the relevant *channel bundle* processes in the kernel.

Of course, the occam-π language does not allow CLAIMs on *undefined* channel bundle variables and this rule is enforced statically by the compiler. So, we do not need to be concerned with such possibilities for these semantics (and this is why `claim` and `release` events do not have to play any part in the `UndefinedBundle` process of our CSP kernel).

4.4 Input/Output on a Channel in a Mobile Channel Bundle

Communications over channel fields in a bundle are modeled following the patterns given in Section 2.

Suppose `client` is a `RESOURCE.LINK+` variable and `server` is `RESOURCE.LINK-`; then communications on `client` follow the directions defined in the declaration:

```
CHAN TYPE RESOURCE.LINK
  MOBILE RECORD
    CHAN RESOURCE.ASK ask?:
    CHAN RESOURCE.ANS ans!:
:
```

but communications on `server` follow the opposite directions.

4.4.1 Positive Communications

We present the mapping for these first, since they are not complicated by the variant protocol defined for `RESOURCE.ASK`. For sending:

```
client[ask] ! size; n
```

⤳

```
write.client.ask!size.n -> ack.client.ask -> SKIP
```

where, in the mapped CSP, `ask` is just the field number (0) in the bundle. The values of `size` and `n` are, respectively, a constant specifying the message type being delivered and the resource size requested. In these semantics (of mobile channel communications), message contents are not relevant – they are only significant for the use made of them by application processes before and after communication. Similarly:

```
client[ask] ! deposit; r
```

⤳

```
write.client.ask!deposit.r -> ack.client.ask -> SKIP
```

On the other hand, for receiving:

```
client[ans] ? r
```

⤳

```
read.client.ans?tmp -> (r := tmp)
```

4.4.2 Negative Communications

The server side communications on **RESOURCE.LINK** are complicated slightly by the *variant* message structure in **RESOURCE.ASK**: the **CASE** input must be dealt with by testing the first (tag) part of the message. This is not significant for these semantics of mobile channel input, which are only concerned with synchronisations between application and kernel processes.

The mappings for sending and receiving do not depend on the polarity of the bundle-ends. Sending is the same as in Section 4.4.1, except that the field names are switched:

```
server[ans] ! r
```

⤳

```
write.server.ans!r -> ack.server.ans -> SKIP
```

Receiving is also as in Section 4.4.1, with switched field names. The complications of the variant protocol, used in this example, are not much of a distraction:

```
server[ask] ? CASE
  INT n:
  size; n
    P (n)
  RESOURCE r:
  deposit; r
    Q (r)
```

⤳

```
read.server.ask?tag.tmp ->
  if tag == size then P' (tmp) else Q' (tmp)
```

4.5 Sending Mobile Channel Bundle Ends

Sending a mobile channel bundle-end depends on whether it is shared. Assume that **client** and **server** are *unshared* and that we have suitable channels **m** and **n** carrying, respectively, **RESOURCE.LINK+** and **RESOURCE.LINK-**. Then:

```
m ! client
```

⤳

```
m!client -> (client := undefined)
```

which introduces the occam-π mobility semantics: *send-and-lose*. The mapping for sending **server** down channel **n** is identical, apart from name changes. These mappings are, of course, for classical (non-mobile) channels **m** and **n**. If these were themselves fields of a mobile channel bundle, the mappings from Section 4.4 would also be applied.

Assume, now, that **client** and **server** are *shared* and that we have suitable channels **m** and **n** carrying, respectively, SHARED RESOURCE.LINK+ and SHARED RESOURCE.LINK-. The communication semantics are now different: the sending process *does not lose* a mobile, if it is shared, and the reference count on the mobile must be incremented (since the receiving process may now engage on it):

```
m ! client
```

↝

```
enroll.client -> m!client -> SKIP
```

Note: it is necessary that the *sender* of the mobile increments the reference count as part of the send, rather than the *receiver*. The receiver could only try to increment that count after receiving it – by which time the sender could have resigned from the bundle itself (perhaps through termination, see Section 4.1, or overwriting of its bundle variable, see below), which might reduce the reference count to zero, terminating the bundle process in the kernel and leaving the receiver deadlocked!

As before, the mapping for sending a *shared* **server** down channel **n** is identical to the above, apart from name changes. Again, these mappings are for classical (non-mobile) channels **m** and **n**. If these were themselves fields of a mobile channel bundle, the mappings from Section 4.4 would also be applied.

4.6 Receiving Mobile Channel Bundle Ends

Receiving a mobile channel bundle-end does *not* depend on whether it is shared. Assume that **client** and **server** are bundle-end variables, *shared* or *unshared*, and that we have suitable channels **m** and **n** for carrying their values. All that must be done is resign from the bundles currently referenced and assign the new references:

```
m ? client
```

↝

```
m?tmp -> resign.client -> (client := tmp)
```

As before, the mapping for receiving a **server**, *shared* or *unshared*, from channel **n** is identical to the above, apart from name changes. Again, both these mappings are for classical (non-mobile) channels **m** and **n**. If these were themselves fields of a mobile channel bundle, the mappings from Section 4.4 would also be applied.

4.7 Assigning Mobile Channel Bundle Ends

Communication and assignment are intimately related in occam-π: an assignment has the same semantics as a communication of some value between variables in the *same* process.

Therefore, assignment of *unshared* mobiles leaves the source variable *undefined* and does not change the reference count on the assigned bundle – though the reference count on the bundle originally referenced by the target variable must decrement. Suppose `a` and `b` are *unshared* bundle-end variables of compatible type. Then:

```
a := b
```
↪
```
resign.a -> (a := b); (b := undefined)
```

However, if `a` and `b` are *shared* bundle-end variables of compatible type. Then:

```
a := b
```
↪
```
(enrol.b -> SKIP ||| resign.a -> SKIP); (a := b)
```

where, as in Section 4.2, we use interleaving to allow as much implementation freedom as possible.

4.8 Forking Processes with Mobile Channel Bundle End Parameters

Section 2 did not review the *forking* mechanism [1, 2, 3] of occam-π. However, passing the ends of mobile channel bundles to forked processes has been widely used in the complex system modelling being developed in our TUNA [13] and CoSMoS [14] projects – so, we do need a proper semantics for it.

Concurrency may be introduced into occam-π processes either by its classical **PAR** construct or by forking. A **PAR** construct does not terminate until *all* its concurrently running component processes terminate (and this maps to the CSP parallel operator). Often, a process needs to construct dynamically a new process, set it running concurrently with itself and continue – this is *forking*. Forking does not technically introduce anything that cannot be done with a **PAR** construct and recursion (which is precisely how its semantics are defined in [10]). However, forking does enable the more direct expression of certain idioms (e.g. a server loop that constructs processes to deal with new clients on demand and concurrently) and, pragmatically, has a more practical implementation (recursion in process components of a **PAR** construct makes an unbounded demand on memory, although a compiler may be able to optimize against that).

For the scope of this paper, we are only concerned about the semantics of passing the mobile ends of channel bundles to a forked process. For static arguments (e.g. data values or shared ends of classical channels), passing is by *communication* along a channel specific to the process type being forked.

Let `P(c)` be a process whose parameter is a mobile channel bundle end and let `forkP` be the channel, specific for `P`, connecting `P`-forking *application* processes to the (**PAR** recursive) `P`-*generator* process (Section 2.4.13 of [10]). The semantics of *forking* simply follows from the semantics of *sending* (Section 4.5 of this paper). If the argument is unshared, the forking process loses it:

```
FORK P (c)
```
↪
```
forkP!c; (c := undefined)
```

More commonly, the argument will be shared, so that the forking process retains it:

```
FORK P (c)

enrol.c -> forkP!c -> SKIP
```

The details of the P-*generator* process at the receiving end of the `forkP` channel are the same as defined for the passing of mobile occam-π *barriers* in Section 2.4.13 of [10].

5. Conclusions and Future Work

The correct, flexible and efficient management of concurrency has growing importance with the advent of multicore processing elements. The mobile extensions to occam-π blend seamlessly into the classical language, maintaining the safety guarantees (e.g. against parallel race hazards), the simple communicating process programming model (which yields compositional semantics) and extreme lightness of overhead (memory and run-time).

This paper has presented a formal operational semantics, in terms of CSP, of all mobile channel mechanisms within occam-π. In turn, this leads to a denotational semantics (traces-failures-divergences [5, 6]) that enables formal reasoning about these aspects of occam-π systems. For use with the FDR model checker [7], the unbounded recursion of the mobile channel generator, `MC`, in Section 3 needs to be bounded – i.e. only a limited number of channel indices can be used, with their associated processes pooled for recycling (instead of terminated) when their reference count hits zero.

Also, these operational semantics specify crucial details of low-level code generation by occam-π compilers and the data structures and algorithms necessary for run-time support. The mobile channel *indices* generated by the real compilers are not the increasing sequence of natural numbers specified in Section 3. Actual memory addresses, dynamically allocated for the bundle structures, are used instead. This is *safe*, since these indices only need to be unique for each existing bundle, and *fast*, since the address gives direct access. Finally, the occam-π kernel is not actually implemented as the parallel collection specified by `MOBILE_CHANNEL_KERNEL` in Section 3, but by logic statically scheduled from it for serial use by the (extremely lightweight and multicore-safe) processes underlying occam-π.

We have not provided semantics for mobile channels in *arbitrary* CSP systems – only those conforming to the classical occam constraint of no output guards in external choice. That constraint was necessary for the modelling of (mobile) channels as processes, which in turn was necessary for the separate modelling of channel ends.

Without that separation, we could not allow *interleaving* of all application processes over the infinite enumeration of potential mobile channels: since if that were allowed, no application process would be able to synchronise with another to send a message down any mobile channel! With separate static channels modelling the *ends* of mobile channels, application processes synchronise with the *kernel* processes (modeling the mobile channels themselves) to communicate with each other and no buffering is introduced (which means that the fully synchronised semantics of CSP channel events are maintained). Application processes synchronise with each other over the *fixed* set of static channels (and other events) through which with they are connected (as normal) or synchronise with the kernel over the infinite, *but still fixed*, set of kernel channels. Either way, the synchronisation sets associated with all parallel operators remain fixed and we have a CSP for mobile channels.

Recently, fast algorithms for implementing external choice whose leading events are multiway synchronisations have been engineered [15, 16] and have been built into the JCSP library [17] and an experimental compiler for exercising CSP [18]. These can be simply applied to allow output guards.

Providing a CSP model for mobile channels both of whose ends can be used as guards in an external choice is an open question. One way may be to drop the separate modelling of channel ends (along with their supporting kernel processes) and divide application processes into two sets: those that receive mobile channel communications and those that send them. The processes in each set interleave with each other over the set of mobile (but actually normal) channels and the two process sets synchronise with each other over the fixed, but infinite, set of mobiles. This is elegant but not very flexible! Even if the division could be defined individually for each mobile channel, it still looks too constraining. In occam-π, mobile channel bundle ends can be passed to *any* processes with the right connections. Some serial processes may hold and use either end of the same channel bundle – though, hopefully, not at the same time! The semantics presented here for mobile channels without output guards has no such constraints.

Acknowledgements

This work started under the TUNA project (EPSRC EP/C516966/1) and is being applied as part of the CoSMoS project (EPSRC grants EP/E053505/1 and EP/E049419/1). We are indebted to all our colleagues in both these projects for their insights, motivation and encouragement. We are also very grateful to the anonymous referees of this paper and to everyone in the Concurrency sub-group at Kent for their many criticisms and corrections to so many crucial points of detail.

References

[1] P.H. Welch and F.R.M. Barnes. Communicating Mobile Processes: Introducing occam-π. In *'25 Years of CSP'*, Lecture Notes in Computer Science vol. 3525, A. Abdallah, C. Jones, and J. Sanders, editors, Springer, pp. 175–210, Apr. 2005.
[2] F.R.M.Barnes and P.H.Welch. Prioritised Dynamic Communicating and Mobile Processes. *IEE Proceedings-Software*, 150(2):121-136, April 2003.
[3] F.R.M. Barnes and P.H. Welch. Prioritised dynamic communicating processes: Part 1. In James Pascoe et al., editors, *Communicating Process Architectures 2002*, volume 60 of *Concurrent Systems Engineering*, pp. 321-352, IOS Press, Amsterdam, The Netherlands, September 2002.
[4] C.A.R. Hoare. Communicating Sequential Processes. In *CACM*, vol. 21-8, pp. 666-677, August, 1978.
[5] C.A.R. Hoare. Communicating Sequential Processes. Prentice-Hall, ISBN 0-13-153271-5, 1985.
[6] A.W. Roscoe. The Theory and Practice of Concurrency. Prentice-Hall, ISBN 0-13-674409-5, 1997
[7] Formal Systems (Europe) Ltd. FDR2 Manual, **www.fsel.com/documentation/fdr2/fdr2manual.pdf**. May 2008.
[8] R. Milner. Communcating and Mobile Systems: the π-Calculus. Cambridge University Press, ISBN-13:9780521658690, 1999.
[9] Inmos Ltd. occam3 Draft Reference Manual, 1991.
[10] P.H. Welch and F.R.M. Barnes. Mobile Barriers for occam-π: Semantics, Implementation and Application. In J.F. Broenink et al., editors, *Communicating Process Architectures 2005*, volume 63 of *Concurrent Systems Engineering Series*, pp. 289-316, IOS Press, The Netherlands, September 2005.
[11] P.H. Welch and J.M.R Martin. Formal Analysis of Concurrent Java Systems. In A.W.P. Bakkers et al., editors, *Communicating Process Architectures 2000*, volume 58 of *Concurrent Systems Engineering*, pp. 275-301. WoTUG, IOS Press (Amsterdam), September 2000.
[12] J.C.P. Woodcock and A.L.C. Cavalcanti. The Semantics of Circus. In *ZB 2002: Formal Specification and Development in Z and B*, volume 2272 of *Lecture Notes in Computer Science*, pp 184-203, Springer-Verlag, 2002.

[13] S. Stepney, J.C.P. Woodcock, F.A.C. Polack, A.L.C. Cavalcanti, S. Schreider, H.E. Treharne and P.H.Welch. TUNA: Theory Underpinning Nanotech Assemblers (Feasibility Study). EPSRC grant EP/C516966/1 (Universities of York, Surrey and Kent). Final report from `www.cs.york.ac.uk/nature/tuna/`, November, 2006.
[14] S. Stepney, F.A.C. Polack, J. Timmis, A.M. Tyrrell, M.A. Bates, P.H. Welch and F.R.M. Barnes. CoSMoS: Complex Systems Modelling and Simulation. EPSRC grant EP/E053505/1 (University of York) and EP/E049419/1 (University of Kent), October 2007 – September 2011.
[15] A.A. McEwan. Concurrent Program Development, DPhil Thesis, University of Oxford, 2006.
[16] P.H. Welch, F.R.M. Barnes, and F. Polack. Communicating Complex Systems. In *Proceedings of the 11th IEEE International Conference on Engineering of Complex Computer Systems (ICECCS-2006)*, pp 107-117, Stanford University, IEEE Society Press, August 2006.
[17] P.H. Welch, N.C.C. Brown, J. Moores, K. Chalmers and B. Sputh. Integrating and Extending JCSP. In Alistair A. McEwan et al., editors, *Communicating Process Architectures 2007*, volume 65 of *Concurrent Systems Engineering Series*, pp. 349-370, Amsterdam, The Netherlands, July 2007. IOS Press.
[18] F.R.M. Barnes. Compiling CSP. In P.H.Welch et al., editors, *Communicating Process Architectures 2006*, volume 64 of *Concurrent Systems Engineering Series*, pp. 377-388, Amsterdam, The Netherlands, September 2006. IOS Press.

Communicating Scala Objects

Bernard SUFRIN

Oxford University Computing Laboratory
and
Worcester College, Oxford, OX1 2HB, England

`Bernard.Sufrin@comlab.ox.ac.uk`

Abstract. In this paper we introduce the core features of CSO (Communicating Scala Objects) – a notationally convenient embedding of the essence of occam in a modern, generically typed, object-oriented programming language that is compiled to Java Virtual Machine (JVM) code. Initially inspired by an early release of JCSP, CSO goes beyond JCSP expressively in some respects, including the provision of a unitary extended rendezvous notation and appropriate treatment of subtype variance in channels and ports. Similarities with recent versions of JCSP include the treatment of channel ends (we call them *ports*) as parameterized types. Ports and channels may be transmitted on channels (including inter-JVM channels), provided that an obvious design rule – the *ownership* rule – is obeyed. Significant differences with recent versions of JCSP include a treatment of network termination that is significantly simpler than the "poisoning" approach (perhaps at the cost of reduced programming convenience), and the provision of a family of type-parameterized channel implementations with performance that obviates the need for the special-purpose scalar-typed channel implementations provided by JCSP. On standard benchmarks such as Commstime, CSO communication performance is close to or better than that of JCSP and Scala's Actors library.

Keywords. occam model, concurrency, Scala, JCSP.

Introduction

On the face of it the Java virtual machine (JVM) is a very attractive platform for realistic concurrent and distributed applications and systems. On the other hand, the warnings from at least parts of the "Java establishment" to neophyte Java programmers who think about using threads are clear:

> If you can get away with it, avoid using threads. Threads can be difficult to use, and they make programs harder to debug.

> It is our basic belief that extreme caution is warranted when designing and building multi-threaded applications ... use of threads can be very deceptive ... in almost all cases they make debugging, testing, and maintenance vastly more difficult and sometimes impossible. Neither the training, experience, or actual practices of most programmers, nor the tools we have to help us, are designed to cope with the non-determinism ... this is particularly true in Java ... we urge you to think twice about using threads in cases where they are not absolutely necessary ...[8]

But over the years JavaPP, JCSP, and CTJ [7,3,4,1,2] have demonstrated that the occam programming model can be used very effectively to provide an intellectually tractable *discipline* of concurrent Java programming that is harder to achieve by those who rely on the lower level, monitor-based, facilities provided by the Java language itself.

So in mid-2006, faced with teaching a new course on concurrent and distributed programming, and wanting to make it a *practical* course that was easily accessible to Java programmers, we decided that this was the way to go about it. We taught the first year of this course using a Java 1.5 library that bore a strong resemblance to the current JCSP library.[1]

Our students' enthusiastic reaction to the occam model was as gratifying as their distaste for the notational weight of its embedding in Java was dismaying. Although we discussed *designs* for our concurrent programs using a CSP-like process-algebra notation and a simplified form of ECSP [5,6], the resulting *coding gap* appeared to be too much for most of the students to stomach.

At this point one of our visiting students introduced us to Scala [9], a modern object-oriented language that generates JVM code, has a more subtle generic type system than Java, and has other features that make it very easy to construct libraries that appear to be notational extensions.

After toying for a while with the idea of using Scala's Actor library [12,13], we decided instead to develop a new Scala library to implement the occam model independently of existing Java libraries,[2] and of Scala's Actor library.[3] Our principal aim was to have a self-contained library we could use to support subsequent delivery of our course (many of whose examples are toy programs designed to illustrate patterns of concurrency), but we also wanted to explore its suitability for structuring larger scale Scala programs.

This paper is an account of the most important features of the core of the Communicating Scala Objects (CSO) library that emerged. We have assumed some familiarity with the conceptual and notational basis of occam and JCSP, but only a little familiarity with Scala.

Readers familiar with JCSP and Scala may be able to get a quick initial impression of the relative notational weights of Scala+CSO and Java+JCSP by inspecting the definitions of FairPlex multiplexer components defined on pages 45 and 54 respectively.

1. Processes

A CSO process is a value with Scala type PROC and is what an experienced object oriented programmer would call a *stereotype* for a thread. When a process is *started* any fresh threads that are necessary for it to run are acquired from a thread pool.[4]

1.1. Process Notation

Processes (p : PROC) are values, denoted by one of:

proc { $expr$ }	A simple process ($expr$ must be a command, *i.e.* have type Unit)
$p_1 \;\|\|\; p_2 \;\|\|\; ... \;\|\|\; p_n$	Parallel composition of n processes (each p_i must have type PROC)
$\|\|$ $collection$	Parallel composition of a finite collection of PROC values. When $collection$ comprises $p_1...p_n$ this is equivalent to $p_1 \;\|\|\; p_2 \;\|\|\; ... \;\|\|\; p_n$.

---

[1]This was derived from an earlier library, written in Generic Java, whose development had been inspired by the appearance of the first public edition of JCSP. The principal differences between that library and the JCSP library were the generically parameterized interfaces, InPort and OutPort akin to modern JCSP channel ends.

[2]Although Scala interoperates with Java, and we could easily have constructed Scala "wrappers" for the JCSP library and for our own derivative library, we wanted to have a pure Scala implementation both to use as part of our instructional material, and to ensure portability to the .NET platform when the Scala .NET compiler became available.

[3]The (admirably ingenious) Actor library implementation is complicated; its performance appears to scale well only for certain styles of use; and it depends for correct functioning on a global timestamp ([13] p183).

[4]The present JVM implementation uses a thread pool from the Java concurrent utility library, though this dependency is really not necessary.

A frequently-occuring pattern of this latter form of composition is one in which the collection is an iterated form, such as: || (**for** (i<−0 until n) **yield** $p(i)$). This form denotes a process equivalent to: $p(0)\ ||\ p(1)\ ||\ ...\ ||\ p(n-1)$,

1.2. Starting and Running Processes

If p is a process, then evaluation of the expression $p()$ runs the process.[5] The following cases are distinguished:

1. p is **proc** $\{\ expr\ \}$
   - $p()$ causes $\{\ expr\ \}$ to be evaluated in the current thread.
   - The process as a whole terminates when the evaluation of $\{\ expr\ \}$ terminates or throws an (uncaught) exception.
   - The behaviour of the expression p() cannot be distinguished from that of the expression {expr}.

2. p is $p_1\ ||\ p_2\ ||\ ...\ ||\ p_n$
   - $p()$ causes all the processes $p_1...p_n$ to be run concurrently.
   - All but one of the processes is run in a new thread; the remaining process is run in the current thread.
   - The process as a whole terminates only when *all* the component p_i have terminated. But if any of the component p_i terminated by throwing an uncaught exception then – *when and only when they have all terminated* – one of those exceptions is chosen nondeterministically and re-thrown; in this case exceptions that are subtypes of cso.Stop are only chosen in the absence of other types of exception.[6]

2. Ports and Channels

2.1. Introduction

Following ECSP [5,6], CSO ports (akin to JCSP channel ends) are generically parameterized, and we define the abbreviations ?[T] and ![T] respectively for InPort[T] and OutPort[T].

The most important method of an ![T] is its write method

    ! (value : T)

and the most important methods of an ?[T] are its read method

    ? () : T

and its *extended rendezvous* method

    ? [U] (body : T => U) : U

The type Chan[T] is the interface implemented by all channels that carry values of type T: it is declared by:

    trait Chan[T] extends InPort[T] with OutPort[T] { ... }

---

[5] A process also has a fork method that runs it in a new thread concurrent with the thread that invoked its fork method. The new thread is recycled when the process terminates.

[6] This is because cso.Stop type exceptions signify anticipated failure, whereas other types signify unexpected failure, and must be propagated, rather than silently ignored. One useful consequence of the special treatment of cso.Stop exceptions is explained in section 4: *Closing Ports and Channels*.

This makes Chan[T] a subtype of both InPort[T] and OutPort[T]. It makes sense to think of a Chan as embodying both an InPort and an OutPort.

The implicit contract of every conventional Chan implementation is that it delivers the data written at its output port to its input port in the order in which the data was written. Different implementations have different synchronization behaviours and different restrictions on the numbers of processes that may access (*i.e.* use the principal methods of) their ports at any time.

The CSO core comes with several predefined channel implementations, the most notable of which for our present purposes are:

- The *synchronous* channels. These all synchronize termination of the execution of a ! at their output port with the termination of the execution of a corresponding ? at their input port.
    * OneOne[T] – No more than one process at a time may access its output port or its input port.[7] This is the classic occam-style point to point channel.
    * ManyOne[T] – No more than one process at a time may access its input port; processes attempting to access its output port get access in nondeterministic order.[8]
    * OneMany[T] – No more than one process at a time may access its output port; processes attempting to access its input port get access in nondeterministic order.
    * ManyMany[T] – Any number of processes may attempt to access either port. Writing processes get access in nondeterministic order, as do reading processes.

- Buf[T](n) – a many-to-many buffer of capacity n.[9]

Access restrictions are enforced by a combination of:

- Type constraints that permit sharing requirements to be enforced statically.
    * All output port implementations that support shared access have types that are subtypes of SharedOutPort.
    * All input port implementations that support shared access have types that are subtypes of SharedInPort.
    * All channel implementations that support shared access to both their ports have types that are subtypes of SharedChannel.
    * Abstractions that need to place sharing requirements on port or channel parameters do so by declaring them with the appropriate type.[10]

- Run-time checks that offer *partial* protection against deadlocks or data loss of the kind that can could otherwise happen if unshareable ports were inadvertently shared.
    * If a read is attempted from a channel with an unshared input port before an earlier read has terminated, then an illegal state exception is thrown.
    * If a write is attempted to a channel with an unshared output port before an earlier write has terminated, then an illegal state exception is thrown.

    These run-time checks are limited in their effectiveness because it is still possible for a single writer process to work fast enough to satisfy illegitimately sharing reader processes without being detected by the former check, and for the dual situation to remain undetected by the latter check.

---

[7]The name is a contraction of "From One writer process to One reader process."

[8]The name is a contraction of "From Many possible writer processes to One reader process." The other forms of synchronous channel are named using the same contraction convention.

[9]We expect that history will soon give way to logic: at that point each form of synchronous channel will be supplemented by an aptly-named form of buffered channel.

[10]See, for example, the component mux2 defined in program 3.

```
def producer(i: int, ![T]) : PROC = ...
def consumer(i: int, ?[T]) : PROC = ...

def mux[T] (ins: Seq[?[T]],    out: ![(int, T)]) : PROC = ...
def dmux[T](in:  ?[(int, T)], outs: Seq[![T]])   : PROC = ...


val left, right = OneOne[T](n)   // 2 arrays of n unshared channels
val mid = OneOne[(int, T)]       // an unshared channel

(  || (for (i<-0 until n) yield producer(i, left(i)))
|| mux(left, mid)
|| dmux(mid, right)
|| || (for (i<-0 until n) yield consumer(i, right(i)))
)()
```

Program 1. A network of producers connected to consumers by a multiplexed channel

```
def producer(i: int, ![T]) : PROC = ...
def consumer(i: int, ?[T]) : PROC = ...

val con = OneOne[T](n)      // an array of n unshared channels

(  || (for (i<-0 until n) yield producer(i, con(i)))
|| || (for (i<-0 until n) yield consumer(i, con(i)))
)()
```

Program 2. A network in which producers are connected directly to consumers

2.2. Examples

In program 1 we show how to connect a sequence of n producers to a sequence of n consumers using a single multiplexed channel that carries values accompanied by the index of their producer to a demultiplexer that dispatches these values to the corresponding consumer. Readers familiar with JCSP may find it useful to compare this with the network illustrated in section 1.5 of [4].

As observed in that paper this isn't the most efficient way of connecting the producers to the consumers within a single JVM; and in program 2 we show a network in which producers and consumers are connected directly.

The signatures of the components producer, consumer, mux, and dmux in programs 1 and 2 specify the types of port (channel end) they require; but the subtype relation between channels and ports means that when connecting these components we can simply provide the connecting channels as parameters, and that the components take the required *views* of them. This means we needn't name the ports explicitly, and significantly reduces the degree of *formal clutter* in the network description.[11]

In program 3 we show how to implement two (unfair) multiplexers and a demultiplexer of the kind that might have been used in program 1.

A multiplexer process generated by mux1 is the concurrent composition of a collection of "labelling" processes, each of which outputs labelled copies of its input, via a ManyOne[(int,T)]

---

[11]The reduction of formal clutter comes at the cost of forcing readers to refer back to the component signatures to ascertain which ports they actually use. The JCSP designers made the tradeoff in the other direction.

```
def mux1[T] (ins: Seq[?[T]], out: ![(int, T)]) : PROC =
{ val mid = ManyOne[(int, T)]
  (  proc { while(true) { out!(mid?) } }
  || (for (i<-0 until ins.length) yield proc { mid!(i, ins(i)?) })
  )
}

def mux2[T] (ins: Seq[?[T]], out: SharedOutPort[(int, T)]) : PROC =
    || (for (i<-0 until ins.length) yield proc { out!(i, ins(i)?) })

def dmux[T](in: ?[(int, T)], outs: Seq[![T]]) : PROC =
proc {
   while (true) { val (n, v) = in?; outs(n)!v  }
}
```

Program 3. Two multiplexers and a demultiplexer

channel, to a forwarding process that writes them to the out port. The forwarding process is necessary because the type-signature of mux1 does not constrain the kind of port that is passed to it as a parameter, so in programming it we must assume that it is not shareable.

On the other hand, mux2 requires that its out parameter is shareable, so it composes a collection of labelling processes that write directly to out.

The function dmux generates demultiplexer processes that forward labelled inputs to the appropriate output ports.

3. Extended Rendezvous

3.1. Introduction

As we explained earlier, the *synchronous* channel implementations ensure that *termination* of a write (!) at their output port is synchronized with the termination of the corresponding read (?) at their input port. Although a standard read terminates once the data is transferred between the writer and the reader process, an *extended rendezvous read* permits a computation on the transferred data to take place *in the reader process*, and it is only when this computation terminates that the read is considered to have terminated and the writing process is permitted to proceed.

The usual form of an extended rendezvous read from in: ?[T] is[12]

        in ? { bv => body }

It is evaluated by transferring a value, v, from the process at the output end of the channel (if necessary waiting for one to become ready), then applying the (anonymous) function { bv => body } to v. The read is considered to have terminated when this application has been completely evaluated. At this point the writing process is permitted to proceed and the result of the application is returned from the read.

3.2. Example: Monitoring Interprocess Traffic

An easily understood rationale for extended rendezvous is given in [4]. We are asked to consider how to monitor the interprocess traffic between a producer process connected to a con-

---
[12]The most general form of extended rendezvous read is in?f where f denotes a function of type T=>U. The type of in?f is then U.

sumer process via a simple channel *without interfering with producer-consumer synchronization*. We want to construct a process that is equivalent to

```
{ val mid = Chan[T]
  producer(mid) || consumer(mid)
}
```

but which also copies traffic on mid to a monitor process of some kind.

A first approximation to such a process is

```
{ val left, mon, right = Chan[T]
  (   producer(left)
   || proc { repeat { val v = left?; mon!v; right!v }
   || consumer(right)
   || monitor(mon)
  )
}
```

But this interferes with producer-consumer synchronization, because once left? has been executed, producer is free to proceed. More specifically, it is free to proceed before consumer reads from right. If the context in which this network of process runs is tolerant of an additional degree of buffering this is not problematic; but if it is not, then we need to be able to synchronize the read from right with the write to left.

The problem is solved by replacing the body of the copying process

```
{ val v = left?; mon!v; right!v }
```

with a body in which the outputs to mon and right are part of an extended rendezvous with the producing process, namely:

```
{ left ? { v => {mon!v; right!v} } }
```

The extended rendezvous is executed by reading a value from left, then applying the function { v => {mon!v; right!v} } to it. Termination of the write to left is synchronized with termination of the evaluation of the function body, so the producer writing to left cannot proceed until the consumer has read from right.

The extended rendezvous doesn't terminate until {mon!v; right!v} has terminated, but delays the output to right until the output to mon has terminated. The following reformulation relaxes the latter constraint, thereby removing a potential source of deadlock:

```
{ left ? { v => {(proc{mon!v} || proc{right!v})()} } }
```

It is a simple matter to abstract this into a reusable component:

```
def tap[T](in: ?[T], out: ![T], mon: ![T]) =
  proc
  { repeat { in ? { v => {(proc{mon!v} || proc{out!v})()} } } }
```

3.3. Example: Simplifying the Implementation of Synchronous inter-JVM Channels

Extended rendezvous is also used to good effect in the implementation of synchronized inter-JVM or cross-network connections, where it keeps the overt intricacy of the code manageable. Here we illustrate the essence of the implementation technique, which employs the two "network adapter" processes.

```
def copyToNet[T](in: ?[T], net: ![T], ack: ?[Unit]) =
    proc { repeat { in ? { v => { net!v; ack? } } } }
```

and

```
def copyFromNet[T](net: ?[T], ack: ![Unit], out: ![T]) =
    proc { repeat { out!(net?); ack!() } }
```

The effect of using the extended rendezvous in copyToNet is to synchronize the termination of a write to in with the reception of the acknowledgement from the network that the value written has been transmitted to out.

At the producer end of the connection, we set up a bidirectional network connection that transmits data and receives acknowledgements. Then we connect the producer to the network via the adapter:

```
def producer(out: ![T]) = ...
val (toNet, fromNet): (![T], ?[Unit]) = ...
val left = OneOne[T]
( producer(left) || copyToNet(left, toNet, fromNet) )()
```

At the consumer end the dual setup is employed

```
def consumer(in: ?[T]) = ...
val (toNet, fromNet): (![Unit], ?[T]) = ...
val right = OneOne[T]
( copyFromNet(fromNet, toNet, right) || consumer(right) )()
```

In reality the CSO networking components deliver their functionality at a higher level of abstraction than this, namely bidirectional client/server connections, and the synchronous implementations piggy-back acknowledgements to client requests on top of server responses.

4. Closing Ports and Channels

4.1. Introduction

A port may be *closed* at any time, *including after it has been closed*. The trait InPort has method

> closein: Unit

whose invocation embodies a promise *on the part of its invoking thread* never again to read from that port.

Similarly, the trait OutPort has method

> closeout: Unit

whose invocation embodies a promise *on the part of its invoking thread* never again to write to that port.

It can sometimes be appropriate to *forbid* a channel to be used for further communication, and the Chan trait has an additional method for that purpose, namely:

> close: Unit

The important design questions that must be considered are:

1. What happens to a process that attempts, or is attempting, to communicate through a port whose peer port is closed, or which closes during the attempt?
2. What does it mean to close a *shared* port?

Our design can be summarised concisely; but we must first explain what it means for a channel to be closed:

Definition: A channel is *closed* if it has been closed at a *non-shared* OutPort by invoking its closeout method, or if it has been closed at a *non-shared* InPort by invoking its closein method, or if it has been closed by invoking its close method.[13]

This means that closing a shared *port* has no effect. The rationale for this is that shared ports are used as "meeting points" for senders and receivers, and that the fact that one sender or receiver has undertaken never to communicate should not result in the right to do so being denied to others.[14]

The effects of closing ports and/or channels now can be summarised as follows:

- Writer behaviour
  1. An attempt to write to a closed channel raises the exception Closed in the writing thread.
  2. Closing a channel whose OutPort is waiting in a write raises the exception Closed in the writing thread.
- Reader behaviour
  1. An attempt to read from a closed channel raises the exception Closed in the reading thread.
  2. Closing a channel whose InPort is waiting in a read raises the exception Closed in the reading thread.

4.2. Termination of Networks and Components

The Closed exception is one of a family of runtime exceptions, the Stop exceptions, that play a special role in ensuring the clean termination of networks of communicating processes.

The form

$$\textbf{repeat} \ (expr_{guard}) \ \{ \ expr_{body} \ \}$$

behaves in exactly the same way as

$$\textbf{while} \ (expr_{guard}) \ \{ \ expr_{body} \ \}$$

except that the raising of a Stop exception during the execution of the $expr_{body}$ causes it to terminate normally. The form **repeat** $\{ \ expr_{body} \ \}$ is equivalent to **repeat** (**true**) $\{ \ expr_{body} \ \}$

The behaviour of **repeat** simplifies the description of cleanly-terminating iterative components that are destined to be part of a network. For example, consider the humble copy component of program 4, which has an iterative copying phase followed by a close-down phase. It is evident that the copying phase terminates if the channel connected to the input port is closed before that connected to the output port. Likewise, if the channel connected to the output port is closed before (or within) a write operation that is attempting to copy a recently-read datum. In either case the component moves into its close-down phase, and this

---

[13] In the case of buffered (non-synchronized) channels, the effect of invoking close is immediate at the InPort, but is delayed at the OutPort until any buffered data has been consumed.

[14] This is a deliberate choice, designed to keep shared channel semantics simple. More complex channel-like abstractions – such as one in which a non-shared end is informed when all subscribers to the shared end have disappeared – can always be layered on top of it.

```
def copy[T](in: ?[T], out: ![T]) =
proc {
  repeat { out!(in?) }                              // copying
  (proc { out.closeout } || proc { in.closein })()  // close-down
}
```

Program 4. A terminating copy component

results in one of the channels being closed again while the other is closed anew. In nearly all situations this behaviour is satisfactory, but it is worth noticing that it can result in a datum being silently lost (in the implicit buffer between the in? and the out!) when a network is closed from "downstream".[15]

In section 1.2 we explained that on termination of the components of a concurrent process: (a) if any of the component processes themselves terminated by throwing an exception then one of those exceptions is chosen nondeterministically and re-thrown; and (b) in making the choice of exception to throw, preference is given to Stop exceptions.

One consequence of (b) is that it is relatively simple to arrange to reach the closedown phase of an iterated component that does concurrent reads and/or writes. For example, the tee component below broadcasts data from its input port to all its output ports concurrently: if the input port closes, or if any output port is closed before or during a broadcast, then the component stops broadcasting and closes all its ports.

```
def tee[T](in: ?[T], outs: Seq[![T]]) =
proc
{ var data      = _   // unspecified initial value
  val broadcast = || for (out<-outs) yield proc { out!data }
  repeat { in ? { d => { data=d; broadcast() }}}
  (|| (for (out<-outs) yield proc { out.closeout }) || in.closein)()
}
```

This is because closing in results in a Closed exception being thrown at the next in?; and because closing an output port causes the corresponding out!data to terminate by throwing a Closed, which is propagated in turn by the || when it terminates.[16]

Careful programming of the closedown phases of communicating components is needed in order to assure the clean termination of networks of interconnected processes, and this is facilitated by the Stop-rethrowing behaviour of ||, and the behaviour of **repeat** when its body Stops.

---

[15]*i.e.* from the out direction. On the face of it it looks like this could be avoided by reprogramming the component with a stronger guard to the iteration, *viz* as: **repeat** (out.open) { out!(in?) } *but this is not so*, because the out.open test and the out! action are not joined atomically, so the channel associated with the output port could be closed between being polled in the guard and being written to in the body of the loop.

[16]Although it is incidental to the theme of this example, it is worth noticing that we construct the concurrent process broadcast before starting the iteration. While this is not strictly necessary, it provides an improvement in efficiency over: **repeat** { in ? { d => {|| (**for** (out<-outs) **yield proc** { out!d })() }}}. This is because the expression: ||(**for** (out<-outs) ...) that constructs the concurrent broadcast process is evaluated only once, rather than being evaluated once per broadcast.

5. Input Alternation

5.1. Introduction

Input alternations are first class values of type Alt. The simplest form of an **alt** is constructed from a sequence of *guarded events*:[17]

$$\textbf{alt} \ (\ port_1 \ (guard_1) \ ==> \ \{ \ cmd_1 \ \}$$
$$| \ \ldots$$
$$| \ port_n \ (guard_n) \ ==> \ \{ \ cmd_n \ \}$$
$$)$$

An event of the form $port \ (guard) \ ==> \ \{ \ cmd \ \}$

- is said to be *enabled*, if $port$ is open and $guard$ evaluates to true
- is said to be *ready* if $port$ is ready to read
- is *fired* by executing its cmd (which *must* read $port$)

If a is an **alt**, then $a()$ starts its execution, which in principle[18] proceeds in phases as follows:

1. All the event guards are evaluated, and then
2. The current thread waits until (at least one) enabled event is ready, and then
3. One of the ready events is chosen and fired.

If no events are enabled after phase 1, or if all the channels associated with the ports close while waiting in phase 2, then the Abort exception (which is also a form of Stop exception) is raised.

If a is an **alt**, then a **repeat** executes these phases repeatedly, but the choices made in phase 3 are made in such a way that if the same group of guards turn out to be ready during successive executions, they will be fired in turn.

For example, the method tagger below constructs a tagging multiplexer that ensures that neither of its input channels gets too far ahead of the other. The tagger terminates cleanly when its output port is closed, or if both its input channels have been closed.

```
def tagger[T](l: ?[T], r: ?[T], out: ![(int, T)]) =
proc
{ var diff = 0
    alt (  l  (!r.open || diff < 5 ) ==> { out!(0, l?); diff+=1 }
        |  r  (!l.open || diff > -5) ==> { out!(1, r?); diff-=1 }
       ) repeat;
    ( proc {l.closein} || proc {r.closein} || proc {out.closeout} )()
}
```

A **prialt** is constructed in the same way as an **alt**, and is executed in nearly the same way, but the choice of which among several ready guards to fire always favours the earliest in the sequence.

5.2. Collections of Guards

Alternations can be composed of collections of guards, as illustrated by the fair multiplexer defined below.[19]

---

[17] Guard expressions must be free of side-effects, and a $(guard)$ that is literally (true) may be omitted.

[18] We say "in principle" because we wish to retain the freedom to use a much more efficient implementation than is described here.

[19] It is perhaps worthwhile comparing this construction with that of the analogous JCSP component shown in program 11 (page 54).

```
def fairPlex[T](ins: Seq[?[T]], out: ![T]) =
    proc { alt (for (in <- ins) yield in ==> { out!(in?) }) repeat }
```

They can also be composed by combining collections and single guards. For example, the following is an extract from a multiplexer than can be dynamically set to favour a specific range of its input ports. It gives priority to its range-setting channels.

```
def primux[T](MIN: ?[int], MAX: ?[int], ins: Seq[?[T]], out: ![T]) =
proc
{ var min = 0
  var max = ins.length - 1
  prialt ( MIN ==> { min = MIN? }
         | MAX ==> { max = MAX? }
         || (for (i <- 0 until ins.length) yield
                    ins(i) (max>=i && i>=min) ==> { out!(ins(i)?) } )
         ) repeat
}
```

5.3. Timed Alternation

An alternation may be qualified with a deadline, after which failure of any of its enabled ports to become ready causes an Abort exception to be thrown. It may also be qualified with code to be executed in case of a timeout – in which case no exception is thrown.[20] We illustrate both of these features with an extended example, that defines the transmitter and receiver ends of an inter-JVM buffer that piggybacks "heartbeat" confirmation to the receiving end that the transmitting end is still alive.

First we define a Scala type Message whose values are of one of the forms Ping or Data(v).

```
trait  Message
case   object    Ping                     extends Message {}
case   class     Data[T](data: T) extends Message {}
```

The transmitter end repeatedly forwards data received from in to out, but intercalates Ping messages whenever it has not received anything for pulse milliseconds.

```
def transmitter[T](pulse: long, in: ?[T], out: ![Message]) =
proc
{ alt (in==>{out!Data(in?)}) before pulse orelse { out!Ping } repeat }
```

The receiver end (whose pulse should be somewhat slower than that of the transmitter) repeatedly reads from in, discarding Ping messages and forwarding ordinary data to out. If (in each iteration) a message has not been received before the timeout, then a message is sent to the fail channel.

```
def receiver[T](pulse: long, in: ?[Message], out: ![T], fail: ![Unit]) =
  proc
  { alt (in ==>
            { in ? match
                { case Ping     => ()
                  case Data(d:T) => out!d
                }
            }
        ) before pulse orelse { fail!() } repeat
  }
```

---

[20]The implementation of this feature is straightforward, and not subject to any potential races.

Though timeout is cheap and safe to implement,[21] the technique used above may not be suitable for use in components where there is a need for more subtle interplay between timing and channel input. But such components can always be constructed (and in a way that may be more familiar to occam programmers) by using periodic timers, such as the simple and straightforward one shown in program 6.

For example, program 5 shows the definition of an alternative transmitter component that "pings" if the periodic timer ticks twice without an intervening input becoming available from in, and "pongs" every two seconds regardless of what else happens.

```
def transmitter2[T](pulse: long, in: ?[T], out: ![Message]) =
proc
{ val tick   = periodicTimer(pulse)
  val tock   = periodicTimer(2000)
  var ticks = 0
  prialt ( tock ==> { out!Pong; tock? }
         | in   ==> { out!Data(in?); ticks = 0 }
         | tick ==> { ticks+=1; if (ticks>1) out!Ping; tick? }
         ) repeat;
  tick.close
  tock.close
}
```

Program 5. A conventionally-programmed transmitter

In the periodic timer of program 6 the fork method of a process is used to start a new thread that runs concurrently with the current thread and periodically writes to the channel whose input port represents the timer. Closing the input port terminates the **repeat** the next time the interval expires, and thereby terminates the thread.

```
def periodicTimer(interval: long) : ?[Unit] =
{ val chan = OneOne[Unit]
  proc { repeat { sleep(interval); chan!() } } . fork
  return chan
}
```

Program 6. A simple periodic timer

6. Port Type Variance

As we have seen, port types are parameterized by the types of value that are expected to be read from (written to) them. In contrast to Java, in which all parameterized type constructors are covariant in their parameter types, Scala lets us specify the variance of the port type constructors precisely. Below we argue that the InPort constructor should be covariant in its type parameter, and the OutPort constructor contravariant in its type parameter. In other words:

1. If T' is a subtype of T, then a $?[T']$ will suffice in a context that requires a $?[T]$; but not *vice-versa*.

---

[21]By imposing an explicit time limit on the wait call that implements the notional second phase of the **alt**.

2. If T' is a subtype of T, then a $![T]$ will suffice in a context that requires a $![T']$; but not *vice-versa*.

Our argument is, as it were, by contradiction. To take a concrete example, suppose that we have an interface Printer which has subtype BonjourPrinter that has an additional method, bonjour.

Suppose also that we have process generators:

```
def printServer    (printers: ![Printer])       : PROC = ...
def bonjourClient(printers: ?[BonjourPrinter]) : PROC = ...
```

Then under the uniformly covariant regime of Java the following program would be type valid, but it would be unsound:

```
val connector = new OneOne[BonjourPrinter]
  (printServer(connector) || printClient(connector))()
```

The problem is that the server could legitimately write a non-bonjour printer that would be of little use to a client that expects to read and use bonjour printers. This would, of course, be trapped as a runtime error by the JVM, but it is, surely, bad engineering practice to rely on this lifeboat if we can avoid launching a doomed ship in the first place![22] And we can: for under CSO's contravariant typing of outports, the type of connector is no longer a subtype of ![Printer], and the expression printServer(connector) would, therefore, be ill-typed.

7. Bidirectional Connections

In order to permit client-server forms of interaction to be described conveniently CSO defines two additional interface traits:

```
trait Connection.Client[Request,Reply] extends OutPort[Reply]
                                       with    InPort[Request]  { ... }

trait Connection.Server[Request,Reply] extends OutPort[Request]
                                       with    InPort[Reply]    { ... }
```

Thus a Server interface is something to which requests are written and from which replies are read, while a Client interface is something from which requests are read and to which replies are written.

A Connection[Request,Reply] has a client interface and a server interface:

```
trait Connection[Request, Reply]
{ def client : Connection.Client[Request,Reply]
  def server : Connection.Server[Request,Reply]
}
```

The implicit contract of a connection implementation is that requests written to its server interface by the code of a client should eventually be readable by the code of the corresponding server in the order in which they were written; likewise responses written to its client interface by the code of a server should eventually be readable by the code of the corresponding client in the order they were written. Different connection implementations implement "eventually" in different ways. The simplest of these is a

---

[22]This difficulty is analogous to the well-known difficulty in Java caused by the covariance of the array constructor.

```
Connection.OneOne[Request, Reply]
```

which connects both directions synchronously.

It is worth noticing that both Client and Server interfaces can be viewed as both an InPort and an OutPort. This lends an air of verisimilitude to the wrong idea that "a connection is a bidirectional channel", but nevertheless contributes to the lack of formal clutter in the programming of clients and servers.

For example, program 7 shows a process farmer component that acquires requests from its in port, and farms them out to servers from which it eventually forwards replies to its out port. This implementation is a little inefficient because we enable *all* the server guards when *any* server is busy.

```
def farmer[Req,Rep]( in:       ?[Req]
                   , out:      ![Rep]
                   , servers:  Seq[Server[Req,Rep]]
                   ) =
proc
{ var busy = 0                       // number of busy servers
  val free = new Queue[![Req]]       // queue of free server connections
  free ++= servers                   // initially all are free
  // INVARIANT: busy+free.length=servers.length
  alt ( | (for (server <- servers) yield
              server (busy>0) ==>
              { out ! (server?)
                free += server
                busy = busy-1
              }
          )
        | in (free.length >0) ==>
          { val server = free.dequeue
            busy = busy+1
            server ! (in?)
          }
      ) repeat
}
```

Program 7. A Process Farmer

8. Performance

The Commstime benchmark has been used as a measure of communication and thread context-swap efficiency for a number of implementations of occam and occam-like languages and library packages. Its core consists of a cyclic network of three processes around which an integer value, initially zero, is circulated. On each cycle the integer is replaced by its successor, and output to a fourth process, Consumer, that reads integers in batches of ten thousand, and records the time per cycle averaged over each batch.

The network is shown diagrammatically in figure 1. Its core components are defined (in two variants) with CSO in program 8, and with Actors in program 9. The SeqCommstime variant writes to Succ and Consumer sequentially. The ParCommstime variant writes to Consumer and Succ concurrently, thereby providing a useful measure of the overhead of starting the additional thread per cycle needed to implement ParDelta.

In table 1 we present the results of running the benchmark for the current releases of Scala Actors, CSO and JCSP using the latest available Sun JVM on each of a range of host

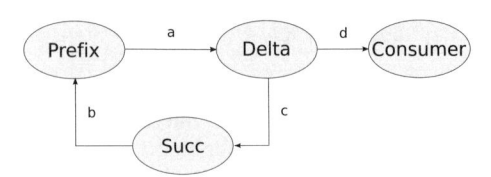

Figure 1. The Commstime network

```
val  a,b,c,d       = OneOne[int]
val  Prefix        = proc { a!0; repeat { a!(b?) }}
val  Succ          = proc { repeat { b!(c?+1) } }
val  SeqDelta      = proc { repeat { val n=a?; c!n; d!n } }
val  SeqCommstime  = (Prefix || SeqDelta || Succ || Consumer)

val  ParDelta      = proc { var n   = a?;
                            val out = proc{c!n} || proc{d!n}
                            repeat { out(); n=a? }
                          }
val  ParCommstime  = (Prefix || ParDelta || Succ || Consumer)
```

Program 8. Parallel and Sequential variants of the Commstime network defined with CSO

```
type Node    = OutputChannel[int]
val  Succ    : Node =
     actor { loop { receive { case n:int => Prefix!(1+n)}}}
val  Prefix  : Node =
     actor { Delta!(0); loop { receive { case n:int => Delta!n}}}
val  Delta   : Node =
     actor { loop { receive { case n:int => {Succ!n; Consume!n}}}}
```

Program 9. The Commstime network defined with Actors

types. The JCSP code we used is a direct analogue of the CSO code: it uses the specialized integer channels provided by the JCSP library. Each entry shows the range of average times per cycle over 10 runs of 10k cycles each.

Table 1. Commstime performance of Actors, CSO and JCSP (Range of Avg. μs per cycle)

Host	JVM	Actors	CSO Seq	JCSP Seq	CSO Par	JCSP Par
4 × 2.66GHz Xeon, OS/X 10.4	1.5	28-32	31-34	44-45	59-66	54-56
2 × 2.4GHz Athlon 64X2 Linux	1.6	25-32	26-39	32-41	24-46	27-46
1 × 1.83GHz Core Duo, OS/X	1.5	62-71	64-66	66-69	90-94	80-89
1 × 1.4GHz Centrino, Linux	1.6	42-46	30-31	28-32	49-58	36-40

8.1. Performance Analysis: CSO v. JCSP

It is worth noting that communication performance of CSO is sufficiently close to that of JCSP that there can be no substantial performance disadvantage to using completely generic component definitions.

```
val a,b,c,d        = Buf[int](4)
val BuffPrefix     = proc { a!0; a!0; a!0; a!0; repeat { a!(b?) } }
...
val BuffCommstime  = (BuffPrefix || SeqDelta || Succ || Consumer)
```

Program 10. Buffered Commstime for Actors v. CSO benchmark

It is also worth noting that process startup overhead of CSO is somewhat higher than that of JCSP. This may well reflect the fact that the JCSP Parallel construct caches the threads used in its first execution, whereas the analogous CSO construct re-acquires threads from its pool on every execution of the parallel construct.

8.2. Performance Analysis: Actors v. Buffered CSO

At first sight it appears that performance of the Actors code is better than that of CSO and JCSP: but this probably reflects the fact that Actors communications are buffered, and communication does not force a context switch. So in order to make a like-for-like comparison of the relative communication efficiency of the Actors and CSO libraries we ran a modified benchmark in which the CSO channels are 4-buffered, and 4 zeros are injected into the network by Prefix to start off each batch. The CSO modifications were to the channel declarations and to Prefix – as shown in program 10; the Actors version of Prefix was modified analogously. The results of running the modified benchmark are provided in table 2.

Host	JVM	Actors	CSO
4×2.66 GHz Xeon, OS/X 10.4	1.5	10-13	10-14
2×2.4 GHz Athlon 64X2 Linux	1.6	16-27	4-11
1×1.83 GHz Core Duo, OS/X 10.4	1.5	27-32	14-21
1×1.4 Ghz Centrino, Linux	1.6	42-45	17-19

Table 2. Buffered Commstime performance of Actors and CSO (Range of Avg. μs per cycle)

Space limitations preclude our presenting the detailed results of the further experiments we conducted, but we noted that even when using an *event-based* variant of the Actors code, the performance of the modified CSO code remains better than that of the Actors code, and becomes increasingly better as the number of initially injected zeros increases. The account of the Actors design and implementation given in [12,13] suggests to us that this may be a consequence of the fact that the network is cyclic.[23]

9. Prospects

We remain committed to the challenge of developing Scala+CSO both as a pedagogical tool and in the implementation of realistic programs. Several small-scale and a few medium-scale case studies on networked multicore machines have given us some confidence that our implementation is sound, though we have neither proofs of this nor a body of successful (*i.e.* non-failed) model checks. The techniques pioneered by Welch and Martin in [10] show the way this could be done.

---

[23]This result reinforces our feeling that the only solution of the scaleability problem addressed by the Actors library is a reduction in the cost of "principled" threading. We are convinced that this reduction could be achieved by (re-)introducing a form of lighter-weight (green) threads, and by providing OS-kernel/JVM collaboration for processor-scheduling.

The open nature of the Scala compiler permits, at least in principle, a variety of compile-time checks on a range of design rules to be enforced. It remains to be seen whether there are any combinations of expressively useful Scala sublanguage and "CSO design rule" that are worth taking the trouble to enforce. We have started our search with an open mind but in some trepidation that the plethora of possibilities for aliasing might render it fruitless – save as an exercise in theory.

Finally, we continue to be inspired and challenged by the work of the JCSP team. We hope that new communication and synchronization components similar to some of those they describe in [4] and a networking framework such as that described in [11] will soon find their way into CSO; and if this happens then the credit will be nearly all theirs.

Acknowledgements

We are grateful to Peter Welch, Gerald Hilderink and their collaborators whose early demonstration of the feasibility of using occam-like concurrency in a Java-like language inspired us in the first place. Also to Martin Odersky and his collaborators in the design and implementation of the Scala language.

Several of our students on the Concurrent and Distributed Programming course at Oxford helped us by testing the present work and by questioning the work that preceded and precipitated it. Itay Neeman drew Scala to our attention and participated in the implementation of a prototype of CSO. Dalizo Zuse and Xie He helped us refine our ideas by building families of Web-servers – using a JCSP-like library and the CSO prototype respectively.

Our colleagues Michael Goldsmith and Gavin Lowe have been a constant source of technical advice and friendly skepticism; and Quentin Miller joined us in research that led to the design and prototype implementation of ECSP [5,6].

The anonymous referees' comments on the first draft of this paper were both challenging and constructive; we are grateful for their help.

Last, but not least, we are grateful to Carsten Heinz and his collaborators for their powerful and versatile LaTeX `listings` package [14].

References

[1] Hilderink G., Broenink J., Vervoort W. and Bakkers A. Communicating Java Threads. In *Proceedings of WoTUG-20: ISBN 90 5199 336 6* (1997) 48–76.
[2] Hilderink G., Bakkers A. and Broenink J. A Distributed Real-Time Java System Based on CSP. In *Proceedings of the Third IEEE International Symposium on Object-Oriented Real-Time Distributed Computing*, (2000), 400–407.
[3] Website for Communicating Sequential Processes for Java, (2008)
http://www.cs.kent.ac.uk/projects/ofa/jcsp/
[4] Welch P. et. al. Integrating and Extending JCSP. In *Communicating Process Architectures* (2007) 48–76.
[5] Miller Q. and Sufrin B. Eclectic CSP: a language of concurrent processes. In *Proceedings of 2000 ACM symposium on Applied computing*, (2000) 840–842.
[6] Sufrin B. and Miller Q. Eclectic CSP. *OUCL Technical Report*, (1998)
http://users.comlab.ox.ac.uk/bernard.sufrin/ECSP/ecsp.pdf
[7] Welch P. et. al. Letter to the Editor. *IEEE Computer*, (1997)
http://www.cs.bris.ac.uk/~alan/Java/ieeelet.html
[8] Muller H. and Walrath K. Threads and Swing. *Sun Developer Network*, (2000)
http://java.sun.com/products/jfc/tsc/articles/threads/threads1.html
[9] Odersky M. et. al. An Overview of the Scala Programming Language. *Technical Report LAMP-REPORT-2006-001, EPFL, 1015 Lausanne, Switzerland* (2006) (also at http://www.scala-lang.org/)
[10] Welch P. and Martin J. Formal Analysis of Concurrent Java Systems. In *Proceedings of CPA 2000 (WoTUG-23): ISBN 1 58603 077 9* (2000) 275–301.

[11] Welch P. CSP Networking for Java (JCSP.net)
http://www.cs.kent.ac.uk/projects/ofa/jcsp/jcsp-net-slides-6up.pdf
[12] Haller P. and Odersky M. Event-based Programming without Inversion of Control. In: Lightfoot D.E. and Szyperski C.A. (Eds) JMLC 2006. LNCS 4228, 4–22. Springer-Verlag, Heidelberg (2006)
(also at http://www.scala-lang.org/)
[13] Haller P. and Odersky M. Actors that unify Threads and Events. In: Murphy A.L. and Vitek J. (Eds) COORDINATION 2007. LNCS 4467, 171–190. Springer-Verlag, Heidelberg (2006)
(also at http://www.scala-lang.org/)
[14] Heinz C. The **listings** Package.
(http://www.ifi.uio.no/it/latex-links/listings-1.3.pdf)

Appendix: Thumbnail Scala and the Coding of CSO

In many respects Scala is a conventional object oriented language semantically very similar to Java, though notationally somewhat different.[24] It has a number of features that have led some to describe it as a *hybrid* functional and object-oriented language, notably

- *Case classes* make it easy to represent free datatypes and to program with them.
- *Functions are first-class values*. The type expression T=>U denotes the type of functions that map values of type T into values of type U. One way of denoting such a function anonymously is { bv => body } (providing body has type U).[25]

The principal novel features of Scala we used in making CSO notationally palatable were:

- *Syntactic extensibility*: objects may have methods whose names are symbolic operators; and an object with an apply method may be "applied" to an argument as if it were a function.
- *Call by Name*: a Scala function or method may have have one or more parameters of type => T, in which case they are given "call by name" semantics and the actual parameter expression is evaluated anew whenever the formal parameter name is mentioned.
- *Code blocks*: an expression of the form {...} may appear as the actual parameter corresponding to a formal parameter of type => T.

The following extracts from the CSO implementation show these features used in the implementation of unguarded repetition and **proc**.

```
// From the CSO module: implementing unguarded repetition
def repeat (cmd: => Unit) : Unit =
{ var go = true;
  while (go) try { cmd } catch { case ox.cso.Stop(_,_) => go=false }
}

// From the CSO module: definition of proc syntax
def proc (body: => Unit) : PROC = new Process (null) (()=>body)
```

Implementation of the guarded event notation of section 5 is more complex. The formation of an InPort.Event from the Scala expression $port(guard) \implies \{cmd\}$ takes place in two stages: first the evaluation of $port(guard)$ yields an intermediate InPort.GuardedEvent object, ev; then the evaluation of $ev \implies \{cmd\}$ yields the required event. An unguarded event is constructed in a simple step.

---

[24]The main distributed Scala implementation translates directly into the JVM; though another compiler translates into the .net CLR. The existence of the latter compiler encouraged us to build a pure Scala CSO library rather than simply providing wrappers for the longer-established JCSP library.

[25]In some contexts fuller type information has to be given, as in: { **case** bv: T => body }. Functions may also be defined by cases over free types; for an example see the match expression within receiver in section 5.3

```
// From the definition of InPort[T]: a regular guarded event
def apply (guard: => boolean) = new InPort.GuardedEvent(this, ()=>guard)

// From the definition of InPort.GuardedEvent[T]
def ==> (cmd: => Unit) = new InPort.Event[T](port, ()=>cmd, guard)

// From the definition of InPort[T]: implementing a true-guarded event
def ==> (cmd: => Unit) = new InPort.Event[T](this, ()=>cmd, ()=>true)
```

Appendix: JCSP Fair Multiplexer

Program 11 shows the JCSP implementation of a fair multiplexer component (taken from [3]) for comparison with the CSO implementation of the component with the same functionality in section 5.2.

```
public final class FairPlex implements CSProcess {
    private final AltingChannelInput[] in;
    private final ChannelOutput        out;
    public FairPlex ( AltingChannelInput[] in, ChannelOutput out)
    { this.in = in; this.out = out; }

    public void run () {
        final Alternative alt = new Alternative (in);
        while (true) { final int i = alt.fairSelect ();
                       out.write (in[i].read ());
        }
    }
}
```

Program 11. Fair Multiplexer Component using JCSP

Combining EDF Scheduling with occam using the Toc Programming Language

Martin KORSGAARD [1] and Sverre HENDSETH

Department of Engineering Cybernetics, Norwegian University of Science and Technology

Abstract. A special feature of the occam programming language is that its concurrency support is at the very base of the language. However, its ability to specify scheduling requirements is insufficient for use in some real-time systems. Toc is an experimental programming language that builds on occam, keeping occam's concurrency mechanisms, while fundamentally changing its concept of time. In Toc, deadlines are specified directly in code, replacing occam's priority constructs as the method for controlling scheduling. Processes are scheduled lazily, in that code is not executed without an associated deadline. The deadlines propagate through channel communications, which means that a task blocked by a channel that is not ready will transfer its deadline through the channel to the dependent task. This allows the deadlines of dependent tasks to be inferred, and also creates a scheduling effect similar to priority inheritance. A compiler and run-time system has been implemented to demonstrate these principles.

Keywords. real-time programming, occam, earliest deadline first, EDF, scheduling.

Introduction

A real-time computer system is a system which success depends not only on the computational results, but also on the time the results are delivered. The typical implementation of a real-time system is also concurrent, meaning that it has tasks that are run in parallel. While there are plenty of programming languages designed with concurrency in mind, for instance Ada [1], Java [2], occam [3] or Erlang [4], there are far fewer based on timing. Synchronous programming languages such as Esterel [5] and Lustre [6] are notable exceptions. The timing support in most other programming languages is limited to delays and access to a clock, and scheduling control is handled by specifying task priorities. A more integrated approach, where timing requirements could be specified directly by using simple language constructs, would make it easier to reason intuitively about the temporal properties of a program.

CSP [7,8] is a process algebra designed to model and reason about concurrent programs and systems. The occam programming language is largely based on CSP. The concurrency handling mechanisms of occam are one of the most fundamental parts of the language. Parallel execution is achieved using a simple `PAR` constructor, and is just as easy as creating serial execution (which requires the mandatory `SEQ` constructor). The channel data type provides native synchronous communication and synchronization between the parallel processes, the only type of inter-process communication allowed. occam has built-in language support for delays and clocks, but lacks a robust way of controlling scheduling, which can cause difficulties when implementing real-time systems [9]. One problem is that priorities are static, disabling online control over scheduling. Another problem is that certain combinations of `PAR`,

---

[1] Corresponding Author: *Martin Korsgaard, Department of Engineering Cybernetics, 7491 Trondheim, Norway.* Tel.: +47 73 59 43 76; Fax: +47 73 59 45 99; E-mail: `martin.korsgaard@itk.ntnu.no`.

PRI PAR, ALT, and PRI ALT can yield intuitively ambiguous programs. In some cases, the addition of prioritized operators can change the logical behaviour of a program [10]. An extended version of occam, called occam-π, adds many new features to occam, including process priority control [11].

Ada [1] is a programming language designed to ensure safe real-time programming for critical systems. It contains both asynchronous and synchronous concurrency functions, the latter which also is influenced by CSP. Ada allows specification of absolute task priorities. Specification of deadlines became possible from Ada 2005. The Ravenscar profile, however, which defines a safer subset of Ada more suitable to critical systems, permits neither asynchronous interruption of tasks nor synchronous concurrency. This is done to increase determinism and to decrease the size and complexity of the run-time system [12].

A recurring problem in designing a programming language that is built on timing is the lack of execution time information. In general, one cannot know in advance how long it will take to execute a given piece of code, which severely limits a system's ability to take preemptive measures to avoid missing deadlines. This also reduces the types of implementable timing requirements: An example is executing a task as late as possible before a given deadline, which is impossible without knowledge of the execution time of the task. Measurements of execution time are inadequate because execution time can vary greatly with input data. Certain features of modern computer architectures, such as caches, pipelining and branch prediction, increases the average performance of a computer, but adds complexity that makes execution times even harder to predict. Nevertheless, safe estimates of the worst-case execution time (WCET) of a program can be found using computerized tools such as aiT [13], which Airbus has used with success [14]. However, the process is inconvenient and computationally expensive, and arguably most suitable for offline schedulability analysis of safety critical systems.

Two common real-time scheduling strategies are rate-monotonic scheduling (RMS) and earliest deadline first (EDF) [15]. In RMS, task priorities are ordered by their inverse periods. If a task's relative deadline is allowed to be shorter than its period, then priorities are ordered by inverse relative deadlines instead and the algorithm is called deadline-monotonic scheduling [16]. EDF scheduling works by always executing the task with the earliest absolute deadline.

If there is not enough time to complete all tasks within their deadlines, the system is said to be overloaded. It is a common misconception that RMS is more predictable during overload that EDF, because tasks will miss deadlines in order of priority. This is not correct [17]. If a task has its execution time extended under RMS, it will affect any or all lower priority tasks in no particular order.

EDF and RMS behave differently under permanent overload. RMS will first execute tasks with a higher priority than the tasks that lead to the overload. This can result in some tasks never being executed, but ensures at least that the higher priority tasks are undisturbed. Note that priority is a function of a task's period and not its importance, so this property is not always useful. Under EDF, a task's absolute deadline will always at some point become the earliest in the system, as tasks will be scheduled even if their deadline has already passed. This means that all tasks will be serviced even if the system is overloaded. EDF has the remarkable property of doing period rescaling, where if the system has a utilisation factor $U > 1$, a task with period t_i will execute under an average period of $U \cdot t_i$ [18]. Which of the overload behaviours is the most suitable will depend on the application. For a thorough comparison of the two scheduling algorithms see [17].

In this paper we present Toc, an experimental programming language where the specification of task deadlines is at the very core of the language. In Toc, a deadline is the only reason for execution, and no code is executed without an associated deadline. Processes do not execute simply because they exist, forcing the programmer to make all assumptions on timing explicit. Scheduling is done earliest deadline first, making use of occam's synchronous channels as a way of propagating a deadline from one task to dependent tasks. Deadlines of dependent processes can thus be inferred, so that the timing requirements specified in code can be the actual requirements stemming from the specifications of the system: If a control system should do something in 10 ms, then that should appear exactly one place in the source code as "TIME 10 MSEC". The occam priority construct "PRI PAR" can then be omitted, and the result is a language that can express many types of timing requirements in an elegant manner, and where the timing requirements are clearly reflected in the source code. A prototype Toc compiler and run-time system has been implemented.

The rest of this paper is organized as follows: Section 1 describes the Toc language and gives a few examples of how to specify timing requirements using Toc. Section 2 discusses scheduling and how it is implemented. The scheduling of an example program is explained. In section 3, some of the implications of writing real-time programs in Toc are considered.

1. The Toc Language

Toc is an experimental programming language based on the idea that specification of tasks and deadlines could be fully integrated in to a procedural programming language, and furthermore, that all functionality of a program should be given an explicit deadline in order to be executed. The idea was first described in [19].

1.1. Language Specification

The language is based on occam 2.1. As with occam, statements are called processes, and are divided into primitive and constructed processes. Primitive processes are assignment, input and output on channels, SKIP and STOP. The constructive processes alter or combine primitive processes, and are made using SEQ, PAR, IF, ALT, CASE or WHILE, where the four first may be replicated using FOR. Definitions of the occam processes can be found in the occam reference manual [3]. Toc has the additional TIME constructor, which takes a relative time t, and a process P. It is defined as follows:

> The construct "TIME t : P" has a minimum allowed execution time and maximum desired execution time of t.

The maximum execution time property of TIME sets a deadline. It is specified only as "desired" to account for the possibility that the actual execution time exceeds t. The minimum execution time property sets a minimum time before the construct is allowed to terminate. In practice this is the earliest possible start time of processes following in sequence. This property is absolute, even in cases where it will cause another deadline to be missed. Furthermore, if a TIME construct follows in sequence to another, then the start time of the latter will be set to the exact termination time of the first, removing drift in ready times between consecutive tasks or task instances. The TIME constructor is used to create deadlines, periods and delays. Since all ALTs that are executed already have a timing requirement associated with them, occam's timer guards in alternations can be replaced by a TIMEOUT guard, which is triggered on the expiration of a deadline. Examples are shown in Table 1 and are discussed in the next section.

Central to Toc is the concept of lazy scheduling, where no code is executed unless given a deadline. With the above terminology in place, a more precise definition of the laziness of Toc can be given:

> In Toc, no primitive processes are executed unless needed to complete a process with a deadline.

Table 1. Use of the TIME constructor

#	Use	Code
1	Set deadline d milliseconds to procedure P. The TIME construct is not allowed to terminate before its deadline.	`TIME d MSEC` `  P()`
2	Delay for 2.5 seconds.	`TIME 2500 MSEC` `  SKIP`

3	Periodic process running procedure P, with deadline and period equal to 1 second.	`WHILE TRUE`
		`  TIME 1 SEC`
		`    P()`

4	Periodic process running procedure P, with deadline d and period t. Assumes d < t.	`WHILE TRUE`
		`  TIME t MSEC`
		`    TIME d MSEC`
		`      P()`

5	Sporadic process running procedure P(value) after receiving input on channel ch from another process with a deadline. The sporadic task is given deadline and minimum period d.	`WHILE TRUE`
		`  INT value:`
		`  SEQ`
		`    ch ? value`
		`    TIME d MSEC`
		`      P(value)`

6	Timeout after t microseconds waiting for input on channel ch	`TIME t USEC`
		`  ALT`
		`    ch ? var`
		`      SKIP`
		`    TIMEOUT`
		`      SKIP`

The restriction to primitive processes means that the outer layers of constructed processes are exempted from the laziness rule, and are allowed to execute until an inner primitive process. This restriction is necessary to allow the `TIME` constructors themselves to be evaluated. A periodic process can then be created by wrapping a `TIME` construct in a `WHILE`, without needing to set a deadline for the `WHILE`.

That only non-primitive processes can execute without a deadline does not imply that code without a deadline only requires insignificant execution time. For example, an arbitrarily complex expression may be used as the condition in a `WHILE` construct. It does mean, however, that no code with side-effects will be executed without a deadline, and consequently, that all functionality requiring input or output from a program will need a deadline.

1.2. Examples

A few examples are given in Table 1. Example one and two in Table 1 are trivial examples of a deadline and a delay, respectively. A simple periodic task with deadline equal to period can be made by enclosing a `TIME` block in a loop, as shown in example three. Here, the maximum time property of the `TIME` constructor gives the enclosed process a deadline, and the minimum time property ensures that the task is repeated with the given period. In this example the period and relative deadline are set to one second, which means that unless P misses a deadline, one instance of P will execute every second. The start time reference of each `TIME` construct is set to the termination time of the previous.

`TIME`-constructors can be nested to specify more complex tasks. The fourth example is a task where the deadline is less than the period. This requires two nested `TIME` constructors; the outermost ensuring the minimum period (here p), and the innermost creating the deadline (d). The innermost maximum desired execution time takes precedence over the outermost, because it is shorter; the outermost minimum allowed execution time takes precedence over the innermost, because it is longer. In general, a timing requirement that takes n different times to specify, will take n different `TIME` constructors to implement.

Combining `TIME` with other processes, for instance an input or an `IF`, makes it possible to create sporadic tasks. Example five shows a sporadic task activated by input on a channel. The task runs `P(value)` with deadline d milliseconds after receiving `value` from channel `ch`. Example six shows a simple timeout, where the `TIMEOUT` guard is triggered on the expiration of the deadline.

2. Scheduling

Because Toc specifies timing requirements as deadlines, EDF was the natural choice of scheduling algorithm. A preemptive scheduler was found necessary to facilitate the response-time of short deadline tasks. A task that has the shortest relative deadline in the system when it starts will never be preempted, somewhat limiting the number of preemptions. This is a property of EDF. When the current task does not have the earliest deadline in the system, it will be preempted when the earliest deadline task becomes ready.

There are two major benefits of using EDF. The first is simply that it is optimal; EDF leads to a higher utilization than any priority-based scheduling algorithm. The second reason is the behaviour in an overload situation: With fixed-priority scheduling, a task with a long relative deadline risks never being run during overload. Because Toc has no way of specifying *importance*, assuming that shorter tasks are more important cannot be justified. EDF gives reduced service to all tasks during overload.

2.1. Discovery

A `TIME` constructor that can be reached without first executing any primitive processes must be evaluated immediately, because it may represent the earliest deadline in the system. The process of evaluating non-primitive processes to look for `TIME` constructs is called *discovery*. Discovery only needs to be run on processes where there is a chance that a `TIME` construct can be reached without executing any primitive processes. This limits the need for discovery to the following situations:

- At the start of a program, on procedure `Main`.
- After a `TIME` construct is completed, on processes in sequence.
- After a channel communication, on processes in sequence at both ends.
- On all branches of a `PAR`.

Discovery stops when a primitive process is encountered.

All `TIME` construct found during the same discovery will have the same ready-time, which is the time of the event that caused the discovery. If the event is the end of an earlier `TIME` block with time t, which deadline was not missed, then the ready time of `TIME` blocks found during the subsequent discovery will be precisely t later than the ready time of the first block. Thus, a periodic process such as example #3 in table 1 will have a period of exactly 1 second.

If a deadline is missed, the termination time of the `TIME` block and the ready time of consequent `TIME` blocks are moved accordingly. This also means that there is no attempt to make up for a missed deadline by executing the next instance of a task faster. This behaviour is a good choice in some real-time systems such as computer control systems or media players, but wrong where synchronization to an absolute clock is intended. In this case the task can re-synchronize itself for instance by skipping a sample and issuing a delay.

2.2. Deadline Propagation

The propagation of deadlines through channels is used to make lazy scheduling and EDF work with dependent processes. If a task with a deadline requires communication with another task, the first task will forward execution to the second to allow the communication to complete, in effect transferring its deadline to the second task. The implementation of this feature relies on a property inherited from the occam language, namely the usage rules. In occam, a variable that is written to by a process cannot be accessed by any other processes in parallel. Likewise, a channel that is used for input or output by a process cannot be used for the same by any other processes in parallel. An occam compiler must enforce these rules at compile-time. The process of enforcing the usage rules also gives the channel-ownerships, defined below:

> The input (/output)-owner of a channel is the process whose execution will lead to the next input (/output) on that channel.

The initial input- and output-owner of a channel is the first process following the declaration of the channel. The channel ownerships are updated at run-time at every PAR and end of PAR, using information gathered by the compiler during the usage rules check. With channel ownership defined, the scheduling mechanism for propagating a deadline over a channel becomes simple:

> If the current task needs to complete an input/output on a channel that is not ready, then forward execution to the output-/input-owner of that channel.

Per definition, the owner is the process whose execution will lead up to the next communication on that channel, so this forwarding is the fastest way of completing the communication. The forwarded process is executed up to the point of communication, where execution then continues from the driving side of the channel.

Usage and channel ownership is fairly simple to evaluate for scalars. With arrays, the indexing expressions may have to be evaluated at compile-time, if correct use cannot be ensured without taking the indices into account. In that case indexes will be limited to expressions of constants, literals and replicators (variables defined by a FOR). The compiler will, if necessary, simulate all necessary replicators to find the correct indices of an array used by a process.

2.3. Alternation

The alternation process ALT executes one of its actions that has a ready guard. A guard consists of a Boolean expression and/or a channel input; a guard is ready if its Boolean expression is TRUE and its channel is ready. Alternation behaves differently in Toc than in occam. In occam, the alternation process will wait if no guards are ready, but in Toc there is always a deadline driving the execution, so if there are no inputs in any guards that are ready, then the ALT must drive the execution of one of them until it, or another, becomes ready. Selecting the best alternative to drive forward is either simple or hard, depending on the deadline that drives the ALT itself.

The simple case is if the execution of the ALT is driven by a deadline from one of the guard channels, and there is no boolean guard that blocks the channel. The alternative that is the source of the deadline is then chosen. This represents programs where the ALT is a server accessed by user processes, and only the users have deadlines.

The choice is less obvious if the ALT is driven by its own deadline, or if the driving channel is not an alternative or is disabled by a boolean guard. Here, the program needs to select an alternative that allows it to proceed, ideally in some way that would aid the earliest deadline task. Unfortunately, predicting the fastest way to e.g. a complex boolean guard may be arbitrarily difficult, thereby making it impossible to find the optimal algorithm for choosing ALT branches.

Because a perfect algorithm is not possible, it is important that the algorithm used is intuitively simple to understand for the programmer, and that the behaviour is predictable. The current implementation uses the following pragmatic decision algorithm:

1. If there are ready inputs as alternatives, choose the one with the earliest deadline. Any input that is ready has a deadline associated with it, or it would not have become ready.
2. If there is a ready guard without a channel input (just a Boolean expression that evaluates to TRUE), then choose it.
3. If there exists a channel input alternative that is not disabled by a boolean guard, forward execution to the output-owner of the channel, but do not select the alternative. This is because execution could require input on another alternative of the same ALT, which would cause a deadlock if the first alternative was already selected. At some point, execution will be forwarded back to the ALT, now with an input ready. That input is then selected.
4. If no alternatives are unguarded, act as a STOP. A STOP in Toc effectively hangs the system, because it never terminates but retains the earliest deadline.

2.4. Deadline Inversion

In general, when synchronizing tasks, a task may be in a blocked state where it is not allowed to continue execution until another task has completed some work. The naïve way to schedule such a system is to ignore any blocked tasks and schedule the rest of the tasks normally. This leads to a timing problem known as unbounded priority inversion.

Say the highest priority task is blocked waiting for the lowest priority task. This is a simple priority inversion, and cannot be avoided as long as high and low-priority tasks share resources. The unbounded priority inversion follows when the scheduler selects the second highest priority task to run, leaving the lowest priority task waiting. Now the highest priority task will remain blocked. In effect, all tasks now take precedence over the one with the highest priority.

A reformulation of the problem is that the scheduler does not actively help to execute its most urgent task. One way to alleviate the problem is to use priority inheritance [20]. Using priority inheritance, if a high priority task is blocked waiting for a lower priority task; the lower priority task will inherit the priority of the blocked task, thus limiting the priority inversion to one level. In a sense, the lower priority task completes its execution on behalf of the higher priority task. Priority inheritance has a number of weaknesses; in particular it does not work well with nested critical regions [21]. Other schemes exist, for instance the priority ceiling protocol [22] and the stack resource policy [23].

The Toc notion of priority inversion — or deadline inversion — is different than the one used in classical scheduling. Classical priority inversion is defined for lockable resources with well-defined owners, where the locking process is always the process that will open it later. This property does not apply to systems synchronized by channels. Also, in Toc, tasks are meant to drive the execution of their dependencies. In principle, when communication is done on channels there is a simple deadline inversion every time the earliest deadline task needs to communicate and the other side is not ready. However, this is an inversion by design, and not an unfortunate event.

A situation more similar to a classical priority inversion is when two tasks communicate with a third server, which could represent a lockable resource. If the earliest deadline task is not ready, then the second task may set the server in a state where it is unable to respond to the earliest deadline task when it becomes ready. A third task, with a deadline between that

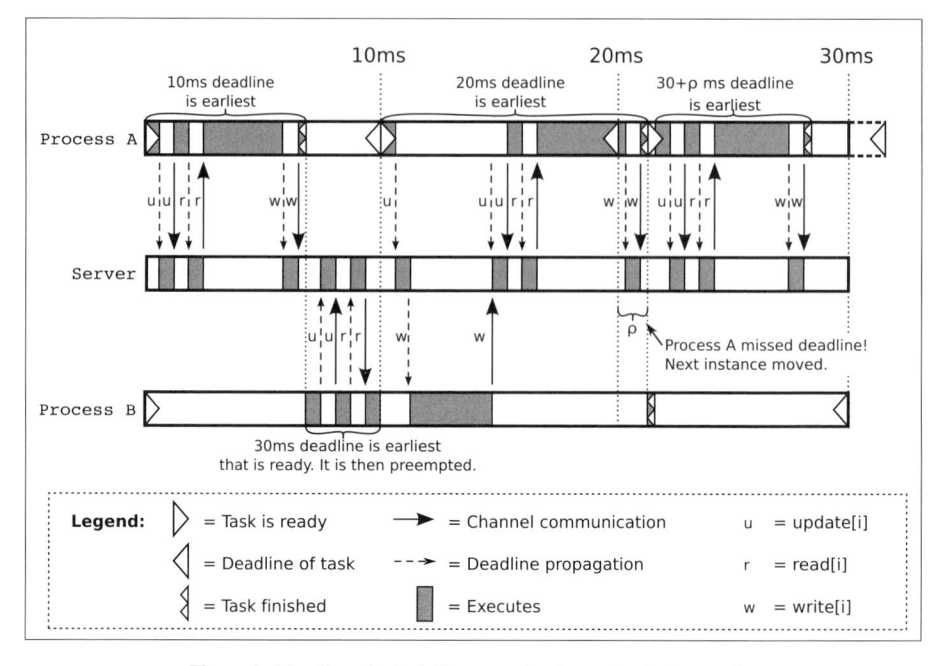

Figure 1. Timeline of scheduling example. Approximate time scale.

of the first and second task will then indirectly block a task with a shorter deadline, possibly leading to unbounded priority inversion.

In Toc, the deadline propagation rule automatically resolves these situations. The earliest deadline process will never be blocked, rather it will transfer its execution and deadline to the blocking processes, so that they execute as if with an earlier deadline. This effect is similar to priority inheritance: With priority inheritance, the blocking process inherits the priority of the blocked process; with deadline propagation, the blocked process will transfer its deadline to the blocking process.

2.5. Scheduling Example

An example of the scheduling of a system with a server and two users is shown in Figures 1 and 2. A server allows a variable to be updated concurrently. Two periodic user tasks access the server. If the server is not ready when the earliest deadline task needs it, then it will finish its current transaction driven by the earliest deadline, the same way it would have executed with a higher priority if priority inheritance was used. A brief explanation is given below:

1. The program starts and executes up to the first primitive processes in all three parallels. Two parallel deadlines are discovered. The user process with the short deadline will be referred to as process A, the other as process B.
2. Process A starts because it has the earliest deadline. It eventually needs to output on channel update[0]. The input-owner of the channel-array is the Server process, so execution is forwarded to the server process through deadline propagation.
3. The server executes up to the communication and zero is sent over the channel.
4. Process A now needs to input from read[0]. First it executes up to the communication and then execution is forwarded to the server. The server executes up to the communication and value is sent over the channel.

```
PROC Server(CHAN[2] INT update?, read!, write?)
  WHILE TRUE
    INT dummy, value:
    ALT i = 0 FOR 2
      update[i] ? dummy
        SEQ
          read[i] ! value
          write[i] ? value
PROC User(VAL INT period, CHAN INT update!, read?, write!)
  WHILE TRUE
    TIME period MSEC
      INT x:
      SEQ
        update ! 0
        read ? x
        WORK 6 MSEC        -- Pseudo-statement for requiring 6 ms of CPU time.
        write ! x+id
        WORK 500 USEC      -- Do some clean-up to finish the task
PROC Main()
  CHAN[2] INT update, read, write:
  PAR
    User(10, update[0], read[0], write[0])    -- Process A
    User(30, update[1], read[1], write[1])    -- Process B
    Server(update, read, write)
```

Figure 2. Code for scheduling example

5. Process A works. Some time later it writes the new value to the server through the `write` channel, and finishes its task.
6. Process B begins in the same way. However, at $t = 10$ms it is preempted by process A, whose task is now ready and has the earliest deadline.
7. Now process B has blocked the server. This is a priority inversion in the classical sense. Process A proceeds as last time, forwarding execution to the input-owner of `update[0]`, which is still the server. To proceed, the server must output on `write[1]`, which is not ready, and forwards execution to the input-owner of that channel (process B).
8. Process B executes up to the output on `write[1]`, driven by the server's need to input on `write[1]`, which again is driven by process A's need to output on `update[0]`. This frees the server and allows process A to continue.
9. Notice that the second instance of process A's task misses its deadline. It has its new period offset accordingly. Also notice that only then is process B allowed to finish.

3. Implications

In occam, a `PRI PAR` or `PRI ALT` can be used to affect the scheduling of parallel processes. This can potentially affect the logical behaviour of a program when modelling the program with CSP, as certain program traces are no longer possible. A `TIME` constructor will also restrict possible traces of a Toc program by always preferring execution of the earliest deadline task to others. The effect of `TIME` constructors on the logical behaviour of a Toc program is much greater than the effect of prioritized constructs in occam programs. For example removing all `TIME` constructors will make any Toc program equal to `STOP`.

Adding a process may also make a program behave as `STOP`. Take the example given in Figure 3. This may be an attempt to initialize a variable before starting a periodic process, but the primitive assignment makes the entire program behave as `STOP`. There is no deadline for

```
PROC Main()
  INT a:
  SEQ
    a := 42
    WHILE TRUE
      TIME 100 MSEC
        P(a)
```

Figure 3. Dead code example. Lack of timing requirements on the assignment makes the program equal to STOP.

```
task body Periodic_Task is                                    WHILE TRUE
  Period      : Time_Span := Milliseconds(30);                  TIME 30 MSEC
  Rel_Deadline : Time_Span := Milliseconds(20);                   TIME 20 MSEC
  Next        : Ada.Real_Time.Time;                                 Action()
begin
  Next := Ada.Real_Time.Clock;
  Set_Deadline(Next+Rel_Deadline);
  loop
    delay until Get_Deadline;
    Action;
    Next := Next + Interval;
    Set_Deadline(Next+Rel_Deadline);
    delay until Next;
  end loop;
end Periodic_Task;
```

Figure 4. Comparison of a periodic process in Ada and Toc. Left: Ada. Right: Toc. Ada example is a modified example from [25].

executing the assignment and therefore it will never happen. The TIME constructor that follows will never be discovered and the periodic process will not start. In Toc, the programmer must consider the deadlines of all functionality in the system, no matter how trivial. Because assignments, like the one in Figure 3, are typically quite fast, not adding a timing requirement signals that the programmer does not care exactly how long time the assignment will take. In Toc, not caring is not an option, and deadlines are always required. The compiler will in many cases issue a warning to avoid such mistakes.

It has been argued that it is more awkward to assign arbitrary deadlines to tasks than to assign arbitrary priorities [24]. This is debatable, but it may anyway be easier to find a *non-arbitrary* deadline than a priority: Many seemingly background tasks can be given sensible deadlines: A key press does not need to give visual feedback faster than it is possible for the human eye to perceive it. A control monitor in a process plant will need some minimum update frequency in order to convey valid information to the operators. Setting a deadline for the latter task, however arbitrary, is a step up from giving it a low priority under fixed-priority scheduling, where a scheduling overload could effectively disable the task.

4. Conclusions and Future Work

This paper presented the language Toc, which allows the specification of deadlines in the source code of a program. This is done using the TIME constructor, which provides elegant language support for specifying deadlines and tasks. A periodic task with a deadline can be implemented with just a few lines of code, compared to the rather more complex construct required in for example Ada, as shown in Figure 4. The combination of EDF and deadline propagation yields a simple and versatile scheduling strategy. Lazy scheduling forces the programmer to consider all timing requirements in the system, not only those that are considered real-time in the classical sense. This may potentially increase awareness of timing requirements for parts of real-time systems for which such requirements were previously ignored, such as sporadic or background tasks and error handling.

The prototype compiler was developed in Haskell, using the parser generator `bnfc` [26]. The compiler generates C code, which can then be compiled with an ordinary C compiler. The current run-time system is written in C, and includes a custom scheduler using POSIX threads. The scheduler is linked into the program, and the resulting executable can be run as an application under another operating system. Both the compiler and run-time system are prototypes under development, and have so far only been used to test small programs, though quite successfully so.

The next task would be to test Toc on a real-time system of some complexity, to see if the `TIME` constructor presented here is practical and suitable for such a task. Of particular interest is seeing how many timing requirements that are actually necessary in the specification of such a system, when no execution is allowed without one.

References

[1] J. D. Ichbiah, B. Krieg-Brueckner, B. A. Wichmann, J. G. P. Barnes, O. Roubine, and J.-C. Heliard, "Rationale for the design of the Ada programming language," *SIGPLAN Not.*, vol. 14, no. 6b, pp. 1–261, 1979.
[2] J. Gosling, B. Joy, G. Steele, and G. Bracha, *The Java Language Specification*, 2000.
[3] SGS-THOMPSON Microelectronics Limited, *occam® 2.1 Reference Manual*, 1995.
[4] J. Armstrong and R. Virding, "ERLANG – an experimental telephony programming language," *Switching Symposium, 1990. XIII International*, vol. 3, 1990.
[5] G. Berry and G. Gonthier, "The ESTEREL synchronous programming language: design, semantics, implementation," *Sci. Comput. Program.*, vol. 19, no. 2, pp. 87–152, 1992.
[6] P. Caspi, D. Pilaud, N. Halbwachs, and J. Plaice, "LUSTRE: A declarative language for programming synchronous systems," *Conference Record of the 14th Annual ACM Symp. on Principles of Programming Languages*, 1987.
[7] C. A. R. Hoare, "Communicating sequential processes," *Communications of the ACM*, vol. 21, pp. 666–677, 1978.
[8] S. Schneider, *Concurrent and Real Time Systems: The CSP Approach*. New York, NY, USA: John Wiley & Sons, Inc., 1999.
[9] D. Q. Z. C. Cecati and E. Chiricozzi, "Some practical issues of the transputer based real-time systems," *Industrial Electronics, Control, Instrumentation, and Automation, 1992. Power Electronics and Motion Control., Proceedings of the 1992 International Conference on*, pp. 1403–1407 vol.3, 9-13 Nov 1992.
[10] C. J. Fidge, "A formal definition of priority in CSP," *ACM Trans. Program. Lang. Syst.*, vol. 15, no. 4, pp. 681–705, 1993.
[11] P. Welch and F. Barnes, "Communicating mobile processes: introducing occam-pi," in *25 Years of CSP* (A. Abdallah, C. Jones, and J. Sanders, eds.), vol. 3525 of *Lecture Notes in Computer Science*, pp. 175–210, Springer Verlag, Apr. 2005.
[12] A. Burns, B. Dobbing, and T. Vardanega, "Guide for the use of the Ada Ravenscar Profile in high integrity systems," *ACM SIGAda Ada Letters*, vol. 24, no. 2, pp. 1–74, 2004.
[13] R. Heckmann and C. Ferdinand, "Worst case execution time prediction by static program analysis," *Parallel and Distributed Processing Symposium, 2004. Proceedings. 18th International*, pp. 125–, 26-30 April 2004.
[14] J. Souyris, E. L. Pavec, G. Himbert, V. Jégu, G. Borios, and R. Heckmann, "Computing the worst-case execution time of an avionics program by abstract interpretation," in *Proceedings of the 5th Intl Workshop on Worst-Case Execution Time (WCET) analysis*, pp. 21–24, 2005.
[15] C. L. Liu and J. W. Layland, "Scheduling algorithms for multiprogramming in a hard-real-time environment," *Journal of the ACM*, vol. 20, no. 1, pp. 46–61, 1973.
[16] J. Leung and J. Whitehead, "On the complexity of fixed-priority scheduling of periodic, real-time tasks," *Performance Evaluation*, vol. 2, no. 4, pp. 237–250, 1982.
[17] G. C. Buttazzo, "Rate monotonic vs. EDF: Judgment day," *Real-Time Syst.*, vol. 29, no. 1, pp. 5–26, 2005.
[18] A. Cervin, J. Eker, B. Bernhardsson, and K.-E. Årzén, "Feedback-feedforward scheduling of control tasks," *Real-Time Systems*, no. 23, pp. 25–53, 2002.

[19] M. Korsgaard, "Introducing time driven programming using CSP/occam and WCET estimates," Master's thesis, Norwegian University of Science and Technology, 2007.
[20] D. Cornhilll, L. Sha, and J. P. Lehoczky, "Limitations of Ada for real-time scheduling," *Ada Lett.*, vol. VII, no. 6, pp. 33–39, 1987.
[21] V. Yodaiken, "Against priority inheritance," tech. rep., FSMLabs, 2002.
[22] L. Sha, R. Rajkumar, and J. P. Lehoczky, "Priority inheritance protocols: an approach to real-time synchronization," *IEEE Transactions on Computers*, vol. 39, no. 9, pp. 1175–1185, Sep 1990.
[23] T. Baker, "A stack-based resource allocation policy for realtime processes," *Real-Time Systems Symposium, 1990. Proceedings., 11th*, pp. 191–200, Dec 1990.
[24] A. Burns and A. Wellings, *Real-Time Systems and Programming Languages*. Essex, England: Pearson Education Limited, third ed., 2001.
[25] A. Burns and A. J. Wellings, "Programming execution-time servers in Ada 2005," *Real-Time Systems Symposium, 2006. RTSS '06. 27th IEEE International*, pp. 47–56, Dec 2006.
[26] M. Pellauer, M. Forsberg, and A. Ranta, "BNF converter: Multilingual front-end generation from labelled BNF grammars," tech. rep., 2004.

Communicating Haskell Processes: Composable Explicit Concurrency using Monads

Neil C.C. BROWN

Computing Laboratory, University of Kent,
Canterbury, Kent, CT2 7NF, England.

neil@twistedsquare.com

Abstract. Writing concurrent programs in languages that lack explicit support for concurrency can often be awkward and difficult. Haskell's monads provide a way to explicitly specify sequence and effects in a functional language, and monadic combinators allow composition of monadic actions, for example via parallelism and choice – two core aspects of Communicating Sequential Processes (CSP). We show how the use of these combinators, and being able to express processes as first-class types (monadic actions) allow for easy and elegant programming of process-oriented concurrency in a new CSP library for Haskell: Communicating Haskell Processes.

Keywords. Communicating Sequential Processes, CSP, Haskell, monads, explicit concurrency.

Introduction

Communicating Sequential Processes (CSP) is a formal algebra encompassing processes, events and synchronous channel communications [1]. CSP is the basis for the occam-π process-oriented programming language [2], and CSP libraries for a wealth of other languages, including: Java [3,4], C++ [5,6], C [7], Python [8] and the .NET languages [9,10].

This paper introduces Communicating Haskell Processes (CHP), a library for Haskell that is also based on CSP. The primary difference between the CHP library and its predecessors is that Haskell is a functional language rather than imperative. However one of Haskell's distinctive features is monads, which allow control over how operations are sequenced and thus allow for imperative-style programming [11].

Haskell's monads and CSP's algebra share an elegant feature: ease of composition. Monadic combinators have been used in the past to implement sequence (and iteration), exceptions, concurrency and choice – CSP allows composition of processes in similar ways. CHP thus marries CSP's process composition with Haskell's monad composition. This allows a level of composition across function/procedure boundaries that is not present in any other CSP-based language or library.

Using our choice operator <-> and our parallelism operator <||> we can write a process that reads from either of its two input channels and sends the value down both its output channels in parallel:

proc (in0, in1) (out0, out1)
  = do x <- readChannel in0 <-> readChannel in1
     writeChannel out0 x <||> writeChannel out1 x

We are then able to compose this process again using choice, sequence, parallelism or iteration[1]. Taking p and q to be processes, and using the notation from Hoare's book [1], the compositions are as follows:

```
p <-> q    -- choice,     CSP: P|Q
p <||> q   -- parallelism, CSP: P||Q
p >> q     -- sequence,   CSP: P ; Q
forever p  -- iteration,  CSP: *P
  where
    p = proc (in0, in1) (out0, out1)
    q = proc (in2, in3) (out2, out3)
```

We will explain the necessary background in Haskell required to be able to understand the examples and concepts in the paper (section 1), and then examine each type of combinator in CHP:

- Sequence (section 2),
- Parallelism (section 3),
- Exception (poison) handling (section 5),
- Choice, often referred to as ALTing (section 6), and
- Iteration constructs (section 7).

We go on to examine building simple processes with several different combinators, and demonstrate the closeness of the Haskell code to the CSP algebra as well as the use of choice over outputs (section 8).

We will also examine how Haskell makes it easy to wire up networks of processes and channels (section 4). Brief details on the implementation of CHP using Software Transactional Memory (section 9) are also provided, as is discussion of related work (section 10).

1. Background – Haskell

Haskell is a statically-typed lazily-evaluated functional programming language. This section provides the necessary background information on Haskell to understand this paper. We explain how to read Haskell types and some basic syntax. Further details are supplied where necessary throughout the paper.

1.1. Types

We precede every Haskell function with its type signature. The format for this is:

functionName :: *typeOfParameter1* -> *typeOfParameter2* -> ... -> *resultType*

The statement $x :: t$ should be read as "x has type t". Each parameter is separated by an arrow, so that $a \rightarrow b$ is "a function that takes a parameter of type a, and returns type b." Any type beginning with an upper-case letter is a specific type, whereas any type beginning a lower-case letter (by convention, single lower-case letters are used) is a parameterised type. This is the type of the *map* function, that applies a transformation to a list:

map :: $(a \rightarrow b)$ -> $[a]$ -> $[b]$

This takes a function transforming a value of type a into type b, and maps from a list of type a to the corresponding list of type b. The a and b types can be the same (and frequently are).

---

[1] Note that in CHP, unlike the formal algebra of CSP, there is no difference in type between an event and a process.

1.2. Functions

A Haskell function definition simply consists of the function name, followed by a label for each of its parameters, then an equals sign and the definition of the function. For example, this function adds the squares of two numbers:

addSquares :: *Int* –> *Int* –> *Int*
addSquares x y = (*x* ∗ *x*) + (*y* ∗ *y*)

There is also an infix notation for functions. The prefix function call *addSquares* 3 5 may also be written 3 `*addSquares*`5; the infix form is created by using backquotes around the function name.

2. Sequential Composition

A monad type in Haskell defines how to compose operations in sequence. Examples of common monads include state monads, error-handling monads (that can short-circuit computation) and the *IO* monad for dealing with input and output.

We will not fully explain the underlying mechanics of monads here, but thankfully Haskell provides a **do** notation that should render our programs readable to those not familiar with Haskell and monads. Consecutive lines in a **do** block are sequential monadic operations. A **do** block follows standard Haskell indentation rules, lasting until the indentation decreases (similar to occam-π's indentation rules), or the end of the expression (e.g. a closing parenthesis that began before the **do** block). The output values of monad operations can be labelled using the <- notation. For example, this program reads in a character and writes it out again twice (then finishes):

main :: *IO* ()
main = **do** *c* <- *getChar*
         putChar c
         putChar c

The type of *getChar* is *IO Char*. This indicates that the function is a monadic action in the *IO* monad that returns a value of type *Char*. The type of the *main* function is *IO* (); the Haskell unit-type ("()", which can be read as an empty tuple) is used to indicate that there is no useful return value, and is very common. Note that there is no difference in types between a **do** block and a single monadic action; the former simply composes several of the latter in sequence, and either or neither may give back a useful value.

Recursion is possible in **do** blocks. We present here the standard *forever* function that repeats a monadic action[2]:

forever :: *Monad m* => *m a* –> *m* ()
forever action = **do** *action*
                    forever action

The type signature of *forever* states that for any monad *m*, *forever* takes a monadic action that returns some value and gives back a monadic action that returns no value. In this instance, the implementation of the function is easier to follow than its type.

CHP defines the *CHP* monad, in which all of its operations take place. Two elementary monadic operations are *readChannel*, which reads from a given channel, and *writeChannel* which writes a given value (second parameter) to a given channel (first parameter). Thus, we can

---
[2]As we will see later on, errors such as poison can break out of a *forever* block.

write the classic *id* process that continually reads from one channel and writes to another, omitting support for poison (see section 5):

idNoPoison :: *Chanin a* −> *Chanout a* −> *CHP* ()
idNoPoison input output
 = *forever* (**do** *x* <− *readChannel input*
      writeChannel output x
)

CHP uses the idea of channel-ends, as most CSP frameworks do. Both channels carry type *a*, which may be any type – but the type carried by the input channel must match the type carried by the output channel; otherwise the compiler will give a type-checking error. Because **in** is a reserved word in Haskell, we use *input* as a variable name instead.

The process could also be written recursively:

idNoPoison input output
 = **do** *x* <− *readChannel input*
    writeChannel output x
    idNoPoison input output

Where possible, we prefer the *forever* idiom, both to shorten definitions and also because it makes clear that no state is carried between iterations of the process. Iteration constructs are examined in more detail in section 7.

The channels in CHP are synchronous. This means that when a process attempts to write to a channel, it must wait until the reader arrives and takes the data before the write operation will complete. Synchronous channels are used in all CSP-based frameworks, and this is one difference between CHP and, for example, Erlang.

3. Parallel Composition

Processes can be composed in parallel using the runParallel function:

runParallel :: [*CHP a*] −> *CHP* [*a*]

Its type can be read as follows: *runParallel* takes a list of processes that return values of type *a*, and composes them into a single process that returns a list of values of type *a*. It gets these values by running the processes in parallel and waiting for them all to complete.

We also provide an operator, <||>, such that *p* <||>*q* is semantically identical to *runParallel* [*p, q*]. The types are slightly different however: the operator returns a pair of values (which can have different types), whereas the function returns a list of identically-typed values. A sum type could be used if heterogenous return types are required.

In contrast to other CSP frameworks, parallelism here supports returning the result values of the sub-processes. This was primarily out of necessity; if we only had the *runParallel_* function that does not return the output of the sub-processes[3]:

*runParallel_* :: [*CHP a*] −> *CHP* ()

Then there would be no easy way to return any values from the parallel processes. Assignment to variables cannot be used because there is no assignment in functional languages, and values could not be communicated back to the parent process because it would be waiting for the sub-processes to finish (and hence deadlock would ensue).

---

[3]The underscore-suffix on a monadic function is a Haskell convention indicating that the output is discarded.

3.1. Forking with Monads

In occam-π, it is possible to use a FORKING block to dynamically start new processes. At the end of the FORKING block, the completion of all the processes is waited for. This idea was carried across to C++CSP2, using the scope of objects to enforce a similar rule – although with an added danger because of the ordering of object destruction [5].

We can again implement this concept in CHP using monads. There is no danger of object destruction, as CHP channels are garbage-collected only when they are no longer in use. We declare a forking monad[4] that gives us the following functions:

forking :: *ForkingCHP a* –> *CHP a*
fork :: *CHP* () –> *ForkingCHP* ()

The *forking* function takes a monadic *ForkingCHP* block and runs it, waiting for all the processes at the end before returning the output. The *fork* function forks off the given *CHP* process from inside the *ForkingCHP* block. Unlike our normal parallelism operators described previously, there is no way for a forked process to directly return a value. Forked processes that need to pass back a value to the parent process may do so using a channel communication.

4. Channel Wiring

In occam-π, PROCedures are not first-class types. A block of monadic code is a first-class type in Haskell, and can be passed around, as we have already seen with our combinators. We can also pass around functions that yield a monadic item: in CHP terms, this is a process that still needs parameters.

We can take advantage of this to provide functions for standard wiring idioms. An obvious example is wiring a list of processes into a pipeline:

pipeline :: [*Chanin a* –> *Chanout a* –> *CHP b*] –> *Chanin a* –> *Chanout a* –> *CHP* [*b*]

This function takes a list of processes that require a reading- and writing-end of a channel carrying type *a*. The *pipeline* function also takes the channel ends to be used at the very beginning and end of the pipeline, and returns the parallel composition of the processes in the pipeline.

The *pipeline* function can be defined in several ways. Here we use an elegant recursive definition of a helper function *wirePipeline* that wires up all the processes and returns them in a list:

```
pipeline procs input output
  = do wiredProcs <- wirePipeline procs input output
       runParallel wiredProcs

wirePipeline :: [Chanin a -> Chanout a -> CHP b]
             -> Chanin a -> Chanout a -> CHP [CHP b]
wirePipeline [p] input output = return [p input output]
wirePipeline (p : ps) input output
  = do c <- newChannel
       rest <- wirePipeline ps (reader c) out
       return ((p input (writer c)) : rest)
```

The first line of *wirePipeline* is a base case, and matches a single-process list. The remaining lines are the recursive step, with a pattern-match to decompose the process list into its

---
[4]Technically, this is a monad transformer that composes the *CHP* monad with a *ForkingT* monad transformer.

head *p* (a single process) and the remainder of the list *ps* (a list of processes). The : constructor is used again in the last line to join an item onto the head of the list *rest*.

Here is an example of using the function:

```
fifoBuffer :: Int -> Chanin a -> Chanout a -> CHP ()
fifoBuffer n input output
  = do pipeline ( replicate n idProcess ) input output
       return ()
```

The *replicate* function takes a replication count and a single item, and returns a list containing the item repeated that many times.

We can also easily define a function for wiring up a cycle of processes, by making the two ends of the pipeline use the same channel:

```
cycle :: [Chanin a -> Chanout a -> CHP b] -> CHP [b]
cycle procs = do c <- newChannel
                 wiredProcs <- wirePipeline (reader chan) (writer chan) procs
                 runParallel wiredProcs
```

It would not be difficult to make general functions for wiring up other common idioms.

4.1. Channel Type Inference

For channels there is a bijective mapping between the two channel-end types and the channel implementation. A *Chanin* and *Shared Chanout* are associated with an any-to-one channel. A *Chanin* and *Chanout* are associated with a one-to-one channel.

This means that if the Haskell type-checker (which uses type inference) knows either the two channel-end types or the channel type, it can infer the other. Typically this is used to allocate a channel using the *newChannel* function, and have the type-checker figure out what type the channel needs to be, based on what processes the ends are passed to. Programmers who prefer to be explicit can still use individual *oneToOneChannel* functions.

The difference in types between the various channel-ends prevents channel-ends being used incorrectly (for example, using a shared channel-end without claiming it), so there is no possibility for error, and it also makes the code simpler. No other CSP framework or language has this capability, because of the lack of such type inference.

Channels do not need to be explicitly destroyed in CHP – instead, they will be garbage-collected when no longer in use (using standard Haskell mechanisms). This removes any worry about correctly nesting the scope of channels.

5. Poison and Exception Handling

Poison is a technique for safely shutting down a process network, without inviting deadlock or forcefully aborting processes [12,13,14]. A channel can either be in a normal operating state, or it can be poisoned. Any attempt to read or write on a poisoned channel will result in a poison exception being thrown.

Poison propagates throughout a network as follows. When a process catches a poison exception, it poisons all its channel-ends. Thus its neighbours (according to channel connections in a process graph) will also get poison thrown, and they will do the same, until all channels in a process network have been poisoned. Once processes have poisoned their channels, they shut down, and thus the process network terminates. This mechanism has previously been incorporated into C++CSP, JCSP and others.

Our discussion here is centred around poison, but the ideas should generalise to any notion of exceptions in process-oriented programs. Haskell supports exceptions in three ways:

in pure code, in the *IO* monad, and in special error monads. The latter approach is the neatest solution. Thus we allow poison exceptions to occur in our *CHP* monad.

The *onPoisonTrap* function may be used (typically infix) to trap and handle poison. For example, here is the identity process with poison handling:

idProcess :: *Chanin a* -> *Chanout a* -> *CHP* ()
idProcess input output
　= (*forever* (**do** *x* <- *readChannel input*
　　　　　　　　　　writeChannel output x
　　　　　　)
　) `onPoisonTrap` (**do** *poison input*
　　　　　　　　　　　　poison output)

It is important that the *forever* combinator is used inside the body of the poison, e.g. (*forever* ...) `onPoisonTrap`(...). If the process was composed as *forever* ((...) `onPoisonTrap` (...)) then it would form an infinite loop, forever catching the poison, handling it, and looping again.

We also add another poison handler (*onPoisonRethrow*) that *always* rethrows after the handler has finished. This handler can be used either inside or outside of the *forever* combinator, without encountering the infinite loop problem. The use of this function is further examined in section 8.

5.1. Parallelism and Poison

One problem with poison has been deciding on the semantics of poison and parallel composition. In short, when p and q are composed in parallel and p exits due to poison, what should happen to q, and what should happen to their parent process?

Forcibly killing off q is an ugly solution that goes against the main principle of poison (allowing for controlled termination). Doing nothing at all is an odd solution, because the parent will not know whether its sub-processes terminated successfully or died because of poison. Consider the following process:

delta2 :: *Chanin a* -> *Chanout a* -> *Chanout a* -> *CHP* ()
delta2 input output0 output1
　= *forever* (**do** *x* <- *readChannel input*
　　　　　　　　writeChannel output0 x <||> *writeChannel output1 x*)
　　`onPoisonRethrow` (**do** *poison input*
　　　　　　　　　　　　　poison output0
　　　　　　　　　　　　　poison output1)

If the parent is never notified about its subprocesses dying of poison, the *delta2* process would continue running if one, or even *both*, of its output channels was poisoned, because the poison exception would be masked by the parallel composition.

The semantics we have chosen are straight-forward. The parent process spawns off all the sub-processes, and waits for them *all* to complete, either normally (no poison) or abnormally (with poison). Once they have all completed, if *any* of the sub-processes exited in a state of exception (with poison), the *runParallel* function (or similar) throws a poison exception in the parent process.

This solution corresponds to the ideas in Hilderink's CSP exception operator [13] and Hoare's concurrent flowcharts [15]. It maintains the associativity of PAR; the following two lines are semantically equivalent:

runParallel [*runParallel* [*p*, *q*], *r*]
runParallel [*p*, *runParallel* [*q*, *r*]]

The other preserved useful property is that running one process in parallel is the same as running the process directly: *runParallel* [*p*] is semantically identical to *p*. Commutativity of PAR is also maintained. It should be noted that the types differ slightly between all the aforementioned examples, but our concern here is only with semantics.

6. Composition using Choice – Alts and Implicit Guards

In occam-π, it is possible to choose between several events using the *ALT* construct, or *PRI ALT* which gives its guards descending priority. Each option has a guard, followed by a body:

```
PRI ALT
  c ? x -- input guard
    d ! x -- body (output)
  SKIP -- guard
    d ! 0 -- body (output)
```

`SKIP` is a guard that is always ready. Thus the above code checks the channel c to see if input is waiting. If some input is waiting it is read and sent out on channel d, otherwise the `SKIP` guard is chosen and the value 0 is sent instead.

Frameworks such as JCSP and C++CSP2 have translated the ALT into a construct that takes an array of guards and returns an integer denoting which guard is ready. The program then follows this up by acting on the guard. For example, the occam-π code above would be written as follows in C++CSP2:

```
Alternative alt (c.guard()) (new SkipGuard);
switch (alt.priSelect()) {
   case 0: {
      c >> x;
      d << x;
   } break;
   case 1: {
      d << 0;
   } break;
}
```

Note how the input must be performed separately from the guard. This was a design decision (taken from JCSP) to easily allow either normal or extended input on the channel after it has been found to be ready by the `Alternative` construct.

6.1. Implicit Guards

In CHP, we are able to integrate the guard and its body. We can write, similar to CSP:

> *alt* [**do** *x* <- *readChannel c*
>   *writeChannel c x*
>   , **do** *skip*
>   *writeChannel d x*]

We say that choice is implicitly available here, because the first action in each body supports choice – such actions are *skip*, a channel read or write (normal or extended), a wait action, a barrier synchronisation or another *alt* (which allows alts to be nested). This is achieved by constructing a special monad that allows us to keep track of the first action (and a hidden associated guard) in any given monadic action. It is possible to supply only one action, such as *skip*, to the alt without a **do** block if no body is required.

In addition to the *alt* and *priAlt* functions, we supply corresponding operators: <-> for choice without priority, and </> for choice with left-bias. That is, the expression *p* <->*q* is identical to *alt* [*p*, *q*] and the expression *p* </>*q* is identical to *priAlt* [*p*, *q*]. The operators are associative. The functions also have the property that *alt* [*p*], *priAlt* [*p*] and *p* are all identical, provided that *p* supports choice (otherwise a run-time error will result). The duality between choice (a sum of processes) and parallelism (a product of processes) is clearer in CHP than it is in occam-π.

An eternal *fairAlt* that cycles priority between the guards is also easy to construct – we choose to represent it here with recursion:

fairAlt :: [*CHP a*] –> *CHP* ()
fairAlt (*g*:*gs*) = **do** *priAlt* (*g*:*gs*)
                          fairAlt (*gs* ++ [*g*])

6.2. Composition of ALTs

ALTs in occam-π are composable to a certain degree. Directly nested `ALT`s are possible:

```
ALT
  ALT
    c ? x
      d ! x
    e ? x
      d ! x
  tim ? AFTER t
    d ! 0
```

The above code chooses between inputs on c and e, and waiting for a timeout (for the time t to occur). The body of each guard is an output on channel d.

However, you cannot pull out guards into a separate procedure:

```
PROC alt.over.all (CHAN INT c?, CHAN INT e?, CHAN INT d!)
  ALT
    c ? x
      d ! x
    e ? x
      d ! x
:
ALT
  alt.over.all (c, e, d)
  tim ? AFTER t
    d ! 0
```

This was a design decision, taken in classical occam, to treat the guards differently. In our Haskell implementation, we only require the first action of any given monadic action to support choice. Since an *alt* supports choice, we can nest them – regardless of function boundaries. Therefore this *is* valid in CHP:

altOverAll :: *Chanin Int* –> *Chanin Int* –> *Chanout Int* –> *CHP* ()
altOverAll c e d = *alt* [**do** *x* <– *readChannel c*
                                  writeChannel d x
                        , **do** *x* <– *readChannel e*
                                  writeChannel d x]

alt [*altOverAll c e d*
    , **do** *waitUntil t*
          writeChannel d 0]

This new composability overcomes one of the shortcomings that Reppy pointed out in the ALT construct when he developed Concurrent ML [16]. He noted that function composition was incompatible with choice. With implicit guards in CHP, this is not the case. This idea would also be possible to build into occam-π or Rain [17], where the presence of choice could be checked at compile-time. The compiler could eliminate the run-time errors that can occur in CHP if you try to choose between something that does not support choice (for example, poisoning a channel).

A further example of using choice can be seen in section 8.2.

7. Iteration

Most processes have repeating behaviour. It is very common to see `WHILE some.condition` or even `WHILE TRUE` at the beginning of occam-π processes. The latter can be expressed with the Haskell combinator *forever*, and can be broken out of using poison or other monadic exception mechanisms.

The *forever* combinator repeatedly runs the same block of code. It does not support easily stopping on a certain condition, or retaining any idea of state between subsequent runs of the same block. For many small processes, such as the identity process, this is acceptable. To demonstrate two different ways state can be implemented in the presence of iteration, we will use the example of a *runningTotal* process that continually reads in numbers and outputs the current total after each one.

The first obvious mechanism is to use recursion. We define the *runningTotal* process as simply setting off another inner process[5]:

```
runningTotal :: Chanin Int -> Chanout Int -> CHP ()
runningTotal input output
  = runningTotal' 0 `onPoisonRethrow` (do poison input
                                          poison output)
  where
    runningTotal' :: Int -> CHP ()
    runningTotal' prevTotal = do x <- readChannel input
                                 let newTotal = prevTotal + x
                                 writeChannel output newTotal
                                 runningTotal' newTotal
```

We take advantage of the scoping of Haskell's **where** clause; *input* and *output* are in scope for *runningTotal'*.

This recursion can get messy if many variables need to be passed to the recursive call. Haskell's state monad-transformer provides another alternative. The state monad-transformer provides *get* and *put* monadic functions for dealing with the state, and a whole block can be evaluated with a given state[6]:

```
runningTotal input output
  = runWithState 0 runningTotal' `onPoisonRethrow` (do poison input
                                                       poison output)
  where
    runningTotal' = forever (do x <- readChannel input
                                prevTotal <- get
                                let newTotal = prevTotal + x
                                put newTotal
                                writeChannel output newTotal
                             )
```

---

[5] We use the suffix ' here: a valid character in Haskell identifiers often used for this purpose.
[6] Technically, our *runWithState* function here is defined as *flip evalStateT*.

With the state monad, the reads and writes to and from the state can be placed more appropriately throughout the code block, rather than having to name all the variables at the start of the function, and pass them all again at the end of the function.

With the recursive method it is possible to control the looping by providing a base case, whereas the state monad has no support for this. However, it is possible to support some more easily controlled looping in Haskell, using yet another monad.

Inspired by Ian East's revival of the DO-WHILE-DO loop (transmuted into his Honey-suckle programming language as repeat-while [18]), CHP offers a loop-while construct using another monad-transformer.

The *loop* function takes a block and executes it. Inside this block may be one or several (or none, to loop forever) *while* statements. As an example of its use, here is a modified identity process that stops (between the input and output) when a certain target value is seen:

```
idUntil :: a -> Chanin a -> Chanout a -> CHP ()
idUntil target input output
  = loop (do x <- readChannel input
             while (x /= target)
             writeChannel output x)
```

This particular process would not be as elegantly expressed using recursion or the state monad. It is possible to combine this looping monad with the state monad.

8. Further Composition

We have now presented five types of composition: sequence, parallelism, choice, exception (poison) handling and iteration (cyclic sequence). All of these compositions can cross function boundaries in Haskell. We first show some general examples of all these types of composition, and also give an example of practical uses while implementing buffers.

8.1. General Composition

In this section we show how to compose several very simple processes. Each process is given both in Haskell code and using CSP notation (with parameters omitted). We borrow Hilderink's exception-handling operator [13]: $P \overset{\rightarrow}{\triangle} Q$ behaves as P, but if P throws a poison exception it behaves instead like Q. We also invent a process, $\Omega(..)$ that poisons all channels passed to it, and THROW that throws a poison exception.

Generally, the smallest composite process in process-oriented programming is the identity process – but this already contains two compositions (sequence and iteration), and three in frameworks with poison such as C++CSP2. We start here with a *forward* process that is one iteration of the identity process:

```
-- CSP: forward = input?x ⟶ output!x ⟶ SKIP
forward :: Chanin a -> Chanout a -> CHP ()
forward input output = do x <- readChannel input
                          writeChannel output x
```

This can then be composed into several other processes:

```
-- CSP: forwardForever = *forward
forwardForever :: Chanin a -> Chanout a -> CHP ()
forwardForever input output = forever (forward input output)
```

$$\text{-- CSP: forwardSealed} = forward \overset{\rightarrow}{\triangle} (\Omega(input, output))$$

```
forwardSealed :: Chanin a -> Chanout a -> CHP ()
forwardSealed input output
  = (forward input output)
    `onPoisonTrap` (do poison input
                       poison output)
```

-- CSP: forwardRethrow = forward $\overrightarrow{\triangle}(\Omega(input, output))$; THROW
```
forwardRethrow :: Chanin a -> Chanout a -> CHP ()
forwardRethrow input output
  = (forward input output)
    `onPoisonRethrow` (do poison input
                          poison output)
```

We include both of the latter two processes so that we can demonstrate their relative composability below. Consider these further-composed processes:

-- CSP: id1 = forwardForever $\overrightarrow{\triangle}(\Omega(input, output))$
```
id1 :: Chanin a -> Chanout a -> CHP ()
id1 input output
  = (forwardForever input output)
    `onPoisonTrap` (do poison input
                       poison output)
```

-- CSP: id2 = forwardForever $\overrightarrow{\triangle}(\Omega(input, output))$; THROW
```
id2 :: Chanin a -> Chanout a -> CHP ()
id2 input output
  = (forwardForever input output)
    `onPoisonRethrow` (do poison input
                          poison output)
```

-- CSP: id3 = *forwardSealed
```
id3 :: Chanin a -> Chanout a -> CHP ()
id3 input output = forever (forwardSealed input output)
```

-- CSP: id4 = *forwardRethrow
```
id4 :: Chanin a -> Chanout a -> CHP ()
id4 input output = forever (forwardWithRethrow input output)
```

-- CSP: id5 = (*forwardRethrow)$\overrightarrow{\triangle}$SKIP
```
id5 :: Chanin a -> Chanout a -> CHP ()
id5 input output
  = (forever (forwardWithRethrow input output))
    `onPoisonTrap` skip
```

Intuitively, *id2* is semantically identical to *id4*, and *id1* is semantically identical to *id5*; proving this is left as an exercise for the reader. We prefer *id4* and *id5*, which locate the poison-handling as close as possible in the composition to the channel-events. Processes *id1* and *id5* are *not* identical to *id3*, as the latter will never terminate, even if its channels are poisoned.

We can see that, pragmatically, the *forwardWithRethrow* function was much more composable than the *forwardSealed* function. The implication in turn is that *id2* and *id4* will prove more composable than their "sealed" counterparts, *id1* and *id5* – and we believe that in practice, processes involving poison should always rethrow in order to make them more composable.

Our example shows that simple CHP programs can be reasoned about. The documentation supplied with CHP contains many useful laws to support such reasoning, for example:

runParallel [*p*] == *p*
throwPoison >> *p* == *throwPoison*
(*p* >> *throwPoison*) <| |> *q* == (*p* <| |> *q*) >> *throwPoison*

The >> operator represents sequence. These laws are similar to the laws of occam presented by Roscoe and Hoare [19].

There is a close correspondence between our extended CSP and the CHP code, especially in the presence of composition. Even if the user of the library knows nothing about CSP, the CHP compositions have inherited the beauty, and some of the reasoning power, of the original CSP calculus.

8.2. Output Guards and Buffers

It is sometimes desirable to introduce buffering between two processes, rather than having direct synchronous communication. A pipeline of N identity processes forms a limited capacity First-In First-Out (FIFO) buffer of size N. Sometimes, more complex buffering is required, such as overwriting buffers. An overwriting buffer also provides a limited capacity FIFO buffer, but when the buffer is full, it continues to accept new data and overwrites the oldest value in the buffer.

It is not possible in occam-π to define an overwriting buffer process with a single input and single output channel. Consider the case of a size-one overwriting buffer process. The process begins by reading in an item of data. If it subsequently writes out the data, it is committed to the write because all writes must be committed to in occam-π. Another piece of data arriving cannot affect this write, and thus the value cannot be overwritten. If the process does not send out the data, it is breaking the semantics of the buffer that the data should be available to be read.

Many frameworks, such as JCSP and C++CSP2 solve this problem by supplied buffered *channels* that encapsulate this behaviour. This complicates the API for channels. In CHP we allow choice over outputs (which no previous framework has done) and we can use this to construct overwriting buffer processes.

Our CHP overwriting buffer process does not have to commit to the output. It therefore chooses between reading a new data item in and writing out an item from the buffer. This allows us to express the correct behaviour:

```
overwritingBuffer  :: Int -> Chanin a -> Chanout a -> CHP ()
overwritingBuffer n input output
  = ( overwritingBuffer' []) `onPoisonRethrow` (do poison input
                                                    poison output)
 where
    overwritingBuffer' :: [a] -> CHP ()
    overwritingBuffer' s | null s        = takeIn
                         | n == length s = takeInReplace <-> sendOut
                         | otherwise     = takeIn <-> sendOut
      where
        takeIn        = do x <- readChannel input
                           over (s ++ [x])
        takeInReplace = do x <- readChannel input
                           over (tail s ++ [x])
        sendOut       = do writeChannel output (head s)
                           over (tail s)
```

We define our buffer as simply setting off an inner process with an empty list[7]. The inner process takes a list of items in the buffer. It then has three guards (indicated by the "|"

---
[7]In our real buffers we use a data structure with $O(1)$ append, but we use lists here for simplicity.

symbol). The first guard that evaluates to *True* is chosen. These are checked in sequential order based on the function arguments, and should not be confused with guards used for alting.

The first guard checks if the buffer is currently empty. If so, the only action should be to take in new data. If the buffer is full (the second guard), the process chooses between taking in new data (and overwriting the oldest existing value) or sending out an item of data. If the buffer is neither empty nor full (the last guard), the process chooses between taking in new data (and adding it to the buffer) and sending out an item of data.

The process behaviours are at the end of the code above. The *head* function picks the first item from a list, and the *tail* function is all of the list *except* for the first item. They use recursion to provide iteration.

Buffers can be written in CHP quite easily because choice is available on channel writes as well as reads; in other frameworks, choice was only available on channel reads because there was not a fast and safe implementation until recently (see the next section).

9. Implementation

CHP's channels and barriers are built on top of Haskell's Software Transactional Memory (STM) library [20]. Channels offer choice at both ends, using Welch's idea for symmetric channels [3]: both ends synchronise on an alting barrier (a multiway synchronisation with choice) then proceed with the communication. The alting barriers use an STM implementation based on the "oracle" mechanism [21,22].

The only way to start an explicit new concurrent process in Haskell is with the *forkIO* function. It takes a process and starts running it. It is a "fork-and-forget" function, providing no means to wait for the completion of the forked process. Thus we implement our parallel operators (that do wait for the completion of processes and return their outputs) by simply having the processes write their result to a shared channel when they complete. The parallel operator thus forks off N processes, then reads from the channel N times, sorts the results and returns them.

10. Related Work

The idea of combining functional programming and process-oriented concurrency is now quite old; Erlang is an obvious successful example [23]. Erlang is based on the actor model [24], and as such has asynchronous communication and untyped addressed mailboxes instead of CSP's synchronous communication and anonymous typed channels. Erlang-style communication has also already been implemented in Haskell [25]. Erlang has explicit sequencing in the language and strict evaluation, in contrast to Haskell's monads and lazy evaluation.

Combining Haskell with some ideas from occam has been done before in Haskell# ("Haskell-hash") [26]. Haskell# began with explicit concurrency and has developed into a separation of computation (expressed in Haskell) and topology/communication (expressed in a separate Haskell Configuration Language), which contrasts to CHP's standing as a Haskell library.

The CHP library is built on top of Software Transactional Memory (STM), a Haskell solution to the problem of explicitly concurrent programming [20]. Most previous concurrent Haskell frameworks were built on top of an older system of shared mutable variables [27]. One of the main advantages of STM over the shared mutable variables is that STM naturally supports choice, which is key to CSP programs, and allows for better composability.

STM allows multiple transactions in the STM monad to be composed via sequence or choice into a single transaction. Parallelism is handled externally to the STM mechanism. STM also allows full choice between sequentially composed transactions; both transactions in sequence must succeed for it to be chosen. This is possible because STM only commits transactions when they are successful. It does not make it possible to rollback a change which can be viewed by another process.

Concurrent ML is another obvious predecessor to CHP [16], and in turn was an influence on STM. It had the notion of an event, and choice between events. Events could be derived from channels by supplying a destination/source for a read/write. Events could be composed via choice, and independently executed later on. This corresponds in CHP to forming a monadic action using choice, and separately executing the action.

Concurrent ML did not feature the idea of poison or anything similar. Poison could probably be built on top of Concurrent ML's primitives, but the idea of poison requires careful thought about the semantics of composing it via parallelism and iteration. ML permits side effects and uses eager evaluation; Haskell's purity and lazy evaluation may offer opportunity for safer programming and different programming methods.

STM and Concurrent ML share some of the elegance in composition of CHP, but we find that CSP is a more comprehensible programming model for explicit concurrency, and in addition provides a formal basis for reasoning about programs.

11. Conclusions

We have presented a library that makes the CSP-based process-oriented programming model available in Haskell, and have shown how natural, powerful and elegant its composition of elements is. Code in CHP can have a strong correspondence to the original CSP algebra, with identical semantics.

We believe that CHP is just as powerful as explicitly concurrent languages such as occam-π for writing concurrent programs, and that parallelism, choice and composing process networks with formulaic wiring are all easy in CHP, with the added feature of support for poison. The occam-π implementation retains a memory and speed advantage over Haskell, the former being able to allocate as little as 32 bytes per process and at least an order of magnitude faster than the latter. CHP contains many of occam-π's other features [2] such as barriers, explicitly-claimed shared channels, extended input, as well as extended output that is not currently present in occam-π.

The examples throughout this paper have also demonstrated how recursion can be used with the CSP model to create small elegant understandable programs. Recursion has not often been used in CSP implementations in the past – previous versions of occam did not support recursion, and other frameworks shy away from it, as non-optimised recursion can lead to needing a large amount of stack memory. In most frameworks, each process requires a separate stack, so efforts are made to minimise the use of the stack.

We hope that this library will prove interesting and useful to a variety of people, including: CSP theoreticians looking for a suitable development platform, process-oriented programmers who wish to use Haskell, and Haskell programmers who want a simple but powerful explicit concurrency framework.

11.1. Future Work

There is currently work at Kent ongoing to write a new occam-π compiler, Tock, in Haskell. One possibility for future work would be to combine Tock with the CHP library in order to produce an occam-π interpreter written in Haskell.

The performance of CHP could also be investigated and benchmarked. Haskell's lazy evaluation means that more thought is required to achieve speed-up through concurrency, so CHP needs to be exercised on parallel processing tasks for this to be tested.

11.2. Practical Details

Like many Haskell programs, the CHP library uses features that are not part of the latest Haskell standard (Haskell 98), but are likely to be part of the next Haskell standard (currently entitled Haskell-Prime). It runs under the latest branch (6.8) of the most popular Haskell compiler, GHC. The library has now been released, and more details on obtaining it can be found at: `http://www.cs.kent.ac.uk/projects/ofa/chp/`.

Acknowledgements

Many thanks are due to Adam Sampson: for his comments on this paper, for his suggestion to try implementing the library on top of STM, and for his suggestion that alts themselves could provide implicit guards, thus allowing alts to be nested as they can be in occam. Thanks are also due to the anonymous reviewers for their helpful comments, and to Claus Reinke for his suggestions regarding a few aspects of the library.

References

[1] C. A. R. Hoare. *Communicating Sequential Processes*. Prentice-Hall, 1985.
[2] Peter H. Welch and Fred R. M. Barnes. Communicating mobile processes: introducing occam-pi. In *25 Years of CSP*, volume 3525 of *Lecture Notes in Computer Science*, pages 175–210. Springer Verlag, 2005.
[3] Peter Welch, Neil Brown, Bernhard Sputh, Kevin Chalmers, and James Moores. Integrating and Extending JCSP. In Alistair A. McEwan, Steve Schneider, Wilson Ifill, and Peter Welch, editors, *Communicating Process Architectures 2007*, pages 349–370, 2007.
[4] Jan F. Broenink, André W. P. Bakkers, and Gerald H. Hilderink. Communicating Threads for Java. In Barry M. Cook, editor, *Proceedings of WoTUG-22: Architectures, Languages and Techniques for Concurrent Systems*, pages 243–262, 1999.
[5] Neil C. C. Brown. C++CSP2: A Many-to-Many Threading Model for Multicore Architectures. In *Communicating Process Architectures 2007*, pages 183–205, 2007.
[6] B. Orlic and J.F. Broenink. Redesign of the C++ Communicating Threads Library for Embedded Control Systems. In *5th Progress Symposium on Embedded Systems*, pages 141–156, 2004.
[7] James Moores. CCSP – A Portable CSP-Based Run-Time System Supporting C and occam. In Barry M. Cook, editor, *Proceedings of WoTUG-22: Architectures, Languages and Techniques for Concurrent Systems*, pages 147–169, 1999.
[8] John M. Bjørndalen, Brian Vinter, and Otto Anshus. PyCSP – Communicating Sequential Processes for Python. In *Communicating Process Architectures 2007*, pages 229–248, 2007.
[9] Kevin Chalmers and Sarah Clayton. CSP for .NET Based on JCSP. In Frederick R. M. Barnes, Jon M. Kerridge, and Peter H. Welch, editors, *Communicating Process Architectures 2006*, pages 59–76, 2006.
[10] Alex Lehmberg and Martin N. Olsen. An Introduction to CSP.NET. In Frederick R. M. Barnes, Jon M. Kerridge, and Peter H. Welch, editors, *Communicating Process Architectures 2006*, pages 13–30, 2006.
[11] Simon L. Peyton Jones and Philip Wadler. Imperative functional programming. In *POPL '93: Proceedings of the 20th ACM SIGPLAN-SIGACT symposium on Principles of Programming Languages*, pages 71–84, New York, NY, USA, 1993. ACM.
[12] Neil C. C. Brown and Peter H. Welch. An Introduction to the Kent C++CSP Library. In Jan F. Broenink and Gerald H. Hilderink, editors, *Communicating Process Architectures 2003*, pages 139–156, 2003.
[13] Gerald H. Hilderink. *Managing Complexity of Control Software through Concurrency*. PhD thesis, Laboratory of Control Engineering, University of Twente, 2005.
[14] Peter H. Welch. Graceful termination – graceful resetting. In André W. P. Bakkers, editor, *OUG-10: Applying Transputer Based Parallel Machines*, pages 310–317, 1989.
[15] C.A.R. Hoare. Fine-grain concurrency. In *Communicating Process Architectures 2007*, pages 1–19, 2007.
[16] John H. Reppy. First-class synchronous operations. In *TPPP '94: Proceedings of the International Workshop on Theory and Practice of Parallel Programming*, pages 235–252. Springer-Verlag, 1995.
[17] Neil C. C. Brown. Rain: A New Concurrent Process-Oriented Programming Language. In *Communicating Process Architectures 2006*, pages 237–251, September 2006.
[18] Ian R. East. The Honeysuckle programming language: an overview. *IEE Proc.-Softw.*, 150(2):95–107, April 2003.

[19] A. W. Roscoe and C. A. R. Hoare. The laws of occam programming. *Theor. Comput. Sci.*, 60(2):177–229, 1988.
[20] Tim Harris, Simon Marlow, Simon Peyton-Jones, and Maurice Herlihy. Composable memory transactions. In *PPoPP '05*, pages 48–60. ACM, 2005.
[21] P.H. Welch. A Fast Resolution of Choice between Multiway Synchronisations (Invited Talk). In Frederick R. M. Barnes, Jon M. Kerridge, and Peter H. Welch, editors, *Communicating Process Architectures 2006*, pages 389–390, 2006.
[22] P.H. Welch, F.R.M. Barnes, and F.A.C. Polack. Communicating complex systems. In Michael G Hinchey, editor, *Proceedings of the 11th IEEE International Conference on Engineering of Complex Computer Systems (ICECCS-2006)*, pages 107–117, Stanford, California, August 2006. IEEE. ISBN: 0-7695-2530-X.
[23] Joe Armstrong, Robert Virding, and Mike Williams. *Concurrent Programming in Erlang*. Prentice Hall, 1993.
[24] Carl Hewitt, Peter Bishop, and Richard Steiger. A Universal Modular ACTOR Formalism for Artificial Intelligence. In *IJCAI*, pages 235–245, 1973.
[25] F. Huch. Erlang-style distributed Haskell. In *11th International Workshop on Implementation of Functional Languages*, September 1999.
[26] Francisco Heron de Carvalho Junior and Rafael Dueire Lins. Haskell#: Parallel programming made simple and efficient. *Journal of Universal Computer Science*, 9(8):776–794, August 2003.
[27] Simon L. Peyton Jones, A. Gordon, and S. Finne. Concurrent Haskell. In *Symposium on Principles of Programming Languages*, pages 295–308. ACM Press, 1996.

Two-Way Protocols for occam-π

Adam T. SAMPSON [1]

Computing Laboratory, University of Kent

Abstract. In the occam-π programming language, the client-server communication pattern is generally implemented using a pair of unidirectional channels. While each channel's protocol can be specified individually, no mechanism is yet provided to indicate the relationship between the two protocols; it is therefore not possible to statically check the safety of client-server communications. This paper proposes *two-way protocols* for individual channels, which would both define the structure of messages and allow the patterns of communication between processes to be specified. We show how conformance to two-way protocols can be statically checked by the occam-π compiler using Honda's session types. These mechanisms would considerably simplify the implementation of complex, dynamic client-server systems.

Keywords. client-server, concurrency, occam-π, protocols, session types.

Introduction

The occam-π process-oriented programming language supports very large numbers of lightweight processes, communicating using channels and synchronising upon barriers. It has roots in CSP and the π-calculus, making it possible to reason formally about the behaviour of programs at all levels from individual processes up to complete systems.

In a process-oriented system, it is very common to have client-server relationships between processes: a server process answers requests from one or more clients, and may itself act as a client to other servers while processing those requests (see figure 1). The *client-server design rules* [1] allow the construction of client-server systems of processes that are guaranteed to be free from deadlock and livelock problems.

The client-server pattern has proved extremely useful when building complex process-oriented systems. Server processes fill approximately the same role as objects in object-oriented languages; indeed, OO languages such as Smalltalk use message-passing terminology to describe method calls between objects. However, process-oriented servers avoid many of the concurrency problems endemic in OO languages, and their interfaces are more powerful: a client-server communication may be a *conversation* containing several messages in both directions, not just a single request-response pair.

Most non-trivial occam-π programs today make some use of the client-server pattern, with communication implemented using channels. However, while occam-π allows the protocol carried over an individual channel to be specified and checked by the compiler, it does not yet provide any facilities for checking the protocols used across two-way communication links such as client-server connections.

In this paper, we will first describe occam-π's existing protocol specification facilities, and how they are currently used to implement client-server communications. We will then examine other implementations of two-way communication protocols, and how they can be

---

[1] Corresponding Author: *Adam Sampson, Computing Laboratory, University of Kent, CT2 7NF, UK.* Tel.: +44 1227 827841; E-mail: `A.T.Sampson@kent.ac.uk`.

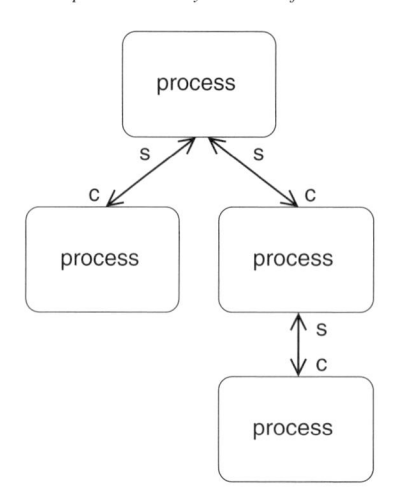

Figure 1. Client-server relationships between processes

formally specified using session types. Finally, we will describe how two-way protocols could be specified and implemented in occam-π, and discuss some possible syntax for them.

1. Unidirectional Protocols

occam-π's channels are unidirectional and unbuffered, and the order of messages permitted over each channel is specified using a protocol. The compiler checks that processes using channels adhere to their protocols. Protocols have been present in the language since occam 2 [2].

A *simple protocol* is just a list of types which will be sent in sequence over the channel.

```
PROTOCOL REPORT IS INT; REAL32:
CHAN REPORT c:
c ! 42; 3.141
```

A *variant protocol* allows choice between several simple protocols, each identified by a unique tag; each communication over the channel is preceded by a tag that indicates the simple protocol to be followed.

```
PROTOCOL COLOUR
  CASE
    rgb; REAL32; REAL32; REAL32
    palettised; INT
:
CHAN COLOUR c:
SEQ
  c ! rgb; 0.3; 0.3; 0.0
  c ! palettised; 42
```

occam-π introduced the idea of *protocol inheritance* [3], which allows the tags from one or more existing variant protocols to be incorporated into a new variant protocol. A reading end of a channel carrying one of the included protocols may be used as if it were a reading end of a channel carrying the new protocol; for writing ends, the opposite applies.

```
PROTOCOL PRINT.COLOUR EXTENDS COLOUR
  CASE
    cmyk; REAL32; REAL32; REAL32; REAL32
:
CHAN PRINT.COLOUR c:
SEQ
  c ! rgb; 0.3; 0.3; 0.0
  c ! cmyk; 0.1; 0.4; 0.1; 0.0
```

Several channels may be grouped into a *channel bundle* [4]. The ends of a channel bundle are *mobile*: they may be sent around between processes, allowing the process network to be dynamically reconfigured at runtime. The channels inside a bundle may be used as if they were regular channels.

```
CHAN TYPE DISPLAY
  MOBILE RECORD
    CHAN COORDS in?:
    CHAN COLOUR out!:
:
DISPLAY! end:
INT c:
SEQ
  end[in] ! 2.4; 6.8
  end[out] ? CASE palettised; c
```

2. Client-Server Communication in occam-π

Client-server communications are currently implemented in occam-π using a pair of channels: one carries *requests* from the client to the server, and the other carries *responses* from the server to the client. The two channels are usually packaged inside a channel bundle. We can use this approach to specify a client-server interface to a random-number generator, which will attempt to roll an N-sided die for you, and either succeed or drop it on the floor:

```
PROTOCOL DIE.REQ
  CASE
    roll; INT
    quit
:
PROTOCOL DIE.RESP
  CASE
    rolled; INT
    dropped
:
CHAN TYPE DIE
  MOBILE RECORD
    CHAN DIE.REQ req?:
    CHAN DIE.RESP resp!:
:
```

The `req` and `resp` channels carry requests and responses respectively, each with their own protocol. In this case, a `roll` message from the client would provoke a `rolled` or `dropped` response from the server; a `quit` message would cause the server to exit with no response. To use this process, a client need only send the appropriate messages over the channels in the bundle:

```
PROC roll.die (DIE! die)
  SEQ
    ... obtain die

    die[req] ! roll; 6
    die[resp] ? CASE
      INT n:
      rolled; n
        ... rolled an 'n'
      dropped
        ... dropped the die

    die[req] ! quit
:
```

The syntax for defining and using client-server interfaces is rather clumsy. Each client-server interface requires two protocols and a channel bundle type to be declared. The protocol names – `DIE.REQ` and `DIE.RESP` in this case – are usually only used within the channel bundle definition. When sending messages over a client-server interface, the name of the channel being used must always be specified, even though it is unambiguous from the direction of communication whether the `req` or `resp` channel should be used.

More seriously, occam-π provides no facility for specifying the relationship between the two protocols in a client-server interface. By convention, the programmer writes a comment saying "replies rolled or dropped" next to the definition of `roll`, but this is only useful to humans. The compiler cannot check that the processes using the channel bundle are correctly ordering messages between channels. For example, a process like this correctly follows the protocol on each individual channel:

```
SEQ
  die[req] ! roll; 6
  die[req] ! roll; 6
```

However, it would deadlock because the server expects to only receive a single `roll` message before sending a response. At the moment, it will be accepted by the compiler without complaint.

The vast majority of channel bundle definitions in existing occam-π code are client-server interfaces like `DIE`. Providing a more convenient language binding for client-server interfaces would not only simplify many programs, but also allow the compiler to detect more programmer errors at compile time.

3. Related Work

Facilities for specifying two-way communication are present in some other process-oriented languages.

The draft occam 3 language specification [5] described a *call channels* mechanism built on top of channel bundles; this provided a way of declaring channel bundles that were used for call-response communications. The declaration of a call channel therefore implicitly defined protocols to carry the parameters and results of a procedure. A call channel named `cosine` with a single input parameter and a single result would be defined as:

```
CALL cosine (RESULT REAL32 result, VAL REAL32 x):
```

The suggested syntax made clients look like procedure calls, and servers look like procedure declarations:

```
PAR
  cosine (cos.pi, 3.141)
  ACCEPT cosine (RESULT REAL32 result, VAL REAL32 x)
    result := COS (x)
```

Since **ACCEPT** is implemented as a channel input for the parameters, followed by a channel output for the results after the block is complete, it is possible to use it as a guard in an **ALT**. The same idea has been implemented in other process-oriented frameworks such as JCSP [6].

Call channels are a useful abstraction for programmers transitioning from the object-oriented world, since they make calls to a server look like method calls upon an object. However, they only allow a single request and response; they do not provide the richer conversations afforded by protocols.

The Honeysuckle language provides facilities for easily composing client-server systems, with interfaces being defined as *services* [7]. Of particular interest here are *compound services*, which allow a server's behaviour to be specified using a subset of Honeysuckle including communication, choice and repetition constructs:

```
service class Console :
{
    ...

    sequence
      receive command
      if command
        write
          acquire String
        read
          sequence
            receive Cardinal
            transfer String
}
```

This notation is very powerful; it allows arbitrary conversations between a client and server to be precisely specified. It is, however, possible to specify a protocol that cannot be statically verified by using repetition with a count obtained from a channel communication. Such protocols may require runtime checks to be inserted by the compiler if the repetition counts cannot be statically determined.

Honeysuckle's services provide a convenient, flexible way of specifying client-server interfaces; we would like to provide a similar facility in occam-π.

4. Session Types

Session types [8] provide a formal approach to the problem of specifying the interactions between multiple processes, by allowing communication protocols to be specified as types. The type of a communication channel therefore describes the sequence of messages that may be sent across it. For example, a channel with the session type

$$foo! \, . \, bar?$$

can be used to send ("!") the message foo, then receive ("?") the message bar; the "." operator sequences communications.

Session types can also specify choice between several labelled variants using the "|" operator. For example,

$$(left! \, . \, INT!) \mid (right! \, . \, BYTE!)$$

can be used to either send $left$ followed by an integer, or $right$ followed by a byte.

When checking the correctness of a process, a session-type-aware compiler will update the type of each channel as communications are performed using it. For example, if a channel's session type is initially $foo! \, . \, bar? \, . \, baz?$, after it is used to send the message foo, its session type will be updated to $bar? \, . \, baz?$.

Session types were originally defined in terms of the π-calculus, but can also be applied to network protocols, operations in distributed systems, and – most interestingly for our purposes – communications between threads in concurrent programming languages.

Neubauer and Thiemann [9] describe an encoding of session types in Haskell's type system, representing communication operations using a continuation-passing approach. Session types may be defined recursively, which is convenient for specifying protocols containing repetition or state progression – for example, a type may be defined as several operations followed by itself again. The specifications are applied to sequences of IO operations, such as communications on a network socket; there is no discussion of their application to local communication, although the same approach could be used to sequence communication between threads.

The L_{doos} language [10] integrates object-oriented programming and session types. Its session type specifications cannot contain branching or selection, but they support arbitrary sequences of communications in both directions, making them more flexible than simple method calls.

Vasconcelos, Ravara and Gay [11] give operational semantics and type-checking rules for a simple functional language with lightweight processes and π-calculus-style channels, where channel protocols are specified using session types. Its session types may be defined recursively, and may include choice between several labelled options. It notes that aliasing of channels can introduce consistency problems, since operations may affect one alias and not update the session type of the others. It demonstrates that session types can be applied effectively to communication between local concurrent processes.

The SJ language [12] extends Java with *session sockets* that are conceptually similar to TCP sockets (and are implemented using TCP), but over which communication takes place according to protocols which are defined using session types. It supports conditional and iteration constructs in which the branch taken is implicitly communicated across the socket by the sending process; this ensures that the two ends cannot get out of step.

Connected session sockets can be passed around between processes, and the SJ system tracks their session types correctly even when they are in mid-communication; this makes it possible to hand off a connected socket to another process to continue the conversation.

5. Two-Way Protocols

We propose generalising occam-π's protocols so that they can specify two-way conversations rather than just sequences of one-way communications.

occam-π's unidirectional protocol specifications can be viewed as a restricted form of session types: all the communications must be in the same direction, only a single choice is permitted at the start of the protocol, and no facilities are provided for iteration and recursion within a protocol (although the same protocol can be used multiple times across the same channel). In order to specify two-way communications, we must relax some of these restrictions.

For example, the `DIE` interface above could be expressed as a single two-way protocol between the client and the server. In this protocol, a client starts a conversation by sending a `roll` or `quit` message; the server will reply to `roll` only with `rolled` or `dropped`. We can specify this as a session type from the client's perspective:

$$(roll! \ . \ INT! \ . \ (rolled? \ . \ INT? \ | \ dropped?)) \ | \ quit?$$

(We do not propose that occam-π programmers should use this syntax for protocol definitions – see section 7.)

Protocols can contain multiple direction changes; for example:

$$move! \ . \ (moved? \ | \ (suspend? \ . \ suspended!))$$

One valid conversation using this protocol would be *move!*, *suspend?*, *suspended!*; another would be *move!*, *moved?*. The conversation *suspended!* would not be valid.

We constrain the first communication in a two-way protocol specification to always be an output; this makes it possible for the compiler to always be able to tell in which direction the next communication is expected to come.

A client-server connection can now just appear as a channel to the occam-π programmer; there is no need to specify whether a particular communication is a request or a response, since that is implicit in the operation being used. For example, our `DIE` client can now be written this way:

```
PROC roll.die (CHAN DIE die!)
  SEQ
    die ! roll; 6
    die ? CASE
      INT n:
      rolled; n
        ... rolled an n
      dropped
        ... dropped the die
:
```

Furthermore, the compiler now has enough information to be able to tell that the following process does not conform to the protocol:

```
SEQ
  die ! roll; 6
  die ! roll; 6
```

Since the session type is now tracked between multiple communications on the same channel, we could allow sequential communications to be split up over multiple communication processes: that is, `die ! roll; 6` would be merely syntactic sugar for:

```
SEQ
  die ! roll
  die ! 6
```

6. Implementation

6.1. Two-Way Channels

Two-way channels could be implemented by the occam-π compiler using a pair of regular channels inside a channel bundle. The transformation required would be very straightforward,

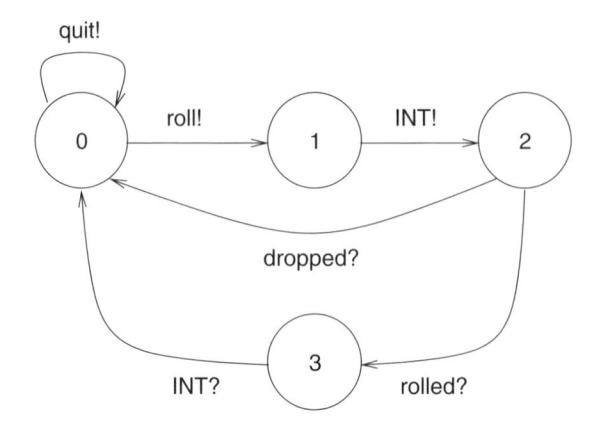

Figure 2. Finite state machine representing the DIE protocol

simply selecting the "request" or "response" channel inside the bundle based on the direction of communication. This approach would allow the translation of occam-π code using two-way channels into code that would be accepted by the existing compiler.

However, all existing implementations of occam-π channels on a single machine allow communication in either directions, provided both users of the channel always agree about the direction of communication they expect. Since session types allow the compiler to reason about the direction of communication whenever a channel is used, we can use the existing occam-π runtime's channels as two-way channels with no additional overhead.

This also offers a small memory saving: one channel can now be used where two and a channel bundle were previously necessary. This may be useful in programs with very large numbers of channels.

6.2. Protocol Checking

Checking that one-way protocols are used correctly is simple: each input or output operation must always perform the complete sequence of communications that the protocol describes, so the compiler can tell what communications should happen from the type of the channel alone. Two-way protocols complicate this somewhat because the protocol may take place across multiple operations. We can solve this problem by representing protocols as session types, and attaching a session type to each channel end.

A common representation for a session type is a finite state machine, with each message being an edge in the state machine's graph (see figure 2). The compiler will translate each protocol definition it sees into a state machine; a session type can then be represented as a pair of a state machine and a state identifier within that machine. Given a channel's current state, this makes it possible to tell whether an operation upon it is valid, and if so what the resulting state is. The same approach is already used in Honeysuckle and in implementations of session types in other languages.

Each protocol has an *initial state* for the start of a conversation. Since occam-π protocols may be repeated arbitrarily as a whole, a message with nothing following it in the protocol specification is recorded as a transition back to the initial state.

To check a program for protocol compliance, it is first transformed into a control flow graph (which occam-π compilers already do in order to perform other sorts of static checks). Each channel end variable is tagged with a state, which is set to the initial state when a channel is first allocated. The control flow graph is traversed; when a communication operation is performed upon a channel end, the current state and the message are checked against the appropriate state machine, and the state is updated. If the communication is not valid for the present state, the compiler can report not only that it's invalid, but also what communications would have been valid at that stage of the protocol.

When two flows of control rejoin, the compiler must check that the state of each channel end is the same in both flows; this ensures that conditionals and loops do not leave channels in an inconsistent state. When a channel is abbreviated – either via an explicit abbreviation, or in a procedure definition – it must be left in the same state at the end of the abbreviation that it was in at the start. (This rule may need to be adjusted to support mid-conversation handoff; see section 7.3.)

Similarly, when a channel end is sent between processes (for example, as part of a channel bundle), its state must be preserved by the communication. As with SJ, it would be perfectly reasonable to hand off a channel end in the middle of a conversation to another process, provided the receiving process agrees what session type it should have. This allows the process network to be dynamically reconfigured without consistency problems.

A `CLAIM` upon a shared channel must start and end with the channel in its initial state, since the channel must be left in a predictable state for its next user. Shared channels are the only case in which channel ends may be aliased in occam-π, so this restriction avoids consistency problems caused by aliasing of session-typed channels.

Note that the channel's state is not tracked at runtime, as with many other session types systems. Two-way protocols incur no runtime overheads.

7. Protocol Specifications

The syntax used so far for session types is hard to read and write, particularly for complex protocols with many choices and direction changes; we would like something more convenient for use in occam-π. We emphasise that we have not yet decided on a final syntax for this; this section describes some of the possibilities we have considered.

7.1. Starting Small

We must preserve the existing syntax for unidirectional protocols in order to avoid breaking existing occam-π code, but since a unidirectional protocol is just a special case of a two-way protocol, we can deprecate the existing syntax in favour of a new one. However, there are some advantages to basing our new syntax on the existing one: the existing syntax has worked well for over twenty years, and occam-π programmers are already familiar with it.

A minimal approach would be to keep the existing syntax for unidirectional protocols – so communications in the same direction are still sequenced using ; – but allow an indented block inside a protocol specification to mean a change of direction. We could then write our `DIE` protocol as:

```
PROTOCOL DIE
  CASE
    roll; INT
      CASE
        rolled; INT
        dropped
    quit
:
```

This protocol is very simple, but if it had more changes of direction (and therefore deeper nesting), it would be harder to tell the direction of each communication. We could require the user to explicit specify the direction of each communication:

```
PROTOCOL DIE
  CASE
    ! roll; INT
      CASE
        ? rolled; INT
        ? dropped
    ! quit
:
```

The directions are specified from the perspective of the client. This is more useful for documentation purposes; a programmer is more likely to be writing a client to somebody else's server than a server to somebody else's client. Since all the communications within a single CASE must be in the same direction, it is somewhat redundant to specify the direction on all of them; it would be possible to apply the direction to the CASE itself instead.

This syntax does not allow the full power of session types, though. It is only possible to have choice at the start of a communication or after a change of direction, which means you cannot send some identifying information followed by a command. Furthermore, there is no way to name and reuse parts of the protocol; you cannot write a protocol containing repetition, or share a response (such as a set of error messages) between several possible commands.

7.2. Protocol Inheritance

We have not yet specified how protocol inheritance would work with these simple two-way protocols. We could allow the inclusion of an existing protocol in a new one by giving the existing protocol's name. The effect would be as if the existing protocol's specification were textually included in the new protocol.

```
PROTOCOL ERROR
  CASE
    ! ok
    ! file.not.found
    ! disk.full
:
PROTOCOL FILE
  CASE
    ! open; FILENAME
      ERROR
    ! write; STRING
      ERROR
:
```

Note that the ERROR protocol's direction has been implicitly reversed when it is included in FILE. In combination with occam-π's existing RECURSIVE keyword, which brings a name into scope for its own definition, this approach would allow protocols containing repetition to be defined recursively:

```
RECURSIVE PROTOCOL ARK
  CASE
    ! animal; ANIMAL
      CASE
        ? ok
          ARK
        ? full
    ! done
:
```

This approach has several downsides, though.

It is only possible to recurse back to the "outside" of a protocol. Mutually recursive protocols – which may be useful when you have a protocol that switches between two or more stable states – cannot be written.

It is also difficult to describe the type of a channel in mid-conversation. The session type of a **CHAN FILE** after an **open** or **write** message has been sent is the same: the expected messages are those from **ERROR**. However, it is not a **CHAN ERROR**, because after the error message has been sent the next communication will be one from **FILE**. This makes it impossible to write a reusable error-handling process.

7.3. Named Subprotocols

A more flexible approach would be to allow the user to define named subprotocols within a single protocol definition – which the compiler will eventually translate into named states within the protocol's state machine. Our **FILE** protocol with errors can now be written using a subprotocol for error reporting:

```
PROTOCOL FILE
  SUBPROTOCOL ERROR
    CASE
       ? ok
       ? file.not.found
       ? disk.full
  :

  CASE
    ! open; FILENAME
       ERROR
    ! write; STRING
       ERROR
:
```

Note that the message directions are now written consistently between the top-level protocol and its subprotocol.

We can now refer to a particular state within a protocol when describing a channel's type, which lets us write abbreviations and procedures expecting a channel in a particular state:

```
PROC handle.error (CHAN FILE[ERROR] c?)
  ...
:

SEQ
  c ! open; "foo.occ"
  handle.error (c?)
```

One problem with this is that, by the checking rules described earlier, the abbreviation of c in the procedure definition would require it to have the same type when the procedure exited – which will not be the case if it has handled the error. To solve this, we could allow the input and output states of a protocol to be specified in an abbreviation's type – for example, **CHAN FILE[ERROR, FILE] c?**. Another option is to just specify **CHAN FILE** and have the compiler infer the input and output states.

The top-level protocol's name is still made available if **RECURSIVE** is used, so **ARK** can be defined as it is above. We could generalise this to permit mutual recursion between subprotocols, which is rather unusual for occam-π; its scoping rules usually forbid mutual recursion. Mutually-recursive subprotocols would allow us to specify a protocol with multiple "stable states": for example, a network socket that may be either connected or disconnected, and supports different sorts of requests in different states.

```
PROTOCOL SOCKET
  SUBPROTOCOL DISCONNECTED
    CASE
      ! connect; ADDRESS
        CONNECTED
  :
  SUBPROTOCOL CONNECTED
    CASE
      ! send; DATA
        CONNECTED
      ! disconnect
        DISCONNECTED
  :

  DISCONNECTED
:

CHAN SOCKET c:
...
SEQ i = 0 FOR SIZE addrs
  SEQ
    c ! connect; addrs[i]
    c ! send; "hello"
    c ! disconnect
```

occam-π protocols are not currently written to be this long-lived: a socket protocol in the current language can only describe a single request. Recursive protocols would allow longer-lasting interactions to be captured.

8. Mobile Channels

At present, occam-π allows channel bundles to be mobile, but not individual channels. In order to make a two-way channel mobile, it would need to be wrapped in a channel bundle. It would be more convenient for most uses of two-way channels if the ends of a channel could simply be declared to be mobile in the same way as data and barriers.

The existing syntax for channel end abbreviations in occam-π appends the ! and ? decorators to the name of the abbreviation. The same syntax could be used for mobile channel ends:

```
MOBILE CHAN FOO out!:
MOBILE CHAN FOO in?:
SEQ
  out, in := MOBILE CHAN FOO
  PAR
    out ! some.foo
    in ? other.foo
```

However, this makes it impossible to write the type of a channel end on its own – for example, if you wanted to declare a protocol carrying channel ends, or define a type alias. Where does the decorator go? It would be simpler to always include the direction as part of the type of a channel end:

```
MOBILE CHAN! FOO out:
MOBILE CHAN? FOO in:
```

9. Conclusion

We have proposed extensions to the occam-π programming language that would significantly extend the expressive power of channel protocols by permitting two-way communication on a single channel. We have shown how session types can be used to specify these protocols, and to check that processes implement the protocols correctly. We have therefore demonstrated that session types can be used to improve the safety of inter-process communication in an existing concurrent programming language.

Two-way protocols would significantly simplify the implementation of client-server systems in occam-π, but it is important to note that they are not tied to the client-server model. For example, it would be possible to build a ring of processes connected by two-way channels, which would be a violation of the client-server design rules, but permissible using I/O-PAR or other approaches. Two-way protocols simply ensure that the communications between any two connected processes proceed in a consistent manner.

Two-way protocols cannot guarantee that a system follows any particular set of design rules; this is, in general, a difficult problem to solve, particularly in the face of dynamically-reconfigurable process networks. Making guarantees about the system as a whole is out of the scope of this proposal, although we hope that the ability to reason formally about channel protocols using session types would make a future implementation of design-rule checking more straightforward.

Acknowledgements

The author would like to thank Neil Brown and the other members of the Concurrency Research Group at the University of Kent for their work on this proposal.

This work was supported by EPSRC grants EP/P50029X/1 and EP/E053505/1.

References

[1] J.M.R. Martin and P.H. Welch. A Design Strategy for Deadlock-Free Concurrent Systems. *Transputer Communications*, 3(4), 1997.
[2] Inmos Limited. occam 2 Reference Manual. Technical report, Inmos Limited, 1988.
[3] Fred Barnes and Peter H. Welch. Prioritised Dynamic Communicating Processes - Part II. In J. Pascoe, R. Loader, and V. Sunderam, editors, *Communicating Process Architectures 2002*, pages 353–370, 2002.
[4] Fred Barnes and Peter H. Welch. Prioritised Dynamic Communicating Processes - Part I. In J. Pascoe, R. Loader, and V. Sunderam, editors, *Communicating Process Architectures 2002*, pages 321–352, 2002.
[5] Geoff Barrett. occam 3 Reference Manual. Technical report, Inmos Limited, March 1992.
[6] Peter H. Welch. Process Oriented Design for Java: Concurrency for All. In P.M.A.Sloot, C.J.K.Tan, J.J.Dongarra, and A.G.Hoekstra, editors, *Computational Science - ICCS 2002*, volume 2330 of *Lecture Notes in Computer Science*, pages 687–687. Springer-Verlag, April 2002. Keynote Tutorial.
[7] Ian R. East. Interfacing with Honeysuckle by Formal Contract. In J.F. Broenink, H.W. Roebbers, J.P.E. Sunter, P.H. Welch, and D.C. Wood, editors, *Communicating Process Architectures 2005*, pages 1–11, September 2005.
[8] Kohei Honda. Types for Dyadic Interaction. In *Proc. CONCUR '93*, number 715 in LNCS, pages 509–523. Springer, 1993.
[9] Matthias Neubauer and Peter Thiemann. An implementation of session types. In Bharat Jayaraman, editor, *PADL*, volume 3057 of *Lecture Notes in Computer Science*, pages 56–70. Springer, 2004.
[10] Mariangiola Dezani-Ciancaglini, Nobuko Yoshida, Alex Ahern, and Sophia Drossopoulou. l_{doos}: a Distributed Object-Oriented language with Session types. In Rocco De Nicola and Davide Sangiorgi, editors, *TGC 2005*, volume 3705 of *LNCS*, pages 299–318. Springer-Verlag, 2005.
[11] V. T. Vasconcelos, Antnio Ravara, and Simon Gay. Session types for functional multithreading. In *CONCUR'04*, number 3170 in LNCS, pages 497–511. Springer-Verlag, 2004.
[12] Raymond Hu, Nobuko Yoshida, and Kohei Honda. Language and Runtime Implementation of Sessions for Java. In Olivier Zendra, Eric Jul, and Michael Cebulla, editors, *ICOOOLPS'2007*, 2007.

Prioritized Service Architecture: Refinement and Visual Design

Ian R. EAST

Dept. for Computing, Oxford Brookes University, Oxford OX33 1HX, England

ireast@brookes.ac.uk

Abstract. Concurrent/reactive systems can be designed free of deadlock using prioritized service architecture (PSA), subject to simple, statically verified, design rules. The Honeysuckle Design Language (HDL) enables such *service-oriented* design to be expressed purely in terms of communication, while affording a process-oriented implementation, using the Honeysuckle Programming Language (HPL). A number of enhancements to the service model for system abstraction are described, along with their utility. Finally, a new graphical counterpart to HDL (HVDL) is introduced that incorporates all these enhancements, and which facilitates interactive stepwise refinement.

Keywords. client-server protocol, deadlock-freedom, programming language, correctness-by-design, visual programming, visual design.

Introduction

Background and Motivation

Every programming language reflects an underlying *model for abstraction*. For example, an imperative language relies upon the procedure, which is simply a composition of commands that act upon objects to update their state. The procedure is itself a command, and may be subject to further composition. Operators dictating the manner of composition for both command and object, together with a set of primitives for each, complete the model.

Three questions arise over any proposed language and its associated model for abstraction: Can the model directly capture all the behaviour of the systems we may wish to build? Is each program then *transparent*? How precisely is the meaning of each program defined, and can we relate one program to another? To put it more simply, how easy are programs both to write and to read, and how can we tell, for example, that one may perform a function indistinguishable from another?

Early languages complicated their model by including reference to the machine level, with pointer and 'goto'. Despite the fact that it has been repeatedly shown to be unnecessary, such things remain in contemporary languages, denying simplicity and transparency of expression. These languages also typically lack a formal semantics, despite valiant attempts to install a foundation after the house has been built. Finally, they fail to capture concurrent and reactive behaviour, or do so in a manner that is unnecessarily complicated and obscure.

For many, occam [1] offered a remedy for all these shortcomings. Its *process-oriented* model could boast a formal foundation in the theory of Communicating Sequential Processes (CSP) [2,3], affording analysis, and a precise interpretation, of every program.

While occam arguably represented a major advance, it left room for further progress. In some ways, it was perhaps too simple. An object could only be copied, and not passed, between processes, giving rise to some inefficiency. Only processes, and not objects, could encapsulate behaviour with information, limiting choice in system abstraction, and suggesting a conflict with (highly popular) object-oriented design. There was no facility for collecting definitions together, according to their utility – something we shall refer to here as *project*, as opposed to system, modularity – in contrast with Java, for example.

Finally, occam offered no intrinsic help in avoiding the additional pathological behaviour inherent with concurrent and reactive behaviour. In particular, one had to rely upon informal design patterns to exclude the threat of deadlock.

The Honeysuckle Programming Language

The Honeysuckle project began as an attempt to establish a successor to occam that builds on its strengths [4]. In particular, it overcomes the additional threat of pathological behaviour introduced when incorporating concurrency and prioritized reactive (event-driven) behaviour. Honeysuckle incorporates formal design rules that guarantee every design and program *a priori* deadlock-free [5]. These are verified 'statically' (upon compilation).

In the Honeysuckle model for abstraction, a system may be reduced into components, which, as in occam, are processes. Unlike occam, the interface between components is expressed as *services* provided or consumed. A service comprises a sequence of communications, each defined by the type of object conveyed (if any), whether ownership, or simply a copy, is passed, and the direction in which conveyance takes place. Each service is oriented according to initiation, which is performed by the client (consumer). At the "sharp end" is the server (provider). A component may either uphold mutual exclusion or *alternate* provision of multiple (aggregated) services according to static prioritization.

It is worth noting that a service is ultimately implemented using at most two synchronous *channels* (one in each direction), and so any Honeysuckle program could be expressed occam. Service architecture may therefore be considered 'superstructure'.

Like occam, Honeysuckle has been built upon a formal foundation, derived from CSP. A formal definition has been provided of both service and *service network component* (SNC), on which a proof of deadlock-freedom has been based [5].

While a component may be specified by its interface, its design is expressed via the declaration of a service network. Interface definition and network declaration form the *context* in which an implementation is later described, and are rendered explicit using a subset of Honeysuckle – the *Honesuckle Design Language* (HDL) – as a 'header' above a process implementation. For example:

```
definition of process p

   ... imports

   process p is
   {
     interface
       provider of s1
       client of s2

     network
       s1 > s2
   }
```

Here, a component is defined that provides one service while consuming another. The service provided *depends* upon the one consumed. Dependency is one way in which services may inter-relate. Each 'input' (service provided at the component interface) may relate to 'output' (service consumed at the interface) via a *chain of dependency*.

While it is possible to express the behaviour of any system solely in terms of service architecture, implementation of every dependency requires the definition of a process.

Dependency *abstracts* process for the purpose of design.

Note that no implementation is given in the above definition. Design and implementation are intended to be separate, with each design verified against rules which guarantee deadlock-freedom [5]. An implementation may be provided later, and is verified only against a design.

Three kinds of *item* must be defined in a Honeysuckle program – process, service, and object class. Definitions may be collected, according to their utility, and traded (imported and exported). The ability to form a *collection* affords "project modularity" – the possibility of sharing (reusing common) definitions between distinct projects.

A Honeysuckle implementation is imperative and process-oriented, and must describe *how* each service is provided or consumed, using the same primitives as occam – SEND, RECEIVE, and ASSIGN – plus two more – TRANSFER and ACQUIRE – which refer to the conveyance of an object, rather than a value. (The distinction between value and object (reference) is made by choice of operator, rather than by modifying an identifier.)

Each process is constructed in much the same way as Pascal, using sequence, selection, and repetition, but with the addition of a parallel construct, as in occam, and one that affords *prioritized alternation*. The latter dictates pre-emptive behaviour similar to that brought about by hardware prioritized vectored interrupts, but subject to a certain discipline [6].

Honeysuckle syntax is intended to result in transparent (highly readable) programs, where simple function is achieved with simple text. For example, a "Hello world!" process can be expressed in just two lines, free of any artifact:

```
process hello is
  send "Hello world!" to console
```

Here, 'console' is the name of a service, not a process.

Overall, it is hoped that Honeysuckle offers even greater simplicity, efficiency, and transparency than did occam, while introducing *a priori* security against deadlock.

The Need for Refinement of Honeysuckle and the PSA *Model for Abstraction*

Honeysuckle incorporates a model for system abstraction that allows a design to be expressed entirely in terms of communication, without reference to process or object. As stated above, the intention is that such a design should offer sufficient information for deadlock to be excluded *before* any implementation is conceived. Unfortunately, with the language thus far reported [7], it remains possible to express a design which would endorse some incorrect implementations – ones that would fail verification of the conditions governing the behaviour of a component of a deadlock-free network [5]. The simplest example is the classic one of "crossed servers", where two clients each seek sequential access to two servers. This would form a hitherto legitimate Honeysuckle design, though any implementation would break at least one condition governing the behaviour of a service network component.

As a result, it became clearly desirable to establish an enhanced service-oriented model in which one could distinguish the correct from the incorrect. Changes to the Honeysuckle Design Language (HDL) are proposed that allow every legitimate design to dictate an implementation that will satisfy all conditions for deadlock-freedom. These changes are documented in Section 1 and 2. Sadly, this does entail complicating the language, but the prize of properly separating verifiable design from implementation is believed well worthwhile.

Interleaving in Context

The refinements to the PSA model needed to address the deficiency mentioned above depend upon a notion of 'interleaving'. In CSP, 'interleaving' refers to that of events identified with processes. Here, we refer to events (communications) identified instead with *services*.

For example, suppose a service comprises the sequence $\{c_i^1\}$, where i indexes progression in the delivery of that service. A second service might be formed by the sequence $\{c_j^2\}$, with progress indexed by j. Should the two services progress together, we might then observe a sequence $c_1^1, c_1^2, c_2^1, c_2^2, \ldots$.

Whenever one service starts before the other is complete, we say that the two *interleave*.

There are three different ways in which services may combine and interleave. Each is capable of endorsing conflict with the formal conditions laid down upon service and component [8,5], and thus aberrant behaviour. To remove such conflict, and establish the possibility of a guarantee of deadlock-freedom, given only a PSA design, certain attributes are introduced to each form of service combination, according to whether interleaving is allowed or denied.

These circumstances and changes are described in Section 2 below.

A Visual Design Language

While investigating these changes, it became increasingly apparent that design could be expressed with considerably greater transparency using pictures. Greater transparency is a key aim in the development of Honeysuckle and the PSA model for system abstraction.

Visual programming languages were explored widely in the 1970s. Particularly successful was the notion of "visual data flow", where processes embody functions repeatedly executed. Each one would await reception on each input port, evaluate the appropriate function, and then transmit the result. (Multiple output ports implied multiple functions evaluated.)

Evidence of the utility of this idea can be found in the huge success of LabView^{TM1}, which facilitates the visual programming of industrial test instrumentation. The dominant commercial appeal appears to be productivity. This is widely believed to result from the visual nature of the language and the consequent ease of expression. Productivity, through ease of expression, is another key aim in the development of PSA and Honeysuckle.

Because it would increase the ease with which a design may be both written and read, a Honeysuckle Visual Design Language is proposed and is described in Section 3.

Summary

Changes to the (textual) Honeysuckle Programming Language (HPL) are described first. Section 1 addresses changes to the abstraction, and syntax for the description, of the component interface. Section 2 deals with changes in the way an internal prioritized service network is conceived and expressed. Finally, Section 3 describes the proposed Honeysuckle Visual Design Language (HVDL).

1. Concurrency and synchronization in the component interface

1.1. Distributed Service Provision

To express the interface of a component, only services provided and consumed need be declared, and not their interdependency. It may be possible for any service offered to be consumed by more than one component simultaneously. Provision must then be *distributed* across more than one internal process [7]. An interface declaration should include the degree of concurrency possible. For example:

```
interface
   provider of [4] publisher
```

---

[1]LabView is a trademark of National Instruments Inc.

1.2. Synchronization of Shared Provision

Any service provided by a component may always be *shared* between multiple consumers. There is no need to advertise the fact, or limit the degree. As the number of consumers rises, beyond the degree of concurrency available, the delay will simply increase.

A queue will form, and service will become sequential.

When sequential service is *synchronized* [9], it is to be shared between a specific number of clients. Each will be served once before the cycle can begin again.

Note that no provision can be both concurrent *and* synchronized, in Honeysuckle.

The description of a component interface is a little different to that of each connection within a network. Because of the possibility of dependency, synchronization may encompass multiple distinct services. Fig. 13 shows an example of how this can occur. The corresponding interface can be expressed as follows:

```
interface
  synchronized
    [2] provider of claim
    provider of request
```

At an interface, it could conceivably make sense for a provision to be both synchronized *and* distributed. One purpose behind synchronization is to expose multiple processes to state whose evolution is perceived in common. Distributed providers could perhaps relay such common state via a single shared service on which they all depend. However, such access would again be sequential, even if such dependency were somehow enforced. Hence, mutual exclusion between distributed (concurrent) and synchronized (sequential) service is to be maintained, in Honeysuckle, at the interface, as within the network.

Once an interface has been established, a design can be developed, which requires the definition of a prioritized service network, declared separately.

2. Interleaving of Internal Service Provision, Dependency, and Consumption

2.1. Interleaved Provision

The Honeysuckle Design Language (HDL) provides a construct by which to express the prioritized interleaving of service *provision* [7]. Delivery of one service may be pre-empted by another, should a client initiate it.

```
network
  interleave
    s1
    s2
```

In the above, should provision of $s2$ be underway when $s1$ is initiated, it will cease to progress until $s1$ provision is complete, whereupon it will resume. Delivery of $s2$ is thus interrupted and pre-empted by that of $s1$. There is no intrinsic necessity for this condition. It has been adopted in order to simplify the model for abstraction, and to remove problems which otherwise arise in forming networks, secure against pathological behaviour.

Note that this behaviour can be stated wholly in terms of *service* provision.

Interleaved provision may be directly implemented using prioritized alternation (the WHEN construct) [6]. Only in implementation do we express behaviour in *process*-oriented terms. Note that alternation is *reactive* (event-driven), but free of concurrency. No two processes composed in this way run concurrently at any time.

Given the semantics applied to interleaved service provision, and its implementation via prioritized alternation, all the conditions placed upon both service conduct and each service network component can be maintained, except for systems where there is feedback.

With feedback, a service of lower priority must *depend* upon one of higher priority.

2.2. Interleaved Dependency

When the provision of one service depends upon the consumption of another, we say that a *dependency* exists. This relation may recur, allowing the formation of a *chain of dependency* of arbitrary length. For example:

```
network
  s1 > s2
  s2 > s3
  ...
```

A chain of dependency may form a feedback path across interleaved provision. To indicate this textually, a hyphen is introduced to the right of the service with the fed-back dependency, and to the left of the one on which it depends. For example:

```
network
  interleave
   -s1
    s2 > s1-
```

Here, provision of $s2$ depends upon consumption of another service ($s1$) with which provision is interleaved, and which has higher priority.

There exists the potential for conflict between the condition upon interleaved provision – that the interrupting service be completed before the interrupted one resumes – and any interleaving across the dependency. Should the consumption of $s2$ interleave with the provision of $s1$ then the former would have to continue while the latter proceeds. As a consequence, we need to refine our notion of prioritized service architecture (PSA) for consistency to prevail.

We shall therefore distinguish two kinds of dependency.

An *interleaved* dependency is one where communications of the service consumed interleave with those of the one provided. An *interstitial* dependency occurs where a service is consumed in its entirety between two communications of the one provided.

Since interstitial dependencies are the safer of the two, they will be assumed, by default. An interleaved dependency must now be indicated, with additional notation. For example:

```
s1 > s2, |s3
```

Here, provision of $s1$ has two dependencies; one interstitial, and one interleaved ($s3$).

The previous interleaving with feedback is now both more precisely defined and free of any conflict between applicable conditions. For the same to be said of any interleaved provision, a simple additional design rule must apply:

Every feedback chain must contain at least one interstitial dependency

A feedback chain of length one has a natural service-oriented interpretation, though is a little troublesome in implementation as it imposes a measure of concurrency upon provider and consumer.

2.3. Interleaved Consumption

Having discussed interleaved provision and interleaved dependency, we now turn to the third and final occasion for interleaving – interleaved *consumption*.

Let us suppose we have a dependency upon two services, as described previously, but where these services are in turn provided subject to mutual exclusion [7][2]:

```
network
  s1 > s2, |s3
  select
    s2
    s3
```

If consumption of $s2$ and $s3$ interleaves then deadlock will surely ensue — a classic example of "crossed servers". For example, an attempt to initiate $s3$ might be made before $s2$ is complete. That attempt will fail because provision of $s2$ and $s3$ has been declared mutually exclusive. On the other hand, if we already know that $s2$ is completed before $s3$ is begun, the configuration is entirely safe.

To ensure security, and retain pure service-oriented program abstraction, we therefore need to distinguish between interleaved and *sequential* consumption. An alternative to the above that is secure against deadlock would be expressed as follows:

```
network
  s1 > s2; |s3
  select
    s2
    s3
```

A semicolon has replaced the comma, indicating that consumption of $s3$ commences only after that of $s2$ is complete.

A comma does not necessarily indicate the converse – that consumption of the two services interleaves. Instead, it merely indicates the *possibility* of interleaving, which would be enough, in this case, to deny any guarantee of the absence of deadlock.

One other potential conflict with security against deadlock arises when chains of dependency cross. A design rule [5] requires that they never do so when fed forward to subsequent interleaved provision (inverting priority). A refinement of this rule is needed to account for crossing over by a feedback chain. We require that any dependency additional to any link in the feedback chain be consumed in sequence, and not interleaved.

3. The Honeysuckle Visual Design Language (HVDL)

3.1. Philosophy

Design, in general, refers to the decomposition of a system into components. Its purpose is to facilitate both analysis and separated development and test. Its outcome must therefore include a precise and complete description of component function and interface. With Honeysuckle, this need refer only to the pattern of communication between components.

PSA is particularly well-suited to *graphical* construction and editing, which is believed capable of improving both insight and productivity. A suitable development tool is needed to both capture a design and, on command, translate it into textual form. In no way does such a tool replace a compiler. It will merely act as an aid to design, though will be capable of verifying, and thus enforcing, both principal design rules [5].

Systems and components, defined purely according to services provided and consumed, may be composed or decomposed, without reference to any process-oriented implementation. Each *dependency* may be reduced, or combined with another. A HVDL editor is expected to afford either operation and thus allow the description of any service architecture by step-

---

[2]The keyword SELECT has been substituted for EXCLUSIVE, used in the earlier publication.

wise refinement. Only when an engineer decides that a system is *fully defined* (*i.e.* ready for implementation in HPL) is a translation into textual form carried out.

HVDL thus constitutes an interactive tool for the designer, whereas HPL serves the implementer (programmer). Design is bound by the Honeysuckle design rules (verified continually by HVDL), while any implementation remains bound by the design.

3.2. Independent Service

A simple, self-contained, system might comprise just the provision and consumption of a single service (Fig. 1).

Figure 1. A single service, both provided and consumed internally.

An arrow represents availability of the service s, and is directed towards the provider.

Provision of s is termed *independent*, by which we mean that it depends upon consumption of no other service. The nature of service s is documented separately.

The thin line at either end of the arrow denotes internal consumption. As depicted, the system therefore has no interface. External consumption or provision may be indicated by the same thin line drawn *across* an arrow, close to one end or the other accordingly (Fig. 2).

Figure 2. External consumption (left) and external provision (right).

Within the corresponding textual description, the distinction is rendered explicit within an accompanying component INTERFACE section [9].

3.3. Dependency and Interface

When provision of one service is dependent upon the consumption of another, an *interface* exists between services, denoted graphically by a '•'. While a process will be required to implement such an interface, there is no need to consider its nature when documenting design.

The nature of the dependency falls into one of two distinct categories (Fig. 3).

Figure 3. A single interstitial (left) and interleaved (right) dependency.

The distinction between interstitial and interleaved dependency was discussed earlier, in Section 2.2. Note that this forms an attribute of the interface, and not of either service.

Dependency upon multiple services requires a distinction between interleaved and sequential consumption (2.3). The latter is indicated graphically by the presence of a thin dividing line (Fig. 4). Absence of such a line indicates the *possibility* of interleaving.

In Fig. 4 (right), the vertical bar, just to the right of the interface symbol, confirms interleaving. Since both consumed services are interstitial, this implies that the two are completed within the same interstice (interval between two communications of the dependent service).

Figure 4. Sequential (left) and interleaved (right) consumption.

3.4. Aggregated Service Provision

As we described in Sections 2.1 and 2.3, service provision may be combined in two ways — mutually exclusive *selection*, and prioritized (pre-emptive) *interleaving* (Fig. 5).

Figure 5. Exclusive selection with shared dependency (left), and prioritized interleaved provision (right).

With exclusive (selective) provision, any dependency of any member is also a dependency of the entire *bunch*. This is because any member may be forced to await consumption. However, it remains useful still to indicate which among the bunch of services possesses a direct dependency. Alignment with the interface connection, or a short horizontal line extending an arrow, is enough (Fig. 6).

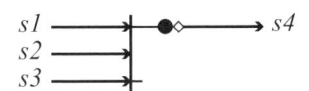

Figure 6. Indicating which mutually exclusive services are directly dependent.

Mutually exclusive selection is both commutative and associative. To eliminate redundant options for expression, nesting is denied. Offering the same service twice is similarly redundant, given *shared provision* (see below), and is thus also denied.

Multiple dependencies, within an interleaved provision, are drawn by extending the corresponding interface vertically (Fig. 7).

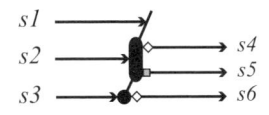

Figure 7. Multiple dependencies within interleaved provision.

Nesting of one INTERLEAVE construction inside another is unnecessary and denied. SELECT within INTERLEAVE is permitted, but not INTERLEAVE within SELECT.

In Section 2.3, we met the constraint on feedback (Fig. 8) across an interleaving that demands that the feedback chain ends with higher-priority provision. A second condition also applies; that, in any feedback chain, at least one dependency must be interstitial. When

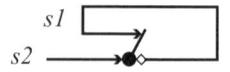

Figure 8. Feedback across interleaved provision.

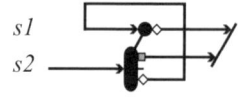

Figure 9. An example of legitimate consumption sequential to a feedback path.

a feedback chain has length one, the dependency is necessarily interstitial, and a measure of concurrency is required between consumer and producer, in implementation.

Any additional dependency of the lower-priority service must be arranged in sequence with the feedback path, when fed forward to further interleaved provision, as in Fig. 9. The order of the sequence is immaterial.

3.5. Sharing and Distribution

Shared provision is depicted via the convergence of multiple arrow shafts towards a single arrow head (Fig. 10).

Figure 10. Shared (left), and synchronized shared (right), provision.

Honeysuckle also affords *synchronized* shared provision, which requires each client to consume the service concerned once before the cycle can begin again [9]. Synchronized sharing is indicated graphically by a loop where arrow shafts join.

Like sharing, synchronized sharing is not an attribute of either service or process. It refers only to how a service is consumed within a particular network design.

While sharing offers many-to-one interconnection, *distributed* shared provision affords many-to-many configurations (Fig. 11).

Figure 11. Distributed shared provision.

It can never make sense to offer more providers than consumers, and so the very possibility is denied. There are therefore no one-to-many connections within any service digraph.

Multiple providers of a dependency may also be expressed (Fig. 12).

All distributed providers must be fully independent of their consumer. No feed-back is allowed. Behaviour is unaffected when a dependency is shared and/or distributed.

Synchronized sharing can only delay, and not deny, consumption.

Both graphical and textual expression of sharing and distribution renders the degree of each explicit. (Recall that, barring asymmetry, sharing denies the need for provision of

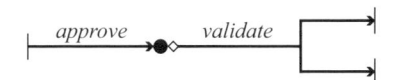

Figure 12. Distributed provision of a dependency.

multiple instances of the same service [7].) The number of clients to a shared service need only be revealed in the description of a complete system. *Synchronized* sharing is ultimately an exception to this rule, since behaviour might depend upon, or limit, its degree.

An example might be found in a component that protects a shared resource by requiring a claim by, say, three separate clients prior to release (Fig. 13).

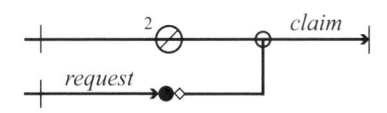

Figure 13. Synchronized sharing for resource protection.

Textual expression of such a system, though concise, is far less transparent:

```
network
  request > claim
  synchronized [3] shared claim
```

Note that an interface declaration says nothing about dependency of 'input' (service provision) upon 'output' (service consumption). As a result, it is impossible to capture any higher level of abstraction than the one implemented. Graphical expression, using HVDL has no such limitation.

An explicit indication of multiplicity in service provision can have two distinct meanings. As we have seen in Fig. 13, it can refer to the number of instances that must be synchronized. (It remains possible for a single client to satisfy this constraint, though perhaps the most common application will be that each instance will be consumed by a separate process.)

In addition, multiplicity can refer to the degree of distributed provision, which equates with the degree of concurrency possible in consumption. An example is depicted in Fig. 14. Here, a system provides up to three concurrent instances of a service. A fourth request will incur a delay until a provider becomes available.

Figure 14. Distributed (concurrent) provision, with a single internal shared dependency.

Note that provision of the dependency is not similarly replicated, and is therefore shared.

4. System Abstraction using HVDL

4.1. Introduction

Honeysuckle was previously able to abstract a system by decomposition into a network of processes, communicating under service protocol. HVDL does nothing to change that. Its effect is rather to provide the means of progressively refining system abstraction purely in terms of priioritized service architecture (PSA). As such, it represents a powerful design tool.

A designer might begin by simply specifying a system via its input and output connections, together with their interdependency. They might subsequently refine this (perhaps by clicking on the graphical representation of an interface) to achieve reduction. At any time, they may choose to subdivide the system into separated components. Refinement might continue until implementation becomes possible. At this point, a command can be given to generate both interface and network definition for each and every component. (A HVDL editor is expected to maintain a description of system/component composition.)

Advantages of this approach include the ability to maintain a hierarchical and graphical depiction of a concurrent/reactive system with any degree of complexity, in a manner that can be guaranteed authentic. Because a HPL compiler must guarantee conformance with interface and network definition, any changes to the abstract system/component specification or design can be similarly guaranteed in the implementation.

Of course, freedom from any possibility of deadlock also remains guaranteed.

4.2. Component Specification and Refinement

A component interface is expressed, using either HDL or HVDL, without including any dependency between 'input' (service provision) and 'output' (consumption). Only information required for connection to other components is provided. Any provision may be shared, but the degree to which it is either synchronized or distributed must be noted. Note that, because of internal structure (*e.g.* Fig. 13), synchronization may extend over services that are distinct, as well as shared. Distributed provision affords concurrent consumption.

Design might begin with only the highest level of abstraction – that of the dependency between component provision and consumption. Refinement might consist of either directly replacing an interface with additional structure (an *expansion*) or of naming that interface and giving it a separate description (an *embedding*).

It is not only an interface that can see its definition refined. A *terminal* (internal provision or consumption) may also be subject to either expansion or naming and embedding.

4.3. An Example

For illustration, let us consider a modern "print on demand" (POD) publishing system, that serves both customers and authors, and which consumes a printing service.

Our design will look a lot more like that of hardware than of software, and will omit certain parameters important in real life, such as the time taken either to complete a service or to progress between steps within it. However, as with a hardware (digital system) design, every detail required for the addition of such constraints to the specification is included.

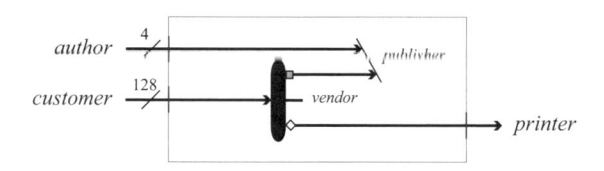

Figure 15. An abstraction of a "print on demand" publishing system.

Fig. 15 depicts an initial design with 'input' and 'output' ports labelled. Such a system can be defined via the following interface:

```
interface
  provider of [128] customer
  provider of [4] author
  client of printer
```

A convention is helpful whereby services are named according to provider. This affords readable use of the Honeysuckle primitives SEND, RECEIVE, TRANSFER, and ACQUIRE [10]. For example:

```
send draft to publisher
```

Client connections ('input' ports) are named only for illustration here. All that is significant is the name of the service provided, and the degree of concurrency available. Up to 128 customers can be served concurrently before one has to wait their turn. A much lower degree of concurrent provision is appropriate for authors.

Fig. 15 depicts the design (prioritized service architecture) as well as the specification (component interface). Labels appropriate to a process implementation have been added. The vendor interface exhibits a dependency of the customer service upon sequential consumption of publishing and printing services. A publisher process alternates service provision between authors and vendor. Note that serving the vendor is attributed a higher priority. (In this illustration, priority increases downwards. As a result, *every* graphic depicting an interleaving must lean to the left. Consistency, in this respect, is essential.)

Before declaring a network, we shall first refine our initial design. Clearly, management of provision (vending), consumption (printing and despatch), and internal accounting and record-keeping, will require significant decomposition into internal services (Fig. 16).

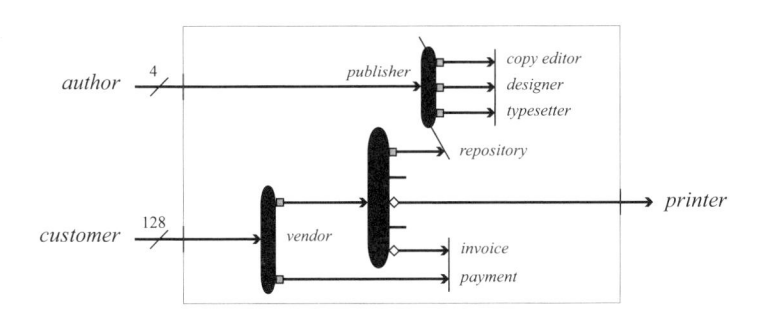

Figure 16. A refinement to the POD system where distribution has been separated.

It may eventually be considered worthwhile to "out-source" production (typesetting, copy-editing, cover and block design, *etc.*). Our design can thus be expressed via the following NETWORK statement:

```
network
  customer > |supply, |payment
  supply > |repository; printer; invoice
  interleave
    recover
    author > |copy_editor, |designer, |typesetter
```

Only prioritized service architecture (PSA) has been required in the description of a design. There has been no need to identify any process, though their location is self-evident.

5. Conclusion

Another step has been taken in the design of a pure service-oriented model for system abstraction, reflecting *protocol* rather than process, together with a corresponding programming language. The predominant aims of easing decomposition and denying pathological behaviour are well-served by an emphasis on communication, which both affords the transparent description of the individual component interface and exposes the overall pattern of interpendency within a complete network, allowing a measure of holistic analysis.

The Honeysuckle Programming Language (HPL) is intended to guarantee conformance with the conditions previously laid down for systems with prioritized service architecture (PSA), which in turn guarantee freedom from deadlock [5]. To this end, it has been necessary to add a little complication to the Honeysuckle Design Language (HDL – a subset of HPL). While HDL is a little less simple than before, it remains transparent and intuitive.

A later paper will offer a formal proof that conformance with the conditions follows conformance with the language, once the latter is finally complete and stable. No significant difficulty is anticipated here since those conditions that apply to each service network component (process) are easily adapted to instead refer to each service interface. No reference to any (process-oriented) implementation is necessary.

A *graphical* counterpart to HDL – the Honeysuckle Visual Design Language (HVDL) has been presented. This offers an alternative to textual expression that affords much greater transparency. A suitable graphical editor could offer hierarchical decomposition (or composition) of systems with PSA, and the automatic generation of interface and network definition that can be subsequently implemented using HPL. It could also offer visualization of existing textual designs, since a one-to-one correspondence exists between text and graphics.

HVDL demonstrates the possibility of employing a formal model for component-based systems to create development tools that combine high productivity with high integrity, free of the need for staff with a scarce combination of mathematical and engineering skill. It also demonstrates the potential for system definition and design using a purely communication-oriented vocabulary and syntax.

Applicability to the engineering of distributed systems with service architecture is obvious, but HDL is by no means limited to such "high-level" abstraction. PSA, and thus Honeysuckle, is equally applicable to engineering systems at *any* level of abstraction, from hardware upward, and seems particularly well-suited to high-integrity embedded systems, and perhaps to 'co-design' of systems to be implemented with a mixture of hardware and software.

References

[1] Inmos. *occam 2 Reference Manual*. Series in Computer Science. Prentice Hall International, 1988.
[2] C. A. R. Hoare. *Communicating Sequential Processes*. Series in Computer Science. Prentice Hall International, 1985.
[3] A. W. Roscoe. *The Theory and Practice of Concurrency*. Series in Computer Science. Prentice-Hall, 1998.
[4] Ian R. East. Towards a successor to occam. In A. Chalmers, M. Mirmehdi, and H. Muller, editors, *Proceedings of Communicating Process Architecture 2001*, pages 231–241, University of Bristol, UK, 2001. IOS Press.
[5] Ian R. East. Prioritised Service Architecture. In I. R. East and J. M. R. Martin et al., editors, *Communicating Process Architectures 2004*, Series in Concurrent Systems Engineering, pages 55–69. IOS Press, 2004.
[6] Ian R. East. Programming prioritized alternation. In H. R. Arabnia, editor, *Parallel and Distributed Processing: Techniques and Applications 2002*, pages 531–537, Las Vegas, Nevada, USA, 2002. CSREA Press.
[7] Ian R. East. Concurrent/reactive system design with honeysuckle. In A. A. McEwan, S. Schneider, W. Ifill, and P. H. Welch, editors, *Proceedings of Communicating Process Architecture 2007*, pages 109–118, University of Surrey, UK, 2007. IOS Press.

[8] Jeremy M. R. Martin. *The Design and Construction of Deadlock-Free Concurrent Systems.* PhD thesis, University of Buckingham, Hunter Street, Buckingham, MK18 1EG, UK, 1996.
[9] Ian R. East. Interfacing with Honeysuckle by formal contract. In J. F. Broenink, H. W. Roebbers, J. P. E. Sunter, P. H. Welch, and D. C. Wood, editors, *Proceedings of Communicating Process Architecture 2005*, pages 1–12, University of Eindhoven, The Netherlands, 2005. IOS Press.
[10] Ian R. East. The Honeysuckle programming language: An overview. *IEE Software*, 150(2):95–107, 2003.

Experiments in Translating CSP‖B to Handel-C

Steve SCHNEIDER [a,1], Helen TREHARNE [a], Alistair McEWAN [b] and Wilson IFILL [c]

[a] *University of Surrey*
[b] *University of Leicester*
[c] *AWE Aldermaston*

Abstract. This paper considers the issues involved in translating specifications described in the CSP‖B formal method into Handel-C. There have previously been approaches to translating CSP descriptions to Handel-C, and the work presented in this paper is part of a programme of work to extend it to include the B component of a CSP‖B description. Handel-C is a suitable target language because of its capability of programming communication and state, and its compilation route to hardware. The paper presents two case studies that investigate aspects of the translation: a buffer case study, and an abstract arbiter case study. These investigations have exposed a number of issues relating to the translation of the B component, and have identified a range of options available, informing more recent work on the development of a style for CSP‖B specifications particularly appropriate to translation to Handel-C.

Keywords. Handel-C, CSP‖B, translation, formal development.

Introduction

This paper investigates the translation of rigorous models written in CSP‖B into the Handel-C programming language, as a route to hardware implementation. CSP‖B [22] is a formal method integrating CSP for the description of control flow and the B-Method for the handling of state. It is supported by industrial-strength tools [9,3,6,12] which enable verification of CSP and B models, separately and together, through model-checking and proof. It is particularly suited for applications where control and data considerations are both critical for correct behaviour. Areas where CSP‖B has been applied include a file transfer protocol [7], dynamic properties of information systems [8], and modelling platelets in blood clot modelling [20]. It has also been used within the AWE funded project 'Future Technologies for System Design' at the University of Surrey, which is concerned with formal approaches to co-design. An element of that project has been the investigation of a development methodology which takes CSP‖B formal models to hardware implementation, via a translation to Handel-C. Handel-C [17,5] is a programming language designed for compilation to hardware, in particular FPGAs. The translation from CSP‖B to Handel-C is a key link in the project's development methodology, which aims to refine requirements down to implementation, via formally verified specifications. CSP‖B is used as the modelling language, since it enables formal analysis with respect to high level properties. Low level programming languages closer to code, such as Handel-C, are not sufficiently abstract to support such analysis, so our ideal is to refine the CSP‖B models (once they are verified) to Handel-C code to provide an appropriate implementation.

---
[1]Corresponding Author: *Steve Schneider, University of Surrey, Guildford, Surrey, GU2 7XH, UK.* E-mail: S.Schneider@surrey.ac.uk.

There is existing work on translating CSP to Handel-C [21,16] and the translation of the CSP controllers of our CSP∥B description is broadly in line with those. There is also work on translating B to Handel-C [11] through use of annotations to indicate the control flow. However, translations of CSP and B combined models have not previously been considered, and this is the gap that this paper begins to address. The novelty of the approach we are taking is the combined translation of control and state, which is not considered in the other authors' previous work, and the new issues being considered are the translation of the state aspects of a CSP∥B description within the context of the CSP translation. Data is often a key aspect of a system description, and our approach introduces the ability to incorporate a significant state component at the abstract level, and through the translation.

The paper describes two exploratory case studies that have been carried out to investigate issues that arise in this translation. The translations so far have been carried out by hand. The first case study is of a buffer, where the control component looks after the input and output to the buffer, and the state component is used to store the data passing through the buffer. The second case study is of an abstract arbiter: a component that tracks a set of requests, and then chooses one element of the set. This case study is inspired by an AMBA bus case study [2,15] which makes use of an arbiter component to resolve contention for the bus. We investigate the use of channels and of signals to carry the requests. Some experimentation was done to see how fast (in terms of clock cycles) the inputting of a signal could be: instead of reading the signal into the CSP controller and then calling the corresponding B operation when appropriate, a further translation was considered in which the line could be read and the operation called within the same clock cycle.

In both case studies we begin with a CSP∥B description, and then consider how it can be translated to Handel-C. The approach taken is relatively systematic in terms of how a control loop should be translated, and where the Handel-C implementations of the B machine should appear. Some parts of the specification have to be translated in a less generic way because of the specific requirements of Handel-C. Indeed, some of the motivation for these investigations is to identify where this becomes necessary.

1. Background

1.1. CSP∥B

The formal method CSP∥B is a combination of the process algebra CSP [10,19] and the state-based B-Method [1], designed to model systems which are rich both in interaction (CSP) and in data (B-Method).

CSP describes systems in terms of *processes*, which interact by synchronising on common events. Processes are defined in terms of the patterns of *events* that they can perform. A process language enables processes to be described. It includes channel input $c?x \to P(x)$, channel output $c!v \to P$, interleaved parallel $P \mid\mid\mid Q$, termination $SKIP$, sequential composition $P; Q$, and recursive definitions $N \triangleq P$. The language also supports choice, synchronised parallel, abstraction, mutual recursion, and channel communications that have both input and output, though these are not used in this paper. Tool support [9,12] enables model-checking for refinement, deadlock, divergence, and temporal logic properties.

The B-Method describes systems in terms of *abstract machines*, which encapsulate state and operations. It is supported by tool support which at the formal modelling level enables proof of internal consistency and refinement [6,3], and model-checking [12]. The tool support also provides for fully formal development. An abstract machine will define state variables, initialisation, invariants on the state, and operations which can be used to read or update the state (or both). Abstract machines can be proven internally consistent with respect to their invariants: that every operation, called within its precondition, is guaranteed to preserve the invariant. Of particular relevance to the work presented here is the use of simultaneous assignments: the construction $x := E \parallel y := F$ evaluates both E and F in the original state, and then simultaneously updates both x and y. This is the same treatment of parallel assignment as provided by Handel-C. Static information (e.g. user defined types, constants, auxiliary functions) can also be defined within abstract machines. The B-method is a design methodology and supports data refinement: changing the way data is represented so the model can be developed in more detail and progress towards implementation. Such changes are written as *refinement machines*, which encapsulate the changed data representation and its relationship to the original abstract machine.

CSP∥B [22] combines CSP processes and B machines by treating B machines as processes that interact through the performance of their operations: an operation op passing a value x will interact with a process which has op as a channel which carries x. The typical unit is a *controlled component* consisting of a CSP process in parallel with a B machine. We think of the CSP process as governing the control flow of the combination, determining the next possible operations or communication; and the B machine maintains the state, and is controlled and updated by the CSP process. For example, in Section 2 we will see a $CELL$ machine (Figure 2) controlled by a process $CONTROLLER1$ (Figure 3). The combination is written $CONTROLLER1 \parallel CELL$. There is also a general approach to establishing consistency between the CSP and the B parts of a description: that operations are called only within their preconditions.

1.2. Handel-C

Handel-C [5] is a clock-synchronous programming language reminiscent of occam [14,18]: it offers co-routines, concurrency and communication, and it is intended for programming applications onto reconfigurable hardware, specifically Field Programmable Gate Arrays (FPGAs). Its aim is to provide an interface to hardware that looks and feels like a traditional imperative programming language, while allowing the programmer to exploit the natural concurrency inherent in a hardware environment. It is not a hardware description language; it is a high level programming language with a clean, intuitive semantics, with a syntax based on C including extensions for concurrency and communication. The extensions are based on similar constructs in occam [14] and have a theoretical basis in CSP.

Handel-C differs from occam in two main respects. Firstly, parallel assignments to state variables are synchronous: all assignments take place on the rising edge of the hardware clock. Secondly, shared access to variables is permitted between processes.

```
a := 0;
b := 1;
par {
  a := 2;
  b := a;
}
```

Figure 1. A parallel assignment

The code fragment in Figure 1 demonstrates how concurrency interacts with the clock in Handel-C. The first two lines of this program are executed sequentially, with the semicolon marking a clock tick: after the second clock tick, a will have the value 0, while b will have the value 1. The program then enters the parallel block. This block takes one clock cycle to execute: both assignments happen simultaneously. Once it has completed, b will have been assigned the *initial* value of a (the value held in a before the rising clock edge). On the rising clock edge, both assignments happen: a takes on the value 2, while b takes on the value 0; the clock then ticks and the statement terminates.

The synchronous nature of assignment gives a very well defined notion of timing in the language: an assignment takes a clock cycle, and expression evaluation takes no time. This is achieved in implementation as follows: when a program is loaded onto an FPGA, a calculation is required to set the clock speed of the hardware. The maximum clock speed permissible by any one application is the maximum combinatorial cost of the longest expression evaluation, along with the assignment cost. Therefore, any one application may be clocked at a different speed from any other on the same FPGA, and the same application may be able to achieve a different speed on a different FPGA.

Shared access to variables also needs to be handled with care. As may be expected in a hardware environment, if two (or more) processes attempt to write to a register at the same time, the result is undefined. There is no resolution offered by Handel-C to this: should a programmer insist on writing co-processes that both write to a shared variable, it is the programmer's responsibility to deal with any non-determinism arising. In the Handel-C development environment DK3.1 [13] used for the experiments reported in this paper, the compiler emits a warning if the same variable appears on the left hand side of an assignment in two or more parallel blocks.

Handel-C provides *channels* which allow concurrent processes to interact through synchronous communication. Channels behave as they do in CSP and occam: communication occurs when both sender and receiver are ready to engage in it, and a value is passed from sender to receiver. The type of the channel defines the possible values that can pass along it.

Handel-C also provides *signals* as a mechanism for communication between concurrent processes. Processes assign values to signals, which can be read by other processes in the same clock cycle. A signal will hold the value assigned to it just for the period of the clock cycle, after which the value will no longer be available. In this sense a signal behaves as a wire. A signal can also have a default value, which it holds when it has not been assigned any other value.

1.3. Translation to Handel-C

When approaching the translation to Handel-C, we integrate the CSP and the B aspects of the description by embedding the code corresponding to the B into the translation of the CSP controller. This can be done in one of three ways:

- include the translation of the operation directly where it appears in the CSP controller. This will involve substituting the actual for the formal parameters directly in the code that is being included;
- use Handel-C macros to encapsulate the translation of the operation;
- use Handel-C functions to encapsulate the translation of the operation.

The first choice is direct, and it may result in replicating code where an operation is called a number of times in the body of a controller. The second choice avoids the replication of code, but at the cost of generating fresh hardware for each instance of a macro call in the code. The main difference between these two choices is to do with readability of the code, and separation of concerns. The third choice provides the best chance of obtaining a separate translation of an entire B machine which is then accessed from the controller. It allows reuse of hardware by generating one implementation of each function. However, multiple instances of a function cannot be called in parallel, in contrast to macros. Hence each of the approaches brings different benefits and it is clear that there will be trade-offs in deciding which to deploy in any given situation.

2. Case Study I: Buffers

The first case study is concerned with simple buffers-style behaviour. This is an artificial example chosen to drive the initial exploration. Thus it includes a B machine to manage some simple state, and control written in CSP to access and update that state. The purpose of this example is to see how such a description translates.

In the first instance, the B machine controlling the information consists of a memory cell in which a single value can be set or retrieved. We consider two controllers for this machine. The first controller simply accepts values from some external source, puts them into the cell, and then reads the value from the cell and outputs it. This case study exposes the issues involved in producing Handel-C corresponding to a CSP∥B specification. Issues emerged on declarations, header files, Handel-C types, and on how to construct an appropriate harness to best interact with the resulting hardware.

The second controller consists of two concurrent processes: an input process responsible for setting the state, and an output one responsible for reading the state and passing it externally. The processes do not synchronise with each other, but both made use of operations of the memory cell.

In the final case study of this family, the B machine tracks not just a single value, but an array (of bits or booleans) in which each element of the array can be flipped independently by the operation called with the corresponding index.

2.1. CSP∥B description of a simple buffer

```
MACHINE          CELL
VARIABLES        xx
INVARIANT        xx : NAT
INITIALISATION   xx := 0
OPERATIONS
   set(yy) = PRE yy : NAT
             THEN xx := yy
             END;

   yy <-- get = BEGIN yy := xx END

END
```

Figure 2. The CELL machine

The first buffer consists of a piece of state in a B machine used to capture the value to be stored in the buffer given in Figure 2, and a CSP controller operating a store/fetch cycle given in Figure 3. Capturing a single piece of state using a B machine, and controlling it with a CSP process, forms a simple example of interaction with state and control in the CSP∥B specification.

$$CONTROLLER1 \triangleq bufin?Value \rightarrow set!Value \rightarrow$$
$$get?Stored \rightarrow bufout!Stored \rightarrow CONTROLLER1$$

Figure 3. CSP controller for the CELL machine

Two operations exist in this B machine. The first, `set`, takes a natural number as a parameter and sets the value of local state. The second, `get`, returns the value of the local state.

The first CSP controller is a recursive process that reads in a value from the environment on the channel *bufin*, and then calls the operation in the B machine to set the value. It then reads the value back from the B machine on the channel `get`, before outputting the value to the environment on the channel *bufout*. The process then recurses.

2.2. Handel-C Translation of a Simple Buffer

The code presented in this section is the Handel-C translation of the CSP‖B specification *CONTROLLER*1 ‖ *CELL*.

The B machine essentially describes the kind of state for which the machine is responsible. Thus its translation to Handel-C will consist of the state declarations (and any other declarations included in the machine definition), and its initialisation to ensure it begins in the correct state. In this case study, we will be including the translation of an operation directly in the place where it is called within the CSP description, so translation of operations will occur within the CSP translations.

The machine has one state variable: `xx`. Variables in Handel-C must have a specified bit width. In this case study, we have decided to use 3-bit values.

```
#define WORD_TYPE unsigned int
#define WORD_WIDTH 3
WORD_TYPE WORD_WIDTH xx;
```

The following channel declarations provide input and output channels.

```
chan WORD_TYPE WORD_WIDTH bufout;
chan WORD_TYPE WORD_WIDTH bufin;
```

The following macros are used to implement atomic statements in the specification. Their use in this case study helps clarify which type of CSP specification statements are implemented, and how they are implemented. The general approach would be to use macros for B operations corresponding to CSP events. Here they are all atomic (input, output, and assignment), but in principle they could be more complex.

The macro procedure `CHANINPUT` takes a channel name X and a variable Y, and performs the Handel-C channel input X?Y. The macro does not type the parameters, although they are checked at compile-time. The macro `CHANOUTPUT` does the same for a CSP channel output.

The macro `BASSIGN` takes a variable Y, and assigns to it the value X; the same type-checking rules apply. Furthermore, it is assumed that the necessary variable parameters have been declared in advance.

```
macro proc CHANINPUT(X,Y) { X?Y; }
macro proc CHANOUTPUT(X,Y) { X!Y; }
macro proc BASSIGN(X,Y){ Y = X; }
```

In the `main` source file, the initialisation xx := 0 from the INITIALISATION clause of the CELL machine (Figure 2) is translated to Handel-C as an assignment. This is followed by the implementation of the CSP controller.

```
void main(void)
{
  /* B machine initialisations */
  BASSIGN(0, xx);
  /* CSP MAIN controller */
  SimpleBuffer(bufin,bufout);
}
```

The implementation of the controller declares two local variables `Stored` and `Value` for the two variables used in the description of *CONTROLLER*1.

$$In \triangleq bufin?Value \rightarrow set!Value \rightarrow In$$
$$Out \triangleq get?Stored \rightarrow bufout!Stored \rightarrow Out$$
$$CONTROLLER2 \triangleq In \ ||| \ Out$$

Figure 4. A second CSP controller for CELL

The CSP loop is translated to Handel-C by translating each of the events in the description of $CONTROLLER1$ in turn:

- $bufin?Value$ is translated as a channel input.
- $set!Value$ represents a call of the set operation in the CELL machine, and so the body of that operation in the B description is translated, as an assignment of Value to xx.
- $get?Stored$ represents a call of the get operation in the CELL machine, and is translated as an assignment of xx to Stored.
- $bufout!Stored$ is translated as a channel output: $bufout$ is a channel rather than an operation call.
- The overall recursive definition is translated to a do loop.

The result is as follows:

```
macro proc SimpleBuffer(bufin, bufout)
{
  /* local CSP state */
  WORD_TYPE WORD_WIDTH Stored;
  WORD_TYPE WORD_WIDTH Value;

  do {
    CHANINPUT(bufin, Value);
    BASSIGN(Value,xx);
    BASSIGN(xx, Stored);
    CHANOUTPUT(bufout, Stored);
    } while(1);
}
```

2.3. A Two-Part Controller

A more complicated case study is to manipulate the state in the B machine using two interleaved controllers: a process In handling input, and a process Out handling output. These are given in Figure 4, together with the resulting controller $CONTROLLER2$ which combines them with interleaving parallel. This does not provide a buffer in CSP$\|$B because the two component controllers can proceed independently, and at different rates. However, our translation to Handel-C will result in their clock-synchronous execution, providing the behaviour of a buffer containing an initial value.

Given that there is no communication and synchronisation between the input and output processes, the implementation of $CONTROLLER2 \ \| \ CELL$ is straightforward. Each is implemented using the same approach as previously, and then the results are run in parallel. We obtain the following:

```
macro proc InterleaveBuffer(bufin, bufout) {
  /* local CSP state */
  WORD_TYPE WORD_WIDTH Stored;
  WORD_TYPE WORD_WIDTH Value;
```

```
    par {
      do {CHANINPUT(bufin,Value);
          BASSIGN(Value, xx);
          } while(1);

      do {BASSIGN(xx, Stored);
          CHANOUTPUT(bufout, Stored);
          } while(1);
    }}
  void main(void)
  {
    BASSIGN(0, xx);
    InterleaveBuffer(bufin,bufout);
  }
```

The CSP description does not in fact necessarily behave as a buffer, because *In* and *Out* are independent and could proceed at different rates. However, their synchronous implementation in Handel-C means that they execute at the same rate, matching outputs to inputs, and yielding a buffer. This is a refinement of the behaviour encapsulated in the CSP∥B description.

2.4. A Boolean Function

This next case study increases the complexity of the B machine. Instead of a natural number, we store a function from natural numbers (indexes) to booleans. This case study introduces more complexity into the implementation as the local state is not just a simple variable. The machine is given in Figure 5. Here the controller accepts bit-wise updates to the array on the input cycles, but outputs the entire array on each output cycle. The associated controller is given in Figure 6.

```
MACHINE BOOLARRAY
VARIABLES    xx
INVARIANT    xx : NAT --> 0..1
INITIALISATION  xx := NAT * { 0 }
OPERATIONS
       set(yy) = PRE yy : NAT
                 THEN xx(yy) := 1 - xx(yy)
                 END;

yy <-- get = BEGIN
                yy := xx
             END
END
```

Figure 5. The BOOLARRAY machine

$$In \mathrel{\hat=} bufin?Value \to set!Value \to In$$
$$Out \mathrel{\hat=} get?Stored \to bufout!Stored \to Out$$
$$CONTROLLER3 \mathrel{\hat=} In \mathbin{|||} Out$$

Figure 6. A controller for the BOOLARRAY machine

In the following implementation of $CONTROLLER3 \parallel BOOLARRAY$, we must restrict to a finite domain for the function: in fact three bit integers, the set 0..7.

```
#define WORD_TYPE unsigned int
#define WORD_WIDTH 3
#define ARRAY_WIDTH 8
unsigned int 1 xx[ARRAY_WIDTH];
```

Input to this buffer is an index, and the bit stored at this index is flipped. The buffer output is the sequence of values stored in the array. As this is sent over a channel it must be a single word of 8 bits. The following code achieves this—but it is not the simple assignment that may have been naïvely expected. An array of 8 bits is of a different type to an 8-bit word so *xx* cannot be assigned to *Stored* directly.

```
macro proc BoolArray(bufin, bufout) {
  /* local CSP state */
  unsigned int WORD_WIDTH Value;
  unsigned int ARRAY_WIDTH Stored;

  par {
    do {
      CHANINPUT(bufin,Value);
      BASSIGN(!xx[Value], xx[Value]);
    } while(1);

    do {
      BASSIGN(xx[7]@xx[6]@xx[5]
              @xx[4]@xx[3]@xx[2]
              @xx[1]@xx[0], Stored);
      CHANOUTPUT(bufout, Stored);
    } while(1);
  }
}
```

2.5. Issues

In this section, we highlight the issues uncovered in implementing this specification.

1. *Variable size in bits.*
   In the B machine, the local state was expressed as a natural number. In a Handel-C program, each variable must be declared in terms of a finite size in bits. This is an issue with any state-based development—and the solution normally is to retrench [4] to the initial abstract specification and rewrite it with acceptable bounds. Retrenchment provides a controlled and traceable way of allowing changes to the specification in the light of implementation considerations. This is the solution that was adopted in this case.
   An orthogonal issue concerns the use of signed bits: a number can be represented in *signed* or *unsigned* format; and the behaviour of program code is highly dependent on this choice. Therefore, we expect that a useful further investigation would be to refine data types in the B machine into bitwise implementations, thereby ensuring that all operations consider signed bit issues.

2. *Preconditions are a specification statement, not an implementation one.*
   Preconditions are specification statements, not implementation ones. In analyzing a CSP∥B specification, one normally proves that the CSP controller never calls a B operation in a state that would violate a precondition. Therefore the precondition need not be implemented: the fact that the precondition is always respected is discharged during analysis and there is no requirement to translate any operation preconditions into Handel-C. Preconditions are therefore dropped during translation.

3. *Main processes.*

   CSP processes are frequently defined—and may easily be analyzed—without declaring a starting point. For instance, the behaviour of the recursive processes $Ex \,\hat{=}\, a \to b \to Ex2$ and $Ex2 \,\hat{=}\, c \to d \to Ex$ have quite clear behaviour; but it is not explicit where *execution* begins. Programs however have a very clearly defined starting point for execution—in Handel-C this is given by a `main` function. When writing a CSP∥B specification for implementation, one must consider where execution will begin and how this can translate into a `main` function. One useful solution to this is the one adopted by the ProB tool [12]—to require that one of the processes is named `Main`.

4. *Variable scope and introduction from channel inputs.*

   Programming languages usually have to be explicit about when they introduce storage and declare variables. However, CSP is not so explicit. Another issue thrown up by this case study concerned the introduction and management of storage introduced by CSP channel inputs.

   The input communication $a?x \to P$ introduces the variable x, and then writes to it from the channel a. This is similar to a `var x` declaration just before the $a?x$ input. The process $P \,\hat{=}\, a?x \to P$ therefore introduces a copy of x each time it recurses, but the scope of the local variable declaration includes all recursive calls. Thus in CSP this amounts to a nesting of local variable declarations. However, the tail recursive nature of the process definition makes it sufficient in an implementation to declare x only once (as would be natural in Handel-C or occam), and to re-use it on each cycle, since each nested declaration supersedes the previous one. Alternatively it could be declared afresh on each cycle. In any case translation to Handel-C would preclude dynamic generation of new local variables, since there would need to be a bound on the hardware state, so a constraint such as the need for recursions to be tail recursive would need to be present to enable translation.

5. *Type conversion.*

   The boolean buffer highlighted an issue of type conversion that became apparent because of the necessity of compile-time constants in bit indexing. We would normally expect that type converters would be introduced at a formal level when data-refining abstract data type; but the appearance of this issue in this case study means that we should be aware of it in further investigations.

3. Case Study II: Simple Arbiter

The AMBA AHB 2.0 specification [2] for the AMBA bus makes use of an arbiter component to choose between competing requests for a bus. This case study is an idealisation of the arbiter function, tracking a set of requests, and then choosing between them. Our focus is on issues such as the storing of local state, interaction of local state with timing, and the distinction between signals and channels, and the impact this has on specifications and implementations.

The case study uses a machine which maintains a *set* of values (4 bit numbers in fact). This is an abstract data structure used in the specification of the AMBA arbiter within the 'Future Technologies for System Design' project, and so it was a natural candidate for exploration. The machine allows two operations: `add` which adds a value to the set; and `choose` which selects an arbitrary element from the set (the statement `yy :: ss` is a nondeterministic assignment of any element of the set `ss` to the variable `yy`), and also resets the set to $\{0\}$. A default element is always in the set. A more abstract specification could be less determined when no element has been added to the set, for example deadlocking, or providing any arbitrary value. However, the AMBA case study specifies the use of a default element, and so we follow this approach.

The translation of the machine will be done it two steps: it first needs to be refined within the B-Method in order to bring it to a point closer to implementation and thus appropriate for translation. The resulting refinement machine is then used in the translation to Handel-C.

The controller for this machine reads bits on a number of lines concurrently, and for any line, if its value is 1 then it should be added to the set. When these have all occurred, then the choice is read and passed as output.

This case study demonstrates the clear difference between treating the input line as a signal, and treating it as a channel. Both translations are provided. The difference in timing behaviour and hence the resulting behaviour between reading signals and reading channels is exposed by the test cases.

Some experimentation was also done to see how fast (in terms of clock cycles) the inputting of a signal could be: instead of reading the signal into the CSP controller and then calling the corresponding B operation when appropriate, a further translation was considered in which the line could be read and the operation called on the same clock cycle.

3.1. CSP∥B Description of a Simple Arbiter

The abstract state contains a set, ss, containing natural numbers in the range 0..15. Elements may be added into the set using add. An element may be chosen from the set non-deterministically using the operation choose, at which time the set is also reset to its default value {0}. The machine is given in Figure 7. Note that the B semantics of parallel assignment means that yy is chosen from the original value of the set ss while ss is simultaneously updated.

```
MACHINE         SetChoice
CONSTANTS       NAT4
PROPERTIES      NAT4 = 0..15
VARIABLES       ss
INVARIANT       ss <: NAT4 & 0 : ss
INITIALISATION  ss := {0}
OPERATIONS
yy <-- choose = BEGIN yy :: ss
                     || ss := { 0 }
                 END;

     add(xx) = PRE xx : NAT4
               THEN ss := ss  \/ { xx }
               END
END;
```

Figure 7. The SetChoice machine

This machine is not ready for implementation in Handel-C because of the presence of non-determinism, and also of the presence of a set-valued variable. The non-determinism in the choose operation needs to be resolved: we decide as a refinement step to choose the maximum value in the set. Furthermore, we need to refine abstract sets into something more concrete, as well as the operations performed on the set. We achieve this as a data refinement mapping the set of values to an array of bits, indexed by the values. The resulting refinement is encapsulated within B as a REFINEMENT machine, and given in Figure 8. Note the linking invariant stating (in ascii) that $ss = arr^{-1}(\!|\ \{1\}\ |\!)$. This states that the set ss is the same as the inverse image of the array (considered as a function) arr on the set $\{1\}$: in other words, the set of all locations in arr that contain the value 1.

```
REFINEMENT SetChoiceR
REFINES    SetChoice
VARIABLES  arr
INVARIANT  arr : 0..15 --> 0..1
           & ss = arr~[{ 1 }] & 0 |-> 1 : arr
INITIALISATION  arr := (1..15 * { 0 })
                       \/ { 0 |-> 1 }
OPERATIONS
yy <-- choose =
       BEGIN yy := max(arr~[{ 1 }] )
                || arr := { 0 |-> 1 }
                       \/ (1..15 * { 0 })
       END;

       add(xx) = BEGIN arr(xx) := 1 END

END;
```

Figure 8. The refinement machine SetChoiceR

In this refinement machine, the set is represented by an array of bits: if a given element is a member of the set, the value of the bit at that index is 1, otherwise is it 0. The first element of the array is always set to 1, implementing the default value. We choose some refinement of the abstract specification of `choose`, and (arbitrarily) select the highest-valued member: the function `max` selects the element with the highest index. This function will be defined in the implementation. Finally, the `add` operation sets the bit of the given parameter to true. The resulting machine is now appropriate for translation to Handel-C.

$$Read(x) = line.x?b \rightarrow \textbf{if } (b = 1)$$
$$\textbf{then } add!x \rightarrow SKIP$$
$$\textbf{else } SKIP$$

$$CONTROLLER = (||| \; x : 0..15 \bullet Read(x));$$
$$choose?y \rightarrow out!y \rightarrow CONTROLLER$$

Figure 9. The CSP controller for SetChoice

The controller is given in Figure 9. The process $Read$ listens on the channel $line.x$, where x is an index. If the boolean value communicated on that line is $true$, the result is stored. The process $CONTROLLER$ replicates $Read$ for each of the lines—there are 16, as the specification is concerned with 4 bit numbers. After reading each line, a candidate is chosen by the B operation `choose`, and output to the environment.

3.2. Handel-C Translation of a Simple Arbiter using Channels

In this section we present Handel-C implementations of the simple arbiter. Several implementations are presented. The first, is a straightforward channel implementation. The second uses signals instead of channels in a naïve way, simply using them in place of the channels in the previous description—but we see that this approach does not work. The third also uses signals, but is an adaptation that performs everything in a single cycle. We discuss what an abstract specification and controller of this version may look like, and use this to demonstrate that a different specification is written for signals than for channels.

The state in the B machine is declared as a 16-element array, where each element is a single bit:

```
unsigned int 1 arr[16];
```

In this implementation, the B machine initialisation and operation *choose* are declared as Handel-C functions. The operation *add* is declared as a macro procedure, which this will in-line the hardware everywhere it is called in source code:

```
void Init();
macro proc add(unsigned int 4 xx);
void choose(unsigned int* yy);
```

The initialisation takes no parameters, and returns no values. The add function takes a 4-bit integer corresponding to the array element to be set to 1. Each interleaved *Read* calls add, and so there will be a copy for each interleaved call. This would be necessary even if add were declared as a function: the possibility of up to 16 simultaneous calls of add would require 16 copies of add to be created. This arises because we wish to call a number of them concurrently within a single cycle.

The choose operation in the B machine takes no parameters, but returns a result. In this implementation we use reference parameters to the function call to return results, instead of a return value of the function implementation. This is because it is possible in a B machine to specify an operation with more than one return type, and the simplest way to implement this in a translation is with reference parameters.

Initialisation sets the value of every element in the array as described in the B machine. All elements are set in parallel so initialisation takes only one clock cycle:

```
void Init() {
  par {
    arr[0] = 1;
    par (i=1; i<16; i++) { arr[i]=0; }
  }
}
```

The add macro is defined as follows:

```
macro proc add(unsigned int 4 xx)
              { arr[xx]=1; }
```

The specification did not state how max would be implemented. Normally a development would include this information. In this case, we have (arbitrarily) implemented max using a number of nested conditional statements that assigns the value stored by the reference parameter:

```
macro proc max(yy) {
  if (arr[15]==1) { *yy=15; } else {
    ⋮
    if (arr[1]==1) { *yy=1; } else {
      *yy = 0;
    } ... }
}
```

The implementation of choose resets the array, in parallel with returning the result of calling max (on the original value of the array). Note here that the reset operation uses the Init function which achieves the resetting of the array, conserving hardware.

```
void choose(unsigned int* yy) {
  par { max(yy);
        Init();
      }
  }
}
```

We now consider the implementation of the controller process. We declare an array of 16 channels corresponding to the input lines into the simple arbiter, and a single output channel:

```
chan unsigned int 1 line[16];
chan unsigned int 4 out;
```

The main process initializes the state in the B machine, and then runs a process monitoring the clock for debugging purposes in parallel with a process SIM_DRIVER (defined below) that drives the Handel-C simulator and the main CSP controller for the case study.

The implementation of the controller replicates one copy of the *Read* process for every input line, then calls the B operation `choose`, outputs the result and iterates:

```
void main() {
  unsigned int 4 y;
  Init();
  par {
    SIM_DRIVER();
    do {
      par (i=0; i<16; i++) { Read(i); }
      choose(&y);
      out!y;
    } while(1);
    /* CHANNELS */
  }
}
```

The *Read* process performs an input on its indexed channel, and branches on a conditional. In this case, the *SKIP* in the specification has been implemented using a `delay` to ensure that both branches of the choice take the same time:

```
macro proc Read(x) {
  unsigned int 1 b;
  line[x]?b; if (b) { add(x); }
             else { delay; }
}
```

The process that drives the simulator reads in an input word and writes indexed bits to individual lines on alternate cycles. In parallel with this, it listens on the output line from the arbiter, and writes this to simulator output, again on alternate cycles:

```
macro proc SIM_DRIVER() {
  unsigned int 16 in;
  unsigned int 4 result;
  par {
    do { input?in;
         par (i=0; i<16; i++) { line[i]!in[i]; }
    } while(1);
    do { out?result; output!result; } while(1);
  }
}
```

3.3. Handel-C (Incorrect) Translation of a Simple Arbiter using Signals

In this section, we describe an attempted implementation of the CSP controller as above; but this implementation uses Handel-C signals in place of channels. This gives different behaviour, and it illustrates that channels cannot simply be replaced by signals.

The only noticeable change to the main function is that the output signal line is assigned to, rather than treated like a channel communication. `Init`, `add`, `max`, and `choose` are unchanged:

```
void main() {
  unsigned int 4 y;
  Init();
  par {
    SIM_DRIVER();
    do {
      par (i=0; i<16; i++) { Read(i); }
      choose(&y);
      out = y;
    } while(1);
  }
}
```

The `Read` process is changed in that it reads from a signal rather than a channel. This has a significant impact on the behaviours of the program. Reading from the signal is guaranteed to take a single cycle, and to take on *only the value written to the signal on that cycle*, whereas the channel communication is guaranteed to block until the channel is written to—and the value written is persistent until it is read:

```
macro proc Read(x) {
  unsigned int 1 b;
  b = line[x]; if(b) {add(x); }
               else { delay; }
}
```

Very contrasting behaviour is observed running this version compared to the previous. Specifically, the channel version produces the expected output, and the signals version does not. This is because there is no persistence in a signal: it only holds a value for single cycle, so unless we can be certain that it will be read on the exact cycle it is written to, the communication cannot be reasoned about. In the above code, the signal is *never* read from on the cycle it is written to.

In the next section, we show an implementation, using signals, that produces the output we would expect. We demonstrate where this is a very different implementation to the above, leading us to conclude that translating to signals requires a different approach to using channels.

3.4. An Improved Signal Implementation

In this section, we present an implementation using signals that produces the expected output. The interesting aspect of this implementation is in the way it does things differently to the channel implementation. The different structure is required to ensure that the value of the input signals are read to, and output signals written to, on every cycle, to ensure that the value is always meaningful to the environment.

The main loop of the program writes to the signal `out` on every cycle. To guarantee this, the `choose` method must take only a single cycle to execute. This was not possible in previous implementations as the signal was assigned, and then local state was assigned, requiring two clock cycles:

```
void main() {
  unsigned int 4 y;
  Init();
  par {
    SIM_DRIVER();
    do {
      par {
        fastchoose(&y);
        out = y;
      }
    } while(1);
  }
}
```

The `choose` method—here named `fastchoose` to distinguish it from the previous choose—must occupy only one clock cycle if the timing statement mentioned above is to hold. To achieve this, we observe that the arbiter can make a decision about the input signals without storing them in local state (the B machine). Losing this assignment saves a clock cycle. The only clock cycle consumed is the one that assigns the result of the operation to the reference parameter representing the return value of the call:

```
void fastchoose(unsigned int* yy) {
  if (line[15]==1) { *yy=15; } else {

    ⋮

  if (line[1]==1) { *yy=1; } else {
    *yy = 0;
  } ... }
}
```

The read process changes also, such that it only consumes a single clock cycle. Instead of storing the value of the signal in local state, the conditional branches on the signal value. A statement like this could not be written in such a way as to consume a single clock cycle if a channel communication were used: a pipelining implementation with a cycle latency would be needed:

```
macro proc Read(x) {
  if (line[x]) { add(x); }
  else { delay; } }
```

The process driving the simulation is also different, and highlights the point above about needing a single cycle latency for a channel communication. The process reads in a value from the simulator, and on the next cycle writes it to the application signals. Signals are written to, and read from, on every cycle:

```
macro proc SIM_DRIVER() {
  unsigned int 16 in;
  par {
    do {input?in;           } while(1);
    do {par (i=0; i<16; i++)
        { line[i] = in[i]; }} while(1);
    do {output!out;         }  while(1);
  }
}
```

3.5. Issues

1. *Macro procedures and function calls: in-lining code.*
   An issue that requires consideration is that of implementing B operations and CSP processes as either function calls or macro procedures. Use of macro procedures is generally more robust as a single piece of hardware is created for each call; whereas use of function calls is more efficient as hardware is reused. Care needs to be taken where parallel access to function calls may happen.

2. *Return values from B operations.*
   A B operation may have several return values. In this case study we have demonstrated that return values can be implemented using reference parameters in Handel-C; therefore functions returning multiple values do not present a problem for us.

3. *Implementation of B methods in single cycle.*
   The model of time assumes that a B operation takes a single cycle to execute. However, it is possible that a specification of a B operation may involve sequential composition; depending on the number of compositions, this can translate into a Handel-C operation that takes some other amount of time. Consideration of the timing semantics of the Handel-C operation is important; our solution to this problem is reflected in the language that we believe suitable for translation.

4. *Resolving underspecification.*
   In the case study above, the implementation of `choose` using `max` was one of many possibilities. This could be seen as a problem of *underspecification*—it was clear what it was supposed to do, but we had not stated *how* it should be achieved. Underspecification issues need to be considered before the translation process can begin because the target language does not contain underspecification constructs.

5. *Implementing $SKIP$.*
   Implementing $SKIP$ is an interesting issue. In this case study, we decided to insert a delay statement (e.g. in the translation of *Read*) to solve combinatorial logic issues where the CSP process terminated, to ensure that the branches either side of a choice took the same time. With more diverse choices, a delay to match the longest branch would be required. However, this may not always be appropriate or necessary.
   Currently, the issue of how to handle $SKIP$ is an open question. Clearly the language suitable for translation needs to include $SKIP$; but it cannot always translate to a delay statement. For instance, in CSP, $SKIP$; $SKIP = SKIP$; however, in Handel-C `delay; delay;` $\neq$ `delay`. The problem here is not just the loss of equivalences; the translation of $SKIP$ will depend on its context.

6. *Timing of signal versus channel.*
   The case study also illustrated the difference between signals and channels. Simply changing channel definitions to signal definitions (and resulting code fragments) produced an implementation that unsurprisingly did not work. A signal implementation that did work was very different in nature to the channel implementation.

4. Discussion

This paper has considered two case studies in translating CSP∥B to Handel-C. The intention was to understand how to treat the state component in the translation process, and to consider a progression of state specifications: single items, arrays, and abstract sets. The CSP controllers we have considered have been in a particular form: recursive sequential processes without choice, and parallel combinations of these. Further CSP operations, including choice, synchronising parallel, and hiding, were not considered in these case studies and remain as subjects for further work.

A number of standard refinement issues arose in the buffer case study, including the need for retrenchment in the implementation of natural numbers, and the dropping of preconditions when implementing operations. More specifically to our translation, the use of local variables in CSP recursive calls has also arisen in the context of CSP to Handel-C translations [21,16], and the approach taken by them is also appropriate in our case in the presence of the B. In the third buffer we have also seen the need for type conversion in refinement in order to achieve a suitable implementation, where we wanted to allow multiple updates to the array within a single clock cycle. In all these case studies we observed how the B component of the specification is translated: to a declaration and initialisation of all the machine state, and with the bodies of the operations translated directly at the point of operation calls.

In the arbiter case study, we have explored the use of channels and of signals to pass information, and have observed the different considerations arising in these two approaches, in particular with regard to timing considerations. In the case of signals, the translation of the controller process had to be hand-crafted to keep all the necessary activity within a single cycle and avoid loss of signals. The case study brought out the question of implementation of operations to macro procedures or to function calls. Implementing operations as macro procedures creates a separate piece of hardware for each call, and so these can be run in parallel; use of function calls only creates one piece of hardware for each operation, allowing reuse for different calls, and hence is more space efficient. However, concurrent calls to the same operation are not possible, so the use of function calls might not always be appropriate, and depends on the context. The case study also demonstrated how outputs to operations can be implemented by using reference parameters, enabling the implementation of multiple outputs allowed for B operations.

These case studies have exposed a considerable amount of detail that needs to be clarified for a translation to Handel-C. The type information in the abstract model needs to be extracted from the various places where it resides: in the B machines' invariants (state), preconditions (operation input), or simply implicit (operation outputs); and in CSP channel declarations (inputs and output). This information needs to be distilled so it can appear (suitably refined) in the Handel-C translation. Other necessary clarifications include the concrete implementation of the abstract types used in the specification; whether CSP communications are to be implemented as signals or channels; whether operations are translated in-line, as macros, or as functions; treatment of $SKIP$ (i.e. should all paths through an operation take the same time); and local variable declaration and scope. This level of detail would need to be provided manually, or extracted in some way from the abstract model, in order to make any automated or machine-assisted translation possible.

This paper has been concerned with initial explorations into the translation, so has not been concerned with questions of formal correctness. Ultimately correctness of the translation would require a relationship to be established between the semantics of CSP‖B and that of Handel-C, and the considerations of the details listed above come into play. The development methodology currently being developed is considering the use of an intermediate CSP‖B description using a restricted subset of CSP‖B which is more readily translatable to Handel-C, and whose semantics is more closely aligned with Handel-C. The benefit of this approach is that the abstract CSP‖B and the restricted CSP‖B are within the same semantic framework, enabling a formal relation to be established between them. It also means that only the restricted subset of CSP‖B needs a translation to Handel-C. To establish correctness of the translation we would aim to establish a simulation relationship between restricted CSP‖B and its Handel-C translation.

The consideration of which aspects of CSP‖B translate most readily into Handel-C has also resulted in the more recent development of a clocked style for CSP‖B specifications particularly appropriate to translation. The core functionality of the machine is bound up in a single operation. Inputs to the machine are either treated separately through input operations to set the state, or else incorporated as inputs to the core operation. Outputs are provided through separate output operations, to make them available to other components independently of the core operation. Our approach now is to develop the machine and identify the appropriate operations in tandem with the development of the control loop, which will typically cycle on input, core operation, and output. This approach to component specification and translation was investigated through the development of a larger case study within the Future Technologies project, and is the subject of ongoing research.

Acknowledgements

This work was funded under the AWE 'Future Technologies for Systems Design' project at the University of Surrey, and was carried out when Alistair McEwan was at the University of Surrey. We are grateful to David Pizarro for advice on Handel-C, and to the anonymous referees for their careful reading and insightful comments.

References

[1] J-R. Abrial. *The B-Book: Assigning Programs to Meaning*. Cambridge University Press, 1996.
[2] ARM. AMBA specification v2.0. Technical Report 0011A, ARM, 1999.
[3] B-Core. *B-Toolkit*.
[4] R. Banach, M. Poppleton, C. Jeske, and S. Stepney. Engineering and theoretical underpinnings of retrenchment. *Science of Computer Programming*, 67:301–329, 2007.
[5] Celoxica Ltd. Handel-C language reference manual. 2004.
[6] Clearsy. *Atelier-B*.
[7] Neil Evans and Helen Treharne. Investigating a file transfer protocol using CSP and B. *Software and System Modeling*, 4(3):258–276, 2005.
[8] Neil Evans, Helen Treharne, Régine Laleau, and Marc Frappier. Applying CSP‖B to information systems. *Software and System Modeling*, 7(1):85–102, 2008.
[9] Formal Systems (Europe) Ltd. *Failures-Divergences Refinement: FDR2 Manual*, 1997.
[10] C. A. R. Hoare. *Communicating Sequential Processes*. Prentice-Hall, 1985.
[11] W. Ifill and S. Schneider. A step towards refining and translating B control annotations to Handel-C. In *Communicating Process Architectures*, pages 399–424. IOS Press, 2007.
[12] M. Leuschel and M. Butler. ProB: A Model Checker for B. In *FM 2003: The 12th International FME Symposium*, pages 855–874, 2003.
[13] Celoxica Ltd. Design suite dk 3.1. 2005.
[14] INMOS Ltd. occam *Programming manual*. Prentice-Hall, 1984.
[15] A. A. McEwan and S. Schneider. Modelling and analysis of the AMBA bus using CSP and B. In *Communicating Process Architectures*, pages 379–398. IOS Press, 2007.
[16] M. Oliveira and J. Woodcock. Automatic generation of verified concurrent hardware. In *International Conference on Formal Engineering Methods, ICFEM 2007*, LNCS 4789, pages 286–306. Springer, 2007.
[17] I. Page. Constructing hardware-software systems from a single description. *Journal of VLSI Signal Processing*, (12(1)):87–107, 1996.
[18] A.W. Roscoe and C.A.R. Hoare. Laws of occam programming. *Theoretical Computer Science*, 60:177 229, 1988.
[19] S. Schneider. *Concurrent and Real-time Systems: The CSP Approach*. John Wiley and Sons, 1999.
[20] S. Schneider, A. Cavalcanti, H. Treharne, and J. Woodcock. A layered behavioural model of platelets. In *ICECCS*, pages 98–106, 2006.
[21] S. Stepney. CSP/FDR2 to Handel-C translation. Technical Report YCS-2002-357, University of York, June 2003.
[22] H. Treharne and S. Schneider. Communicating B machines. In *ZB2002: 2nd International Conference of B and Z users*, LNCS 2272, pages 416–435. Springer, 2002.

FPGA based Control of a Production Cell System

Marcel A. GROOTHUIS, Jasper J.P. VAN ZUIJLEN and Jan F. BROENINK

Control Engineering, Faculty EEMCS, University of Twente,
P.O. Box 217 7500 AE Enschede, The Netherlands.

{M.A.Groothuis,J.F.Broenink}@utwente.nl,J.J.P.vanzuijlen@alumnus.utwente.nl

Abstract. Most motion control systems for mechatronic systems are implemented on digital computers. In this paper we present an FPGA based solution implemented on a low cost Xilinx Spartan III FPGA. A Production Cell setup with multiple parallel operating units is chosen as a test case. The embedded control software for this system is designed in gCSP using a reusable layered CSP based software structure. gCSP is extended with automatic Handel-C code generation for configuring the FPGA. Many motion control systems use floating point calculations for the loop controllers. Low cost general purpose FPGAs do not implement hardware-based floating point units. The loop controllers for this system are converted from floating point to integer based calculations using a stepwise refinement approach. The result is a complete FPGA based motion control system with better performance figures than previous CPU based implementations.

Keywords. embedded systems, CSP, FPGA, Handel-C, gCSP, 20-sim, motion control, PID, code generation.

Introduction

Nowadays, most motion controllers are implemented on programmable logic controllers (PLCs) or PCs. Typical features of motion controllers are the hard real-time timing requirements (loop frequencies of up to 10 kHz). Running multiple controllers in parallel on a single PC can result in missing deadlines when the system load is becoming too high. This paper describes the results of a feasibility study on using a Xilinx Spartan III 3s1500 FPGA for motion control together with a CSP based software framework.

FPGAs are programmable devices that can be used to implement functionality that is normally implemented in dedicated electronic hardware, but can also be used to execute tasks that run normally on CPU based systems. Having a general purpose FPGA as motion control platform compared to CPU based implementations has several advantages:

- Parallel execution: no Von Neumann bottleneck and no performance degradation under high system load due to large scale parallelism;
- Implementation flexibility: from simple glue-logic to soft-core CPUs;
- Timing: FPGAs can give the exact timing necessary for motion controllers;
- High speed: directly implementing the motion controller algorithms in hardware allows for high speed calculations and fast response times. Although not directly required for the chosen system this can, for example, be beneficial for *hardware-in-the-loop* simulation systems. Typical PC based solutions can reach up to 20-40 kHz sampling frequencies, while FPGA based solutions can reach multi-MHz sampling frequencies.

A main disadvantage is that a general purpose FPGA is not natively capable of doing floating point calculations, which are commonly used in motion control systems. For more information on FPGAs and their internal structure, see [1].

One of our industrial partners in embedded control systems is moving from their standardised CPU + FPGA platform towards an FPGA-only platform. A soft-core CPU implemented on the FPGA is used to execute the motion controllers. This approach however still suffers from the Von Neumann bottleneck and the implementation of a soft-core CPU requires a large FPGA.

The target for this feasibility study is a mock-up of a Production Cell system (see figure 1) based on an industrial plastic molding machine. This system consists of 6 moving robots that are each controlled by a motion controller. The previous implementation of the system software was running on an embedded PC. The motion controllers in this implementation suffer from performance degradation when the system is under high load (when all moving robots are active at the same time). The Production Cell system already contains an FPGA. It is currently only used as an I/O board (PWM generators, quadrature encoder interfaces and digital I/O), to interface the embedded PC with the hardware.

The problems with the software implementation, the possible benefits of using an FPGA and the move towards FPGA-only platforms resulted in this feasibility study in which we wanted to implement a motion control system inside an FPGA without using a soft-core CPU.

We have used a model based design approach to realize the FPGA based motion control implementation for this setup. The tools 20-sim [2] and gCSP [3] are used to design the loop-controllers and the embedded control software. The CSP process algebra and the Handel-C hardware description language [4] are used in combination with code-generation from 20-sim and gCSP for the design and implementation of the embedded control software.

Section 1 gives more background information on the production cell setup, our previous experiments, motion control and our model based design method. Section 2 describes the designed software framework and section 3 describes the consequences for the design of the loop controllers when running them on an FPGA. This paper concludes with the results (section 4) and conclusions of this feasibility study and future work.

1. Background

1.1. Production Cell

An industrial Production Cell system is a production line system consisting of a series of actors that are coordinated to fulfill together a production step in a factory. The production cell system that is used for this feasibility study is a mock-up designed to resemble a plastics molding machine that creates buckets from plastic substrate. The system consists of several devices that operate in parallel [5]. Its purpose is to serve as a demonstrator for CSP based software, distributed control and to prototype embedded software architectures. Figure 1 shows an overview of the setup.

The setup is a circular system that consists of 6 robots that operate simultaneously and need to synchronize to pass along metal blocks. In this paper each of these robots is called a Production Cell Unit, or PCU. Each PCU is named after its function in the system (see also figure 1). The operation sequence begins by inserting a metal block (real system: plastic substrate) at the *feeder belt*. This causes the feeder belt to transport the block to the *feeder* which, in turn, pushes the block against the closed *molder door*. At this point, the actual molding (real system: creating a bucket from the plastic substrate) takes place. The feeder retracts and the molder door opens. The *extraction robot* can now extract the block (real system: bucket) from the molder. The block is placed on the *extraction belt* which transports it to the *rotation robot*. The rotation robot picks up the block from the extraction belt and puts

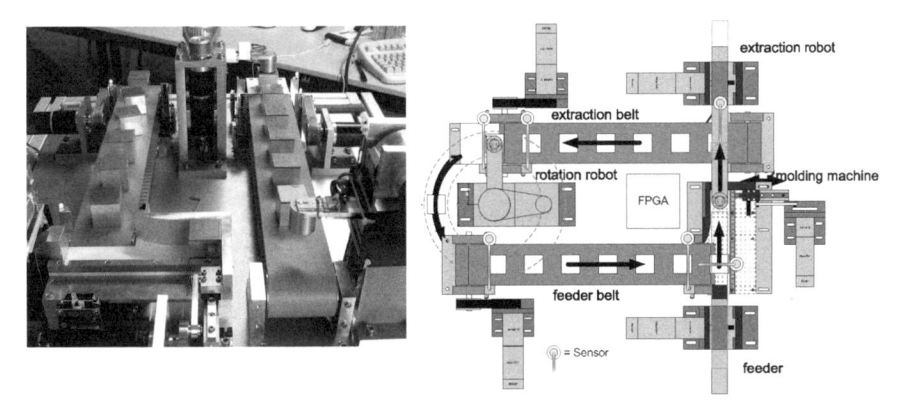

Figure 1. The Production Cell setup

it again on the feeder belt to get a loop in this demonstration setup. This loop can also result in a nice (for teaching purposes) deadlock when 8 or more blocks are in the system. This deadlock occurs when all sensor positions are occupied with blocks, resulting in the situation that all robots are waiting for a free position (at the next sensor), in order to move their block forward.

The belts allow for multiple blocks to be buffered so that every PCU can be provided with a block at all times, allowing all PCUs to operate simultaneously. The blocks are picked up using electromagnets mounted on the extraction robot and the rotation robot. Infrared detectors (sensors in figure 1) are used for detection of the blocks in the system. They are positioned before and after each PCU.

1.2. Previous Experiments

Several other software based solutions have been made in the past to control the Production Cell setup. The first implementation [6] is made using gCSP [7] in combination with our CTC++ library [8] and RTAI (real-time) Linux. 20-sim [2] is used to model the system dynamics and to derive the control laws needed for the movements in the system. Its purpose was to evaluate gCSP/CTC++ for controlling a complex mechatronic setup. This software implementation operates correctly when the system is not overloaded with too many blocks. When all 6 PCUs are active and many sensors are triggered at the same time, the CPU load reaches 100%, resulting in a serious degradation of system performance, unsafe operation, and sometimes even in a completely malfunctioning system. Another implementation [9] is made using the Parallel Object Oriented Specification Language (POOSL [10], based on Milner's CCS [11]). The main focus for this implementation was on the combination of discrete event and continuous time software, the design method and predictable code generation. The properties (e.g. timing, order of execution) of the software model should be preserved during the transformation from model to code. This implementation also could not guarantee meeting deadlines for control loops under high system load. Furthermore, neither implementation incorporates safety features in its design.

1.3. gCSP

gCSP is our graphical CSP tool [7] based on the graphical notation for CSP proposed by Hilderink [12]. gCSP diagrams contain information about compositional relationships (SEQ, PAR, PRI-PAR, ALT and PRI-ALT) and communication relationships (rendezvous channels).

An example of a gCSP diagram with channels, processes and SEQ and PAR compositions is given in figure 3. From these diagrams gCSP is able to generate CSPm code (for deadlock and livelock checking with FDR2/ProBE), occam code and CTC++ code [8]. Recent additions to gCSP are the Handel-C code generation feature (see section 4) and animation/simulation facilities [13].

1.4. Handel-C

Handel-C [4] is an ANSI C based hardware description language born out of the idea to create a way to map occam programs onto an FPGA. Handel-C uses a subset of ANSI C, extended with CSP concepts like channels and constructs. Its built-in support for massive parallelism and the timing semantics (single clock tick assignments) are the strongest features of Handel-C. The close resemblance with the C programming language makes it a suitable target for tools with C based code generation facilities. This was one of the reasons that Rem et al. [14] used Handel-C as a code generation language together with MATLAB/Simulink to design an FPGA based motion controller. While Simulink can be used to generate FPGA optimized motion controller code by using Handel-C templates for each library block, it does not support the design of a software framework with multiple parallel processes containing these motion controllers (targeted for FPGA usage). gCSP is more suited for this purpose.

1.5. Motion Control

Typical motion control systems consist of motion profiles (the trajectory to follow) and loop controllers. Their purpose is to control precisely the position, velocity and acceleration of rotational or translational moving devices, resulting in a smooth movement. The control laws for the loop controllers require a periodic time schedule in which jitter and latency are undesired. Hard real-time behaviour is required for the software implementation, to assure predictable timing behaviour with low latency and jitter. Missing deadlines may result in a catastrophic system failure. The embedded control software of a motion control system often contains a layered structure [15] as shown in Fig. 2 .

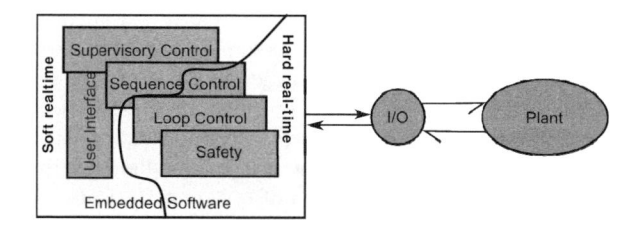

Figure 2. Embedded control system software structure

The typical software layers in motion control systems are:

- Man-machine/user interface;
- Supervisory control;
- Sequence control;
- Loop control;
- Data analysis;
- Measurements and actuation.

Besides a functional division in layers from a control engineering point of view, a division can also be made between hard real-time and soft real-time behaviour: the closer the software layer is to the machine or plant, the more strict the timing must be. Hence, the supervisory control and parts of the sequence control are soft real-time, and mostly run at lower sampling frequencies. In the case of the Production Cell, at least loop controllers (including motion profiles) and sequence controllers (to determine the order of actions) are needed.

1.6. Design Method

To structure the design process for these kind of systems, we use the following design approach:

- Abstraction;
- Top-down design;
- Model-based design;
- Stepwise refinement, local and predictable, aspect oriented

For the system software this means that we start with a top-level abstraction of the system that is refined towards the final implementation. During these stepwise refinements we focus on different aspects (e.g. concurrency, interactions between models of computation, timing, predictable code generation) of the system. To design the loop controller, we follow a similar stepwise refinement approach. The first step is *physical system modelling*: model and understand the plant dynamics. The second step is *control law design*: design a proper control law for the required plant movements. The third step is the *embedded control system implementation* phase in which relevant details about the target are incorporated in the model. These include the non-idealness of the interfaces with the outside world (sampling, discretization, signal delays, scaling), target details (CPU, FPGA). This step ends with code generation and integration of the loop controllers into the systems embedded software. Verifications by simulation are used after the first three steps. Validation and testing are done on the last step *realization*.

In the following two sections, the above design method is applied on the production cell FPGA design.

2. Structure and Communication

This section describes the design and implementation of the CSP based structural and communication (S&C) framework in which the loop controllers are embedded. First, requirements are formulated, after which the design is constructed in a top-down way.

2.1. Requirements

To focus the experiments and tests, the following requirements are formulated:

- Decentralised design to allow distribution across multiple FPGAs (or CPUs) if needed. This means that each PCU must be able to operate independently and that a central supervisory controller is missing.
- CSP based. Exploit parallelism. The setup consists of parallel operating robots, so the natural parallelism of the set up will be exploited.
- Generic. It should be usable for both a software and hardware implementation and for other mechatronic setups.
- Layered structure. This setup should be representative for industrial-sized machine control. Support for hierarchy using the layered structure is inevitable.
- Safety software distinguished from the normal operation. Handling faults can best be separated from the normal operation. This better structures the software, so that parts can be tested individually. Furthermore, design patterns about fault handling strategies can be used here.

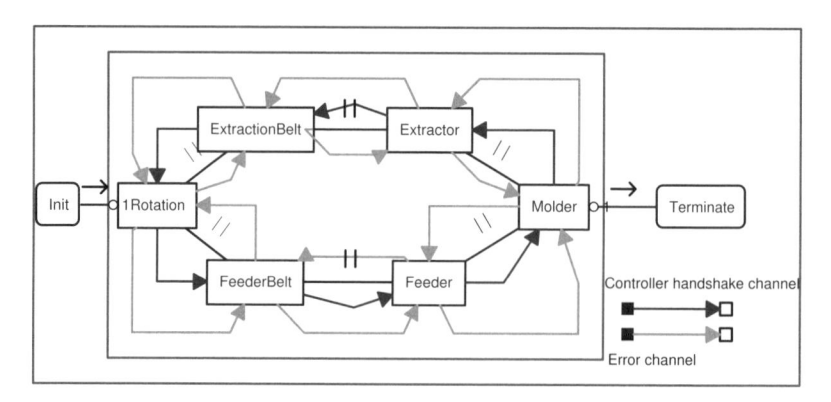

Figure 3. Top-level gCSP diagram

2.2. Top Level Design

We have chosen to implement the 'software' in a layered structure, taking into account the above requirements. The resulting top-level design is shown in figure 3. It shows an abstract view of the Production Cell system with PCUs implemented as parallel running CSP processes. Each PCU is connected to its neighbours using rendezvous channels. No central supervisory process exists and the PCUs are designed such that they are self sustaining. Since the production cell setup has a fixed direction for the blocks (*feeder belt > feeder > molder door > extractor > extraction belt > rotation*), normal communication is only necessary with the next PCU. The communication is a handshake (CSP rendezvous) for transporting a block. This normal communication will be called the *normal-flow* of the system. When a failure occurs, communication with both neighbours is required. For instance, when the feeder is stuck, not only should the molder door be opened, but also the feeder belt should be stopped in order to stop the flow of blocks. The next sections describe the design of the PCUs in more detail.

2.3. Production Cell Unit Design

A PCU is designed such that most of its operation is independent of the other PCUs. Each PCU can be seen as an independent motion control system. Communication with its neighbours is only needed for delivering a block to the next PCU, or in case of local failures that need to be communicated to both neighbours. Based on the layered structure described in section 1.5 and other implementations [16,15,17] a generic CSP model is made for all 6 PCUs. Figure 4 shows this PCU model, containing three parallel running processes. The controller process implements the motion controller intelligence via a sequence controller and a loop controller. Section 2.5 describes the controller in more detail. The command process implements the Man-machine interface for controlling a PCU from a PC (e.g. to request status info or to command the controller). The safety process contains data analysis intelligence to detect failures and unsafe commands. All communication between the setup, the command interface and the controller is inspected for exceptional conditions. The safety layer will be explained in more detail in the next section. The low-level hardware process contains the measurement and actuation part. Quadrature encoder interfaces for position measurement of the motors, digital I/O for the magnets and the block sensors and PWM generators to steer the DC motors are implemented here.

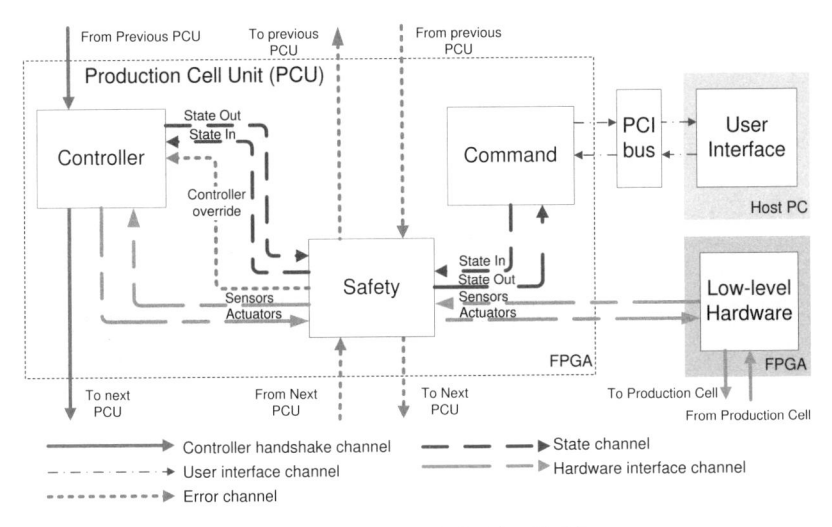

Figure 4. Production Cell Unit – Model

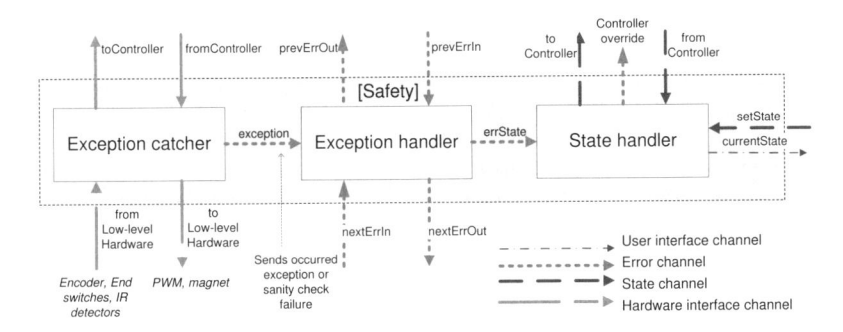

Figure 5. Production Cell Unit – Safety

2.4. Safety

The safety process implements a safety layer following the general architecture of protection systems [18] and the work of Wijbrans [19]. The safety consists of three stages: the *exception catcher*, the *exception handler* and the *state handler* (see figure 5). The *exception catcher* process catches exceptions (hardware to *controller* errors) as well as sanity check failures (*controller* to hardware errors). It sends an error message to the *exception handler*, which converts the error message into three state change messages:

- Its own (safe) controller state via the *errState* channel;
- A safe controller state for the previous PCU in the chain;
- A safe controller state for the next PCU in the chain.

The *state handler* process controls the states in a PCU and is the link between the *normal flow* and the error flow. Here the decision is made what state is being sent to the controller process (figure 4). It receives state information from the *Exception handler* process, the *Controller* process and the *User interface*.

The highest priority channel is the *errState* state channel from the exception handler. This channel transports the 'safe' state from the exception handler to the state handler when a failure has occurred. Once this channel is activated, this state will always be sent to the *Con-*

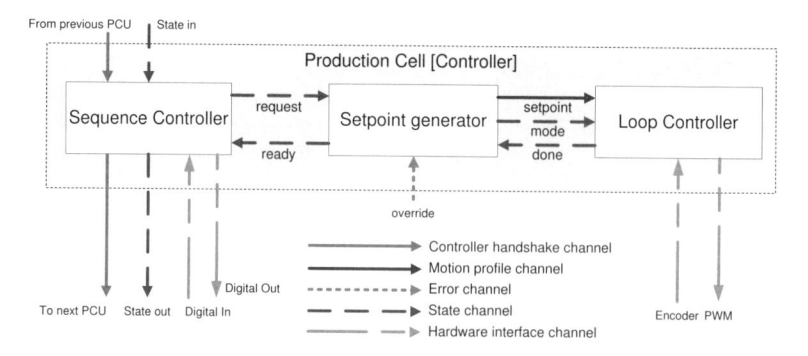

Figure 6. Production Cell Unit – Controller operation

troller (figure 4). The *override* channel is activated as well in order to keep the neighbouring PCU in its state until the error has been resolved.

2.5. Controller

Figure 6 shows the internals of the controller process. The controller process consists of a *sequence controller*, a *setpoint generator* and a *loop controller*. The sequence controller acts on the block sensor inputs and the rendezvous messages from the previous PCU. It determines which movement the PCU should make. It controls the *setpoint generator* that contains setpoints for stationary positions and it is able to generate motion profiles for movements between these stationary positions. The *loop controller* receives setpoints from the generator. Dependent on the mode set by the *setpoint generator* the loop controller is able to:

- Run a homing profile;
- Execute a regulator control algorithm (to maintain a stationary position);
- Execute a servo control algorithm (to track motion profiles).

The homing profile mode is needed to initialize the quadrature encoder position sensors on the motor at start-up. The design of the loop controller algorithm is explained in section 3.

2.6. Communication Sequence

The process framework is now almost complete. The framework is now capable of communicating with other PCUs and it can safely control a single PCU. This section briefly describes the interactions between the PCUs: the startup phase, handshaking and communication.

2.6.1. Normal Flow

When the hardware setup is turned on, all PCUs execute their homing action for sensor initialization and as a startup test. After the homing phase, the system is idle until a block is introduced and triggers a sensor. Figure 7 shows an example of the communication between the PCUs when a single block makes one round starting at the *feeder belt*.

2.6.2. Error Flow

In case a failure occurs, for example a block is stuck at the *feeder*, the local *exception catcher* and *exception handler* will put the *feeder* in a safe state and communicate to the neighbours (the feeder belt and the molder door) that it has a problem. The *molder door* will open and the *feeder belt* will stop supplying new blocks.

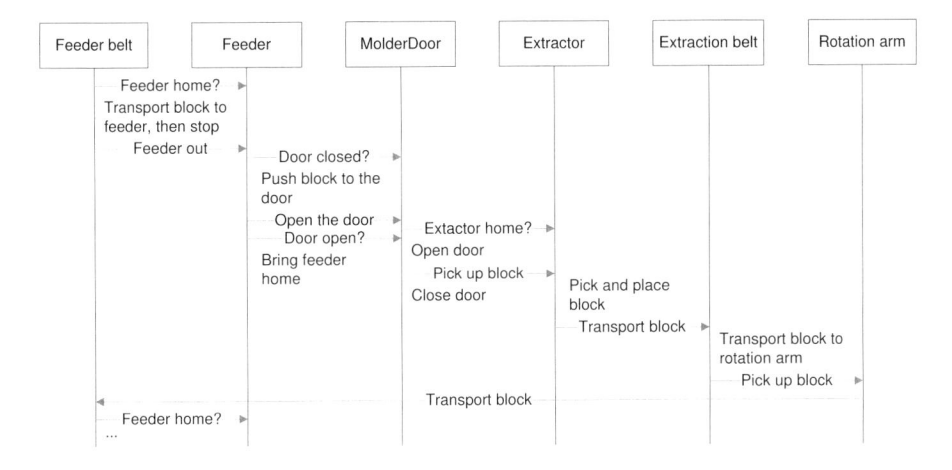

Figure 7. Production Cell – Normal operation sequence diagram

3. Loop Controller Design

An important part of this feasibility study is the implementation of loop-controllers in an FPGA. The control laws for these loop-controller are designed via step-wise refinement in 20-sim using a model of the plant behaviour. The design of these loop-controllers was originally done for software implementations. The resulting discrete-time PID[1] loop controllers and motion profiles used floating point calculations. The PID controller [20] is based on a computer algorithm that relies on the floating point data type (see listing 1 for the algorithm). Its purpose is to minimize the *error* between the current position and the desired position. This error is typically a small value (for the PCUs at most 0.0004m) so some calculation accuracy is needed here. The chosen FPGA has no on-board floating point unit (FPU), so another solution is needed here.

3.1. Floating Point to Integer

Table 1 shows some alternatives to using a floating-point data type.

Table 1. Alternatives to using the floating point data type on an FPGA

	Alternative	Benefit	Drawback
1.	Floating point library	High precision; re-use existing controller	Very high logic utilization because each calculation gets its own hardware
2.	Fixed point library	Acceptable precision	High logic utilization because each calculation gets its own hardware
3.	External FPU	High precision; re-use existing controller	Only available on high end FPGAs; expensive
4.	Soft-core CPU+FPU	High precision; re-use existing controller	High logic utilization unless stripped
5.	Soft-core FPU	High precision; re-use existing controller	Scheduling / resource manager required
6.	Integer	Native datatype	Low precision in small ranges; adaptation of the controllers needed

The numerical precision is coupled to the logic cell utilization, resulting in a design trade-off between numerical precision and FPGA utilization [21]. Agility, a provider of em-

---
[1] Proportional, Integral, Derivative.

```
factor = 1 / (sampletime + tauD * beta);
uD = factor * (tauD * previous(uD) * beta + tauD * kp * (error -
                previous(error)) + sampletime * kp * error);
uI = previous(uI) +  sampletime * uD / tauI;
output = uI + uD;
```

Listing 1. The PID loop-controller algorithm

bedded systems solutions formed from the merger of Catalytic Inc. and Celoxica's ESL business, delivers Handel-C libraries for both floating point and fixed point calculation. A main drawback of the first two options is that the resulting FPGA implementations have very high logic utilization because each calculation gets its own hardware. This is not a viable alternative for the chosen FPGA (a small test with 1 PID controller resulted in a completely filled FPGA for floating point). The third option requires a high-end FPGA with DSP facilities. The fourth option to use a soft-core CPU with a floating point unit (e.g. a Xilinx Microblaze CPU with single precision FPU which costs around 1800 LUTs[2]). The advantage is that we still can use our existing loop controllers (from the previous software version). The drawback is that the design becomes more complicated, due to the combination of Handel-C hardware and soft-core CPU software. Furthermore, we need a scheduler if all 6 PID controllers should run on the same soft-core CPU. The soft-core CPU solution is excessive for just a PID controller. An FPU-only soft-core is a better choice here, but still some scheduling is needed. The last option, integer based calculation, is the most suitable for efficient FPGA usage. However, this requires a redesign of the PID controllers. Despite the disadvantages of switching to an integer based PID controller, we have chosen this solution because the first three options are unfeasible for our FPGA and our goal was to not use a soft-core CPU.

To make the PID algorithm suitable for integer based control, taking into account the needed numerical precision, the following conversions are necessary:

- Integer based parameters;
- Integer based mathematics;
- Proper scaling to reduce the significance of fractional numbers;
- Take into account the quantization effects of neglecting the fractional. numbers

The original controllers used SI-units for I/O and parameters, resulting in many fractional numbers. All signals and parameters are now properly scaled, matching the value ranges of the I/O hardware (PWM, encoder). The conversions mentioned earlier are executed via step-wise refinement in 20-sim using simulations and a side-by-side comparison with the original floating point controllers. The new integer based controllers are validated on the real setup using the CPU based solution, to make sure that the resulting FPGA based integer PID controllers have a similar behaviour compared to the original floating-point version. Some accuracy is lost due to the switch to integer mathematics, resulting in a slightly larger error (0.00046m).

4. Realization and Results

The embedded control software structure for the production cell setup from section 2 was first checked for deadlocks using a separate gCSP model. The top-level structure in figure 3 is extended with an extra block inserter process. An FDR2 test shows indeed a deadlock with 8 blocks or more in the system as described in section 1.1.

For the Handel-C implementation, the gCSP model of figure 3 is refined in a systematic way to a version suitable for automatic code generation. To support automatic code gener-

---

[2]Look-Up Table, and a measure of FPGA size/utilization, with one LUT for each logic block in the FPGA.

```
void Rotation(chan* eb2ro_err, chan* ro2eb_err, chan* fb2ro_err, chan* ro2fb_err,
              chan* eb2ro, chan* ro2fb)
{
  /* Declarations */
  chan int cnt0_w encoder_in;
  chan int 12 pwm_out;
  chan int 2 endsw_in;
  chan int 1 magnet_out;
  chan int state_w setState;
  chan int state_w currentState;
  chan int state_w saf2ctrl;
  chan int state_w override;
  chan int 12 ctrl2hw;
  chan int state_w ctrl2saf;
  chan int cnt0_w hw2ctrl;
  chan int 1 magnet_saf;

  /* Process Body */
  par {
    LowLevel_hw(&encoder_in, &pwm_out, &endsw_in, &magnet_out);
    seq {
      Init(&encoder_in, &magnet_out, &pwm_out);
      par {
        Command(&setState, &currentState);
        Safety(&eb2ro_err, &saf2ctrl, &ro2eb_err, &override, &encoder_in,
               &fb2ro_err, &pwm_out, &setState, &ro2fb_err, &ctrl2hw,
               &currentState, &ctrl2saf, &hw2ctrl);
        Controller(&saf2ctrl, &override, &eb2ro, &ctrl2hw, &ctrl2saf,
                   &ro2fb, &hw2ctrl, &magnet_saf);
      }
      Terminate(&encoder_in, &magnet_out, &pwm_out);
    }
  }
}
```

Listing 2. Generated Handel-C code for the Rotation PCU

ation, gCSP is extended with Handel-C code generation capabilities. Due to the CSP foundation of gCSP, mapping the gCSP diagrams to Handel-C code was rather straightforward. Because Handel-C does not support the ALT and PRI-PAR constructs (only PRI-ALT and PAR are supported) some drawing restrictions were added. Furthermore, gCSP was extended with the possibilities to add non-standard datatypes to be able to use integer datatypes of a specific width. Listing 2 shows an example of the gCSP generated Handel-C code for the *rotation* PCU. This PCU is implemented in gCSP using the design shown figure 4 (the *Init* and *Terminate* blocks for the hardware are not shown in this figure).

The loop-controllers are implemented using a manually adapted version of the code that 20-sim has generated. Currently, 20-sim generates only ANSI-C floating-point based code. The Handel-C integer PID controller is first tested stand-alone in a one-to-one comparison with an integer PID running on the PC containing our FPGA card.

To be able to see what is happening inside the FPGA and to test the PID controllers and the *Command* process, we have implemented a PCI bus interface process (see also figure 4) to communicate between our development PC and the FPGA. This has proved to be a useful debugging tool during the implementation phase. Currently we are using the PCI debugging interface in cooperation with a Linux GUI program to show the internal status of the PCUs and to manually send commands to the FPGA.

Table 2 shows some characteristics of the realized FPGA implementation to get an idea of the estimated FPGA usage for this system. The total FPGA utilization for the Spartan III 1500 is 43% (measured in slices).

The behaviour of the Production Cell setup is similar to the existing software implementations. Compared to the existing CPU based solutions, the FPGA implementation shows

Table 2. Estimated FPGA usage for the Production Cell Motion Controller

Element	LUTs (amount)	Flipflops (amount)	Memory
PID controllers	13.5% (4038)	0.4% (126)	0.0%
Motion profiles	0.9% (278)	0.2% (72)	0.0%
I/O + PCI	3.6% (1090)	1.6% (471)	2.3%
S&C Framework	10.3% (3089)	8.7% (2616)	0.6%
Available	71.7% (21457)	89.1% (26667)	97.1%

perfect performance results under high system load (many blocks in the system) and all hard real-time constraints are met. The controller calculations are finished long before the deadline. Usage of the Handel-C timing semantics to reach our deadlines is not needed with a deadline of 1 ms (sampling frequency of 1 kHz). The PID algorithm itself requires only 464 ns (maximum frequency 2.1 MHz).

The performance of the FPGA based loop controllers is comparable to the CPU based versions. No visible differences in the PCU movements are observed and the measured position tracking errors remain well within limits. An additional feature of the FPGA solution is the implementation of a safety layer, which was missing in the software solutions.

5. Conclusions and Future work

The result of this feasibility study is a running production cell setup where the embedded control software is completely and successfully implemented in a low-cost Xilinx Spartan III XC3s1500 FPGA, using Handel-C as a hardware description language. The resulting software framework is designed such that it is generic and re-usable in other FPGA based or CPU based motion control applications. An FPGA only motion control solution is feasible, without using a soft-core CPU solution. The switch from CPU based implementations towards an FPGA based solution resulted in a much better performance with respect to the timing and the system load. However, the design process for the loop controllers requires more design iterations to ensure that a switch from floating-point calculations to integer based calculations results in correct behaviour.

The potential for FPGA based motion control systems running multiple parallel controllers is not limited to our production cell system. It is also a suitable alternative for our humanoid (walking) soccer robot that contains 12 controllers and a stereo vision system [22].

Although not needed for this setup, the implemented PID controller can reach frequencies of up to 2.1 MHz, which is impossible to achieve on a PC (maximum 40 kHz). This means that other applications requiring high controller frequencies can benefit from an FPGA based controllers.

While this feasibility study shows the potential of using a low-cost FPGA for complex motion control systems, there is still room for improvement and further investigation.

Table 2 shows that the PID controllers take almost half of the required FPGA cells. We have now implemented 6 dedicated PID controllers. A possible optimization would be to implement one PID process and schedule the calculations. We have enough time left to serialise the calculations. However, this conflicts with our goal of exploiting parallelism within the FPGA.

The process for designing integer based motion controllers should be simplified. 20-sim currently has too little support for assisting in the design of integer based motion controllers. Research on the topic of integer based control systems could potentially result in better design methods. Besides this, it would also be a good idea to evaluate the other implementation possibilities from table 1 (especially the soft-core with FPU option), to compare and explore these design space choices. In this way we can better advise on what to use for FPGA based motion control systems in which situations.

Integer based control systems need further research from the control engineering point of view. Especially with respect to accuracy and scaling effects. This is not only needed for FPGA based designs but also for microcontroller targets and soft-core CPUs without an FPU.

While the software framework was successfully designed using gCSP and its new Handel-C code generation output, there are opportunities for improvement in order to facilitate the future design of production cell-like systems. The structure and communication framework can be re-used, so having the option of using library blocks or gCSP design templates would speed-up the design process. Furthermore, only a subset of the Handel-C and the gCSP language (GML) is supported by the code-generation module. This should be extended.

References

[1] Clive Maxfield. *The Design Warriors Guide to FPGAs, Devices, Tools, and Flows*. Mentor Graphics Corp., 2004. www.mentor.com.
[2] Controllab Products B.V. 20-sim, 2008. http://www.20sim.com.
[3] Dusko S. Jovanovic, Bojan Orlic, Geert K. Liet, and Jan F. Broenink. gCSP: A Graphical Tool for Designing CSP systems. In Ian East, Jeremy Martin, Peter H. Welch, David Duce, and Mark Green, editors, *Communicating Process Architectures 2004*, pages 233–251. IOS press, Oxford, UK, 2004.
[4] Agility Design Systems. Handel-C, 2008. http://www.agilityds.com.
[5] L.S. van den Berg. Design of a production cell setup. MSc Thesis 016CE2006, University of Twente, 2006.
[6] Pieter Maljaars. Control of the production cell setup. MSc Thesis 039CE2006, University of Twente, 2006.
[7] D.S. Jovanovic. *Designing dependable process-oriented software, a CSP approach*. PhD thesis, University of Twente, Enschede, NL, 2006.
[8] Bojan Orlic and Jan F. Broenink. Redesign of the C++ Communicating Threads library for embedded control systems. In Frank Karelse, editor, *5th PROGRESS Symposium on Embedded Systems*, pages 141–156. STW, Nieuwegein, NL, 2004.
[9] Jinfeng Huang, Jeroen P.M. Voeten, Marcel A. Groothuis, Jan F. Broenink, and Henk Corporaal. A model-driven approach for mechatronic systems. In *IEEE International Conference on Applications of Concurrency to System Design, ACSD2007*, page 10, Bratislava, Slovakia, 2007. IEEE.
[10] Bart D. Theelen, Oana Florescu, M.C.W. Geilen, Jinfeng Huang, J.P.H.A van der Putten, and Jeroen P.M. Voeten. Software / hardware engineering with the parallel object-oriented specification language. In *ACM-IEEE International Conference on Formal Methods and Models for Codesign (MEMOCODE2007)*, pages 139–148, Nice, France, 2007.
[11] Robin Milner. *Communication and Concurrency*. Prentice-Hall, Englewood Cliffs, 1989.
[12] Gerald H. Hilderink. *Managing Complexity of Control Software through Concurrency*. PhD thesis, University of Twente, Netherlands, 2005.
[13] T.T.J. van der Steen. Design of animation and debug facilities for gCSP. MSc Thesis 020CE2008, University of Twente, 2008.
[14] B Rem, A Gopalakrishnan, T.J.H. Geelen, and H.W. Roebbers. Automatic Handel-C generation from MATLAB and simulink for motion control with an FPGA. In Jan F. Broenink, Herman W. Roebbers, Johan P.E. Sunter, Peter H. Welch, and David C. Wood, editors, *Communicating Process Architectures CPA 2005*, pages 43–69. IOS Press, Eindhoven, NL, 2005.
[15] S. Bennett. *Real-Time Computer Control: An Introduction*. Prentice-Hall, London, UK, 1988.
[16] Herman Bruyninckx. Project Orocos. Technical report, Katholieke Universiteit Leuven, 2000.
[17] S. Bennett and D.A. Linkens. *Real-Time Computer Control*. Peter Peregrinus, 1984.
[18] P. A Lee and T. Anderson. *Fault tolerance, principles and practice*. Springer-Verlag, New York, NY, 1990.
[19] K.C.J. Wijbrans. *Twente Hierarchical Embedded Systems Implementation by Simulation (THESIS)*. Universiteit van Twente, 1993.
[20] Karl J. Åström and T. Hagglund. *PID Controllers: Theory, Design and Tuning*. ISA, second edition, 1995.

[21] Michael J. Beauchamp, Scott Hauck, Keith D. Underwood, and K. Scott Hemmert. Embedded floating-point units in FPGAs. In Steven J. E. Wilton and André DeHon, editors, *FPGA*, pages 12–20. ACM, 2006.

[22] Dutch Robotics. 3TU humanoid soccer robot, TUlip, 2008. http://www.dutchrobotics.net.

Shared-Clock Methodology for Time-Triggered Multi-Cores

Keith F. ATHAIDE [a], Michael J. PONT [a] and Devaraj AYAVOO [b]

[a] *Embedded Systems Laboratory, University of Leicester*
[b] *TTE Systems Ltd, 106 New Walk, Leicester LE1 7EA*

Abstract. The co-operative design methodology has significant advantages when used in safety-related systems. Coupled with the time-triggered architecture, the methodology can result in robust and predictable systems. Nevertheless, use of a co-operative design methodology may not always be appropriate especially when the system possesses tight resource and cost constraints. Under relaxed constraints, it might be possible to maintain a co-operative design by introducing additional software processing cores to the same chip. The resultant multi-core microcontroller then requires suitable design methodologies to ensure that the advantages of time-triggered co-operative design are maintained as far as possible. This paper explores the application of a time-triggered distributed-systems protocol, called "shared-clock", on an eight-core microcontroller. The cores are connected in a mesh topology with no hardware broadcast capabilities and three implementations of the shared-clock protocol are examined. The custom multi-core system and the network interfaces used for the study are also described. The network interfaces share higher level serialising logic amongst channels, resulting in low hardware overhead when increasing the number of channels.

Keywords. co-operative, shared-clock, multi-core, multiprocessor, MPSoC.

Introduction

In the majority of embedded systems, some form of scheduler may be employed to decide when tasks should be executed. These decisions may be made in an "event-triggered" fashion (i.e. in response to sporadic events) [1] or in a "time-triggered" fashion (i.e. in response to pre-determined lapses in time) [2]. When a task is due to be executed, the scheduler can *pre-empt* the currently executing task or wait for the executing task to relinquish control *co-operatively*.

Co-operative schedulers have a number of desirable features, particularly for use in safety-related systems [1, 3-5]. Compared to a pre-emptive scheduler, co-operative schedulers can be identified as being simpler, having lower overheads, being easier to test and having greater support from certification authorities [4]. Resource sharing in co-operative schedulers is also a straightforward process, requiring no special design considerations as is the case with pre-emptive systems [6, 7]. The simplicity may suggest better predictability while simultaneously necessitating a careful design to realise the theoretical predictions in practice.

One of the simplest implementations of a co-operative scheduler is a cyclic executive [8, 9]: this is one form of a broad class of time triggered, co-operative (TTC) architectures. With appropriate implementations, TTC architectures are a good match for a wide range of applications, such as automotive applications [10, 11], wireless (ECG) monitoring systems [12], various control applications [13-15], data acquisition systems, washing-machine control and monitoring of liquid flow rates [16].

Despite having many excellent characteristics, a TTC solution will not always be appropriate. Since tasks cannot interrupt each other, those with long execution times can increase the amount of time it takes for the system to respond to changes in the environment. This then imposes a constraint that all tasks must have short execution times in order to improve system response times [3]. A change in system specifications (e.g. higher sampling rates) post-implementation could require a re-evaluation of all system properties and validation that the static schedule still holds.

In this paper, we consider ways in which – by adapting the underlying processor hardware – we can make it easier to employ TTC architectures in embedded systems. From the outset we should note that there is a mismatch between generic processor architectures and time-triggered software designs. For example, most processors support a wide range of interrupts, while the use of a (pure) time-triggered software architecture generally requires that only a single interrupt is active on each processor. This leads to design "guidelines", such as the "one interrupt per microcontroller rule" [17]. Such guidelines can be supported when appropriate tools are used for software creation (e.g. [18, 19]). However, it is still possible for changes to be made (for example, during software maintenance or upgrades) that lead to the creation of unreliable systems.

The present paper represents the first step in a new research programme in which we are exploring an alternative solution to this problem. Specifically, we are seeking to develop a novel "System-on-chip" (SoC) architecture, which is designed to support TTC software. This approach has become possible since the advent of the reduced cost of field-programmable gate array (FPGA) chips with increasing gate numbers [20].

Following the wider use of FPGAs, SoC integrated circuits have been used in embedded systems, from consumer devices to industrial systems. These complex circuits are an assembly upon a single silicon die from several simpler components such as instruction set processors, memories, specialised logic, etc. A SoC with more than one instruction set processor (or simply "processor") is referred to as a multi-core or multiprocessor system-on-chip (MPSoC).

MPSoCs running decomposed single-processor software as TTC software may have time-triggered tasks running concurrently on separate, simpler, heterogeneous cores, with the tasks synchronising and exchanging timing information through some form of network. In this configuration, MPSoCs resemble micro versions of distributed systems, without the large amount of cabling and high installation and maintenance costs. Like a distributed system, an MPSoC requires software that is reliable and operates in real-time.

Previously, shared-clock schedulers have been found to be a simple and effective means of applying TTC concepts to distributed systems communicating on a single shared channel [17, 21, 22]. This paper explores the use of shared-clock schedulers for a custom MPSoC where communication may take place on multiple channels.

The remainder of this paper is organised as follows. In Section 1, previous work on design methodologies for software on MPSoCs is reviewed, while Section 2 proposes an enhancement to the shared-clock design. Section 3 then describes the MPSoC and network interface module being used in this work. In Section 4, the results of applying the shared-clock design enhancement to the MPSoC are presented. Finally, our conclusions are delivered in Section 5.

1. Previous work

Previous research on MPSoC software design has advocated a modular approach, either composing a system from pre-existing modules or producing an executable from a high-level model, with the entire system running on either a bare-bones or a fully fledged real-time operating system (RTOS) [23, 24].

Virtuoso is a RTOS with a pre-emptive microkernel for heterogeneous multiprocessor signal-processing systems [25]. Communication between processors is packetised with packets inheriting the priority of the generating task. Tasks share no common memory and communicate and synchronise via message-passing. This RTOS allows an MPSoC to be programmed as a virtual single processor, allowing for processor and communication topology independence but at the expense of a larger code base and greater communication overheads.

Alternatively, the multiprocessor nature can be exposed to the designer, as in [26] where a model based design process is proposed for an eight processor MPSoC running time-triggered software. The MPSoC is considered to be composed of various chunks communicating through a network-on-chip (NoC) partitioned into channels using a global time-division-multiple-access (TDMA) schedule. The TDMA scheme which is employed is similar to that of the time-triggered protocol (TTP) used in distributed systems [27], where cores synchronise by comparing their locally maintained time with the global time maintained on the network. This synchronisation is crucial to ensuring that cores transmit only in their allocated timeslots. However, this global TDMA scheme may be constrained by transmission times to distant nodes on certain topologies, requiring unsuitably large timeslots for such nodes.

Another implementation of a NoC for time-triggered software also uses the TTP method of synchronising the attached processors [28]. The broadcast requirements of this implementation limit it to processors connected in a unidirectional ring topology. It also faces the same global TDMA problem described above – as the number of nodes increases, the schedule table quickly becomes full of delay slots.

In the domain of distributed systems, shared-clock scheduling (SCS) is a simple and effective method of applying the TTP principles [17]. In contrast to TTP where all nodes have the same status, SCS follows a single master-multiple slave model, with the global time maintained at the master. The global time is propagated by the master using an accurate time source to generate regular *Ticks* or messages, which then trigger schedulers in the slave nodes, which send back *Acknowledgements* (**Figure 1**).

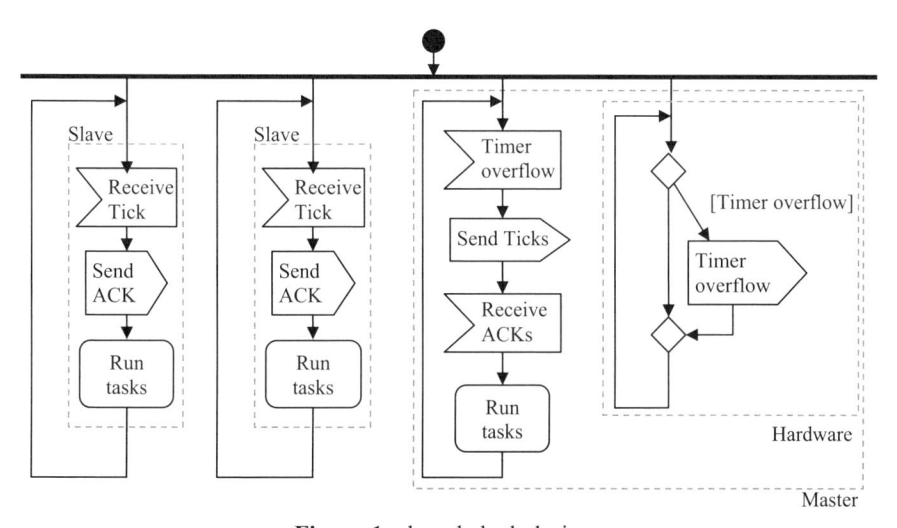

Figure 1: shared clock design.

The Ticks and Acknowledgements can also be used for master-to-slave and slave-to-master communication. Slave-to-slave communication can be carried out through the master or by snooping (if allowed by the communication media). The healthiness of a node is indicated by the presence and contents of its Acknowledgement message. A shared-bus is most often used as the communication medium.

SCS also faces the global TDMA problem resulting in either limited Tick intervals and/or slow slave responses. This paper looks at a way of alleviating this problem by taking advantage of underlying topologies with an SCS design for multiple communication channels.

2. A Multi-Channel Shared-Clock Scheduler

The shared-clock scheme works well because of the implicit broadcast nature of a bus: anything placed on the bus is visible to all connected nodes. However, to maintain cost effectiveness, buffer-heavy features such as broadcasts may be omitted from the NoC in an MPSoC [29].

Without hardware broadcast functionality, the master is then required to address each node separately for each Tick. One method to achieve this would be to send a Tick to a slave, wait for its Acknowledgement and then contact the next one. However, it is easily seen that this method will suffer from excessive jitter as the Acknowledgement time may well vary from Tick to Tick. The time spent communicating will increase as the number of nodes increase, eventually leaving very little time to do anything else.

Another method is to send all Ticks immediately either using a hardware buffer or a software polling loop. Again, however, scaling to a large number of nodes would either require a very large hardware buffer or would consume too much processor time. Transmission times might also mean that it is never possible to send all of one Tick's messages before the next Tick needs to be signalled.

An alternative is to broadcast the Tick in a tree-like manner, having more than one master in the network [30]. In this scheme, the slave of one master acts as the master of other nodes, propagating Tick messages, or multiples thereof through the network. In this scheme, a master only receives Acknowledgements from the immediate slaves and is unaware of the presence of the other nodes.

An example of this scheme is seen in the mesh arrangement in Figure 2, where for illustrative purposes, the initial master is located at the bottom left corner.

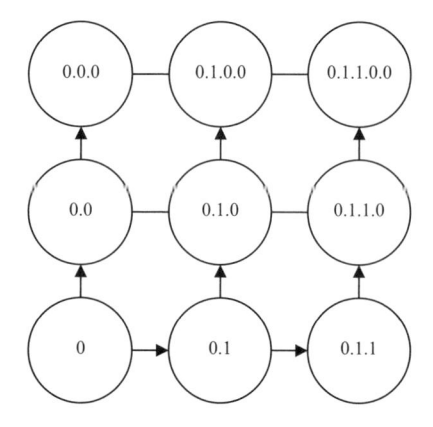

Figure 2: example of tree propagation.

The number in the node is obtained by joining the number of the node that sent it the Tick and the order in which it received the Tick. For example, the initial master first sends its Tick north (to 0.0), then east (to 0.1). The east node then sends its own Tick north (to 0.1.0) and east (to 0.1.1) and so on. The lines represent existing physical connections with the arrowheads showing the direction of propagation of messages. Lines without arrowheads are unused by the scheduling scheme and may be used for additional application-dependent inter-slave communication.

This decentralisation resembles a broadcast done through software, however it is more predictable as the "broadcast" message will always come from a pre-determined node down a pre-determined channel (with static routing). The overhead on a single master can be kept within limits, as opposed to a scheme that only possesses a single master. However, it will cause distant nodes to lag behind the initial master in Ticks. For example, if transmission time equalled half a Tick, a sample Tick propagation of the Figure 2 network is seen in Figure 3. The most distant node eventually lags two Ticks behind the initial master, and any data it generates can reach the initial master only two further Ticks later. This lag needs to be considered during system design and node placement.

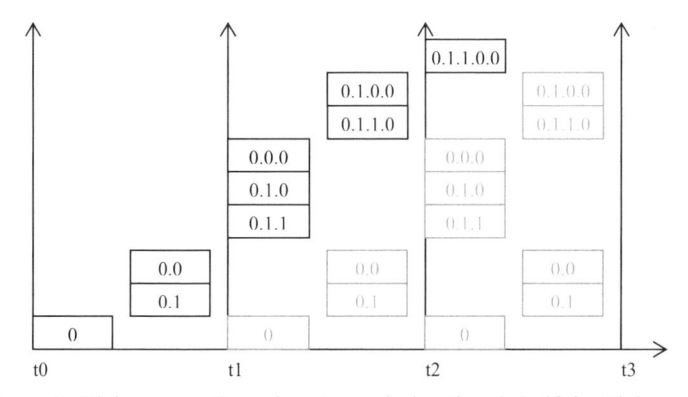

Figure 3: Tick propagation when transmission time is half the Tick period.

3. The TT MPSoC

The MPSoC is formed from several processor clusters linked together by a network-on-chip (**Figure 4**).

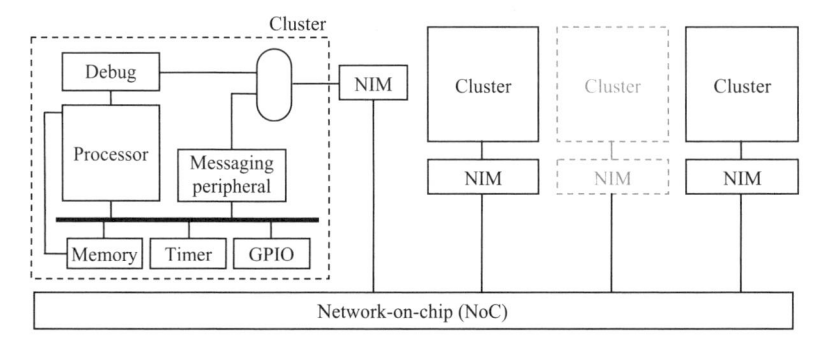

Figure 4: MPSoC architecture.

Each processor cluster contains one processor and associated peripherals. A cluster interfaces to the NoC using a network interface module (NIM).

A development TT MPSoC will always contain a special debug cluster that is used as the interface between a developer's computer and the MPSoC. This debug cluster allows processor cores to be paused, stopped, etc as well as allowing memory in the cluster to be read from or written to. These capabilities are provided by a separate debug node in each cluster connected to the processor and memory in that cluster. Special logic in the cluster recognises when a debug message has been received and multiplexes it appropriately.

The MPSoC has currently been designed for a 32-bit processor called the PH processor [31]. The PH processor was designed for time-triggered software and has a RISC design which is compatible with the MIPS I ISA (excluding patented instructions): it has a Harvard architecture, 32 registers, a 5-stage pipeline and support for precise exceptions. In the present (MPSoC) design, the number and type of peripherals connected to each processor is independently configurable, as is the frequency at which each processor is clocked.

A "messaging" peripheral is present with all clusters. This peripheral allows the processor to receive and transmit messages on the NoC. The PH processor is sensitive to a single event multiplexed from various sources. In this implementation, events can be generated by the messaging peripheral and timer peripherals, if present.

An interactive GUI is used to specify the set of files to be compiled for each processor, generating one binary per processor. These binaries are then uploaded via a JTAG connection to the debug cluster which transmits them onward to the specified processor.

3.1 The Network Interface Module (NIM)

A network interface module (NIM) forms the interface between a processor cluster and the NoC. The NIMs use static routing, store-and-forward switching, send data in non-interruptible packet streams and have no broadcast facilities. Communication is performed asynchronously with respect to software executing on the processors.

The NIM was designed to be flexible (for research purposes) and supports any number of channels of communication. By varying the number of channels and the links between them, various topologies can be obtained.

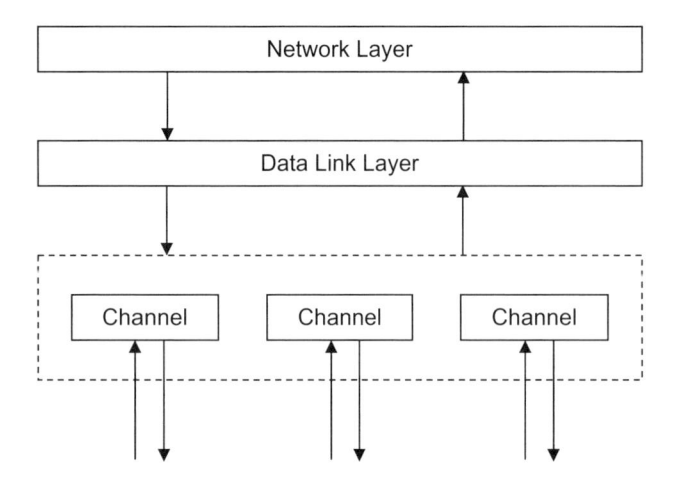

Figure 5: implemented OSI layers.

The channels constitute the physical layer of the Open Systems Interconnection (OSI) model [32], with the data link and network layers shared between channels (**Error! Reference source not found.**). Each channel uses four handshaking control lines and any number of bidirectional data lines, as dictated by the application (**Figure 6**). A channel is insensitive to the frequency at which the other channel is running. This insensitivity comes at the cost of a constant overhead (it is independent of the bit-width of the data line) per data transfer. A channel with n data lines may be referred to as an n-bit channel.

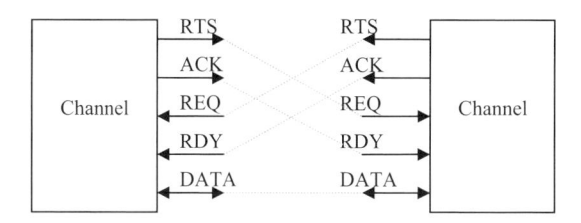

Figure 6: channel OSI layers.

The packet structure for a 6-bit channel is shown in **Figure 7**. The preamble size was chosen to match the physical layer data size, with the cluster ID and node ID sizes chosen so that there is a maximum of 2^{20} clusters each with a maximum of sixteen nodes. Data was fixed at ninety-six bits to obtain reasonable bandwidth. Fixed-length packets were chosen to eliminate transmission-time jitter. Each packet is affixed with a 12-bit cyclic redundancy checksum (CRC)[1], which was used for error detection. This checksum is capable of detecting single bit errors, odd number of errors and burst errors up to twelve bits. There is no automatic error correction in the present implementation.

137	132 131	112 111	108 107	12 11	0
Preamble	Cluster ID	Node ID	Data	CRC	

Figure 7: Network packet structure.

This packet structure has a theoretical transmission efficiency of approximately 70%. The packet has a size of 138 bits, which equates to 23 "chunks" with a 6-bit channel. If the channel uses 7 cycles for each chunk, then the packet in total takes 161 cycles to reach the destination. At 25 MHz, this equates to a throughput of ~15 Mbps, which increases to ~60 Mbps at 100 MHz. This performance can be improved by increasing either the communication frequency and/or the channel bit width.

3.2 Hardware Implementation

A prototype with 6-bit wide data channels was implemented on multiple Spartan 3 1000K FPGAs, using the Digilent Nexys development board [33]. Using one FPGA for each processor, as opposed to a single one containing all processors, allows for lower synthesis times and greater design exploration space in addition to supplying a number of useful debugging peripherals (switches, LEDs, etc.) to each processor.

The hardware usage for the NIM is shown in **Table 1** with varying numbers of channels and data widths. The values were obtained from the synthesis report generated by the Xilinx ISE WebPACK [34] with the NIM selected as the sole module to be synthesised.

---

[1] Using the standard CRC-12 polynomial: $x^{12} + x^{11} + x^3 + x^2 + x + 1$

The 8-bit and 16-bit channel implementations have a 144-bit data packet with an 8-bit preamble and 4-bits of padding added to the 12-bit CRC field.

As can be seen in the table on the next page, logic consumption sees a very minor increase as the number of channels is increased. There is an initial hump when moving from one to two channels as multi-channel logic that is optimised away in the single-channel case (by the synthesis software) is reinserted, but beyond two channels, the trend appears steady. The number of slices counterintuitively reduces when moving from a 6-bit data width to an 8-bit one because certain addition/subtraction logic is optimised into simpler shifts.

Table 1: logic utilisation of the network interface module.

Data width (bits)		Number of channels			
		1	2	3	4
6	Slices used	637	681	699	723
	Percentage used	8.3%	8.9%	9.1%	9.4%
	Percentage increase	—	6.5%	2.6%	3.3%
8	Slices used	613	653	676	706
	Percentage used	8.0%	8.5%	8.8%	9.2%
	Percentage increase	—	6.1%	3.4%	4.2%
16	Slices used	659	709	751	786
	Percentage used	8.6%	9.2%	9.8%	10.2%
	Percentage increase	—	7.1%	5.6%	4.5%

Simulations with ModelSim [35] suggest the interval from transmission request on one channel to reception notification on another channel to be 2.43 μs, 4.59 μs and 5.79 μs for the 16-bit, 8-bit and 6-bit data channels respectively. These simulations assumed ideal propagation times.

Compared to the lightweight circuit-switched PNoC [36], the NIM has a low logic cost for additional channels. PNoC has also been designed for flexibility and suitability for any topology and the 8-bit wide, 4-channel implementation uses 249 slices on the Xilinx Virtex-II Pro FPGA [37]. This is a 66% increase over a 2-channel implementation and 78% less than an 8-channel implementation. In contrast, the 8-bit wide 4-channel NIM when synthesised for this FPGA uses 707 slices while the 8-channel version uses 722 slices, a relatively miniscule increase (2%). Compared to the PNoC equivalents, the 4-channel NIM is about 65% larger while the 8-channel version is about 35% smaller.

The time-triggered NoC in [28] is tied to a unidirectional ring topology. It is not designed to serialise data transfer and completes its 128-bit data transfer in a single cycle. On the Cyclone II (EP2C70) FPGA [38], it uses 480 logic cells. The 16-bit wide, 2-channel NIM when synthesised for this FPGA using the Quartus II Web Edition [39], uses 1161 logic cells, about 2.4 times the amount.

Both the above implementations use on-chip memory for buffer storage, while for the moment the NIM employs registers. The logic utilisation report from the WebPACK suggests that about 50% of the logic in the NIM is spent on registers, further suggesting possible logic savings by moving to on-chip memory for storage.

4. Case Study

To illustrate the operation of this MPSoC design, the MPSoC described in Section 3 was configured to have eight processors and one debug node, arranged in a mesh structure as seen in **Figure 8**, with each processor and the NoC clocked at 25 MHz. In this arrangement, there is no direct path between certain cluster pairs. The XY routing strategy is employed where a message is propagated first horizontally and then vertically in order to reach the targeted processor.

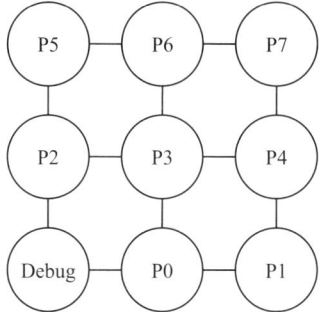

Figure 8: testbed topology.

Three types of schedulers were implemented:

- In the first design (SCH1), P1 (i.e. the processor on cluster 1) acts as a master to all other processors, sending Ticks to them in the same Tick interval, sending a Tick to a processor only when the previous one has sent its Acknowledgement.

- In the second design (SCH2), P1 again acts as a master, but this time sends Ticks to the processors in turns (one Tick per timer overflow).

- In the third design (SCH3), each processor acts as a master to its immediate slaves (**Figure 9**), according to the technique described in Section 2. Ticks are sent to a slave only when the previous one has sent its Acknowledgement.

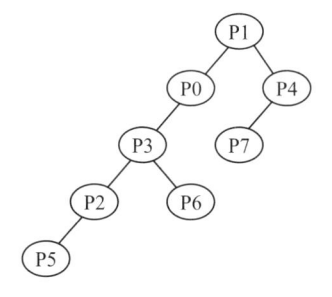

Figure 9: master-slave tree.

Slave processors toggle a pin when the messaging peripheral raises an interrupt. Master processors toggle pins on transmitting Ticks or receiving Acknowledgements. All these pulses are run into a common pulse detection device (with a resolution of 20 ns), providing relative timing on each of these signals. The Tick transmission times from the immediate master are shown in **Figure 10**.

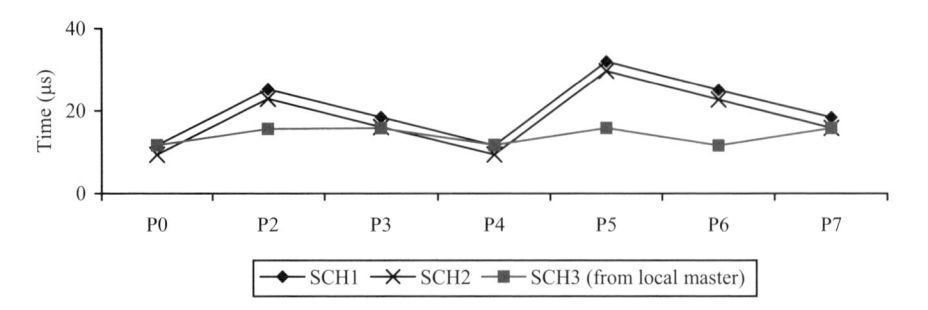

Figure 10: Time taken for the transmission of a Tick from P1.

As expected, the Tick transmission times increase as the number of hops increase. The first and second implementations show near identical transmission times (the difference between them is due to software overhead). Tick transmission times for SCH3 measured from the immediate master are the lowest since all transmissions are single hops. The slight variations observed in the SCH3 line may be due to noise in the transmission medium and has a range of 4.22 µs. The single hop transmission times averaged at about 6.79 µs, not including software overheads.

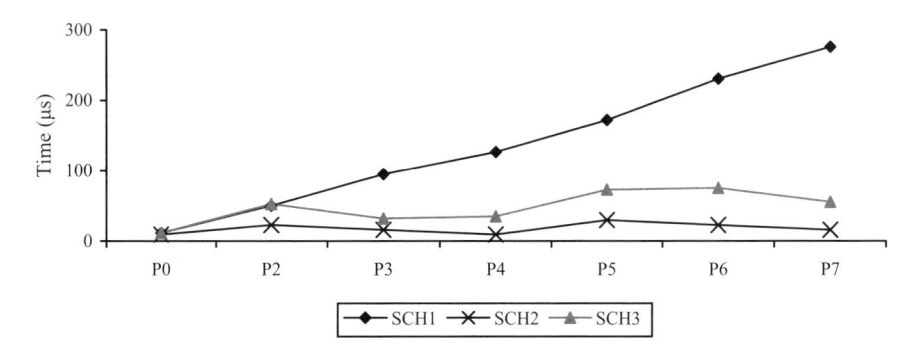

Figure 11: Time from P1 timer overflow to Tick receipt.

Figure 11 shows the time from when the initial master, P1, receives notification of a timer overflow to when a node receives notification of a Tick message. SCH1 has a high overhead as the number of nodes increases, while SCH2 and SCH3 maintain a more or less steady overhead. The overhead for SCH2 depends solely on the number of hops required to reach the node while for SCH3 it is dependent on the number of slaves and the distance from the initial master.

The SCH3 Tick receipt trend might be better understood by comparing the master-slave chain in **Figure 9** and **Figure 12**. In **Figure 12**, the times have been plotted in the order in which the nodes receive their Ticks, each master-slave chain plotted separately. The order is based on the averages, but since chains can be independent (for example, the P0-P3 and the P4-P7 chains), the order could vary in particular Tick instances.

In SCH3, masters wait for slave Acknowledgements before contacting other slaves. In that implementation, when the timer overflows, P1 sends a Tick to P0 and waits. On receiving the Tick, P0 sends an Acknowledgement to P1 and then immediately sends a Tick to P3. However, since the channels share their higher layers, the Acknowledgement must be sent before Tick transmission can commence. Because of this, the Acknowledgement reaches P1 at about the same time that P0 starts to send the Tick to P3. A few microseconds later (spent in software processing), P1 sends a Tick to P4. Since both are single hops, the Tick to P4 reaches slightly later than the Tick to P3. The same trend can be observed for P5 and P6.

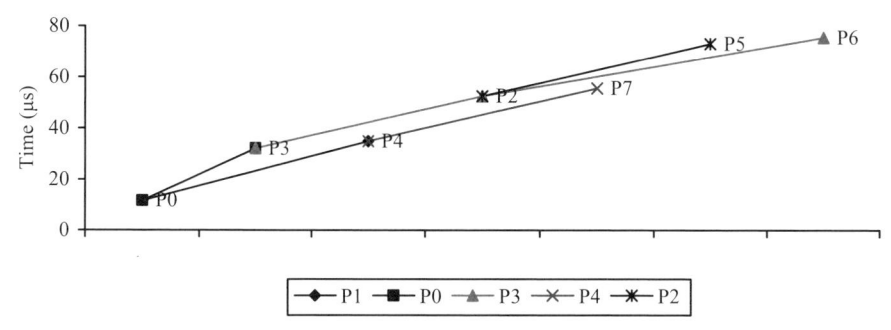

Figure 12: SCH3 times from P1 timer overflow in order of Tick receipt.

The jitter in transmission times can be seen in **Figure 13**. The jitter increases as the number of hops increases, most easily seen in the graphs of SCH1 and SCH2. SCH1 and SCH2 start out with the same amount of jitter, though it quickly starts to increase for SCH1. This is because SCH1 waits for Acknowledgements before sending further Ticks, so the jitter in receiving Acknowledgements is added to that of Ticks sent later.

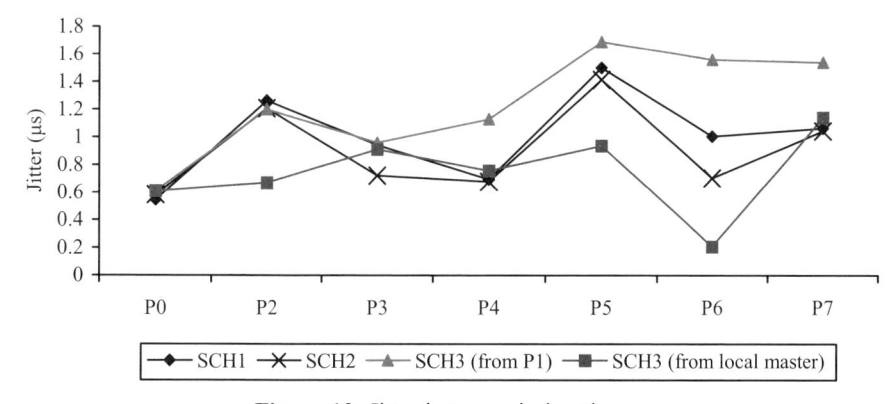

Figure 13: Jitter in transmission times.

The jitter for SCH3 is shown both for transmission from the immediate master and from P1, though the jitter when measured from P1 is more important as it is a measure of the variation in sensing the timer overflow. The jitter when measured from P1 follows the same trend as the transmission time measured in the same way – it increases when travelling down the master-slave chains (**Figure 14**).

The SCH3 jitter measurement is further affected by the channels sharing higher layers. That is why the jitter for P6 is very low; no other messages have been queued for transmission from P3 at that time.

While SCH2 appears to be a better choice in most of the comparisons, it is worth noting that since it sends a message to only one node per timer overflow, the frequency of

Ticks a node receives will decrease as the number of nodes increases. A mix of SCH1 and SCH2 where certain nodes are can be sent Ticks more frequently might prove better. It could also be mixed with SCH3 to introduce a degree of scalability.

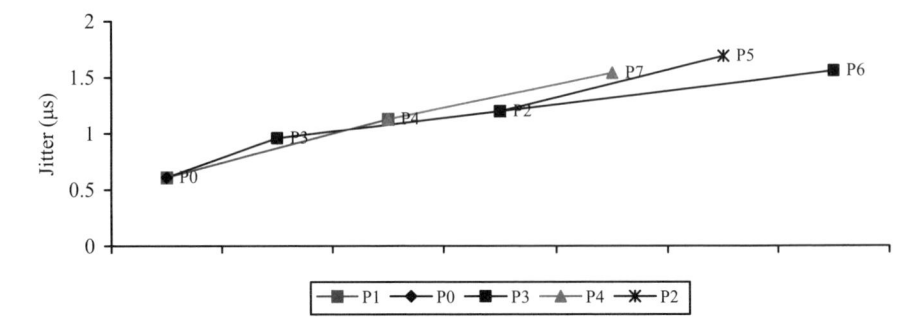

Figure 14: SCH3 jitter from P1 in the order of Tick receipt.

While SCH2 appears to be a better choice in most of the comparisons, it is worth noting that since it sends a message to only one node per timer overflow, the frequency of Ticks a node receives will decrease as the number of nodes increases. A mix of SCH1 and SCH2 where certain nodes are can be sent Ticks more frequently might prove better. It could also be mixed with SCH3 to introduce a degree of scalability.

Choosing between scheduler implementations or a mix of all three scheduler implementations will be highly application dependent and requires a deeper analysis into the rationale behind such choices.

5. Conclusions

This paper has presented results from a study which has explored the use of the time-triggered shared-clock distributed protocol on a non-broadcast, statically routed NoC employed in an MPSoC for time-triggered software. A scheduler design that used a tree-based form of broadcasting clock information was described and implemented on an MPSoC with eight processors connected in a mesh. This design offered a fast response and low overhead, promising to scale well to a large number of nodes, though at the cost of increased jitter and latency when compared to the other implementations.

Acknowledgments

This project is supported by the University of Leicester (Open Scholarship award) and TTE Systems Ltd.

References

[1] N. Nissanke, *Realtime Systems*: Prentice-Hall, 1997.
[2] H. Kopetz, "Time Triggered Architecture," *ERCIM NEWS,* pp. 24-25, Jan 2002.
[3] S. T. Allworth, *Introduction to Real-Time Software Design*: Macmillan, 1981.
[4] I. J. Bate, "Introduction to scheduling and timing analysis," in *The Use of Ada in Real-Time Systems*, 2000.

[5] N. J. Ward, "The static analysis of a safety-critical avionics control system," in *Air Transport Safety: Proceedings of the Safety and Reliability Society Spring Conference*, 1991.
[6] H. Wang, M. J. Pont, and S. Kurian, "Patterns which help to avoid conflicts over shared resources in time-triggered embedded systems which employ a pre-emptive scheduler," in *12th European Conference on Pattern Languages of Programs (EuroPLoP 2007)* Irsee Monastery, Bavaria, Germany, 2007.
[7] A. Maaita and M. J. Pont, "Using 'planned pre-emption' to reduce levels of task jitter in a time-triggered hybrid scheduler," in *Proceedings of the Second UK Embedded Forum*, Birmingham, UK, 2005, pp. 18-35.
[8] T. Baker and A. Shaw, "The cyclic executive model and Ada," *Real-Time Systems,* vol. 1, pp. 7-25, 1989.
[9] C. D. Locke, "Software architecture for hard real-time applications: cyclic executives vs. fixed priority executives," *Real-Time Systems,* vol. 4, pp. 37-53, 1992.
[10] D. Ayavoo, M. J. Pont, and S. Parker, "Using simulation to support the design of distributed embedded control systems: a case study," in *Proceedings of the UK Embedded Forum 2004*, Birmingham, UK, 2004, pp. 54 - 65.
[11] M. Short and M. J. Pont, "Hardware in the loop simulation of embedded automotive control system," in *Proceedings of the 8th IEEE International Conference on Intelligent Transportation Systems (IEEE ITSC 2005)*, 2005, pp. 426 - 431.
[12] T. Phatrapornnant and M. J. Pont, "Reducing jitter in embedded systems employing a time-triggered software architecture and dynamic voltage scaling," *IEEE Transactions on Computers,* vol. 55, pp. 113-124, 2006.
[13] R. Bautista, M. J. Pont, and T. Edwards, "Comparing the performance and resource requirements of 'PID' and 'LQR' algorithms when used in a practical embedded control system: A pilot study," in *Proceedings of the Second UK Embedded Forum*, Birmingham, UK, 2005.
[14] T. Edwards, M. J. Pont, P. Scotson, and S. Crumpler, "A test-bed for evaluating and comparing designs for embedded control systems," in *Proceedings of the first UK Embedded Forum*, Birmingham, UK, 2004, pp. 106-126.
[15] S. Key and M. J. Pont, "Implementing PID control systems using resource-limited embedded processors," in *Proceedings of the first UK Embedded Forum*, Birmingham, UK, 2004, pp. 76-92.
[16] M. J. Pont, *Embedded C*. London: Addison-Wesley, 2002.
[17] M. J. Pont and Association for Computing Machinery, *Patterns for time-triggered embedded systems : building reliable applications with the 8051 family of microcontrollers*. Harlow: Addison-Wesley, 2001.
[18] C. Mwelwa, M. J. Pont, and D. Ward, "Code generation supported by a pattern-based design methodology," in *Proceedings of the first UK Embedded Forum*, Birmingham, UK, 2004, pp. 36-55.
[19] C. Mwelwa, K. Athaide, D. Mearns, M. J. Pont, and D. Ward, "Rapid software development for reliable embedded systems using a pattern-based code generation tool," in *Society of Automotive Engineers (SAE) World Congress*, Detroit, Michigan, USA, 2006.
[20] J. Gray, "Designing a Simple FPGA-Optimized RISC CPU and System-on-a-Chip," Gray Research LLC, 2000.
[21] D. Ayavoo, M. J. Pont, M. Short, and S. Parker, "Two novel shared-clock scheduling algorithms for use with 'Controller Area Network' and related protocols," *Microprocessors & Microsystems,* vol. 31, pp. 326-334, 2007.
[22] M. Nahas, M. J. Pont, and A. Jain, "Reducing task jitter in shared-clock embedded systems using CAN," in *Proceedings of the UK Embedded Forum 2004*, A. Koelmans, A. Bystrov, and M. J. Pont, Eds. Birmingham, UK: Published by University of Newcastle upon Tyne, 2004, pp. 184-194.
[23] L. Benini and G. De Micheli, *Networks on Chips: Technology and Tools*: Morgan Kaufmann, 2006.
[24] A. A. Jerraya and W. H. Wolf, *Multiprocessor systems-on-chips*: Morgan Kaufmann, 2005.
[25] E. Verhulst, "The Rationale for Distributed Semantics as a Topology Independent Embedded Systems Design Methodology and its Implementation in the Virtuoso RTOS," *Design Automation for Embedded Systems,* vol. 6, pp. 277-294, 2002.
[26] H. Kopetz, R. Obermaisser, C. E. Salloum, and B. Huber, "Automotive Software Development for a Multi-Core System-on-a-Chip," in *Proceedings of the 4th International Workshop on Software Engineering for Automotive Systems*: IEEE Computer Society, 2007.
[27] H. Kopetz and G. Grünsteidl, "TTP - A Protocol for Fault-Tolerant Real-Time Systems," *Computer,* vol. 27, pp. 14-23, 1994.
[28] M. Schoeberl, "A Time-Triggered Network-on-Chip," in *International Conference on Field-Programmable Logic and Applications (FPL 2007)*, Amsterdam, Netherlands, 2007, pp. 377-382.

[29] A. Rădulescu and K. Goossens, "Communication services for networks on silicon," in *Domain-Specific Processors: Systems, Architectures, Modeling, and Simulation*, S. S. Bhattacharyya, E. Deprettere, and J. Teich, Eds.: Marcel Dekker, 2002.
[30] T. Kielmann, R. F. H. Hofman, H. E. Bal, A. Plaat, and R. A. F. Bhoedjang, "MagPIe: MPI's collective communication operations for clustered wide area systems," in *Proceedings of the Seventh ACM SIGPLAN Symposium on Principles and Practice of Parallel Programming (PPoPP'99)*. vol. 34 Atlanta, GA, 1999, pp. 131-140.
[31] Z. M. Hughes, M. J. Pont, and H. L. R. Ong, "The PH Processor: A soft embedded core for use in university research and teaching," in *Proceedings of the Second UK Embedded Forum*, Birmingham, UK, 2005, pp. 224-245.
[32] H. Zimmermann, "OSI Reference Model - The ISO Model of Architecture for Open Systems Interconnection," *IEEE Transactions on Communications*, vol. 28, pp. 425- 432, April 1980.
[33] Digilent Inc., "Nexys," http://www.digilentinc.com/Products/Detail.cfm?Prod=NEXYS, 2006.
[34] Xilinx, "Xilinx ISE WebPACK," http://www.xilinx.com/ise/logic_design_prod/webpack.htm.
[35] Mentor Graphics, "ModelSim," http://www.model.com/.
[36] C. Hilton and B. Nelson, "PNoC: a flexible circuit-switched NoC for FPGA-based systems," *IEE Proceedings of Computers and Digital Techniques*, vol. 153, pp. 181-188, 2006.
[37] Xilinx, "Virtex-II Pro FPGAs," http://www.xilinx.com/products/silicon_solutions/fpgas/virtex/virtex_ii_pro_fpgas/index.htm.
[38] Altera Corporation, "Cyclone II FPGAs," http://www.altera.com/products/devices/cyclone2/cy2-index.jsp.
[39] Altera Corporation, "Quartus II Web Edition Software," http://www.altera.com/products/software/producots/quartus2web/sof-quarwebmain.html.

Transfer Request Broker: Resolving Input-Output Choice

Oliver FAUST, Bernhard H. C. SPUTH and Alastair R. ALLEN

Department of Engineering, University of Aberdeen, Aberdeen AB24 3UE, UK

{o.faust, b.sputh, a.allen}@abdn.ac.uk

Abstract. The refinement of a theoretical model which includes external choice over output and input of a channel transaction into an implementation model is a long-standing problem. In the theory of communicating sequential processes this type of external choice translates to resolving input and output guards. The problem arises from the fact that most implementation models incorporate only input guard resolution, known as alternation choice. In this paper we present the transaction request broker process which allows the designer to achieve external choice over channel ends by using only alternation. The resolution of input and output guards is refined into the resolution of input guards only. To support this statement we created two models. The first model requires resolving input and output guards to achieve the desired functionality. The second model incorporates the transaction request broker to achieve the same functionality by resolving only input guards. We use automated model checking to prove that both models are trace equivalent. The transfer request broker is a single entity which resolves the communication between multiple transmitter and receiver processes.

Keywords. guard resolution, hardware process networks, CSP, relation matrix.

Introduction

Communicating processes offer a natural and scalable architecture for distributed systems, because they make it possible to design networks within networks and the dependencies within these networks are always explicit. The process algebra CSP provides a formal basis to describe the processes and the communication between them. CSP semantics are powerful enough to allow the analysis of systems in order that failures such as non-determinism, deadlock and livelock can be detected. Within the CSP algebra external choice over input and output of a channel transaction are an important abstraction for the design of distributed systems. These constructs allow natural specification of many programs, as in programs with subroutines that contain call-by-result parameters. Furthermore, they ensure that the externally visible effect and behaviour of every parallel command could be modeled by some sequential command [1,2].

However, problems arise during the transition from theoretical CSP models to practical implementations. To ease this transition, Hoare [1] introduced the concept of channels. Channels are used for communication in only one direction and between only two processes. The combination of message and channel constitutes the output guard for the sender and the input guard for the receiver. This combination makes up a CSP event that constitutes the theoretical concept of a guard. External choice over input and output of a channel transaction describes the situation when sender processes offer sets of output guards and receiver processes offer sets of input guards for the environment to choose from. This situation is explicitly allowed by the CSP process algebra and indeed necessary to ensure that every parallel command can be modelled by some sequential command [3]. Very early in the history of CSP Bernstein [4] pointed out the practical difficulties in the resolution of directed guards. In general external choice is possible between both channel inputs and outputs, but in this paper we consider only external choice of channel inputs for the implementation model. This concession is necessary to keep the solution simple enough so that it can be implemented in hardware logic.

As stated in the previous paragraph, to resolve output and input guards is an old problem. There are many proposed solutions for sequential or semi-parallel processing machines. For example, Buckley and Silberschatz [2] claim that they did an effective implementation for the generalized input-output construct of CSP. However, programming languages (such as occam-π [5]) and libraries (CTJ [6], JCSP [7,8], libCSP [9], C++CSP [10], etc), offering CSP primitives and operators do not incorporate these solutions. The main reason for this absence is the high computational overhead which results from the complexity of the algorithm used to resolve input and output guards [11]. To avoid this overhead, these programming environments provide synchronous message passing mechanisms to conduct both communication and synchronisation between two processes. Under this single phase commit mechanism, a sender explicitly performs an output of a message on a channel while the corresponding receiver inputs the message on the same channel. Such a communication will be delayed until both participants are ready. Choice is only possible over input channels; an outputting process always commits. In other words, only input guards can be resolved.

More recent attempts to solve choice problems involve an external entity. Welch et al. propose an external entity (oracle) which resolves choice between arbitrary multiway synchronisation events [12,13]. The advantage is that this method can be implemented with the binary handshake mechanism. The method was proposed for machine architectures in the form of an extension to programming languages and libraries. These extensions allow the implementation of sophisticated synchronisation concepts, such as the alting barrier [14]. However, the oracle uses lists to resolve the choice between arbitrary multiway sync events. These lists are not practical in hardware implementations, because in hardware process networks there is no dynamic allocation of memory. Similarly, Parrow and Sjödin propose a solution to the problem of implementing multiway synchronisation in a distributed environment [15]. Their solution involves a central synchroniser which resolves the choice.

This paper presents a solution for the choice resolution problem, suitable for implementation in hardware logic, that only supports input choice resolution. We introduce the Transfer Request Broker (TRB), an external entity which resolves output and input guards. The system relies entirely on a single phase commit mechanism. Therefore, no fundamental changes are required on the communication protocol. This solution is particularly sound for process networks implemented in hardware logic. These hardware processes are never descheduled and they never fail for reasons other than being wrongly programmed. Furthermore, all hardware processes are executed in parallel, therefore the TRB concept does not introduce delay caused by scheduling.

We approach the discussion in classical CSP style. Section 1.1 specifies the problem and Section 1.2 introduces a specification model. This model is a process network which incorporates the resolution of input and output guards. Section 1.3 discusses the choice resolution problem which results from the specification model with input and output guards. We provide evidence that input and output choice is not possible with a single phase commit mechanism. Therefore, the next step is to refine the specification into an implementation model which uses only input choice resolution. This refinement makes the implementation model less expressive but easier to implement. From a theoretical perspective, we have to increase the model complexity to compensate for this lack of expression. Section 1.4 details the implementation model which incorporates an additional TRB process. On the down side, the TRB process makes the implementation model more complex. On the upside, only choice resolution of

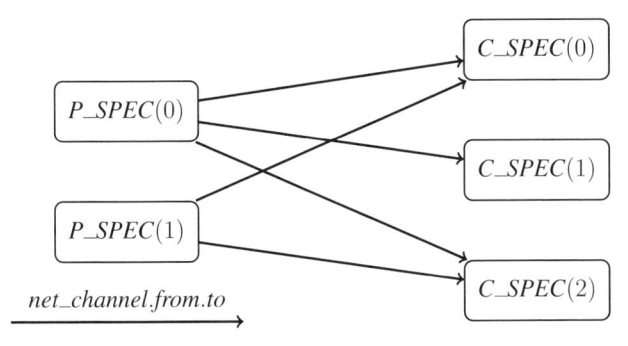

Figure 1. SPECIFICATION process network

input guards is required and this resolution can be achieved with the single phase commit mechanism. Having specification and implementation models leads us naturally to refinement checks. In Section 2 we prove that specification and implementation are trace equivalent. However, we could not establish the same equivalence for failures and divergences. This point is addressed in the discussion in Section 3. Also in the discussion section we introduce a practical example which shows the benefits of having the ability to allow external choice over input and output of a channel transaction in an implementation model. This also shows the practicality of the TRB concept. The paper finishes with concluding remarks.

1. Transfer Request Broker

The idea behind the TRB concept is to have a mediator or broker, which matches multiple inputs against multiple outputs, resolving those choices in a way that guarantees progress in the overall system where possible. The big advantage of this concept arises from the fact that all entities or nodes involved in the system need only comply with a simple binary handshake mechanism.

1.1. Problem Specification

We illustrate the problem by specifying an example process network which involves the resolution of input and output guards. The network incorporates receiver and transmitter processes which communicate over channels. External choice ensures that the environment can choose to transfer a message from one of multiple transmitters to one of multiple receivers.

In the specification environment a sender process can transfer messages over one of multiple output channels. Similarly, a receiver process is able to receive messages from one of multiple input channels. The *SPECIFICATION* process network, shown in Figure 1, depicts such a scenario. The process network contains two producer processes, *P_SPEC*(0) and *P_SPEC*(1). These producer processes generate messages and offer the environment the choice over which channels these are transferred to consumer processes. The *SPECIFICATION* process network contains three consumer processes, *C_SPEC*(0), *C_SPEC*(1) and *C_SPEC*(2). Figure 1 shows five *net_channel*s. These channels transfer messages from producer to consumer. The following list describes the network setup:

1. *net_channel*.0.0 connects *P_SPEC*(0) to *C_SPEC*(0);
2. *net_channel*.0.1 connects *P_SPEC*(0) to *C_SPEC*(1);
3. *net_channel*.0.2 connects *P_SPEC*(0) to *C_SPEC*(2);
4. *net_channel*.1.0 connects *P_SPEC*(1) to *C_SPEC*(0);
5. *net_channel*.1.2 connects *P_SPEC*(1) to *C_SPEC*(2).

Mathematically, the *net_channels* establish a relation between producers and consumers. A relation exists when a channel between producer and consumer exists. We can express such a relation as a matrix where a '1' indicates a relation exists and a '0' indicates that no relation exists. We construct a matrix which relates producers to consumers in the following way: the 0s and 1s in a particular matrix row describe the connection of one producer. Similarly, the entries in a column describe the connection of a particular consumer. The network, shown in Figure 1, contains 2 producers and 3 consumers, therefore the relation matrix is 2 by 3. We define: row 1 contains the connections of producer $P\_SPEC(0)$ and row 2 contains the connections of producer $P\_SPEC(1)$. Furthermore, the connections of consumer $C\_SPEC(0)$ are described in column 1. Similarly, the connections of $C\_SPEC(1)$ and $C\_SPEC(2)$ are described in columns 2 and 3 respectively. These definitions result in the following relation matrix:

$$\begin{array}{c|ccc} & C\_SPEC(0) & C\_SPEC(1) & C\_SPEC(2) \\ \hline P\_SPEC(0) & 1 & 1 & 1 \\ P\_SPEC(1) & 1 & 0 & 1 \end{array} \quad (1)$$

This relationship matrix provides the key to an elegant solution of the output and input guard resolution problem. But, before we attempt to explain the problem and solution in greater detail, the following section provides a formal model of the functionality we want to achieve.

1.2. Specification Model

A specification model is entirely focused on functionality. Therefore, we use the full set of abstract constructs provided by the CSP theory. To be specific, we use external choice for reading and writing. This leads to a very compact specification model which describes the desired functionality. This model serves two purposes. First, it allows the designer to focus on functionality. Second, the model provides a specification with which an implementation model can be compared.

The CSP model starts with the definition of the relationship matrix. This matrix describes the complete network, therefore it is called *net*. Equation 2 states the relation matrix for the network shown in Figure 1.

$$net = \begin{bmatrix} 1 & 1 & 1 \\ 1 & 0 & 1 \end{bmatrix} \quad (2)$$

The *net* matrix is the only parameter in the model, therefore all other constants are derived from this matrix. Equation 3 defines *data* as the set of messages which can be transferred over the *net_channels*. To keep the complexity of the model low, the set contains only two messages $(0, 1)$.

$$data = \{0..1\} \quad (3)$$

The following two equations define the constants n as the number of rows and m as the number of columns of the *net* matrix respectively.

$$n = \dim\_n(net) \quad (4)$$

$$m = \dim\_m(net) \quad (5)$$

where the function $\dim\_n(A)$ extracts the number of rows from a matrix A and $\dim\_m(A)$ extracts the number of columns of a matrix A.

Next, we interpret the process network, shown in Figure 1, in a process-centric way. Each producer is connected to a set of consumers. This consumer set is entirely governed by the indexes of the consumers, therefore it is sufficient to state only the consumer IDs in the index set. The following function extracts the connection set of a particular producer process $P(i)$.

$$p\_set(i) = \text{locate}(\text{get\_n}(i, net), 1) \tag{6}$$

where the function get_n(i, A) extracts the row vector i from matrix A and the function locate(vec, val) returns a set with the positions of val in the vector vec.

The following function extracts the connection set of a particular consumer process $C(j)$.

$$c\_set(j) = \text{locate}(\text{get\_m}(j, net), 1) \tag{7}$$

where the function get_m(j, A) extracts the column vector j for the matrix A.

After having defined all necessary constants and helper sets, we start with the process definitions. The first process to be defined is $P\_SPEC(i)$, where i is the process index. The process reads a message x from the input channel $in.i$. After that, $P\_SPEC(i)$ is willing to send the message over one of the $net\_channels$ which connects the producer process to a consumer process. After the choice is resolved and the message is sent, the process recurses.

$$P\_SPEC(i) = in.i?x \rightarrow \underbrace{\Box_{j \in p\_set(i)} net\_channel.i.j!x}_{\text{output guards}} \rightarrow P\_SPEC(i) \tag{8}$$

where $\Box_{j \in p\_set(i)} net\_channel.i.j!x$ indicates an indexed external choice over all connected $net\_channels$. The channel-messages which are sent out constitute the output guards for particular P_SPEC processes.

The $C\_SPEC(j)$ process waits for a message from one of its connected $net\_channels$. After having received a message x the process is willing to send it on via the output channel $out.j$ before it recurses.

$$C\_SPEC(j) = \underbrace{\Box_{i \in c\_set(j)} net\_channel.i.j?x}_{\text{input guards}} \rightarrow out.j!x \rightarrow C\_SPEC(j) \tag{9}$$

The channel-messages which are sent out constitute the input guards for particular C_SPEC processes.

The specification system composition follows the process network diagram shown in Figure 1. We start by observing that the producers, in this case $P\_SPEC(0)$ and $P\_SPEC(1)$, do not share a channel. That means they can make progress independently from one another. In Equation 10 we use the interleave operator '|||' to model this independence. Similarly, the consumers exchange no messages. Therefore, Equation 11 combines all consumer processes $C\_SPEC(j)$ into the $CONSUMER\_SPEC$ process with the interleave operator.

$$PRODUCER\_SPEC = |||_{i \in \{0..n-1\}} P\_SPEC(i) \tag{10}$$

$$CONSUMER\_SPEC = |||_{j \in \{0..m-1\}} C\_SPEC(j) \tag{11}$$

where $|||_{i \in \{0..1\}} P\_SPEC(i)$ represents the indexed interleave operator which expresses: $P\_SPEC(0) ||| P\_SPEC(1)$.

The $SPECIFICATION$ process combines $CONSUMER\_SPEC$ and $PRODUCER\_SPEC$. In this case the processes which are combined depend on one another to make progress. To be specific, the $PRODUCER\_SPEC$ process sends messages to the $CONSUMER\_SPEC$ process via the $net\_channels$. We say that $PRODUCER\_SPEC$ and $CONSUMER\_SPEC$ execute in parallel, however they have to agree upon all messages which can be transferred via the $net\_channels$. Equation 12 models this behaviour by combining $PRODUCER\_SPEC$ and $CONSUMER\_SPEC$ via the alphabetised parallel operator.

$$SPECIFICATION = CONSUMER\_SPEC \underset{\{|net\_channel|\}}{\|} PRODUCER\_SPEC \quad (12)$$

where $\|_{\{|net\_channel|\}}$ is the alphabetised parallel operator. The expression $\{|\ net\_channel\ |\}$ indicates the set of all events which can be transferred over *net_channel*, i.e.:

$$\{|\ net\_channel\ |\} = \{net\_channel.i.j.x \mid i \in \{0..1\}, j \in \{0..2\}, x \in \{0..1\} \wedge \neg((i=1) \wedge (j=1))\}$$

The definition of the *SPECIFICATION* process concludes our work on the functionality model. We have now a model against which we can measure any implementation model.

1.3. Transfer Request Broker Model

In the specification model external choice allows the environment to choose over which channel $P\_SPEC(i)$ outputs a message and the same operator enables the environment to choose over which channel $C\_SPEC(i)$ inputs a message. The fundamental problem for an implementation system is: Who resolves this choice? Neither producers (P_SPEC) nor consumers (C_SPEC) are able to perform this task with a single phase commit algorithm. To support this point, we analyse three different transfer scenarios which could occur in the process network shown in Figure 1.

First, we assume that the choice is resolved by the producer and the consumer engages in any one transfer which is offered. We dismiss this approach with the following counter example. The producer process $P\_SPEC(0)$ has resolved that it wants to send a message to consumer process $C\_SPEC(2)$. At the same time the producer process $P\_SPEC(1)$ has resolved that it wants to send a message to the same consumer process $C\_SPEC(2)$. Now, according to our assumption $C\_SPEC(2)$ is not able to resolve the choice between the messages offered by $P\_SPEC(0)$ and $P\_SPEC(1)$, at the same time it is impossible to accept both messages. Clearly, this counter example breaks the approach that the producers resolve the choice.

The second approach is that the consumer resolves the choice and the producer reacts to the choice. In this case, the counter example is constructed as follows: $C\_SPEC(0)$ and $C\_SPEC(1)$ have resolved to receive a message from $P\_SPEC(1)$. Now, $P\_SPEC(1)$ faces a similar dilemma as $C\_SPEC(2)$ in the previous example. According to the assumption that the consumer resolves the choice, $P\_SPEC(1)$ is not allowed to choose and it is also not allowed to service both requests.

The last scenario gives both consumer and producer the ability to resolve the choice. We break this scheme with yet another counter example. $P\_SPEC(0)$ tries to send to $C\_SPEC(0)$. However, $C\_SPEC(0)$ blocks this message, because it has resolved the input choice such that it wants to receive a message from $P\_SPEC(1)$. In this example, $P\_SPEC(1)$ is unwilling to send a message to $C\_SPEC(0)$, because it has resolved to output a message to $C\_SPEC(2)$. Unfortunately, $C\_SPEC(2)$ has decided to wait for a message from $P\_SPEC(0)$. However, this message will never come, because $P\_SPEC(0)$ tries to send to $C\_SPEC(0)$. *A classical deadlock*.

This small discussion shows that there is no *simple* solution to the choice resolution problem. The fundamental problem is that individual processes are not aware of what happens around them in the network. In other words, a process with an overview is missing. These considerations lead to the proposal of a *TRB* process. This *TRB* process is aware of the network state. This additional process allows us to use the simple receiver choice resolution method. That means no process in the network incorporates external choice for writing.

For example, the *SPECIFICATION* process network can be in one of up to $243 (= 3^5)$ different states (5 fairly independent processes, each with 3 states). To achieve the awareness of the network state the *TRB* process has to know the state of the individual producer and

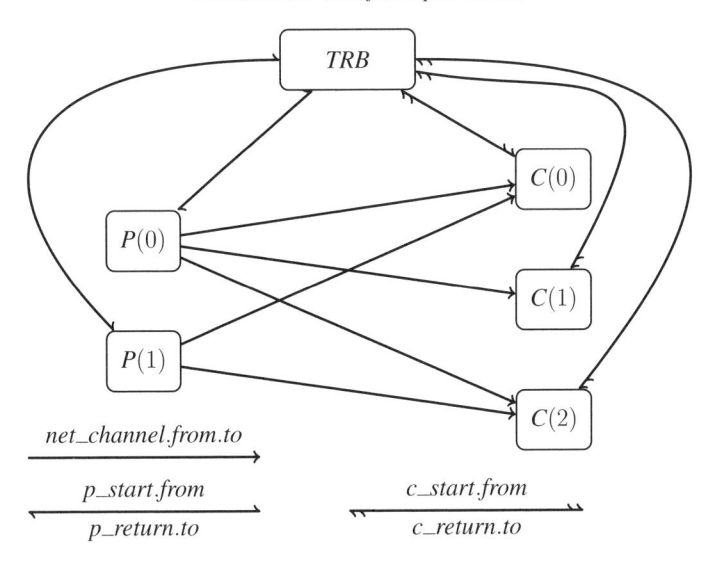

Figure 2. IMPLEMENTATION process network

consumer processes. In the following text the group of consumers and producers are referred to as clients of the *TRB*. The clients transfer their state to the *TRB* via *p_start.from* and *c_start.from* channels. The *TRB* communicates its decision via *p_return.to* and *c_return.to* channels.

Figure 2 shows the *IMPLEMENTATION* process network. Each client process is connected to the *TRB* via a *start.from* and a *return.to* channel. The message passing channels *net_channel.from.to* connect producer processes with consumer processes. The relation matrix of Equation 1 describes the network setup.

1.4. CSP Implementation Model

Similar to the definition of the specification model, we adopt a bottom up design approach for the definition of the CSP implementation model. First, we define the individual processes, shown in Figure 2, then we assemble the *IMPLEMENTATION* process network. The model uses the network matrix, defined in Equation 2, and the constants defined in Equations 3 to 5.

The first process to be defined is the producer process $P(i)$. Initially, this process waits for a message from the input channel *in.i*. After having received a message x the process registers with the *TRB*. The registration is done by sending a vector of dimension m to the *TRB* via the *p_start.i* channel. The entries in this vector are either 0 or 1. A 1 indicates that the producer process wants to send a message. The position of the 1 within the vector indicates the consumer address. The *TRB* resolves the request and returns the index of the selected consumer via a message over the *p_return.i* channel. After having received the index (*select*) of the consumer process the $P(i)$ process is ready to send out the message x over the channel *p_return.i?select* before it recurses. The following equation defines the producer process $P(i)$.

$$P(i) = in.i?x \rightarrow p\_start.i!\text{get\_n}(i, net) \\ \rightarrow p\_return.i?select \rightarrow net\_channel.i.select!x \rightarrow P(i) \quad (13)$$

where $\text{get\_n}(i, A)$ returns the ith row of matrix A.

Initially, the consumer process $C(j)$ is ready to register with the *TRB*. In this case registration means sending the jth column vector of the network matrix *net*. This indicates that

$C(j)$ is willing to receive a message from any one of the connected input channels. The *TRB* resolves this request and sends the result back to the $C(j)$ process via the *c_return.j* channel. The result is the index (*select*) of the producer which is selected to send a message. Now, the consumer process waits on the particular *net_channel.select.j* channel. After having received a message from this channel the consumer sends out the message over the *out* channel before it recurses. The following equation defines the consumer process $C(j)$.

$$C(j) = c\_start.j!\text{get\_m}(j, net) \\ \rightarrow c\_return.j?select \rightarrow net\_channel.select.j?x \rightarrow out.j!x \rightarrow C(j) \qquad (14)$$

The last and most complex process defines the TRB functionality. Indeed this task is so complex that it is broken into 5 processes. The first of these processes is $TRB(p\_array, c\_array)$. The *p_array* parameter is an $n \times m$ matrix which holds the information about the registered producers. Similarly, the *c_array* parameter is an $n \times m$ matrix which holds the information about the registered consumers. Initially, the *TRB* process is ready to receive the registration information from any one of its clients. This is modelled as external choice over two indexed external choices. The first indexed external choice chooses between *p_start* messages and the second indexed external choice chooses between *c_start* messages. If the complete choice construct is resolved and a message is transferred via the *p_start* channel, then *p_array* is updated with the registration information before the process behaves like $EVAL(p\_array, c\_array, 0, 0)$. To be specific, *p_array* is updated by replacing the *i*th row, where *i* is the index of the producer which registered, with the registration vector *p_vector*. As a matter of fact, *p_vector* and the channel index represent the registration information. Similarly, if the external choice over two indexed external choices is resolved in favour of a *c_start* message then *c_array* is updated with the registration information before the process behaves like $EVAL(p\_array, c\_array, 0, 0)$.

$$TRB(p\_array, c\_array) = \\ \begin{pmatrix} \Box_{i \in \{0..n-1\}} p\_start.i?p\_vector \rightarrow EVAL(\text{set\_n}(i, p\_vector, p\_array), c\_array, 0, 0) \\ \Box \\ \Box_{j \in \{0..m-1\}} c\_start.j?c\_vector \rightarrow EVAL(p\_array, \text{set\_m}(j, c\_vector, c\_array), 0, 0) \end{pmatrix} \qquad (15)$$

where the function $\text{set\_n}(i, vec, A)$ replaces the *i*th row vector of matrix A with *vec* and the function $\text{set\_m}(j, vec, A)$ replaces the *j*th column vector of matrix A with *vec*.

The $EVAL(p\_array, c\_array, i, j)$ process compares the entries in *p_array* with the entries in *c_array*. The parameters *i* and *j* describe row and column of the entry to be compared. If both *p_array* and *c_array* have a 1 at position *i, j* then the process resolves the choice such that producer $P(i)$ has to send a message to consumer $C(j)$. We model this case with the process $RETURN\_P(p\_array, c\_array, i, j)$. In all other cases the process continues to compare entries at other matrix positions. This is done by checking whether or not *j* defines the position of the last entry in a row, i.e. $j = m - 1$. If this is the case then the process behaves like $TRB'(p\_array, c\_array, i, j)$, in all other circumstances the column index is incremented ($j = j + 1$) before the process recurses.

$$EVAL(p\_array, c\_array, i, j) = \\ \begin{pmatrix} \textbf{if } \text{get\_nm}(i, j, c\_array) = 1 \textbf{ and } \text{get\_nm}(i, j, p\_array) = 1 \textbf{ then} \\ \quad RETURN\_P(p\_array, c\_array, i, j) \\ \textbf{else if } j = m - 1 \textbf{ then} \\ \quad TRB'(p\_array, c\_array, i, j) \\ \textbf{else} \\ \quad EVAL(p\_array, c\_array, i, j+1) \end{pmatrix} \qquad (16)$$

where the function get_nm(i, j, A) returns the entry $a_{i,j}$ of matrix A.

The *RETURN_P*(p_array, c_array, i, j) sends out the message j, which is the resolution result, to the producer $P(i)$ via the channel *p_return.i*. After the message is transferred the process behaves like *RETURN_C*(p_array, c_array, i, j). In effect, clearing all 1s in row i indicates that $P(i)$ is unable to send out any more messages in this resolution round.

$$\begin{aligned} RETURN\_P(p\_array, c\_array, i, j) = \\ p\_return.i!j \rightarrow RETURN\_C(\text{set\_n}(i, \text{vzeros}(m), p\_array), c\_array, i, j) \end{aligned} \quad (17)$$

where the function vzeros(m) generates an m dimensional zero vector.

The *RETURN_C*(p_array, c_array, i, j) sends out the message i, which is the resolution result, to the consumer $C(j)$ via the channel *c_return.i*. After the message is transferred the process clears all 1s in column j of *c_array* and behaves like *TRB'*(p_array, c_array, i, j). In effect, clearing all 1s in column j indicates that $C(j)$ is unable to receive any further messages in this resolution round.

$$\begin{aligned} RETURN\_C(p\_array, c\_array, i, j) = \\ c\_return.j!i \rightarrow TRB'(p\_array, \text{set\_m}(j, \text{vzeros}(n), c\_array), i, j) \end{aligned} \quad (18)$$

The *TRB'*(p_array, c_array, i, j) detects whether or not the entries in the last row were checked. If this is the case, then the process behaves like *TRB*(p_array, c_array). In all other circumstances the row index i is incremented and the column index j is set to 0 before the process recurses to *EVAL*$(...)$ for another round of entry checking.

$$TRB'(p\_array, c\_array, i, j) = \begin{pmatrix} \textbf{if } i = n-1 \textbf{ then} \\ \quad TRB(p\_array, c\_array) \\ \textbf{else} \\ \quad EVAL(p\_array, c\_array, i+1, 0) \end{pmatrix} \quad (19)$$

The *TRB* definition concludes the description of the individual processes in the *IMPLEMENTATION* process network. In the next step we connect the individual processes such that they form the process network, as shown in Figure 2. From this figure we observe that the *TRB* process and the additional channels do not alter the fact that the producers do not communicate among themselves. Therefore, they form an independent group of processes. This group is modelled as the *PRODUCER* process in Equation 20. Similarly, the consumers form a group of independent processes. This group is established as the *CONSUMER* process in Equation 21.

$$PRODUCER = \big|\big|\big|_{i \in \{0..n-1\}} P(i) \quad (20)$$

$$CONSUMER = \big|\big|\big|_{j \in \{0..m-1\}} C(j) \quad (21)$$

The parallel combination of *PRODUCER* and *CONSUMER* forms the *NETWORK* process. The two processes, which engage in the parallel construct, must agree on all messages sent over the *net_channels*. This is modelled in the following equation:

$$NETWORK = PRODUCER \underset{\{|net\_channel|\}}{\big\|} CONSUMER \quad (22)$$

Now, we are able to define the *IMP* process as the parallel combination of *NETWORK* and *TRB* processes. Equation 23 shows that the *TRB* process is initialised with two zero matrices, i.e. both *p_array* and *c_array* are initialised with $n \times m$ zero matrices.

$$IMP = NETWORK \underset{\{|p\_start,c\_start,p\_return,c\_return|\}}{\|} TRB(\text{zeros}(n,m), \text{zeros}(n,m)) \qquad (23)$$

where the function zeros(n, m) returns an $n \times m$ zero matrix. The alphabetised parallel operator ensures that all clients can register with and receive a solution from the *TRB* process.

$$IMPLEMENTATION = IMP \setminus \{| \; p\_start, c\_start, p\_return, c\_return \; |\} \qquad (24)$$

where $P \setminus \{a\}$ is the hiding operation which makes all events (messages) a internal to P. Hiding the communication to and from the *TRB* is necessary for model checking.

For the sake of clarity this section introduced particular scenarios, i.e. network setups to explain the TRB concept. These particular network setups do not prove that the concept works with other network configurations. To build up trust, the algorithm was tested with other network setups. The tests are described in the following section. These tests were successful, therefore confidence is growing that the algorithm works for arbitrary network configurations which can be described by a relation matrix.

2. Automated Refinement Checks

We use the FDR tool [16] to establish that both *SPECIFICATION* and *IMPLEMENTATION* do not contain any pathological behaviours. This is done in the following section. Furthermore, we use FDR to compare *SPECIFICATION* and *IMPLEMENTATION*. This is done in Section 2.2.

2.1. Basic Checks: Deadlock, Divergence and Non-determinism

Parallel and concurrent systems can exhibit pathological behaviours such as deadlocks and livelocks. These problems arise from the fact that in such systems independent entities communicate. To be specific, a deadlock occurs if two or more independent entities prevent each other from making progress. A livelock occurs when a system can make indefinite progress without engaging with the outside world. CSP models the independent entities as processes which communicate over channels. Equation 25 instructs the automated model checker FDR to verify that the *SPECIFICATION* process is deadlock and divergence (livelock) free.

$$\begin{array}{l} \textbf{assert } SPECIFICATION :[\text{ deadlock free [F] }] \\ \textbf{assert } SPECIFICATION :[\text{ divergence free }] \end{array} \qquad (25)$$

In the CSP community it is custom to publish the output of the model checker to support the claim that a particular test was successful. Figure 3 shows the output of the model checker. A ✓ indicates that this particular test was successful.

In Equation 26 we instruct FDR to verify that the *IMPLEMENTATION* process is deadlock and divergence free.

$$\begin{array}{l} \textbf{assert } IMPLEMENTATION :[\text{ deadlock free [F] }] \\ \textbf{assert } IMPLEMENTATION :[\text{ divergence free }] \end{array} \qquad (26)$$

Figure 3 shows that both tests were successful.

As a final test on individual processes we establish whether or not a process is deterministic. A process is deterministic if it reacts in the same way to the same input. In the following equation we instruct FDR to verify that both *SPECIFICATION* and *IMPLEMENTATION* processes are deterministic.

$$\textbf{assert } SPECIFICATION :[\text{ deterministic [FD] }] \tag{27}$$
$$\textbf{assert } IMPLEMENTATION :[\text{ deterministic [FD] }]$$

Figure 3 shows that the *IMPLEMENTATION* process is not deterministic. Hiding the communication to and from the *TRB* process causes this behaviour. To support this statement, in Equation 28 we test whether or not the *IMP* process is deterministic. The *IMP* process is the *IMPLEMENTATION* with no hidden communication.

$$\textbf{assert } IMP :[\text{ deterministic [FD] }] \tag{28}$$

Figure 3 shows that *IMP* is deterministic.

2.2. Trace Refinement

The refinement operation relates two processes. There are different types of refinement, such as trace and failure refinement. In this section we consider only trace refinement. Trace refinement tests the safety of a particular process. A trace is a sequence of events observed by the outside world (outside of the process itself). A system is safe if and only if the *IMPLEMENTATION* can only exhibit a subset of the traces from the *SECIFICATION*. In other words, the *IMLEMENTATION* refines the *SPECIFICATION*.

In this particular case we establish that the *IMPLEMENTATION* process can exhibit the same traces as the *SPECIFICATION* process. Therefore, the *IMPLEMENTATION* is also safe. With Equation 29, we test whether or not the *SPECIFICATION* process is refined by the *IMPLEMENTATION* process.

$$\textbf{assert } SPECIFICATION \sqsubseteq_T IMPLEMENTATION \tag{29}$$

where $\sqsubseteq_T$ is the trace refinement operator.

Next, we establish that the *SPECIFICATION* process exhibits a subset of all *IMPLEMENTATION* traces.

$$\textbf{assert } IMPLEMENTATION \sqsubseteq_T SPECIFICATION \tag{30}$$

Figure 3 shows that both refinement tests were successful. That means, the *IMPLEMENTATION* process is able to exhibit a subset of the *SPECIFICATION* traces and the *SPECIFICATION* process exhibits a subset of the *IMPLEMENTATION* traces. This implies that both processes exhibit exactly the same traces. This is a very important result, because it establishes that the *SPECIFICATION* is trace equivalent with the *IMPLEMENTATION*. The *SPECIFICATION* process models the desired functionality, therefore we have established that the implementation model, i.e. the *IMPLEMENTATION* process complies with the specification on the level of traces.

3. Discussion

This section discusses three issues raised in the main body of the paper. First, we examine the theoretical concept of failure refinement and its application to the TRB model. We highlight practical difficulties and point out possible solutions. The second issue under discussion is the practical application of the TRB concept. Even though input and output choice resolution appears to be such a natural concept, there are not many text book applications. One reason for this absence is that in the past such systems were difficult to implement and therefore text book authors focused on structures which did not include such constructs. For example, many text books discuss structures such as: farmer-worker and client-server. The final issue is the relevance of the TRB concept to hardware logic implementation.

```
✓  SPECIFICATION deadlock free [F]
✓  SPECIFICATION livelock free
✓  IMPLEMENTATION deadlock free [F]
✓  IMPLEMENTATION livelock free
✓  SPECIFICATION deterministic [FD]
✗• IMPLEMENTATION deterministic [FD]
✓  IMP deterministic [FD]
✓  SPECIFICATION [T= IMPLEMENTATION
✓  IMPLEMENTATION [T= SPECIFICATION
```

Figure 3. Output of the FDR model checker.

3.1. Failure Refinement

Failure (deadlock or/and livelock) refinement ensures that a particular process exhibits only a subset of the allowed traces and a subset of allowed stable failures associated with a particular trace. Section 1.1 defines the *SPECIFICATION* process and Section 1.4 defines the *IMPLEMENTATION* or implementation process. The *IMPLEMENTATION* process does not refine the *SPECIFICATION* in terms of stable failures. In this case, *IMPLEMENTATION* exhibits more stable failures then *SPECIFICATION*. The reason for this failure to refine is the sequential nature of the resolution operation in the *TRB* process. Sequential nature means that the *TRB* establishes a fixed sequence in which the individual sender and receiver can communicate. This sequence is enforced by stepping through Equations 16, 17 and 18. In contrast the *SPECIFICATION* process is free from such prioritisation and therefore it has fewer stable failures.

One way to overcome this difficulty is to introduce prioritisation into the *SPECIFICATION*. The idea is to show that a specification with prioritised external choice can be refined into an implementation which incorporates the TRB concept and therefore requires only prioritised alternation (PRI ALT). However, this paper aims to introduce the TRB concept. Therefore, we limited the scope of the automated checks to trace refinement. Failure refinement would have shifted the scope away from this introduction, because prioritised alternation requires extensive introduction.

3.2. Application

As mentioned before, there are not many textbook examples of systems which require input and output choice resolution. One of the few examples is an N lift system in a building with F floors. This example was presented by Forman [17] to justify the introduction of multiparty interactions in Raddle87. The lift problem concerns the logic to move lifts between floors. We limit the discussion to the following constraints:

1. Each lift has a set of buttons, one button for each floor. These illuminate when pressed and cause the lift to visit the corresponding floor. The illumination is cancelled when the corresponding floor is visited (i.e. stopped at) by the lift.
2. Each floor has two buttons (except ground and top), one to request an up-lift and one to request a down-lift. These buttons illuminate when pressed. The buttons are cancelled when a lift visits the floor and is either travelling in the desired direction, or visiting the floor with no requests outstanding. In the latter case, if both floor request buttons are illuminated, only one should be cancelled. The algorithm used to decide which floor to service should minimise the waiting time for both requests.
3. When a lift has no requests to service, it should remain at its final destination with its doors closed and await further requests (or model a "holding" j floor).

4. All requests for lifts from floors must be serviced eventually, with all floors given equal priority.
5. All requests for floors within lifts must be serviced eventually, with floors being serviced sequentially in the direction of travel.

From these constraints it is clear that there is a need for an external entity which manages the requests from the various buttons. The TRB concept can be used to solve this problem. All we have to do is to reinterpret the *IMPLEMENTATION* process network shown in Figure 2. The producer processes *P* model the buttons of the lift system and the consumer *C* processes model the lifts. Now, whenever a lift is free it registers with the TRB and whenever a button is pressed it illuminates and registers with the TRB. The TRB matches the requests from the buttons with the available lifts. After the choice is resolved, the consumer (lift) knows where to go and the producer (button) knows where to send the message to. The message is only exchanged when the lift stops at the desired floor. After the message is transferred, the button illumination is cancelled.

The TRB functionality, defined in Section 1.4 is a very crude solution to this particular problem, because it does not incorporate any algorithms to minimise the waiting time. However, the *IMPLEMENTATION*, shown in Figure 2, is proven to be deadlock and divergence free and transmitter processes can send messages to one of multiple receiver processes. Furthermore, it offers the same traces as a specification system which resolves input and output guards.

3.3. On Hardware Logic Implementation

In the introduction of his article about output guards and nondeterminism in CSP Bernstein mentions that an external entity which resolves input and output guards is impractical [4]. His argument was basically that this external entity would create a serious bottleneck, because of all the context switches necessary to schedule all processes involved in the message exchange. Now, this is only true for systems which have a scheduler. Say, all machine architectures with a multitasking operating system have such a scheduler. However, there is no scheduler in hardware process networks. In such systems all processes are executed in parallel, there are no context switches necessary. From this perspective the TRB concept does not introduce a performance penalty.

Another big advantage of hardware logic implementation is the trade-off relationship between execution time and silicon area. The *TRB* process, defined in Equations 15 to 19, has a complexity of roughly $O(TRB) = n \times m$ where n is the number of producer processes and m is the number of consumer processes. However, in hardware logic systems this does not mean that the processing time increases with $n \times m$. Depending on the implementation, the processing time might stay constant and the silicon area increases with $n \times m$. Therefore, the processing time bottleneck which is introduced by the TRB concept depends on the particular implementation.

A last point which highlights the practicality of the *IMPPLEMENTATION* model of the TRB concept is the efficiency with which binary matrices and vectors can be implemented in hardware logic. All hardware description languages provide direct support for such constructs. Due to this direct support there is no resource waste. Say, the 3 dimensional binary vector, which is transferred over the *p_start* channels, is represented by only 3+2 signals. The two additional signals are necessary for the channel functionality. This efficiency keeps the routing overhead manageable.

4. Conclusion

In this paper we present a solution to the long-standing problem of refining a model with input and output guards into a model which contains only input guards. The problem has practical relevance, because most implementation models incorporate only the resolution of input guards. We approach the solution in classic CSP style by defining a specification model which requires the resolution of input and output guards. This specification model provides the opportunity to test the functionality of the implementation model. The implementation model resolves only input guards. In other words the implementation model utilises only alternation. We achieved this by incorporating the transfer request broker process. The TRB concept is an elegant solution to the arbitration problem. The concept comes from fusing graph theory with CSP and introducing matrix manipulation functionality to CSP_M. The combination of these three points makes the TRB concept unique.

Automated model checking establishes that specification as well as implementation models are deadlock and divergence free. Furthermore, we prove that specification and implementation are trace equivalent. This is a strong statement in terms of safety, i.e. if the specification exhibits only safe traces then the implementation exhibits also only safe traces. However, trace equivalence is a weak statement about the functionality of the implementation, because it implies that the implementation is able to exhibit the same traces as the specification. The phrase 'is able to' indicates that the implementation can choose not to exhibit a particular event, even if it is within its traces. In other words, the implementation may have more stable failures then the specification. To resolve this problem may require prioritised external choice in the specification model, but these considerations would shift the focus away from the introduction of the TRB concept, therefore we leave this for further work.

The TRB concept allows us to extend the set of standard implementation models with systems which require the resolution of input and output guards. One of these examples is the N lift system. The TRB is used to mediate or broker between the requests from various buttons in the system and the individual lifts. Apart from introducing additional standard problems, the TRB concept eases also the refinement of specification into implementation models. Many specification models incorporate the resolution of output and input guards. With the TRB concept such models can be refined into implementation models which have to resolve only input guards without loss of functionality. These implementation models can be easily mapped into communicating sequential hardware processes executed by flexible logic devices. Under the assumption that there was no error in the mapping operation, the resulting implementation has the same properties as the implementation model.

The implementation model is particularly relevant for process networks in hardware logic. All hardware processes are executed in parallel, this limits the performance penalty incurred by introducing the additional TRB process. This is a big difference to software processes where scheduling problems might make the TRB impracticable. One last thought on system speed. The most expensive operation in the TRB process is the matrix checking. This is a logic operation therefore it can be done with a registered logic circuit which takes only one clock cycle to do the matrix checking. However, the area requirement for the logic circuit is of the order $O(n \times m)$ where n is the number of senders and m is the number of receivers. This makes the TRB concept very fast but not area efficient. But this constitutes no general problem for practical systems, because it is always possible to exchange processing speed and chip area. Therefore, the TRB concept benefits the design of many practical systems.

Acknowledgements

This work was supported by the European FP6 project "WARMER" (contract no.: FP6-034472).

References

[1] C. A. R. Hoare. *Communicating Sequential Processes*. Prentice Hall, Upper Saddle River, New Jersey 07485 United States of America, first edition, 1985. http://www.usingcsp.com/cspbook.pdf Accessed Feb. 2007.
[2] G. N. Buckley and Abraham Silberschatz. An Effective Implementation for the Generalized Input-Output Construct of CSP. *ACM Trans. Program. Lang. Syst.*, 5(2):223–235, 1983.
[3] C. A. R. Hoare. Communicating sequential processes. *Commun. ACM*, 21(8):666–677, 1978.
[4] Arthur Bernstein. Output Guards and Nondeterminism in "Communicating Sequential Processes". *ACM Trans. Program. Lang. Syst.*, 2(2):234–238, 1980.
[5] David C. Wood and Peter H. Welch. The Kent Retargettable occam Compiler. In Brian C. O'Neill, editor, *Proceedings of WoTUG-19: Parallel Processing Developments*, pages 143–166, feb 1996.
[6] Jan F. Broenink, Andrè W. P. Bakkers, and Gerald H. Hilderink. Communicating Threads for Java. In Barry M. Cook, editor, *Proceedings of WoTUG-22: Architectures, Languages and Techniques for Concurrent Systems*, pages 243–262, mar 1999.
[7] P H Welch. Communicating Processes, Components and Scaleable Systems. Communicating Sequential Processes for Java (JCSP) web page, http://www.cs.kent.ac.uk/projects/ofa/jcsp/. Accessed Feb. 2007.
[8] Peter H. Welch and Jeremy M. R. Martin. A csp model for java multithreading. In *PDSE '00: Proceedings of the International Symposium on Software Engineering for Parallel and Distributed Systems*, page 114, Washington, DC, USA, 2000. IEEE Computer Society.
[9] Rick D. Beton. libcsp - a Building mechanism for CSP Communication and Synchronisation in Multithreaded C Programs. In Peter H. Welch and Andrè W. P. Bakkers, editors, *Communicating Process Architectures 2000*, pages 239–250, sep 2000.
[10] Neil C. Brown and Peter H. Welch. An Introduction to the Kent C++CSP Library. In Jan F. Broenink and Gerald H. Hilderink, editors, *Communicating Process Architectures 2003*, pages 139–156, sep 2003.
[11] Geraint Jones. On guards. In Traian Muntean, editor, *OUG-7: Parallel Programming of Transputer Based Machines*, pages 15–24, sep 1987.
[12] Peter H. Welch. A Fast Resolution of Choice between Multiway Synchronisations. In Frederick R. M. Barnes, Jon M. Kerridge, and Peter H. Welch, editors, *Communicating Process Architectures 2006*, page 389, sep 2006.
[13] Peter H. Welch, Frederick R. M. Barnes, and Fiona A. C. Polack. Communicating complex systems. In *ICECCS '06: Proceedings of the 11th IEEE International Conference on Engineering of Complex Computer Systems*, pages 107–120, Washington, DC, USA, 2006. IEEE Computer Society.
[14] Peter H. Welch, Neil C. Brown, James Moores, Kevin Chalmers, and Bernhard Sputh. Integrating and Extending JCSP. In Alistair A. McEwan, Wilson Ifill, and Peter H. Welch, editors, *Communicating Process Architectures 2007*, pages 349–369, jul 2007.
[15] Parrow and Sjödin. Multiway Synchronization Verified with Coupled Simulation. In *CONCUR: 3rd International Conference on Concurrency Theory*. LNCS, Springer-Verlag, 1992.
[16] Formal Systems (Europe) Ltd., 26 Temple Street, Oxford OX4 1JS England. *Failures-Divergence Refinement: FDR Manual*, 1997.
[17] I. R. Forman. Design by decomposition of multiparty interactions in Raddle87. In *IWSSD '89: Proceedings of the 5th international workshop on Software specification and design*, pages 2–10, New York, NY, USA, 1989. ACM.

Mechanical Verification of a Two-Way Sliding Window Protocol

Bahareh BADBAN [a,*], Wan FOKKINK [b] and Jaco VAN DE POL [c]

[a] *University of Konstanz, Department of Computer and Information Science*
   `badban@inf.uni-konstanz.de`
[b] *Vrije Universiteit Amsterdam, Department of Computer Science*
   `wanf@cs.vu.nl`
[c] *University of Twente, Department of EEMCS, Formal Methods and Tools*
   `j.c.vandepol@ewi.utwente.nl`

> **Abstract.** We prove the correctness of a two-way sliding window protocol with piggybacking, where the acknowledgements of the latest received data are attached to the next data transmitted back into the channel. The window sizes of both parties are considered to be finite, though they can be different. We show that this protocol is equivalent (branching bisimilar) to a *pair of FIFO* queues of finite capacities. The protocol is first modeled and manually proved for its correctness in the process algebraic language of μCRL. We use the theorem prover PVS to formalize and mechanically prove the correctness of the protocol. This implies both safety and liveness (under the assumption of fairness).
>
> **Keywords.** μCRL, FIFO queues, PVS, sliding window protocols.

Introduction

A sliding window protocol [7] (SWP) ensures successful transmission of messages from a sender to a receiver through a medium in which messages may get lost. Its main characteristic is that the sender does not wait for incoming acknowledgments before sending next messages, for optimal use of bandwidth. Many data communication systems include a SWP, in one of its many variations.

In SWPs, both the sender and the receiver maintain a buffer. We consider a *two-way* SWP, in which both parties can both send and receive data elements from each other. One way of achieving full-duplex data transmission is to have two separate communication channels and use each one for simplex data traffic (in different directions). Then there are two separate physical circuits, each with a forward channel (for data) and a reverse channel (for acknowledgments). In both cases the bandwidth of the reverse channel is almost entirely wasted. In effect, the user is paying for two circuits but using the capacity of one. A better idea is to use the same circuit in both directions. Each party maintains two buffers, for storing the two opposite data streams. In this two-way version of the SWP, an acknowledgment that is sent from one party to the other may get a free ride by attaching it to a data element. This method for efficiently passing acknowledgments and data elements through a channel in the same direction, which is known as *piggybacking*, is used broadly in transmission control protocols,

---

[*] Corresponding Author: *Bahareh Badban, University of Konstanz, Department of Computer and Information Science, Universitätsstr. 10, P.O. Box D67, 78457 Konstanz, Germany.* Tel.: +49 (0)7531 88 4079; Fax: +49 (0)7531 88 3577; E-mail: `badban@inf.uni-konstanz.de`.

see [39]. The main advantage of piggybacking is a better use of available bandwidth. The extra acknowledgment field in the data frame costs only a few bits, whereas a separate acknowledgment would need a header and a checksum. In addition, fewer frames sent means fewer 'frame arrived' interrupts.

The current paper builds on a verification of a one-way version of the SWP in [1, 9]. The protocol is specified in μCRL [15], which is a language based on process algebra and abstract data types. The verification is formalized in the theorem prover PVS [29]. The correctness proof is based on the so-called *cones and foci* method [10, 18], which is a symbolic approach towards establishing a branching bisimulation relation. The starting point of the cones and foci method are two μCRL specifications, expressing the implementation and the desired external behavior of a system. A *state mapping* ϕ relates each state of the implementation to a state of the desired external behavior. Furthermore, the user must declare which states in the implementation are *focus points*, whereby each reachable state of the implementation should be able to get to a focus point by a sequence of hidden transitions, carrying the label τ. If a number of *matching criteria* are met, consisting of equations between data objects, then states s and $\phi(s)$ are branching bisimilar. Roughly, the matching criteria are: (1) if $s \xrightarrow{\tau} s'$ then $\phi(s) = \phi(s')$, (2) each transition $s \xrightarrow{a} s'$ with $a \neq \tau$ must be matched by a transition $\phi(s) \xrightarrow{a} \phi(s')$, and (3) if s is a focus point, then each transition of $\phi(s)$ must be matched by a transition of s.

The crux of the cones and foci method is that the matching criteria are formulated syntactically, in terms of relations between data terms. Thus, one obtains clear proof obligations, which can be verified with a theorem prover. The cones and foci method provides a general verification approach, which can be applied to a wide range of communication protocols and distributed algorithms.

The main motivations for the current research is to provide a mechanized correctness proof of the most complicated version of the SWP in [39], including the piggybacking mechanism. Here we model buffers (more realistically) as ordered lists, without multiple occurrences of the same index. Therefore two buffers are equal only if they are identical. That is, any swapping or repetition of elements results in a different buffer. It was mainly this shift to ordered lists without duplications (i.e. each buffer is uniquely represented with no more that once occurrence of each index), that made this verification exercise hard work. Proving that each reachable state can get to a focus point by a sequence of τ-transitions appeared to be considerably hard (mainly because communication steps of the two data streams can happen simultaneously).

The medium between the sender and the receiver is modeled as a lossy queue of capacity one. With buffers of sizes $2n_1$ and $2n_2$, and windows of sizes n_1 and n_2, respectively, we manually (paper-and-pencil) prove that the external behavior of this protocol is branching bisimilar [43] to a pair of FIFO queues of capacity $2n_1$ and $2n_2$. This implies both safety and liveness of the protocol (the latter under the assumption of fairness, which intuitively states that no message gets lost infinitely often).

The structure of the proof is as follows. First, we linearize the specification, meaning that we get rid of parallel operators. Moreover, communication actions are stripped from their data parameters. Then we eliminate modulo arithmetic, using an idea from Schoone [35]. Finally, we apply the cones and foci technique, to prove that the linear specification without modulo arithmetic is branching bisimilar to a pair of FIFO queues of capacity $2n_1$ and $2n_2$. The lemmas for the data types, the invariants, the transformations and the matching criteria have all been checked using PVS 2.3. The PVS files are available via http://www.cs.utwente.nl/~vdpol/piggybacking.html.

The remainder of this paper is set up as follows. In Section 1 the μCRL language is explained. In Section 2 the data types needed for specifying the protocol are presented. Section

3 features the μCRL specifications of the two-way SWP with piggybacking, and its external behavior. In Section 4, three consecutive transformations are applied to the specification of the SWP, to linearize the specification, eliminate arguments of communication actions, and get rid of modulo arithmetic. In Section 5, properties of the data types and invariants of the transformed specification are formulated; their proofs are in [2]. In Section 6, it is proved that the three transformations preserve branching bisimilarity, and that the transformed specification behaves as a pair of FIFO queues. In Section 7, we present the formalization of the verification of the SWP in PVS. We conclude the paper in Section 8.

Related Work

Sliding window protocols have attracted considerable interest from the formal verification community. In this section we present an overview. Many of these verifications deal with unbounded sequence numbers, in which case modulo arithmetic is avoided, or with a fixed finite buffer and window size at the sender and the receiver. Case studies that do treat arbitrary finite buffer and window sizes mostly restrict to safety properties.

Unbounded sequence numbers Stenning [38] studied a SWP with unbounded sequence numbers and an infinite window size, in which messages can be lost, duplicated or reordered. A timeout mechanism is used to trigger retransmission. Stenning gave informal manual proofs of some safety properties. Knuth [25] examined more general principles behind Stenning's protocol, and manually verified some safety properties. Hailpern [19] used temporal logic to formulate safety and liveness properties for Stenning's protocol, and established their validity by informal reasoning. Jonsson [22] also verified safety and liveness properties of the protocol, using temporal logic and a manual compositional verification technique. Rusu [34] used the theorem prover PVS to verify safety and liveness properties for a SWP with unbounded sequence numbers.

Fixed finite window size Vaandrager [40], Groenveld [12], van Wamel [44] and Bezem and Groote [4] manually verified in process algebra a SWP with window size one. Richier *et al.* [32] specified a SWP in a process algebra based language Estelle/R, and verified safety properties for window size up to eight using the model checker Xesar. Madelaine and Vergamini [28] specified a SWP in Lotos, with the help of the simulation environment Lite, and proved some safety properties for window size six. Holzmann [20, 21] used the Spin model checker to verify safety and liveness properties of a SWP with sequence numbers up to five. Kaivola [24] verified safety and liveness properties using model checking for a SWP with window size up to seven. Godefroid and Long [11] specified a full duplex SWP in a guarded command language, and verified the protocol for window size two using a model checker based on Queue BDDs. Stahl *et al.* [37] used a combination of abstraction, data independence, compositional reasoning and model checking to verify safety and liveness properties for a SWP with window size up to sixteen. The protocol was specified in Promela, the input language for the Spin model checker. Smith and Klarlund [36] specified a SWP in the high-level language IOA, and used the theorem prover MONA to verify a safety property for unbounded sequence numbers with window size up to 256. Jonsson and Nilsson [23] used an automated reachability analysis to verify safety properties for a SWP with a receiving window of size one. Latvala [26] modeled a SWP using Coloured Petri nets. A liveness property was model checked with fairness constraints for window size up to eleven.

Arbitrary finite window size Cardell-Oliver [6] specified a SWP using higher order logic, and manually proved and mechanically checked safety properties using HOL. (Van de Snepscheut [41] noted that what Cardell-Oliver claims to be a liveness property is in fact a safety property.) Schoone [35] manually proved safety properties for several SWPs using assertional verification. Van de Snepscheut [41] gave a correctness proof of a SWP as a sequence of correctness preserving transformations of a sequential program. Paliwoda and Sanders [30] specified a reduced version of what they call a SWP (but which is in fact very similar to the bakery protocol from [13]) in the process algebra CSP, and verified a safety property modulo trace semantics. Röckl and Esparza [33] verified the correctness of this bakery protocol modulo weak bisimilarity using Isabelle/HOL, by explicitly checking a bisimulation relation. Chkliaev et al. [8] used a timed state machine in PVS to specify a SWP with a timeout mechanism and proved some safety properties with the mechanical support of PVS; correctness is based on the timeout mechanism, which allows messages in the mediums to be reordered.

1. μCRL

μCRL [15] (see also [17]) is a language for specifying distributed systems and protocols in an algebraic style. It is based on the process algebra ACP [3] extended with equational abstract data types [27]. We will use $\approx$ for equality between process terms and $=$ for equality between data terms.

A μCRL specification of data types consists of two parts: a signature of function symbols from which one can build data terms, and axioms that induce an equality relation on data terms of the same type. They provide a loose semantics, meaning that it is allowed to have multiple models. The data types needed for our μCRL specification of a SWP are presented in Section 2. In particular we have the data sort of booleans $Bool$ with constants true and false, and the usual connectives $\wedge, \vee, \neg, \rightarrow$ and $\leftrightarrow$. For a boolean b, we abbreviate $b = $ true to b and $b = $ false to $\neg b$.

The process part of μCRL is specified using a number of pre-defined process algebraic operators, which we will present below. From these operators one can build process terms, which describe the order in which the atomic actions from a set $\mathcal{A}$ may happen. A process term consists of actions and recursion variables combined by the process algebraic operators. Actions and recursion variables may carry data parameters. There are two predefined actions outside $\mathcal{A}$: δ represents deadlock, and τ a hidden action. These two actions never carry data parameters.

Two elementary operators to construct processes are *sequential composition*, written $p \cdot q$, and *alternative composition*, written $p+q$. The process $p \cdot q$ first executes p, until p terminates, and then continues with executing q. The process $p+q$ non-deterministically behaves as either p or q. *Summation* $\sum_{d:D} p(d)$ provides the possibly infinite non-deterministic choice over a data type D. For example, $\sum_{n:\mathbb{N}} a(n)$ can perform the action $a(n)$ for all natural numbers n. The *conditional* construct $p \triangleleft b \triangleright q$, with b a data term of sort $Bool$, behaves as p if b and as q if $\neg b$. *Parallel composition* $p \parallel q$ performs the processes p and q in parallel; in other words, it consists of the arbitrary interleaving of actions of the processes p and q. For example, if there is no communication possible between actions a and b, then $a \parallel b$ behaves as $(a \cdot b) + (b \cdot a)$. Moreover, actions from p and q may also synchronize to a communication action, when this is explicitly allowed by a predefined *communication function*; two actions can only synchronize if their data parameters are equal. *Encapsulation* $\partial_{\mathcal{H}}(p)$, which renames all occurrences in p of actions from the set $\mathcal{H}$ into δ, can be used to force actions into communication. For example, if actions a and b communicate to c, then $\partial_{\{a,b\}}(a \parallel b) \approx c$. *Hiding* $\tau_{\mathcal{I}}(p)$ renames all occurrences in p of actions from the set $\mathcal{I}$ into τ. Finally, processes can be specified by means of recursive equations

$$X(d_1:D_1, \ldots, d_n:D_n) \approx p$$

where X is a recursion variable, d_i a data parameter of type D_i for $i = 1, \ldots, n$, and p a process term (possibly containing recursion variables and the parameters d_i). For example, let $X(n:\mathbb{N}) \approx a(n) \cdot X(n+1)$; then $X(0)$ can execute the infinite sequence of actions $a(0) \cdot a(1) \cdot a(2) \cdot \cdots$.

Definition 1 (Linear process equation) *A recursive specification is a linear process equation (LPE) if it is of the form*

$$X(d{:}D) \approx \sum_{j \in J} \sum_{e_j : E_j} a_j(f_j(d, e_j)) \cdot X(g_j(d, e_j)) \triangleleft h_j(d, e_j) \triangleright \delta$$

with J a finite index set, $f_j : D \times E_j \to D_j$, $g_j : D \times E_j \to D$, and $h_j : D \times E_j \to Bool$.

Note that an LPE does not contain parallel composition, encapsulation and hiding, and uses only one recursion variable. Groote, Ponse and Usenko [16] presented a linearization algorithm that transforms μCRL specifications into LPEs.

To each μCRL specification belongs a directed graph, called a labeled transition system. In this labeled transition system, the states are process terms, and the edges are labeled with parameterized actions. For example, given the μCRL specification $X(n{:}\mathbb{N}) \approx a(n) \cdot X(n+1)$, we have transitions $X(n) \xrightarrow{a(n)} X(n+1)$. Branching bisimilarity $\underline{\leftrightarrow}_b$ [43] and strong bisimilarity $\underline{\leftrightarrow}$ [31] are two well-established equivalence relations on states in labeled transition systems.[1] Conveniently, strong bisimilarity implies branching bisimilarity. The proof theory of μCRL from [14] is sound with respect to branching bisimilarity, meaning that if $p \approx q$ can be derived from it then $p \underline{\leftrightarrow}_b q$.

Definition 2 (Branching bisimulation) *Given a labeled transition system. A strong bisimulation relation $\mathcal{B}$ is a symmetric binary relation on states such that if $s \mathcal{B} t$ and $s \xrightarrow{\ell} s'$, then there exists t' such that $t \xrightarrow{\ell} t'$ and $s' \mathcal{B} t'$. Two states s and t are strongly bisimilar, denoted by $s \underline{\leftrightarrow} t$, if there is a strong bisimulation relation $\mathcal{B}$ such that $s \mathcal{B} t$.*

A strong and branching bisimulation relation $\mathcal{B}$ is a symmetric binary relation on states such that if $s \mathcal{B} t$ and $s \xrightarrow{\ell} s'$, then
- *either $\ell = \tau$ and $s' \mathcal{B} t$;*
- *or there is a sequence of (zero or more) τ-transitions $t \xrightarrow{\tau} \cdots \xrightarrow{\tau} \hat{t}$ such that $s \mathcal{B} \hat{t}$ and $\hat{t} \xrightarrow{\ell} t'$ with $s' \mathcal{B} t'$.*

Two states s and t are branching bisimilar, denoted by $s \underline{\leftrightarrow}_b t$, if there is a branching bisimulation relation $\mathcal{B}$ such that $s \mathcal{B} t$.

See [42] for a lucid exposition on why branching bisimilarity constitutes a sensible equivalence relation for concurrent processes.

The goal of this section is to prove that the initial state of the forthcoming μCRL specification of a two-way SWP is branching bisimilar to a pair of FIFO queues. In the proof of this fact, in Section 6, we will use three proof techniques to derive that two μCRL specifications are branching (or even strongly) bisimilar: invariants, bisimulation criteria, and cones and foci. An *invariant* $I : D \to Bool$ [5] characterizes the set of reachable states of an LPE $X(d{:}D)$. That is, if $I(d) = \texttt{true}$ and X can evolve from d to d' in zero or more transitions, then $I(d') = \texttt{true}$.

Definition 3 (Invariant) $I : D \to Bool$ *is an invariant for an LPE in Definition 1 if for all $d{:}D$, $j \in J$ and $e_j{:}E_j$. $(I(d) \wedge h_j(d, e_j)) \to I(g_j(d, e_j))$.*

If I holds in a state d and $X(d)$ can perform a transition, meaning that $h_j(d, e_j) = \texttt{true}$ for some $e_j{:}E$, then it is ensured by the definition above that I holds in the resulting state $g_j(d, e_j)$.

---
[1] The definitions of these relations often take into account a special predicate on states to denote successful termination. This predicate is missing here, as successful termination does not play a role in our SWP specification.

Bisimulation criteria rephrase the question whether $X(d)$ and $Y(d')$ are strongly bisimilar in terms of data equalities, where $X(d{:}D)$ and $Y(d'{:}D')$ are LPEs. A *state mapping* ϕ relates each state in $X(d)$ to a state in $Y(d')$. If a number of bisimulation criteria are satisfied, then ϕ establishes a strong bisimulation relation between terms $X(d)$ and $Y(\phi(d))$.

Definition 4 (Bisimulation criteria) *Given two LPEs,*

$$X(d{:}D) \approx \sum_{j \in J} \sum_{e_j:E_j} a_j(f_j(d,e_j)){\cdot}X(g_j(d,e_j)) \triangleleft h_j(d,e_j) \triangleright \delta$$
$$Y(d'{:}D') \approx \sum_{j \in J} \sum_{e'_j:E'_j} a_j(f'_j(d',e'_j)){\cdot}X(g'_j(d',e'_j)) \triangleleft h'_j(d',e'_j) \triangleright \delta$$

and an invariant $I : D \to \text{Bool}$ *for* X. *A state mapping* $\phi : D \to D'$ *and local mappings* $\psi_j : E_j \to E'_j$ *for* $j \in J$ *satisfy the* bisimulation criteria *if for all states* $d \in D$ *in which invariant* I *holds:*

I $\quad \forall j \in J \forall e_j{:}E_j\ (h_j(d,e_j) \leftrightarrow h'_j(\phi(d),\psi_j(e_j)))$,
II $\quad \forall j \in J \forall e_j{:}E_j\ (h_j(d,e_j) \land I(d)) \to (a_j(f_j(d,e_j)) = a_j(f'_j(\phi(d),\psi_j(e_j))))$,
III $\quad \forall j \in J \forall e_j{:}E_j\ (h_j(d,e_j) \land I(d)) \to (\phi(g_j(d,e_j)) = g'_j(\phi(d),\psi_j(e_j)))$.

Criterion I expresses that at each summand i, the corresponding guard of X holds if and only if the corresponding guard of Y holds with parameters $(\phi(d),\psi_j(e_j))$. Criterion II (III) states that at any summand i, the corresponding action (next state, after applying ϕ on it) of X could be equated to the corresponding action (next state) of Y with parameters $(\phi(d),\psi_j(e_j))$.

Theorem 5 (Bisimulation criteria) *Given two LPEs* $X(d{:}D)$ *and* $Y(d'{:}D')$ *written as in Definition 4, and* $I : D \to \text{Bool}$ *an invariant for* X. *Let* $\phi : D \to D'$ *and* $\psi_j : E_j \to E'_j$ *for* $j \in J$ *satisfy the bisimulation criteria in Definition 4. Then* $X(d) \underline{\leftrightarrow} Y(\phi(d))$ *for all* $d \in D$ *in which* I *holds.*

This theorem has been proved in PVS. The proof is available at http://www.cs.utwente.nl/~vdpol/piggybacking.html.

The *cones and foci* method from [10, 18] rephrases the question whether $\tau_\mathcal{I}(X(d))$ and $Y(d')$ are branching bisimilar in terms of data equalities, where $X(d{:}D)$ and $Y(d'{:}D')$ are LPEs, and the latter LPE does not contain actions from some set $\mathcal{I}$ of internal actions. A *state mapping* ϕ relates each state in $X(d)$ to a state in $Y(d')$. Furthermore, some $d{:}D$ are declared to be *focus points*. The *cone* of a focus point consists of the states in $X(d)$ that can reach this focus point by a string of actions from $\mathcal{I}$. It is required that each reachable state in $X(d)$ is in the cone of a focus point. If a number of *matching criteria* are satisfied, then ϕ establishes a branching bisimulation relation between terms $\tau_\mathcal{I}(X(d))$ and $Y(\phi(d))$.

Definition 6 (Matching criteria) *Given two LPEs:*

$$X(d{:}D) \approx \sum_{j \in J} \sum_{e_j:E_j} a_j(f_j(d,e_j)){\cdot}X(g_j(d,e_j)) \triangleleft h_j(d,e_j) \triangleright \delta$$
$$Y(d'{:}D') \approx \sum_{\{j \in J | a_j \notin \mathcal{I}\}} \sum_{e_j:E_j} a_j(f'_j(d',e_j)){\cdot}Y(g'_j(d',e_j)) \triangleleft h'_j(d',e_j) \triangleright \delta$$

Let $FC: D \to \text{Bool}$ *be a predicate which designates the focus points, and* $\mathcal{I} \subset \{a_j \mid j \in J\}$. *A state mapping* $\phi : D \to D'$ *satisfies the* matching criteria *for* $d{:}D$ *if for all* $j \in J$ *with* $a_j \notin \mathcal{I}$ *and all* $k \in J$ *with* $a_k \in \mathcal{I}$:

I. $\forall e_k{:}E_k\ (h_k(d,e_k) \to \phi(d) = \phi(g_k(d,e_k)))$;
II. $\forall e_j{:}E_j\ (h_j(d,e_j) \to h'_j(\phi(d),e_j))$;
III $FC(d) \to \forall e_j{:}E_j\ (h'_j(\phi(d),e_j) \to h_j(d,e_j))$;
IV $\forall e_j{:}E_j\ (h_j(d,e_j) \to f_j(d,e_j) = f'_j(\phi(d),e_j))$;
V $\forall e_j{:}E_j\ (h_j(d,e_j) \to \phi(g_j(d,e_j)) = g'_j(\phi(d),e_j))$.

Matching criterion I requires that the internal transitions at d are inert, meaning that d and $g_k(d, e_k)$ are branching bisimilar. Criteria II, IV and V express that each external transition of d can be simulated by $\phi(d)$. Finally, criterion III expresses that if d is a focus point, then each external transition of $\phi(d)$ can be simulated by d.

Theorem 7 (Cones and foci) *Given LPEs $X(d{:}D)$ and $Y(d'{:}D')$ written as in Definition 6. Let $I : D \to Bool$ be an invariant for X. Suppose that for all $d{:}D$ with $I(d)$:*

1. *$\phi : D \to D'$ satisfies the matching criteria for d; and*
2. *there is a $\hat{d}{:}D$ such that $FC(\hat{d})$ and X can perform transitions $d \xrightarrow{c_1} \cdots \xrightarrow{c_k} \hat{d}$ with $c_1, \ldots, c_k \in \mathcal{I}$.*

Then for all $d{:}D$ with $I(d)$, $\tau_{\mathcal{I}}(X(d)) \underline{\leftrightarrow}_b Y(\phi(d))$.

PVS proof of this is in [10]. For example, consider the LPEs $X(b{:}Bool) \approx a{\cdot}X(b) \triangleleft b \triangleright \delta + c{\cdot}X(\neg b) \triangleleft \neg b \triangleright \delta$ and $Y(d'{:}D') \approx a{\cdot}Y(d')$, with $\mathcal{I} = \{c\}$ and focus point $\texttt{true}$. Moreover, $X(\texttt{false}) \xrightarrow{c} X(\texttt{true})$, i.e., $\texttt{false}$ can reach the focus point in a single c-transition. For any $d'{:}D'$, the state mapping $\phi(b) = d'$ for $b{:}Bool$ satisfies the matching criteria.

Given an invariant I, only $d{:}D$ with $I(d) = \texttt{true}$ need to be in the cone of a focus point, and we only need to satisfy the matching criteria for $d{:}D$ with $I(d) = \texttt{true}$.

2. Data Types

In this section, the data types used in the μCRL specification of the two-way SWP are presented: booleans, natural numbers supplied with modulo arithmetic, buffers, and lists. Furthermore, basic properties are given for the operations defined on these data types. The μCRL specification of the data types, and of the process part are presented in here.

Booleans. We introduce constant functions $\texttt{true}$, $\texttt{false}$ of type $Bool$. $\wedge$ and $\vee$ both of type $Bool \times Bool \to Bool$ represent conjunction and disjunction operators, also $\to$ and $\leftrightarrow$ of the same exact type, denote implication and bi-implication, and $\neg : Bool \to Bool$ denotes negation. For any given sort D we consider a function $if : Bool \times D \times D \to D$ which functions an If-Then-Else operation, and also a mapping $eq : D \times D \to Bool$ such that $eq(d, e)$ holds if and only if $d = e$. For notational convenience we take the liberty to write $d = e$ instead of $eq(d, e)$.

Natural Numbers. $0: \to \mathbb{N}$ denotes zero and $S{:}\mathbb{N} \to \mathbb{N}$ the successor function. The infix operations $+$, $\dot{-}$ and $\cdot$ of type $\mathbb{N} \times \mathbb{N} \to \mathbb{N}$ represent addition, monus (also called cut-off subtraction) and multiplication, respectively. The infix operations $\leq, <, \geq$ and $>$ of type $\mathbb{N} \times \mathbb{N} \to Bool$ are the less-than(-or-equal) and greater-than(-or-equal) operations. $|, div$ of type $\mathbb{N} \times \mathbb{N} \to \mathbb{N}$ are modulo (some natural number) and dividing functions respectively. The rewrite rules applied over this data type, are explained in detail in Section 2 in [2].

Since the buffers at the sender and the receiver in the SWP are of finite size, modulo calculations will play an important role. $i|_n$ denotes i modulo n, while $i\, div\, n$ denotes i integer divided by n. In the proofs we will take notational liberties like omitting the sign for multiplication, and abbreviating $\neg(i = j)$ to $i \neq j$, $(k < \ell) \wedge (\ell < m)$ to $k < \ell < m$, $S(0)$ to 1, and $S(S(0))$ to 2. We will also use the standard induction rule to prove some properties.

Buffers. Each party in the two-way SWP will both maintain two buffers containing the sending and the receiving window (outside these windows both buffers will be empty).

$[\,] :\to Buf;\quad inb, add : \Delta \times \mathbb{N} \times Buf \to Buf;$
$|, \| : Buf \times \mathbb{N} \to Buf;$
$smaller, test : \mathbb{N} \times Buf \to Bool;\quad sorted : Buf \to Bool;$
$retrieve : \mathbb{N} \times Buf \to \Delta;\quad remove : \mathbb{N} \times Buf \to Buf;$

$release, release|_n : \mathbb{N} \times \mathbb{N} \times Buf \to Buf$;
$next\text{-}empty, next\text{-}empty|_n : \mathbb{N} \times Buf \to \mathbb{N}$;
$in\text{-}window : \mathbb{N} \times \mathbb{N} \times \mathbb{N} \to Bool$ and
$max : Buf \to \mathbb{N}$

are the functions we use for buffers. And, the rewrite rules are:

$add(d, i, []) = inb(d, i, [])$
$add(d, i, inb(e, j, q)) = \text{if}(i{>}j, inb(e, j, add(d, i, q)),$
$\qquad\qquad\qquad\qquad\qquad\qquad inb(d, i, remove(i, inb(e, j, q))))$
$[]|_n = []$ and $inb(d, i, q)|_n = inb(d, i|_n, q|_n)$
$[]\|_n = []$ and $inb(d, i, q)\|_n = add(d, i|_n, q\|_n)$
$smaller(i, []) = \texttt{true}$ and $smaller(i, inb(d, j, q)) = i < j \wedge smaller(i, q)$
$sorted([]) = \texttt{true}$ and $sorted(inb(d, j, q)) = smaller(j, q) \wedge sorted(q)$
$test(i, []) = \texttt{false}$ and $test(i, inb(d, j, q)) = i{=}j \vee test(i, q)$
$retrieve(i, inb(d, j, q)) = \text{if}(i{=}j, d, retrieve(i, q))$
$remove(i, []) = []$
$remove(i, inb(d, j, q)) = \text{if}(i{=}j, remove(i, q), inb(d, j, remove(i, q)))$
$release(i, j, q) = \text{if}(i \geq j, q, release(S(i), j, remove(i, q)))$
$release|_n(i, j, q) = \text{if}(i|_n{=}j|_n, q, release|_n(S(i), j, remove(i|_n, q)))$
$next\text{-}empty(i, q) = \text{if}(test(i, q), next\text{-}empty(S(i), q), i)$
$next\text{-}empty|_n(i, q) = \text{if}(next\text{-}empty(i|_n, q) < n, next\text{-}empty(i|_n, q),$
$\qquad\qquad\qquad\qquad \text{if}(next\text{-}empty(0, q) < n, next\text{-}empty(0, q), n))$
$in\text{-}window(i, j, k) = i \leq j < k \vee k < i \leq j \vee j < k < i$
$max([])=0$ and $max(inb(d, i, q))=\text{if}(i \geq max(q), i, max(q))$

More explanation on this is in [2] Section 2.

Δ represents the set of data elements that can be communicated between the two parties. The buffers are modeled as a list of pairs (d, i) with $d{:}\Delta$ and $i{:}\mathbb{N}$, representing that cell (or sequence number) i of the buffer is occupied by datum d; cells for which no datum is specified are empty. The empty buffer is denoted by $[]$, and $inb(d, i, q)$ is the buffer that is obtained from q by simply putting (d, i) on top of the buffer q.

add inserts data into the queue, while keeping it sorted (if the queue itself is so) and avoiding duplications. $q|_n$ is taking the sequence numbers in q of modulo n, and With $q\|_n$ the resulting buffer is further sorted out. *sorted* announces whether or not a buffer is sorted. *smaller* makes sure that the first data in the queue is having the smallest index number.

$test(i, q)$ is true if and only if the ith location in q is occupied. $retrieve(i, q)$ reveals q's ith element [2] $remove(i, q)$ wipes the ith element out. $release(i, j, q)$ empties ith to jth locations, where $release|_n(i, j, q)$ does the analogous modulo n. $next\text{-}empty(i, q)$ reveals the first empty cell in q as of i, where $next\text{-}empty|_n(i, q)$ operates the same modulo n. $in\text{-}window(i, j, k)$ is true if and only if $i \leq j \leq k \dot{-} 1$, modulo n. Finally, $max(q)$ reports the greatest occupied place in q.

Lists. $List$ is used for the specification of the external behavior of the protocol. $\langle\rangle :\to List$, $inl : \Delta \times List \to List$, $length : List \to \mathbb{N}$, $top : List \to \Delta$, $tail : List \to List$, $append : \Delta \times List \to List$, $\text{++} : List \times List \to List$ and $\lambda, \lambda' : List$ represent the functions, where $\langle\rangle$ denotes the empty list, and $inl(d, \lambda)$ adds datum d at the top of list λ. A special datum d_0 is specified to serve as a dummy value for data parameters. $length(\lambda)$ denotes the length of λ, $top(\lambda)$ produces the datum that resides at the top of λ, $tail(\lambda)$ is obtained by removing

---

[2]Note that $retrieve(i, [])$ is undefined. One could choose to equate it to a default value in Δ, or to a fresh error element in Δ. However, with the first approach an occurrence of $retrieve(i, [])$ might remain undetected, and the second approach would needlessly complicate the data type Δ. We prefer to work with an under-specified version of *retrieve*, which is allowed in μCRL, since data types have a loose semantics. All operations in μCRL data models, however, are total; partial operations lead to the existence of multiple models.

the top position in λ, $append(d, \lambda)$ adds datum d at the end of λ, and $\lambda{+}{+}\lambda'$ represents list concatenation. Finally, $q[i..j\rangle$ is the list containing the elements in buffer q at positions i up to but not including j. The rewrite rules which are being used are:

$length(\langle\rangle) = 0$ and $length(inl(d, \lambda)) = S(length(\lambda))$
$top(inl(d, \lambda)) = d, \quad tail(inl(d, \lambda)) = \lambda$
$append(d, \langle\rangle) = inl(d, \langle\rangle)$
$append(d, inl(e, \lambda)) = inl(e, append(d, \lambda))$
$\langle\rangle{+}{+}\lambda = \lambda$ and $inl(d, \lambda){+}{+}\lambda' = inl(d, \lambda{+}{+}\lambda')$
$q[i..j\rangle = if(i \geq j, \langle\rangle, inl(retrieve(i, q), q[S(i)..j\rangle))$

Detailed description on this data type is written in [2] Section 2.

3. Two-Way SWP with Piggybacking

This section contains the specification of the protocol in μCRL. Figure 1 illustrates the the protocol we work on (i.e. a two-way SWP with piggybacking). In this protocol sender (S/R) stores data elements that it receives via channel A in a buffer of size $2n$, in the order in which they are received. It can send a datum, together with its sequence number in the buffer, to a receiver R/S via a medium that behaves as a lossy queue of capacity one, represented by the medium K and the channels B and C. Upon receipt, the receiver may store the datum in its buffer, where its position in the buffer is dictated by the attached sequence number. In order to avoid a possible overlap between the sequence numbers of different data elements in the buffers of sender and receiver, no more than one half of each of these two buffers may be occupied at any time; these halves are called the sending and the receiving window, respectively. The receiver can pass on a datum that is located at the first cell in its window via channel D; in that case the receiving window slides forward by one cell. Furthermore, the receiver can send the sequence number of the first empty cell in (or just outside) its window as an acknowledgment to the sender via a medium that behaves as a lossy queue of capacity one, represented by the medium L and the channels E and F. If the sender receives this acknowledgment, its window slides forward accordingly. In a two-way SWP, data streams

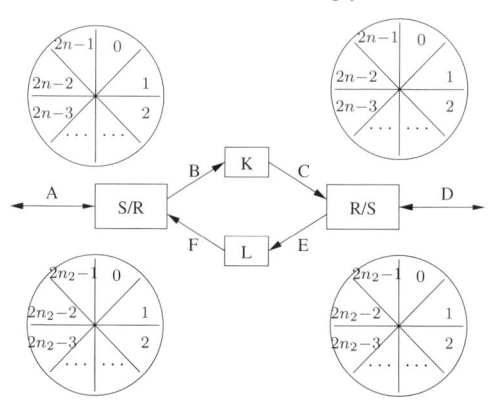

Figure 1. A two sided Sliding window protocol

are in both directions, meaning that S/R and R/S both act as sender and receiver at the same time. In addition to this, in our protocol when a datum arrives, the receiver may either send an acknowledgment back to the channel or it might instead wait until the network layer passes on the next datum. In latter case, once this new datum is to be sent into the channel, the awaited acknowledgment can be attached to it, and hence get a free ride. This technique is known as *piggybacking*.

3.1. Specification

The sender/receiver $\mathbf{S}/\mathbf{R}$ is modeled by the process $\mathbf{S}/\mathbf{R}(\ell, m, q, q'_2, \ell'_2)$, where q is its sending buffer of size $2n$, ℓ is the first cell in the window of q, and m the first empty cell in (or just outside) this window. Furthermore, q'_2 is the receiving buffer of size $2n_2$, and ℓ'_2 is the first cell in the window of q_2.

The μCRL specification of $\mathbf{S}/\mathbf{R}$ consists of seven clauses. The first clause of the specification expresses that $\mathbf{S}/\mathbf{R}$ can receive a datum via channel A and place it in its sending window, under the condition that this window is not yet full. The next two clauses specify that $\mathbf{S}/\mathbf{R}$ can receive a datum/acknowledgment pair via channel F; the data part is either added to q_2 if it is within the receiving window (second clause), or ignored if it is outside this window (third clause). In both clauses, q is emptied from ℓ up to but not including the received acknowledgment. The fourth clause specifies the reception of a single (i.e., non-piggybacked) acknowledgment. According to the fifth clause, data elements for transmission via channel B are taken (at random) from the filled part of the sending window; the first empty position in (or just outside) the receiving window is attached to this datum as an acknowledgment. In the sixth clause, $\mathbf{S}/\mathbf{R}$ sends a single acknowledgment. Finally, clause seven expresses that if the first cell in the receiving window is occupied, then $\mathbf{S}/\mathbf{R}$ can send this datum into channel A, after which the cell is emptied.

$\mathbf{S}/\mathbf{R}(\ell{:}\mathbb{N}, m{:}\mathbb{N}, n{:}\mathbb{N}, n_2{:}\mathbb{N}, q{:}Buf, q'_2{:}Buf, \ell'_2{:}\mathbb{N})$
$\approx \sum_{d:\Delta} r_A(d){\cdot}\mathbf{S}/\mathbf{R}(\ell, S(m)|_{2n}, add(d, m, q), q'_2, \ell'_2) \triangleleft \textit{in-window}(\ell, m, (\ell+n)|_{2n}) \triangleright \delta$
$+ \sum_{d:\Delta} \sum_{i:\mathbb{N}} \sum_{k:\mathbb{N}} r_F(d, i, k){\cdot}\mathbf{S}/\mathbf{R}(k, m, \textit{release}|_{2n}(\ell, k, q), add(d, i, q'_2), \ell'_2)$
$\qquad \triangleleft \textit{in-window}(\ell'_2, i, (\ell'_2+n_2)|_{2n_2}) \triangleright \delta$
$+ \sum_{d:\Delta} \sum_{i:\mathbb{N}} \sum_{k:\mathbb{N}} r_F(d, i, k){\cdot}\mathbf{S}/\mathbf{R}(k, m, \textit{release}|_{2n}(\ell, k, q), q'_2, \ell'_2)$
$\qquad \triangleleft \neg\textit{in-window}(\ell'_2, i, (\ell'_2+n_2)|_{2n_2}) \triangleright \delta$
$+ \sum_{k:\mathbb{N}} r_F(k){\cdot}\mathbf{S}/\mathbf{R}(k, m, \textit{release}|_{2n}(\ell, k, q), q'_2, \ell'_2)$
$+ \sum_{k:\mathbb{N}} s_B(\textit{retrieve}(k, q), k, \textit{next-empty}|_{2n_2}(\ell'_2, q'_2)){\cdot}\mathbf{S}/\mathbf{R}(\ell, m, q, q'_2, \ell'_2) \triangleleft \textit{test}(k, q) \triangleright \delta$
$+ s_B(\textit{next-empty}|_{2n_2}(\ell'_2, q'_2)){\cdot}\mathbf{S}/\mathbf{R}(\ell, m, q, q'_2, \ell'_2)$
$+ s_A(\textit{retrieve}(\ell'_2, q'_2)){\cdot}\mathbf{S}/\mathbf{R}(\ell, m, q, \textit{remove}(\ell'_2, q'_2), S(\ell'_2)|_{2n_2}) \triangleleft \textit{test}(\ell'_2, q'_2) \triangleright \delta$

The μCRL specification of $\mathbf{R}/\mathbf{S}$ (in [2] Appendix A) is symmetrical to the one of $\mathbf{S}/\mathbf{R}$. In the process $\mathbf{R}/\mathbf{S}(\ell_2, m_2, q_2, q', \ell')$, q' is the receiving buffer of size $2n$, and ℓ' is the first position in the window of q. Furthermore, q_2 is the sending buffer of size $2n_2$, ℓ_2 is the first position in the window of q_2, and m_2 the first empty position in (or just outside) this window.

Mediums $\mathbf{K}$ and $\mathbf{L}$, introduced below, are of capacity one. These mediums are specified in a way that they may lose frames or acknowledgments:

$\mathbf{K} \approx \sum_{d:\Delta} \sum_{k:\mathbb{N}} \sum_{i:\mathbb{N}} r_B(d, k, i){\cdot}(j{\cdot}s_C(d, k, i) + j){\cdot}\mathbf{K} + \sum_{i:\mathbb{N}} r_B(i){\cdot}(j{\cdot}s_C(i) + j){\cdot}\mathbf{K}$

$\mathbf{L} \approx \sum_{d:\Delta} \sum_{k:\mathbb{N}} \sum_{i:\mathbb{N}} r_E(d, k, i){\cdot}(j{\cdot}s_F(d, k, i) + j){\cdot}\mathbf{L} + \sum_{i:\mathbb{N}} r_E(i){\cdot}(j{\cdot}s_F(i) + j){\cdot}\mathbf{L}$

For each channel $i \in \{B, C, E, F\}$, actions s_i and r_i can communicate, resulting in the action c_i. The *initial state* of the SWP is expressed by $\tau_\mathcal{I}(\partial_\mathcal{H}(\mathbf{S}/\mathbf{R}(0, 0, [], [], 0) \parallel \mathbf{R}/\mathbf{S}(0, 0, [], [], 0) \parallel \mathbf{K} \parallel \mathbf{L}))$ where the set $\mathcal{H}$ consists of the read and send actions over the internal channels B, C, E, and F, namely $\mathcal{H} = \{s_B, r_B, s_C, r_C, s_E, r_E, s_F, r_F\}$ while the set $\mathcal{I}$ consists of the communication actions over these internal channels together with j, namely $\mathcal{I} = \{c_B, c_C, c_E, c_F, j\}$.

3.2. External Behavior

Data elements that are read from channel A should be sent into channel D in the same order, and vice versa data elements that are read from channel D should be sent into channel A in the same order. No data elements should be lost. In other words, the SWP is intended to be a solution for the following linear μCRL specification, representing *a pair of* FIFO queues of capacity $2n$ and $2n_2$.

$$\mathbf{Z}(\lambda_1{:}List, \lambda_2{:}List) \approx \sum_{d:\Delta} r_A(d) \cdot \mathbf{Z}(append(d, \lambda_1), \lambda_2) \triangleleft length(\lambda_1) < 2n \triangleright \delta$$
$$+ s_D(top(\lambda_1)) \cdot \mathbf{Z}(tail(\lambda_1), \lambda_2) \triangleleft length(\lambda_1) > 0 \triangleright \delta$$
$$+ \sum_{d:\Delta} r_D(d) \cdot \mathbf{Z}(\lambda_1, append(d, \lambda_2)) \triangleleft length(\lambda_2) < 2n_2 \triangleright \delta$$
$$+ s_A(top(\lambda_2)) \cdot \mathbf{Z}(\lambda_1, tail(\lambda_2)) \triangleleft length(\lambda_2) > 0 \triangleright \delta$$

Note that $r_A(d)$ can be performed until the list λ_1 contains $2n$ elements, because in that situation the sending window of **S/R** and the receiving window of **R/S** will be filled. Furthermore, $s_D(top(\lambda_1))$ can only be performed if λ_1 is not empty. Likewise, $r_D(d)$ can be performed until the list λ_2 contains $2n_2$ elements, and $s_A(top(\lambda_2))$ can only be performed if λ_2 is not empty.

4. Modifying the Specification

This section witnesses three transformations, one to eliminate parallel operators, one to eliminate arguments of communication actions, and one to eliminate modulo arithmetic.

Linearization. The starting point of our correctness proof is a linear specification $\mathbf{M}_{mod}$, in which no parallel composition, encapsulation and hiding operators occur. $\mathbf{M}_{mod}$ can be obtained from the μCRL specification of the SWP without the hiding operator, i.e., $\partial_{\mathcal{H}}(\mathbf{S/R}(0,0,[],[],0) \parallel \mathbf{R/S}(0,0,[],[],0) \parallel \mathbf{K} \parallel \mathbf{L})$ by means of the linearization algorithm presented in [16]; and according to [16], the following result can be obtained:

Proposition 8 $\partial_{\mathcal{H}}(\mathbf{S/R}(0,0,[],[],0) \parallel \mathbf{R/S}(0,0,[],[],0) \parallel \mathbf{K} \parallel \mathbf{L}) \leftrightarrow$
$\mathbf{M}_{mod}(0,0,[],[],0,5,0,d_0,0,5,0,d_0,0,0,0,[],[],0)$.

$\mathbf{M}_{mod}$ contains eight extra parameters: $e, e_2{:}D$ and $g, g', h, h', h_2, h'_2{:}\mathbb{N}$. Intuitively, g is 5 when medium **K** is inactive, is 4 or 2 when **K** just received a data frame or a single acknowledgment, respectively, and is 3 or 1 when **K** has decided to pass on this data frame or acknowledgment, respectively. The parameters e, h and h'_2 represent the memory of **K**, meaning that they can store the datum that is being sent from **S/R** to **R/S**, the position of this datum in q, and the first empty position in the window of q'_2, respectively. Initially, or when medium **K** is inactive, g, e, h and h'_2 have the values 5, d_0, 0 and 0. Likewise, g' captures the five states of medium **L**, and e_2, h_2 and h' represent the memory of **L**.

The linear specification $\mathbf{M}_{mod}$ of the SWP, with encapsulation but without hiding, is written below. For the sake of presentation, in states that results after a transition we only present parameters whose values have changed. In this specification

- The first summand describes that a datum d can be received by **S/R** through channel A, if q's window is not full (*in-window*$(\ell, m, (\ell+n)|_{2n})$). This datum is then placed in the first empty cell of q's window ($q{:=}add(d,m,q)$), and the next cell becomes the first empty cell of this window ($m{:=}S(m)|_{2n}$).
- By the 2nd summand, a frame (*retrieve*$(k,q), k,$ *next-empty*$|_{2n_2}(\ell'_2, q'_2)$) can be communicated to **K**, if cell k in q's window is occupied (*test*(k,q)). And by the 19th summand, an acknowledgment *next-empty*$|_{2n_2}(\ell'_2, q'_2)$ can be communicated to **K**.
- The fifth and third summand describe that medium **K** decides to pass on a frame or acknowledgment, respectively. The fourth summand describes that **K** decides to lose this frame or acknowledgment.
- The sixth and seventh summand describe that the frame in medium **K** is communicated to **R/S**. In the sixth summand the frame is within the window of q' (*in-window*$(\ell', h, (\ell'+n)|_{2n})$), so it is included ($q'{:=}add(e, h, q')$). In the seventh summand the frame is outside the window of q', so it is omitted. In both cases, the first cell of the window of q' is moved forward to h'_2 ($\ell_2{:=}h'_2$), and the cells before h'_2 are emptied ($q_2{:=}release|_{2n_2}(\ell_2, h'_2, q_2)$).

- The twentieth and last summand describes that the acknowledgment in medium **K** is communicated to **R/S**. Then the first cell of the window of q' is moved forward to h'_2, and the cells before h'_2 are emptied.
- By the eighth summand, **R/S** can send the datum at the first cell in the window of q' (*retrieve*(ℓ', q')) through channel D, if this cell is occupied (*test*(ℓ', q')). This cell is then emptied ($q':=$*remove*(ℓ', q')), and the first cell of the window of q' is moved forward by one ($\ell':=S(\ell')|_{2n}$).
- Other summands are symmetric counterparts to the ones described above.

$\mathbf{M}_{mod}(\ell, m, q, q'_2, \ell'_2, g, h, e, h'_2, g', h_2, e_2, h', \ell_2, m_2, q_2, q', \ell')$

$\approx \sum_{d:\Delta} r_A(d)\cdot \mathbf{M}_{mod}(m:=S(m)|_{2n}, q:=add(d,m,q)) \triangleleft \textit{in-window}(\ell, m, (\ell+n)|_{2n}) \triangleright \delta$ \hfill $(A1)$

$+ \sum_{k:\mathbb{N}} c_B(\textit{retrieve}(k,q), k, \textit{next-empty}|_{2n}(\ell'_2, q'_2))\cdot \mathbf{M}_{mod}(g:=4, e:=\textit{retrieve}(k,q), h:=k,$
$\quad h'_2:=\textit{next-empty}|_{2n_2}(\ell'_2, q'_2)) \triangleleft \textit{test}(k,q) \wedge g=5 \triangleright \delta$ \hfill $(B1)$

$+ j\cdot \mathbf{M}_{mod}(g:=1, e:=d_0, h:=0) \triangleleft g=2 \triangleright \delta$ \hfill $(C1)$

$+ j\cdot \mathbf{M}_{mod}(g:=5, e:=d_0, h:=0, h_2:=0) \triangleleft g=2 \vee g=4 \triangleright \delta$ \hfill $(D1)$

$+ j\cdot \mathbf{M}_{mod}(g:=3) \triangleleft g=4 \triangleright \delta$ \hfill $(E1)$

$+ c_C(e,h,h'_2)\cdot \mathbf{M}_{mod}(\ell_2:=h'_2, q':=add(e,h,q'), g:=5, e:=d_0, h:=0, h'_2:=0,$
$\quad q_2:=release|_{2n_2}(\ell_2, h'_2, q_2)) \triangleleft \textit{in-window}(\ell', h, (\ell'+n)|_{2n}) \wedge g=3 \triangleright \delta$ \hfill $(F1)$

$+ c_C(e,h,h'_2)\cdot \mathbf{M}_{mod}(\ell_2:=h'_2, g:=5, e:=d_0, h:=0, h'_2:=0, q_2:=release|_{2n_2}(\ell_2, h'_2, q_2))$
$\quad \triangleleft \neg \textit{in-window}(\ell', h, (\ell'+n)|_{2n}) \wedge g=3 \triangleright \delta$ \hfill $(G1)$

$+ s_D(\textit{retrieve}(\ell', q'))\cdot \mathbf{M}_{mod}(\ell':=S(\ell')|_{2n}, q':=remove(\ell', q')) \triangleleft \textit{test}(\ell', q') \triangleright \delta$ \hfill $(H1)$

$+ c_E(\textit{next-empty}|_{2n}(\ell', q'))\cdot \mathbf{M}_{mod}(g':=2, h_2:=0, h':=\textit{next-empty}|_{2n}(\ell', q')) \triangleleft g'=5 \triangleright \delta$ \hfill $(I1)$

$+ j\cdot \mathbf{M}_{mod}(g':=1, e_2:=d_0, h_2:=0) \triangleleft g'=2 \triangleright \delta$ \hfill $(J1)$

$+ j\cdot \mathbf{M}_{mod}(g':=5, h_2:=0, e_2:=d_0, h':=0) \triangleleft g'=2 \vee g'=4 \triangleright \delta$ \hfill $(K1)$

$+ j\cdot \mathbf{M}_{mod}(g':=3) \triangleleft g'=4 \triangleright \delta$ \hfill $(L1)$

$+ c_F(h')\cdot \mathbf{M}_{mod}(\ell:=h', q:=release|_{2n}(\ell, h', q), g':=5, h_2:=0, e_2:=d_0, h':=0) \triangleleft g'=1 \triangleright \delta$ \hfill $(M1)$

$+ \sum_{d:\Delta} r_D(d)\cdot \mathbf{M}_{mod}(m_2:=S(m_2)|_{2n}, q_2:=add(d, m_2, q_2))$
$\quad \triangleleft \textit{in-window}(\ell_2, m_2, (\ell_2+n_2)|_{2n_2}) \triangleright \delta$ \hfill $(N1)$

$+ \sum_{k:\mathbb{N}} c_E(\textit{retrieve}(k, q_2), k, \textit{next-empty}|_{2n}(\ell', q'))\cdot \mathbf{M}_{mod}(g':=4, e_2:=\textit{retrieve}(k, q_2),$
$\quad h_2:=k, h':=\textit{next-empty}|_{2n}(\ell', q')) \triangleleft \textit{test}(k, q_2) \wedge g'=5 \triangleright \delta$ \hfill $(O1)$

$+ c_F(e_2, h_2, h')\cdot \mathbf{M}_{mod}(\ell:=h', q'_2:=add(e_2, h_2, q'_2), g':=5, e_2:=d_0, h_2:=0, h':=0,$
$\quad q:=release|_{2n}(\ell, h', q)) \triangleleft \textit{in-window}(\ell'_2, h_2, (\ell'_2+n_2)|_{2n_2}) \wedge g'=3 \triangleright \delta$ \hfill $(P1)$

$+ c_F(e_2, h_2, h')\cdot \mathbf{M}_{mod}(\ell:=h', g':=5, e_2:=d_0, h_2:=0, h':=0, q:=release|_{2n}(\ell, h', q))$
$\quad \triangleleft \neg \textit{in-window}(\ell'_2, h_2, (\ell'_2+n_2)|_{2n_2}) \wedge g'=3 \triangleright \delta$ \hfill $(Q1)$

$+ s_A(\textit{retrieve}(\ell'_2, q'_2))\cdot \mathbf{M}_{mod}(\ell'_2:=S(\ell'_2)|_{2n_2}, q'_2:=remove(\ell'_2, q'_2)) \triangleleft \textit{test}(\ell'_2, q'_2) \triangleright \delta$ \hfill $(R1)$

$+ c_B(\textit{next-empty}|_{2n_2}(\ell'_2, q'_2))\cdot \mathbf{M}_{mod}(g:=2, h:=0, h'_2:=\textit{next-empty}|_{2n_2}(\ell'_2, q'_2)) \triangleleft g=5 \triangleright \delta$ \hfill $(S1)$

$+ c_C(h'_2)\cdot \mathbf{M}_{mod}(\ell_2:=h'_2, q_2:=release|_{2n_2}(\ell_2, h'_2, q_2), g:=5, h:=0, e:=d_0, h'_2:=0) \triangleleft g=1 \triangleright \delta$ \hfill $(T1)$

$\mathbf{N}_{mod}$: **No Communication Action's Arguments.**

The linear specification $\mathbf{N}_{mod}$ (Written in [2] Appendix A) is obtained from $\mathbf{M}_{mod}$ by renaming all arguments from communication actions (e.g. $c_F(e_2, h_2, h')$) to a fresh action c. Since we want to show that the "external" behavior of this protocol is branching bisimilar to *a pair of* FIFO queues (of capacity $2n$ and $2n_2$), the internal actions can be removed. The following proposition is then a trivial result of this renaming:

Proposition 9 $\tau_\mathcal{I}(\mathbf{M}_{mod}(0,0,[],[],0,5,0,d_0,0,5,0,d_0,0,0,0,[],[],0)) \underline{\leftrightarrow}$
$\tau_{\{c,j\}}(\mathbf{N}_{mod}(0,0,[],[],0,5,0,d_0,0,5,0,d_0,0,0,0,[],[],0)).$

$\mathbf{N}_{nonmod}$: No Modulo Arithmetic.

The specification of $\mathbf{N}_{nonmod}$ is obtained by eliminating all occurrences of $|_{2n}$ (resp. $|_{2n_2}$) from $\mathbf{N}_{mod}$, and replacing all guards of the form *in-window*$(i,j,(i+k)|_{2n})$ (respectively *in-window*$(i,j,(i+k)|_{2n_2})$) with $i{\leq}j{<}i+n$ (respectively $i{\leq}j{<}i+n_2$). According to what just mentioned, only $A1$, $F1$, $G1$, $N1$, $P1$ and $Q1$ whose guards are of this form, will be subjected to change. We name each new clause after its corresponding one by removing the index 1 from it, that is e.g. $A1$ will become A, and so forth. As an example we show this clause below, the whole specification of $\mathbf{N}_{nonmod}$ is in [2] Appendix A.

$$\sum_{d:\Delta} r_A(d)\cdot\mathbf{N}_{nonmod}(m{:=}S(m), q{:=}add(d,m,q)) \triangleleft l<m<\ell+n \triangleright \delta \qquad (A)$$

In Section 6.1, we will prove that $\mathbf{N}_{nonmod}$ and $\mathbf{N}_{mod}$ are strongly bisimilar. In order to demonstrate the correctness of $\mathbf{N}_{nonmod}$ (see Section 6.2) there will be a number of properties on the Data Types which should be investigated first. In the next section we list these properties, and thereafter, in its following section, we will prove the correctness.

5. Properties of Data Types

This section presents some properties of the data types and the *ordered* buffers, also some invariants of the final specification of the system; all proofs are in [2] Appendix B.

5.1. Basic Properties

These properties contain some mathematical reasoning over the functions in our specification of the system, with/without modulo arithmetic. One of them for example is: $test(k,q) \rightarrow add(retrieve(k,q), k, q)[i..j\rangle = q[i..j\rangle$. The entire list is in [2] Appendix B.1.

5.2. Ordered Buffers

Lemma 10 *Some properties on add(.,.) function:*

1. $test(i,q) \rightarrow test(i, add(d,j,q))$
2. $\textit{next-empty}(i, add(d,j,q)) \geq \textit{next-empty}(i,q)$
3. $test(i, add(d,j,q)) = (i{=}j \vee test(i,q))$
4. $retrieve(i, add(d,j,q)) = \textit{if}(i{=}j, d, retrieve(i,q))$
5. $remove(i, add(d,i,q)) = remove(i,q)$
6. $j \neq \textit{next-empty}(i,q) \rightarrow \textit{next-empty}(i, add(d,j,q)) = \textit{next-empty}(i,q)$
7. $\textit{next-empty}(i, add(d, \textit{next-empty}(i,q),q)) = \textit{next-empty}(S(\textit{next-empty}(i,q)),q)$
8. $i<j \rightarrow remove(i, add(d,j,q)) = add(d,j, remove(i,q))$
9. $i \neq j \rightarrow add(e,i, add(d,j,q)) = add(d,j, add(e,i,q))$

Lemma 11 *Ordered buffers maintain the following properties:*

1. $smaller(i,q) \rightarrow smaller(i, remove(j,q))$
2. $i<j \wedge smaller(i,q) \rightarrow smaller(i, add(d,j,q))$
3. $smaller(i,q) \rightarrow remove(i,q) = q$
4. $i<j \wedge smaller(j,q) \rightarrow smaller(i,q)$
5. $sorted(q) \rightarrow sorted(add(d,i,q))$
6. $smaller(i,q) \rightarrow add(d,i,q) = inb(d,i,q)$
7. $sorted(q) \wedge j<i \rightarrow remove(i, add(d,j,q)) = add(d,j, remove(i,q))$
8. $sorted(q) \rightarrow add(d,i,q) = add(d,i, remove(i,q))$

Lemma 12 For $n > 0$, the following results hold on $q\|_n$.

1. $sorted(q\|_n)$
2. $test(i, q|_n) = test(i, q\|_n)$
3. $retrieve(i|_n, q|_n) = retrieve(i|_n, q\|_n)$
4. $j \neq i \rightarrow remove(i, add(d, j, q\|_n)) = add(d, j, remove(i, q\|_n))$
5. $\forall j{:}\mathbb{N}(test(j, q) \rightarrow i \leq j < i + n) \wedge i \leq k \leq i + n \rightarrow$
    $next\text{-}empty|_{2n}(k|_{2n}, q|_{2n}) = next\text{-}empty|_{2n}(k|_{2n}, q\|_{2n})$
6. $\forall j{:}\mathbb{N}(test(j,q) \rightarrow i \leq j < i+n) \wedge i \leq k \leq i+n \rightarrow remove(k, q)\|_{2n} = remove(k|_{2n}, q|_{2n})$
7. $\forall j{:}\mathbb{N}(test(j,q) \rightarrow i \leq j < i+n) \wedge i \leq k \leq i+n \rightarrow release(i, k, q)\|_{2n} =$
    $release|_{2n}(i|_{2n}, k|_{2n}, q\|_{2n})$
8. $\forall j{:}\mathbb{N}(test(j,q) \rightarrow i \leq j < i+n) \wedge i \leq k \leq i+n \rightarrow add(d, k, q)\|_{2n} = add(d, k|_{2n}, q\|_{2n})$

All the abovementioned lemmas are proved in detail in [2] Appendix B.2.

5.3. Invariants

Invariants of a system are properties of data that are satisfied throughout the reachable state space of the system (see Definition 3). Lemma 13 collects 19 invariants of $\mathbf{N}_{nonmod}$ (and their symmetric counterparts). Occurrences of variables $i, j{:}\mathbb{N}$ in an invariant are always implicitly universally quantified at the outside of the invariant.

Invariants 6, 8, 15 and 17 are only needed in the derivation of other invariants. We provide some intuition for the (first of each pair of) invariants that will be used in the correctness proofs in Section 6 and in the derivations of the data lemmas. Invariants 4, 11, 12, 13 express that the sending window of $\mathbf{S/R}$ is filled from ℓ up to but not including m, and that it has size n. Invariants 7, 10 express that the receiving window of $\mathbf{R/S}$ starts at ℓ' and stops at $\ell' + n$. Invariant 2 expresses that $\mathbf{S/R}$ cannot receive acknowledgments beyond $next\text{-}empty(\ell', q')$, and Invariant 9 that $\mathbf{R/S}$ cannot receive frames beyond $m \dot{-} 1$. Invariants 16, 18, 19 are based on the fact that the sending window of $\mathbf{S/R}$, the receiving window of $\mathbf{R/S}$, and $\mathbf{K}$ (when active) coincide on occupied cells and frames with the same sequence number. Invariants 1, 3, 5 and 14 give bounds on the parameters h and h' of mediums $\mathbf{K}$ and $\mathbf{L}$.

Lemma 13 $\mathbf{N}_{nonmod}(\ell, m, q, q'_2, \ell'_2, g, h, e, h'_2, g', h_2, e_2, h', \ell_2, m_2, q_2, q', \ell')$ satisfies the following invariants.

1. $h' \leq next\text{-}empty(\ell', q')$ and $h'_2 \leq next\text{-}empty(\ell'_2, q'_2)$
2. $\ell \leq next\text{-}empty(\ell', q')$ and $\ell_2 \leq next\text{-}empty(\ell'_2, q'_2)$
3. $g' \neq 5 \rightarrow \ell \leq h'$ and $g \neq 5 \rightarrow \ell_2 \leq h'_2$
4. $test(i, q) \rightarrow i < m$ and $test(i, q_2) \rightarrow i < m_2$
5. $(g = 3 \vee g = 4) \rightarrow h < m$ and $(g' = 3 \vee g' = 4) \rightarrow h_2 < m_2$
6. $test(i, q') \rightarrow i < m$ and $test(i, q'_2) \rightarrow i < m_2$
7. $test(i, q') \rightarrow \ell' \leq i < \ell' + n$ and $test(i, q'_2) \rightarrow \ell'_2 \leq i < \ell'_2 + n_2$
8. $\ell' \leq m$ and $\ell'_2 \leq m_2$
9. $next\text{-}empty(\ell', q') \leq m$ and $next\text{-}empty(\ell'_2, q'_2) \leq m_2$
10. $next\text{-}empty(\ell', q') \leq \ell' + n$ and $next\text{-}empty(\ell'_2, q'_2) \leq \ell'_2 + n_2$
11. $test(i, q) \rightarrow \ell \leq i$ and $test(i, q_2) \rightarrow \ell_2 \leq i$
12. $\ell \leq i < m \rightarrow test(i, q)$ and $\ell_2 \leq i < m_2 \rightarrow test(i, q_2)$
13. $m \leq \ell + n$ and $m_2 \leq \ell_2 + n_2$
14. $(g = 3 \vee g = 4) \rightarrow next\text{-}empty(\ell', q') \leq h + n$ and
    $(g' = 3 \vee g' = 4) \rightarrow next\text{-}empty(\ell'_2, q'_2) \leq h_2 + n_2$
15. $\ell' \leq i < h' \rightarrow test(i, q')$ and $\ell'_2 \leq i < h'_2 \rightarrow test(i, q'_2)$
16. $(g = 3 \vee g = 4) \wedge test(h, q) \rightarrow retrieve(h, q) = e$ and
    $(g' = 3 \vee g' = 4) \wedge test(h_2, q_2) \rightarrow retrieve(h_2, q_2) = e_2$

17. $(test(i,q) \wedge test(i,q')) \rightarrow retrieve(i,q) = retrieve(i,q')$ and
    $(test(i,q_2) \wedge test(i,q'_2)) \rightarrow retrieve(i,q_2) = retrieve(i,q'_2)$
18. $((g = 3 \vee g = 4) \wedge test(h,q')) \rightarrow retrieve(h,q') = e$ and
    $((g' = 3 \vee g' = 4) \wedge test(h_2,q'_2)) \rightarrow retrieve(h_2,q'_2) = e_2$
19. $(\ell \leq i \wedge j \leq \text{next-empty}(i,q')) \rightarrow q[i..j\rangle = q'[i..j\rangle$ and
    $(\ell_2 \leq i \wedge j \leq \text{next-empty}(i,q'_2)) \rightarrow q_2[i..j\rangle = q'_2[i..j\rangle$

In the initial state $\mathbf{N}_{nonmod}(0,0,[],[],0,5,0,d_0,0,5,0,d_0,0,0,0,[],[],0)$ all these invariants are satisfied. Also, all invariants are preserved by all summands. So they are satisfied in all reachable states of $\mathbf{N}_{nonmod}$. For a proof of this lemma see [2] Appendix B.3.

6. Correctness of $\mathbf{N}_{mod}$

In Section 6.1, we establish the strong bisimilarity of $\mathbf{N}_{mod}$ and $\mathbf{N}_{nonmod}$. In order to prove this, we show that the bisimulation criteria in Definition 4 hold. Then according to Theorem 5, proof is complete. Section 6.2 demonstrates that $\mathbf{N}_{nonmod}$ behaves like *a pair of* FIFO queues. Finally, the correctness of the two-way SWP is established in Section 6.3.

6.1. Equality of $\mathbf{N}_{mod}$ and $\mathbf{N}_{nonmod}$

Proposition 14 $\mathbf{N}_{nonmod}(0,0,[],[],0,5,0,d_0,0,5,0,d_0,0,0,0,[],[],0) \leftrightarrow \mathbf{N}_{mod}(0,0,[],[],0,5,0,d_0,0,5,0,d_0,0,0,0,[],[],0)$.

Proof. By Theorem 5, it suffices to define a state mapping ϕ and local mappings ψ_j for $j = 1, 2, \ldots, 20$ that satisfy the bisimulation criteria in Definition 4, with respect to the invariants in Lemma 13.

Let Ξ abbreviate $\mathbb{N} \times \mathbb{N} \times Buf \times Buf \times \mathbb{N} \times \mathbb{N} \times \mathbb{N} \times \Delta \times \mathbb{N} \times \mathbb{N} \times \mathbb{N} \times \Delta \times \mathbb{N} \times \mathbb{N} \times \mathbb{N} \times Buf \times Buf \times \mathbb{N}$. We use $\xi{:}\Xi$ to abbreviate $(\ell, m, q, q'_2, \ell'_2, g, h, e, h'_2, g', h_2, e_2, h', \ell_2, m_2, q_2, q', \ell')$, then we define $\phi : \Xi \rightarrow \Xi$ by:

$$\phi(\xi) = (\ell|_{2n}, m|_{2n}, q\|_{2n}, q'_2\|_{2n_2}, \ell'_2|_{2n_2}, g, h|_{2n}, e, h'_2|_{2n_2},$$
$$g', h_2|_{2n_2}, e_2, h'|_{2n}, \ell_2|_{2n_2}, m_2|_{2n_2}, q_2\|_{2n_2}, q'\|_{2n}, \ell'|_{2n})$$

Furthermore, $\psi_2 : \mathbb{N} \rightarrow \mathbb{N}$ maps k to $k|_{2n}$, and $\psi_{15} : \mathbb{N} \rightarrow \mathbb{N}$ maps k to $k|_{2n_2}$; the other 18 local mappings are simply the identity. We show that ϕ and the ψ_j satisfy the bisimulation criteria. For each summand, we list (and prove) the non-trivial bisimulation criteria that it induces. For a detailed proof, see [2] Appendix C. ∎

6.2. Correctness of $\mathbf{N}_{nonmod}$

We prove that $\mathbf{N}_{nonmod}$ is branching bisimilar to the pair of FIFO queues $\mathbf{Z}$ (see Section 3.2), using cones and foci (see Theorem 7).

The state mapping $\phi : \Xi \rightarrow List \times List$, which maps states of $\mathbf{N}_{nonmod}$ to states of $\mathbf{Z}$, is defined by:

$$\phi(\xi) = (\phi_1(m, q, \ell', q'), \phi_2(m_2, q_2, \ell'_2, q'_2))$$

where

$$\phi_1(m, q, \ell', q') = q'[\ell'..\text{next-empty}(\ell', q')\rangle + q[\text{next-empty}(\ell', q')..m\rangle$$
$$\phi_2(m_2, q_2, \ell'_2, q'_2) = q'_2[\ell'_2..\text{next-empty}(\ell'_2, q'_2)\rangle + q_2[\text{next-empty}(\ell'_2, q'_2)..m_2\rangle$$

Intuitively, ϕ_1 collects data elements in the sending window of **S/R** and the receiving window of **R/S**, starting at the first cell in the receiving window (i.e., ℓ') until the first empty cell in this window, and then continuing in the sending window until the first empty cell in that window (i.e., m). Likewise, ϕ_2 collects data elements in the sending window of **R/S** and the receiving window of **S/R**.

The focus points are states where in the direction from **S/R** to **R/S**, either the sending window of **S/R** is empty (meaning that $\ell = m$), or the receiving window from **R/S** is full and all data elements in this receiving window have been acknowledged (meaning that $\ell = \ell' + n$). Likewise for the direction from **R/S** to **S/R**. That is, the focus condition reads

$$FC(\xi) := (\ell = m \vee \ell = \ell' + n) \wedge (\ell_2 = m_2 \vee \ell_2 = \ell'_2 + n_2)$$

Lemma 15 *For each $\xi{:}\Xi$ with $\mathbf{N}_{nonmod}(\xi)$ reachable from the initial state, there is a $\hat{\xi}{:}\Xi$ with $FC(\hat{\xi})$ such that $\mathbf{N}_{nonmod}(\xi) \xrightarrow{c_1} \cdots \xrightarrow{c_n} \mathbf{N}_{nonmod}(\hat{\xi})$, where $c_1, \ldots, c_n \in \mathcal{I}$.*

Proof. We prove (see [2] Appendix C) that for each $\xi{:}\Xi$ where the invariants in Lemma 13 hold, there is a finite sequence of internal actions which ends in a state where $(\ell = m \vee \ell = \ell' + n) \wedge (\ell_2 = m_2 \vee \ell_2 = \ell'_2 + n_2)$. ∎

Proposition 16 $\tau_{\{c,j\}}(\mathbf{N}_{nonmod}(0, 0, [], [], 0, 5, 0, d_0, 0, 5, 0, d_0, 0, 0, 0, [], [], 0)) \underline{\leftrightarrow}_b \mathbf{Z}(\langle\rangle, \langle\rangle)$.

Proof. We prove this using cones and foci method. See [2] Appendix C. ∎

6.3. Correctness of the Two-Way Sliding Window Protocol

Finally, we can prove the main result of our specification which is:

Theorem 17 (Correctness)
$\tau_{\mathcal{I}}(\partial_{\mathcal{H}}(\mathbf{S/R}(0, 0, [], [], 0) \parallel \mathbf{R/S}(0, 0, [], [], 0) \parallel \mathbf{K} \parallel \mathbf{L})) \underline{\leftrightarrow}_b \mathbf{Z}(\langle\rangle, \langle\rangle)$

Proof. We combine the equivalences that have been obtained so far:

$\tau_I(\partial_H(\mathbf{S/R}(0, 0, [], [], 0) \parallel \mathbf{K} \parallel \mathbf{R/S}(0, 0, [], [], 0) \parallel \mathbf{L}))$
$\underline{\leftrightarrow} \ \tau_I(\mathbf{M}_{mod}(0, 0, [], [], 0, 5, 0, d_0, 0, 5, 0, d_0, 0, 0, 0, [], [], 0))$ (Proposition 8)
$\underline{\leftrightarrow} \ \tau_{\{c,j\}}(\mathbf{N}_{mod}(0, 0, [], [], 0, 5, 0, d_0, 0, 5, 0, d_0, 0, 0, 0, [], [], 0))$ (Proposition 9)
$\underline{\leftrightarrow} \ \tau_{\{c,j\}}(\mathbf{N}_{nonmod}(0, 0, [], [], 0, 5, 0, d_0, 0, 5, 0, d_0, 0, 0, 0, [], [], 0))$ (Proposition 14)
$\underline{\leftrightarrow}_b \ \mathbf{Z}(\langle\rangle, \langle\rangle)$ (Proposition 16)
∎

7. Formalization in PVS

In this section we show the formalization and verification of the correctness proof of the SWP with piggybacking in PVS [29].

The PVS specification language is based on simply typed higher-order logic. Its type system contains basic types such as *boolean, nat, integer, real,* etc. and type constructors such as *set, tuple, record,* and *function*. Tuple types have the form [T1, ..., Tn], where Ti are type expressions. A record is a finite list of fields of the form R:TYPE=[# F1·T1, , En·Tn #], where Ei are *record accessor functions*. A function type constructor has the form F:TYPE=[T1, ..., Tn->R], where F is a function with domain D=T1×...×Tn and range R.

A PVS specification can be structured through a hierarchy of *theories*. Each theory consists of a *signature* for the type names and constants introduced in the theory, and a number of axioms, definitions and theorems associated with the signature. A PVS theory can be parametric in certain specified types and values, which are placed between [ ] after the theory name.

In μCRL, the semantics of a data specification is the set of all its models. Incomplete data specifications may have multiple models. Even worse, it is possible to have inconsistent data specifications for which no models exist. Here the necessity of specification with PVS emerges, because of this probable incompleteness and inconsistency which exists when working with µCRL. Moreover, PVS was used to search for omissions and errors in the manual µCRL proof of the SWP with piggybacking.

In Section 7.1 we show examples of the original specification of some data functions, then we introduce the modified forms of them. Moreover, we show how measure functions are used to detect the termination of recursive definitions. In Section 7.2 and 7.3 we represent the LPEs and invariants of the SWP with piggybacking in PVS. Section 7.4 presents the equality of µCRL specification of the SWP with piggybacking with and without modulo arithmetic. Section 7.5 explains how the cones and foci method is used to formalize the main theorem, that is the µCRL specification of the SWP with piggybacking is branching bisimilar to a FIFO queue of size $2n$. Finally, Section 7.6 is dedicated to some remarks on the verification in PVS.

7.1. Data Specifications in PVS

In PVS, all the definitions are first type checked, which generates some *proof obligations*. Proving all these obligations ascertains that our data specification is complete and consistent.

To achieve this, having total definitions is required. So in the first place, partially defined functions need to be extended to total ones. Below there are some examples of partial definitions in the original data specification of the SWP with piggybacking, which we changed into total ones. Second, to guarantee totality of recursive definitions, PVS requires the user to define a so-called *measure function*. Doing this usually requires time and effort, but the advantage is that recursive definitions are guaranteed to be well-founded. PVS enabled us to find non-terminating definitions in the original data specification of the SWP with piggybacking, which were not detected within the framework of µCRL. After finding these non-terminating definitions with PVS, we searched for new definition which can express the operation we look for. Then we replaced the old definitions with new terminating ones in our µCRL framework. Below we show some of the most interesting examples.

Example 18 *We defined a function next-empty which seeks for the first empty position in q from a given position i. This function is identified as:*

$$\text{next-empty}(i, q) = \text{if}(\text{test}(i, q), \text{next-empty}(S(i), q), i).$$

We also need to have next-empty$|_n(i, q)$ *as a function which produces the first empty position in q modulo n, from position i. It looked reasonable to define it as:*

$$\text{next-empty}|_n(i, q) = \text{if}(\text{test}(i, q), \text{next-empty}|_n(S(i)|_n, q), i)$$

Although the definition looks total and well-founded, this was one of the undetected potential errors that PVS detected during the type checking process. Below we bring an example to show what happens. Let $q = [(d_0, 0), (d_1, 1), (d_2, 2), (d_3, 3), (d_5, 5)]$, $n = 4$, $i = 5$ *then*

$$\text{next-empty}|_4(5, q) = \text{next-empty}|_4(6|_4, q) = \text{next-empty}|_4(2, q) = \text{next-empty}|_4(3, q)$$
$$= \text{next-empty}|_4(0, q) = \text{next-empty}|_4(1, q) = \text{next-empty}|_4(2, q) = \ldots$$

which will never terminate. The problem is that modulo n all the places in q are occupied, and since $0 \leq i|_n < n$ hence test(i, q) will always be true. *Hence each position will call for its immediate next position and so on. Therefore the calls will never stop.*

At the end we replaced it with the following definition, which is terminating and operates the way as we expect.

$$\text{next-empty}|_n(i, q) = \text{if}(\text{next-empty}(i|_n, q) < n, \text{next-empty}(i|_n, q),$$
$$\text{if}(\text{next-empty}(0, q) < n, \text{next-empty}(0, q), n))$$

```
...
D:nonempty_type
Buf:type=list[[D,nat]]
x,i,j,k,l,n: VAR nat
...
dm(i,j,n): nat =
        IF mod(i,n)<=mod(j,n)
        THEN mod(j,n)-mod(i,n)
        ELSE n+mod(j,n)-mod(i,n)
        ENDIF
...
release(n)(i,j,q): RECURSIVE Buf=
        IF mod(i,n)=mod(j,n) THEN q
        ELSE release(n)(mod(i+1,n),j,remove(mod(i,n),q))
        ENDIF
        measure dm(i,j,n)
...
```

Figure 2. An example of data specification in PVS

This function first checks whether there is any empty place after $i|_n$ (incl. $i|_n$ itself). If this is the case then that position would be the result, otherwise using next-empty$(0, q)$ it will check if there is any empty position in the buffer modulo n. If so then that position would be the value of the function since next-empty$(i|_n, q)$ will reach it. If all the buffer modulo n is full then n would be the result, because n is bigger that all the possible values for the function (i.e. $i|_n$ at most) and moreover it indicates that the buffer is full modulo n.

In [2] Appendix D there are similar examples for $release(i, j, q)$ and $release|_n(i, j, q)$, detected errors by PVS, and also our ultimate solutions for them.

We represented the μCRL abstract data types directly by PVS types. This enables us to reuse the PVS library for definitions and theorems of "standard" data types. Figure 2 illustrates part of a PVS theory defining $release|_n$. There D is an unspecified but non-empty type which represents the set of all data elements that can be communicated between the sender and the receiver. Buf is list of pairs of type $D \times \mathbb{N}$ defined as $list[[D, nat]]$. Here we used *list* to identify the type of lists, which is defined in the prelude in PVS. Therefore we simply use it without any need to define it explicitly. This figure also represents $release|_n(i, j, q)$ in PVS. Since it is defined recursively, in order to establish its termination (or totality), it is required by PVS to have a measure function. We define a measure function called dm which is decreasing and non-recursive. Here, PVS uses its type-checker to check the validity of dm. It generates two type-check proof obligations: if $i|_n < j|_n$ then $j|_n - i|_n \geq 0$ and if $i|_n \geq j|_n$ then $n + j|_n - i|_n \geq 0$. The first proof obligation is proved in one trivial step. The second one is proved using Lemma 19.

In [2] Appendix D, we also list the extra data lemmas which had to be proved in PVS while they are considered to be trivial in the manual proof.

7.2. Representing LPEs

We now reuse [10] to show how the μCRL specification of the SWP with piggybacking (an LPE) can be represented in PVS. The main distinction will be that we have assumed so far that LPEs are *clustered*. This means that each action label occurs in at most one summand, so that the set of summands could be indexed by the set of action labels. This is no limitation, because any LPE can be transformed in clustered form, basically by replacing $+$ by $\sum$ over

```
LPE[Act,State,Local:TYPE,n:nat]: THEORY BEGIN
    SUMMAND:TYPE= [State,Local-> [#act:Act,guard:bool,next:State#] ]
    LPE:TYPE= [#init:State,sums:[below(n)->SUMMAND]#]
END LPE
```

Figure 3. Definition of LPE in PVS

finite types. Clustered LPEs enable a notationally smoother presentation of the theory. However, when working with concrete LPEs this restriction is not convenient, so we avoid it in the PVS framework: an arbitrarily sized index set $\{0,\ldots,n-1\}$ will be used, represented by the PVS type `below(n)`. A second deviation is that we will assume from now on that every summand has the same set of local variables. Again this is no limitation, because void summations can always be added (i.e. $p = \sum_{d:D} p$, when d does not occur in p). This restriction is needed to avoid the use of polymorphism, which does not exist in PVS. The third deviation is that we do not distinguish action labels from action data parameters. We simply work with one type of expressions for actions. This allows that a summand can generate transitions with various labels. This generalization makes the formalization a bit smoother, but was not really exploited.

So an LPE is parameterized by sets of actions (`Act`), global parameters (`State`) and local variables (`Local`), and by the size of its index set (`n`). Note that the guard, action and next-state of a summand depend on the global parameters $d : State$ and on local variables $e : Local$. This dependency is represented in the definition SUMMAND by a PVS function type. In Figure 3 an LPE consists of an initial state and a list of summands indexed by `below(n)`.

A concrete LPE by a fragment of the linear specification $\mathbf{N}_{mod}$ of SWP with piggybacking in PVS (see Figure 6 in Appendix D in [2]) is introduced as an `lpe` of a set of actions: `Nnonmod_act`, states: `State`, local variables: `Local`, and a digit: `20` referring to the number of summands. The LPE is identified as a pair, called `init` and `sums`, where `init` is introducing the initial state of $\mathbf{N}_{mod}$ and `sums` the summands. The first LAMBDA maps each number to the corresponding summand in $\mathbf{N}_{mod}$. The second LAMBDA is representing the summands as functions over `State` and `Local`. Here, `State` is the set of states and `Local` is the data type $D \times \mathbb{N}$ of all pairs (d, k) of the summation variables, which is considered as a global variable regarding the property: $p = \sum_{(d,k):local} p$, which is mentioned before.

7.3. Representing Invariants

Invariants are boolean functions over the set of states. In Figure 4, we explain how to represent an invariant of the μCRL specification, in PVS. This figure illustrates the (first part of the) Invariant 13.9 from Section 5.3

7.4. Equality of $\mathbf{N}_{mod}$ and $\mathbf{N}_{nonmod}$

Strong bisimilarity of $\mathbf{N}_{mod}$ and $\mathbf{N}_{nonmod}$ (Proposition 14) is depicted in Figure 5. `state_f` and `local_f` are introduced to construct the state mapping between $\mathbf{N}_{nonmod}$ and $\mathbf{N}_{mod}$. In PVS we introduce the state mapping (`state_f, local_f`) from the set of states and local variables of $\mathbf{N}_{nonmod}$ to those of $\mathbf{N}_{mod}$. Then we use the corresponding relation to this state mapping, and we show that this relation is a bisimulation between $\mathbf{N}_{nonmod}$ and $\mathbf{N}_{mod}$.

In PVS we defined an LPE as a list of summands (not as a recursive equation), equipped with the standard LTS semantics. It could be proved directly that state mappings preserve strong bisimulation.

```
...
l,m,l12,g,h,h12,g1,h2,h1,l2,m2,l1: var nat
q,q1,q2,q12  : var Buf
e,e2: var D
...
inv(l,m,q,q12,l12,g,h,e,h12,g1,h2,e2,h1,l2,m2,q2,q1,l1): bool= next_empty(l1,q1)<=m
...
```

Figure 4. An example of representing invariants in PVS

```
...
state_f(l,m,q,q12,l12,g,h,e,h12,g1,h2,e2,h1,l2,m2,q2,q1,l1): State=
      (mod(l,2*n),mod(m,2*n),modulo2(q,2*n),modulo2(q12,2*n2),mod(l12,2*n2),
       g,mod(h,2*n),e,mod(h1,2*n2),g1,mod(h2,2*n2),e2,mod(h1,2*n),
       mod(l2,2*n2),mod(m2,2*n2),modulo2(q2,2*n2),modulo2(q1,2*n),
       mod(l1,2*n)),
local_f(l:Local,i:below(20)): Local=
       LET  (e,k)=l IN
       IF i=4 THEN (e,mod(k,2*n)) ELSE (IF i=9 THEN (e,mod(k,2*n2)) ELSE(e,k)) ENDIF
...
Propsimilaosition_6_22: proposition bisimilar (lpe2lts(Nnonmod),lpe2lts(Nmod))
...
```

Figure 5. Equality of N_{mod} and N_{nonmod} in PVS

By contrast, the manual proof that N_{mod} and N_{nonmod} are strongly bisimilar is based on the proof principle CL-RSP [5], which states that each LPE has a unique solution, modulo strong bisimilarity. An advantage of this approach is that by using algebraic principles only, the stated equivalence also holds in non-standard models for process algebra + CL-RSP. We did not formalize CL-RSP in PVS because it depends on recursive process equations; this would have required a laborious embedding of μCRL in PVS, which would complicate the formalization too much.

7.5. Correctness of N_{mod}

The branching bisimilarity verification of N_{mod} and Z (Theorem 17) is pictured in Figure 6. The function `fc(l,m,q,q12,l12,g,h,e,h12,g1,h2,e2,h1,l2,m2,q2,q1,l1)` defines the focus condition for $N_{nonmod}(\ell, m, q, q'_2, \ell'_2, g, h, e, h'_2, g', h_2, e_2, h', \ell_2, m_2, q_2, q', \ell')$ as a boolean function on set of states. `qlist(q,i,j)` is used to describe the function $q[i..j\rangle$, which is defined as an application on triples. The state mapping h maps states of N_{nonmod} to states of Z, which is called $\phi : \Xi \rightarrow List \times List$ in Section 6.2. k is a Boolean function which is used to match each external action of N_{nonmod} to the corresponding one of Z. This is done by corresponding the number of each summand of N_{nonmod} to one of Z. As PVS requires, this function must be total, therefore without loss of generality we map all the summands with an internal action, from N_{nonmod}'s specification, to the second summand of Z's specification.

According to cones and foci proof method [10], to derive that N_{nonmod} and N_{mod} are branching bisimilar, it is enough to check the matching criteria and the reachability of focus points. The two conditions of the cones and foci proof method are represented by mc and WN, namely matching criteria and the reachability of focus points, respectively. mc establishes that

```
...
fc(l,m,q,q12,l12,g,h,e,h12,g1,h2,e2,h1,l2,m2,q2,q1,l1): bool =
        (l=m OR l=l1+n) AND (l2=m2 OR l2=l12+n2)
k(i): below(2)= IF i=18 THEN 0 ELSE
        IF i=10 THEN 1 ELSE
        IF i=11 THEN 2 ELSE 3 ENDIF ENDIF ENDIF
h(l,m,q,q12,l12,g,h,e,h12,g1,h2,e2,h1,l2,m2,q2,q1,l1): [List_,List_]=
        (concat(qlist(q1,l1,next_empty(l1,q1)),qlist(q,next_empty(l1,q1),m)),
         concat(qlist(q12,l12,next_empty(l12,q12)),qlist(q2,next_empty(l12,q12),m2)))
mc: THEOREM FORALL d: reachable(Nnonmod)(d) IMPLIES MC(Nnonmod,Z,k,h,fc)(d)
WN: LEMMA FORALL S: reachable(Nnonmod)(S) IMPLIES WN(Nnonmod,fc)(S)
main: THEOREM brbisimilar(lpe2lts(Nmod),lpe2lts(Z))
...
```

Figure 6. Correctness of N_{mod} in PVS

all the matching criteria (see Section 1) hold for every reachable state d in Nnonmod, with the aforementioned h, k and fc functions. WN represents the fact that from all reachable states S in Nnonmod, a focus point can be reached by a finite series of internal actions. The function lpe2lts provides the Labeled Transition System semantics of an LPE (see [10]).

7.6. Remarks on the Verification in PVS

We used PVS to find the omissions and undetected potential errors that have been ignored in the manual μCRL proofs; some of them have been shown as examples in Section 7.1. PVS guided us to find some important invariants. We affirmed the termination of recursive definitions by means of various measure functions. We represented LPEs in PVS and then introduced N_{mod} and N_{nonmod} as LPEs. We verified the bisimulation of N_{nonmod} and N_{mod}. Finally we used the cones and foci proof method [10], to prove that N_{mod} and the external behavior of the SWP with piggybacking, represented by Z, are branching bisimilar.

8. Conclusions

In this paper we verify a two-way sliding window protocol which has the acknowledgments piggybacked on data. This way acknowledgments take a free ride in the channel. As a result the available bandwidth is used better. We present a specification of a sliding window protocol with piggybacking in μCRL, and then verify the specification with the PVS theorem prover.

An important aim of this paper is to show how one can incrementally extend a PVS verification effort, in this case the one described in [1]. PVS verification can be reused to check modifications of the SWP nearly automatically. We benefited from the PVS formalizations and lemmas in [1], e.g. properties of data types and those invariants which are not directly working with the internal structure of buffers (i.e. ordered lists). These are also mentioned in [2]. Note that a large part of the complete formalization consists of developing the meta theory. This part is split in generic PVS files with proofs. This generic part can be reused for the correctness proof of many other protocols. In particular, the generic part consists of the definition of an LTS, various forms of bisimulation (with proofs that they form equivalence relations), the definition of LPEs, their operational semantics, the notions of state mappings between LPEs, the notion of an invariant of an LPE (and its relation with reachable states), the proof rules for tau-reachability (with a soundness proof), and the matching criteria (including the proof of the theorem, that from the cones and foci method one may conclude branching bisimilarity).

For a specific protocol verification one must formalize the used data types (or find them in PVS's prelude), define LPEs for the specification and implementation, list the invariants, the focus conditions and the state mapping. From this, all proof obligations (like invariants and matching criteria) are generated automatically. Most obligations can be discharged automatically, but still many must be proven manually. Also, tau-reachability must typically be proven manually, using the predefined proof rules. However, some steps remain protocol-specific, such as the transition from modulo to full arithmetic in the case of the Sliding Window Protocol.

Here, we model the medium between the sending and receiving window as a queue of capacity one. So a possible extension of this work would be to verify this protocol with mediums of unbounded size, i.e. we can define the mediums as lists of pairs (d, i) by:

cons: $[\,] :\to Medium$
func: $add : \Delta \times \mathbb{N} \times Medium \to Medium$

Acknowledgements

We would like to thank Jun Pang for the helpful discussions and for his μCRL files on the one-way SWP.

References

[1] B. Badban, W. Fokkink, J. Groote, J. Pang, and J. van de Pol. Verifying a sliding window protocol in μCRL and PVS. *Formal Aspects of Computing*, 17(3):342–388, 2005.
[2] B. Badban, W. Fokkink, and J. van de Pol. Mechanical verification of a two-way sliding window protocol (full version including proofs). Technical Report TR-CTIT-08-45, University of Twente, CTIT, July 2008.
[3] J. Bergstra and J. Klop. Process algebra for synchronous communication. *Information and Control*, 60(1-3):109–137, 1984.
[4] M. Bezem and J. Groote. A correctness proof of a one bit sliding window protocol in μCRL. *The Computer Journal*, 37(4):289–307, 1994.
[5] M. Bezem and J. Groote. Invariants in process algebra with data. In B. Jonsson and J. Parrow, editors, *Proc. 5th Conference on Concurrency Theory*, LNCS 836, pages 401–416, 1994.
[6] R. Cardell-Oliver. Using higher order logic for modeling real-time protocols. In J. Diaz and F. Orejas, editors, *Proc. 4th Joint Conference on Theory and Practice of Software Development*, Lecture Notes in Computer Science 494, pages 259–282, 1991.
[7] V. Cerf and R. Kahn. A protocol for packet network intercommunication. *IEEE Transactions on Communications*, COM-22:637–648, 1974.
[8] D. Chkliaev, J. Hooman, and E. de Vink. Verification and improvement of the sliding window protocol. In H. Garavel and J. Hatcliff, editors, *Proc. 9th Conference on Tools and Algorithms for the Construction and Analysis of Systems*, Lecture Notes in Computer Science 2619, pages 113–127, 2003.
[9] W. Fokkink, J. Groote, J. Pang, B. Badban, and J. van de Pol. Verifying a sliding window protocol in μCRL. In S. M. C. Rattray and C. Shankland, editors, *Proc. 10th Conference on Algebraic Methodology and Software Technology*, Lecture Notes in Computer Science 3116, pages 148–163, 2004.
[10] W. Fokkink, J. Pang, and J. van de Pol. Cones and foci: A mechanical framework for protocol verification. *Formal Methods in System Design*, 29(1):1–31, 2006.
[11] P. Godefroid and D. Long. Symbolic protocol verification with Queue BDDs. *Formal Methods and System Design*, 14(3):257–271, 1999.
[12] R. Groenveld. Verification of a sliding window protocol by means of process algebra. Technical Report P8701, University of Amsterdam, 1987.
[13] J. Groote and H. Korver. Correctness proof of the bakery protocol in μCRL. In A. Ponse, C. Verhoef, and S. v. Vlijmen, editors, *Proc. 1st Workshop on Algebra of Communicating Processes '94*, Workshops in Computing Series, pages 63–86. Springer-Verlag, 1995.
[14] J. Groote and A. Ponse. Proof theory for μCRL: A language for processes with data. In D. Andrews, J. Groote, and C. Middelburg, editors, *Proc. Workshop on Semantics of Specification Languages*, Workshops in Computing Series, pages 231–250, 1994.

[15] J. Groote and A. Ponse. The syntax and semantics of µCRL. In A. Ponse, C. Verhoef, and S. v. Vlijmen, editors, *Proc. 1st Workshop on Algebra of Communicating Processes '94*, Workshops in Computing Series, pages 26–62. Springer-Verlag, 1995.
[16] J. Groote, A. Ponse, and Y. Usenko. Linearization in parallel pCRL. *Journal of Logic and Algebraic Programming*, 48(1-2):39–72, 2001.
[17] J. Groote and M. Reniers. Algebraic process verification. In J. Bergstra, A. Ponse, and S. Smolka, editors, *Handbook of Process Algebra*, chapter 17, pages 1151–1208. Elsevier, 2001.
[18] J. Groote and J. Springintveld. Focus points and convergent process operators. a proof strategy for protocol verification. *Journal of Logic and Algebraic Programming*, 49(1-2):31–60, 2001.
[19] B. Hailpern. *Verifying Concurrent Processes Using Temporal Logic*. LNCS 129. Springer-Verlag, 1982.
[20] G. Holzmann. *Design and Validation of Computer Protocols*. Prentice Hall, 1991.
[21] G. Holzmann. The model checker Spin. *IEEE Transactions on Software Engineering*, 23:279–295, 1997.
[22] B. Jonsson. *Compositional Verification of Distributed Systems*. PhD thesis, Department of Computer Science, Uppsala University, 1987.
[23] B. Jonsson and M. Nilson. Transitive closures of regular relations for verifying infinite-state systems. In S. Graf and M. Schwartzbach, editors, *Proc. 6th Conference on Tools and Algorithms for Construction and Analysis of Systems*, Lecture Notes in Computer Science 1785, pages 220–234. Springer-Verlag, 2000.
[24] R. Kaivola. Using compositional preorders in the verification of sliding window protocol. In *Proc. 9th Conference on Computer Aided Verification*, LNCS 1254, pages 48–59, 1997.
[25] D. Knuth. Verification of link-level protocols. *BIT*, 21:21–36, 1981.
[26] T. Latvala. Model checking LTL properties of high-level Petri nets with fairness constraints. In J. Colom and M. Koutny, editors, *Proc. 21st Conference on Application and Theory of Petri Nets*, Lecture Notes in Computer Science 2075, pages 242–262. Springer-Verlag, 2001.
[27] J. Loeckx, H. Ehrich, and M. Wolf. *Specification of Abstract Data Types*. Wiley/Teubner, 1996.
[28] E. Madelaine and D. Vergamini. Specification and verification of a sliding window protocol in LOTOS. In E. Knuth and L. Wegner, editors, *Proc. 4th Conference on Formal Description Techniques for Distributed Systems and Communication Protocols*, IFIP Transactions (C-2), pages 495–510. North-Holland, 1991.
[29] S. Owre, S. Rajan, J. Rushby, N. Shankar, and M. Srivas. PVS: Combining specification, proof checking, and model checking. In R. Alur and T.A. Henzinger, editors, *Proc. 8th Conference on Computer Aided Verification*, Lecture Notes in Computer Science 1102, pages 411–414. Springer-Verlag, 1996.
[30] K. Paliwoda and J. Sanders. An incremental specification of the sliding-window protocol. *Distributed Computing*, 5(2):83–94, 1991.
[31] D. Park. Concurrency and automata on infinite sequences. In P. Deussen, editor, *Proc. 5th GI-Conference on Theoretical Computer Science*, LNCS 104, pages 167–183, 1981.
[32] J. Richier, C. Rodriguez, J. Sifakis, and J. Voiron. Verification in Xesar of the sliding window protocol. In H. Rudin and C. West, editors, *Proc. 7th Conference on Protocol Specification, Testing and Verification*, pages 235–248. North-Holland, 1987.
[33] C. Röckl and J. Esparza. Proof-checking protocols using bisimulations. In J. Baeten and S. Mauw, editors, *Proc. 10th Conference on Concurrency Theory*, LNCS 1664, pages 525–540, 1999.
[34] V. Rusu. Verifying a sliding-window protocol using PVS. In M. Kim, B. Chin, S. Kang, and D. Lee, editors, *Proc. 21st Conference on Formal Techniques for Networked and Distributed Systems*, IFIP Conference Proceedings 197, pages 251–268. Kluwer Academic, 2001.
[35] A. Schoone. *Assertional Verification in Distributed Computing*. PhD thesis, Utrecht University, 1991.
[36] M. Smith and N. Klarlund. Verification of a sliding window protocol using IOA and MONA. In T. Bolognesi and D. Latella, editors, *Proc. 20th Joint Conference on Formal Description Techniques for Distributed Systems and Communication Protocols*, pages 19–34. Kluwer Academic Publishers, 2000.
[37] K. Stahl, K. Baukus, Y. Lakhnech, and M. Steffen. Divide, abstract, and model-check. In D. Dams, R. Gerth, S. Leue, and M. Massink, editors, *Proc. 6th SPIN Workshop on Practical Aspects of Model Checking*, Lecture Notes in Computer Science 1680, pages 57–76. Springer-Verlag, 1999.
[38] N. Stenning. A data transfer protocol. *Computer Networks*, 1:99–110, 1976.
[39] A. Tanenbaum. *Computer Networks*. Prentice Hall, 1981.
[40] F. Vaandrager. Verification of two communication protocols by means of process algebra. Technical Report Report CS-R8608, CWI, 1986.
[41] J. van de Snepscheut. The sliding window protocol revisted. *Formal Aspects of Computing*, 7(1):3–170, 1995.
[42] R. van Glabbeek. What is branching time and why to use it? *The Concurrency Column, Bulletin of the EATCS*, 53:190–198, 1994.
[43] R. van Glabbeek and W. Weijland. Branching time and abstraction in bisimulation semantics. *Journal of the ACM*, 43(3):555–600, 1996.

[44] J. van Wamel. A study of a one bit sliding window protocol in ACP. Technical Report P9212, University of Amsterdam, 1992.

RRABP: Point-to-Point Communication over Unreliable Components

Bernhard H.C. SPUTH, Oliver FAUST and Alastair R. ALLEN

School of Engineering, University of Aberdeen, Aberdeen, AB24 3UE, UK
`bernhard@erg.abdn.ac.uk, {o.faust, a.allen}@abdn.ac.uk`

Abstract. This paper establishes the security, stability and functionality of the resettable receiver alternating bit protocol. This protocol creates a reliable and blocking channel between sender and receiver over unreliable non-blocking communication channels. Furthermore, this protocol permits the sender to be replaced at any time, but not under all conditions without losing a message. The protocol is an extension to the alternating bit protocol with the ability for the sender to synchronise the receiver and restart the transmission. The resulting protocol uses as few messages as possible to fulfil its duty, which makes its implementation lightweight and suitable for embedded systems. An unexpected outcome of this work is the large number of different messages needed to reset the receiver reliably.

Keywords. point-to-point communication, reliability, unstable sender.

Introduction

Over the past years CSP process networks have become more and more dynamic. Such dynamic process networks adapt to a changing environment during runtime. The start of this dynamicalisation of process networks was the introduction of mobile channels and mobile processes to occam-π [1,2]. Mobile channels reconfigure the network connections. Similarly, mobile processes reconfigure the functionality of a process network. Both techniques allow the designer to influence the network functionality during runtime. These mobility techniques have since been added to the Communicating Sequential Processes for Java (JCSP) library, by the jcsp.mobile package [3]. This package takes mobility a step further by making processes and channels mobile over the nodes in a TCP/IP network.

Another technique to adjust process networks to changes in the environment is process subnetwork termination by localised graceful-termination [4], also known as localised poisoning. It is no problem to come up with scenarios where a change in the environment renders a complete subnetwork obsolete. One example where subnetwork termination is helpful is a network server. Such as server assigns each connecting client a dedicated process subnetwork. If a client disconnects from the server, its dedicated process subnetwork becomes obsolete. It is highly desirable to terminate these subnetworks, because this releases resources and prevents errors. Localised poisoning means that the termination message (poison) does not leave the subnetwork to be terminated. This containment of poison within a subnetwork makes it possible to terminate clearly defined parts of a process network. The localised poisoning technique was first presented for JCSP in the form of JCSP-Poison [5]. Since then it has been refined and integrated into the JCSP library [6]. The ability to perform a localised graceful-termination is also available in C++CSP [7]. Another approach to terminating subnetworks and even replacing them is the exception handling mechanism in Communicating Threads for Java (CTJ) [8].

Both techniques, mobility and poisoning rely on an entity controlling or initiating the process network change. This controlling entity ensures that the environment outside the terminating process subnetwork knows about this termination and can prepare for it. However, there are situations in which no entity controls the change of the process network. This is, for instance, the case when a process network is spread over multiple unreliable nodes, which communicate over unreliable communication channels. In such systems, nodes may suddenly decide to terminate and then restart later. Furthermore, unreliable communication channels lose and replicate messages in an unpredictable way. An example for such a system is the water quality monitoring system of the WARMER (**WA**ter **R**isk **M**anagement in **EuR**ope) consortium [9,10]. This system consists of multiple in-situ monitoring stations (IMS) and one data centre. The IMSs periodically transmit their measurement data to the data centre, where the data gets accumulated and stored. There are multiple unreliable components involved in this system. The first unreliable component is the communication channel between the IMSs and the data centre. Apart from dropping completely, this unreliable communication channel has other undesirable properties such as losing or replicating data-messages which an IMS entrusts to it. Furthermore, the individual IMS might be unstable, for instance it runs out of energy, or is severely damaged or sunk due to harsh weather conditions. The data centre has to be classified as unstable as well, because its hardware may fail.

The unreliable channel does not pose a very big problem, because there are protocols which can deal with this, such as Bartlett *et. al.* Alternating Bit Protocol (ABP) [11]. However, the ABP does not handle unstable senders and receivers. An unstable receiver may cause a message duplication, while an unstable sender can cause a message loss. A message duplication is unpleasant, but can be lived with and on the receiver side even dealt with, because the receiver is aware that it just started. However, a potential message lost is clearly unacceptable.

The Resettable Receiver Alternating Bit Protocol (RRABP) solves the problems which unstable senders introduce. This is achieved by extending the ABP with a mechanism with which the sender can reset the receiver and thus avoid the potential message loss. This was achieved by introducing three new messages, instead of only a single reset message as we anticipated. However, model checking revealed undesired resets of the receiver and message losses when only a single reset message was in use.

The next section gives an introduction to the Alternating Bit Protocol, which is the foundation to the Resettable Receiver Alternating Bit Protocol. Section 2 develops the formal specification of the protocol, followed by the formal model of the RRABP in Section 3. Section 4 establishes security, stability, and functionality of the protocol. The paper closes with conclusions and further work.

1. Materials and Methods

The RRABP is designed as an extension of the alternating bit protocol. This section explains how the ABP overcomes unreliable communication channels. After detailing the ABP this section gives a model for an unreliable communication channel. These details are based upon Roscoe's description and model of the alternating bit protocol in [12, page 130ff].

1.1. Alternating Bit Protocol

The alternating bit protocol is designed for point-to-point communication systems like the one illustrated in Figure 1. This system consists of the sender S, the unreliable bidirectional communication channel *CHAN* and the receiver R. To overcome the unreliable communication channel, S appends a tag-bit to each data-message, input from the channel *in*. This tag-bit alternates between 0 and 1 for each data-message. This allows R to identify replications

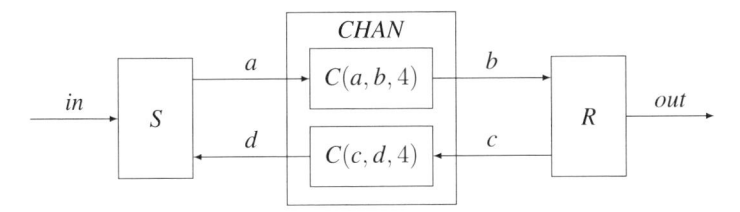

Figure 1. Process network structure used to model a point-to-point communication system.

of the data-message caused by the channel. To counter the loss of data-messages introduced by the unreliable communication channel, the sender S sends multiple copies of each data-message until it receives an acknowledgement from R, with the correct tag-bit. R has to cope with CHAN losing acknowledgments and therefore sends acknowledgements for the last received data-message until it receives a new data-message. The resulting protocol uses as few different messages as possible to fulfil its duty, which makes its implementation lightweight and suitable for embedded systems. An unexpected outcome of this work is the large number of messages necessary to reliably reset the receiver. This has far reaching consequences, because we can show that a certain complexity is necessary to meet the specification. All models or implementations which are less complex do not meet the specification.

1.2. Modelling an Unreliable Channel

The unreliable bidirectional communication channel CHAN (Equation 1) consists of two unreliable communication channels $C(i, o, n)$ (Equation 2). These processes input messages from the channel i and may output them on the channel o. The parameter n defines the maximum length of a burst error, *i.e.* how many consecutive messages get dropped (lost) at most and the maximum number of additional copies of the original message. In this paper the maximum burst error length is set to four. This means that the protocol must be able to deal with at most four lost messages in a row, or four duplications of a message.

$$CHAN = C(a,b,4) \;|||\; C(c,d,4) \tag{1}$$

with

$$C(i,o,n) = C'(i,o,n,n)$$
$$C'(i,o,n,r) = i?x \rightarrow C''(i,o,x,n,r)$$
$$C''(i,o,x,n,r) = \begin{pmatrix} \textbf{if } r = 0 \textbf{ then} \\ \quad o!x \rightarrow C'(i,o,n,n) \\ \textbf{else} \\ \quad C'''(i,o,x,n,r) \end{pmatrix} \tag{2}$$
$$C'''(i,o,x,n,r) = o!x \rightarrow C'(i,o,n,n)$$
$$\sqcap\; o!x \rightarrow C''(i,o,x,n,r-1)$$
$$\sqcap\; C'(i,o,n,r-1)$$

1.2.1. Properties of CHAN

We expect the unreliable communication channel CHAN (Equation 1) to be non-deterministic, but deadlock and livelock free. The FDR [13] output shown in Figure 2 shows that our expectations are correct. The non-determinism of the process is caused by the fact that the process chooses internally whether it drops or replicates a message, unless it has reached the maximum number of message drops or replications, upon which it must send the current message and then reset its drop / replication counter.

> ✗• CHAN deterministic [FD]
> ✓ CHAN deadlock free [F]
> ✓ CHAN livelock free

Figure 2. FDR output after checking determinism, deadlock and livelock properties of *CHAN*.

2. The Resettable Receiver Alternating Bit Protocol Specification

The Resettable Receiver Alternating Bit Protocol (RRABP) uses the idea of a tag-bit, introduced by the alternating bit protocol, to overcome the undesirable properties of the unreliable communication channel. Furthermore, it is a sender driven protocol which means that the receiver generates acknowledgements only as a response to a message received from the sender. This prevents surplus messages in case this acknowledgement has been received already, but the sender has no new data-messages to send.

The RRABP ensures that the first message sent by an unreliable sender, *i.e.* a sender which suddenly fails and is subsequently replaced, is not lost. In a system which uses the ABP protocol the first sent data-message gets lost when the receiver classifies it as a duplication of a previous data-message. This happens when the replaced sender uses the same value for the tag-bit which was used by the last transmitted data-message. Clearly, this is undesirable, therefore a protocol extension is necessary which allows the sender to inform the receiver about its replacement. The RRABP achieves this functionality by resetting the receiver. The sender performs this receiver reset during its initialisation. Resetting the receiver forces it into a predefined state, *i.e.* the tag-bit is set to its default value.

The protocol must ensure that no messages are duplicated or lost, when the unreliable sender fails and subsequently is replaced. However, there are two conditions in which it is out of the hands of the protocol and data-messages may be lost nevertheless:

1. The sender dies before it could send the data-message to the receiver. Hence, the message is definitely lost.
2. The sender dies after sending the data-message but before receiving an acknowledgment for the message. This does not necessarily mean that the message was not received by the receiver, but the message may have been lost by the communication channel! It is hard if not impossible to handle this, because the new instance of the sender does not know about the last message which its predecessor tried to send.

Both instances could be handled by introducing a reliable process in front of the sender which waits for an acknowledgment by the sender before supplying a new message to the sender process. However, in the scenario where a complete IMS loses power, this process would terminate as well, making this process unreliable as well.

This section concentrates on showing that the RRABP works in two scenarios:

1. Normal operation: sender and receiver are stable. The corresponding specification is the process *COPY*.
2. Unstable sender together with a stable receiver, with the corresponding specification *SD_SPEC*, where *SD* stands for 'Sender Dies'.

2.1. Normal Operation

The specification model for normal operation behaves like the so called *COPY* process (Equation 3). This means every message on the channel *in* gets output on channel *out*.

$$COPY = in?x \rightarrow out!x \rightarrow COPY \qquad (3)$$

> ✓ COPY deterministic [FD]
> ✓ COPY deadlock free [F]
> ✓ COPY livelock free

Figure 3. FDR output after checking determinism, deadlock and livelock properties of *COPY*.

2.1.1. Properties of COPY

We expect the specification model for normal operation (*COPY*) to be deterministic, deadlock and livelock free. The FDR output of Figure 3 shows that our expectations are correct.

2.2. Unreliable Sender

The RRABP ensures that the receiver keeps the first data-message sent by a replaced sender. Furthermore, once the restart has been completed the system should not lose or duplicate any data-messages.

SD_SPEC (*SD* stands for Sender Dies), Equation 4, specifies how the communication system should behave when a sender suddenly dies. The process is a parallel composition of the sender specification (process *SD_S_SPEC*, Equation 6) and the receiver specification (process *SD_R_SPEC*, Equation 9). The *sender_dies* event represents a failure of the sender, which results in a replacement of the sender. The specification exposes this event to the environment for refinement checks with the implementation model presented in Section 3.3. Figure 4 illustrates the relationship between *SD_S_SPEC* and *SD_R_SPEC* processes. The two processes exchange messages over the channel *tc*. They synchronise on the *ready* event: this event represents a successful synchronisation between sender and receiver processes. Furthermore, they synchronise on the event *sender_dies*, for reasons explained in the discussion of the receiver side specification in Section 2.2.2.

$$SD\_SPEC = SD\_S\_SPEC \underset{\alpha SD\_SYNC}{\|} SD\_R\_SPEC \setminus \{| \, ready, tc \, |\} \quad (4)$$

with

$$\alpha SD\_SYNC = \{| \, ready, tc, sender\_dies \, |\} \quad (5)$$

2.2.1. Sender Side Specification

Process *SD_S_SPEC* (Equation 6) represents, together with *SD_S_SPEC'* (Equation 7) and *SD_S_SPEC''(m)* (Equation 8), the replaceable sender. Initially, *SD_S_SPEC* waits for the occurrence of a *ready* event, or a *sender_dies* event. The *ready* event signals that the receiver accepts data-messages now. After synchronising on the *ready* event, the process transits to the state *SD_S_SPEC'* (Equation 7).

$$\begin{aligned} SD\_S\_SPEC = \; & ready \rightarrow SD\_S\_SPEC' \\ & \square \; sender\_dies \rightarrow SD\_S\_SPEC \end{aligned} \quad (6)$$

SD_S_SPEC' accepts the events: *sender_dies* and *in.m*. When it encounters *sender_dies* it transits to state *SD_S_SPEC*, to model the restart of the sender. In case of *in?m* the process reads in the message from the channel and then transits to the state *SD_S_SPEC''(m)*.

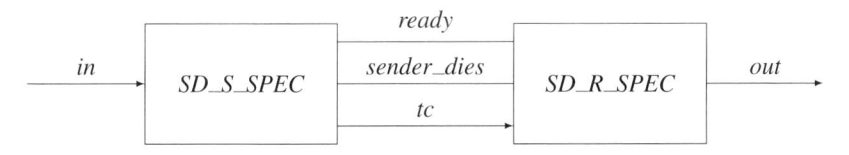

Figure 4. The *SD_SPEC* process detailled

$$SD\_S\_SPEC' = in?m \to SD\_S\_SPEC''(m) \qquad (7)$$
$$\Box \; sender\_dies \to SD\_S\_SPEC$$

$SD\_S\_SPEC''(m)$ offers to output the message m on channel tc, after which it transits to SD_S_SPEC. When it encounters *sender_dies* it transits to SD_S_SPEC, losing its message. This models the replacement of the sender.

$$SD\_S\_SPEC''(m) = tc!m \to SD\_S\_SPEC \qquad (8)$$
$$\Box \; sender\_dies \to SD\_S\_SPEC$$

The sender side specification is now complete.

2.2.2. Receiver side specification

SD_R_SPEC (Equation 9) is the entry point of the receiver specification. Unless the event *sender_dies* occurs, this process signals that it is *ready* and then transits to state SD_R_SPEC' (Equation 9).

$$SD\_R\_SPEC = ready \to SD\_R\_SPEC' \qquad (9)$$
$$\Box \; sender\_dies \to SD\_R\_SPEC$$

SD_R_SPEC' offers to read a data-message from channel tc. After this it transits to $SD\_R\_SPEC''(m)$ (Equation 11). If instead *sender_dies* occurs the process transits to SD_R_SPEC. This models the case that the receiver expects to be reset by the sender.

$$SD\_R\_SPEC' = tc?m \to SD\_R\_SPEC''(m) \qquad (10)$$
$$\Box \; sender\_dies \to SD\_R\_SPEC$$

$SD\_R\_SPEC''(m)$ (Equation 11), outputs the data-message m to the receiver backend, by outputting it on channel *out*. Upon successfully outputting the data-message the process transits to SD_R_SPEC, this completes an uninterrupted message transfer. If instead *sender_dies* occurs the receiver still outputs the previously received data-message. This is modelled by $SD\_R\_SPEC''(m)$ recursing upon engaging in *sender_dies*.

$$SD\_R\_SPEC''(m) = out!m \to SD\_R\_SPEC \qquad (11)$$
$$\Box \; sender\_dies \to SD\_R\_SPEC''(m)$$

This completes the specification of how the RRABP system should behave to the outside world, when an unstable sender is in use.

2.2.3. Properties of SD_SPEC

The process SD_SPEC (Equation 4) is expected to be non-deterministic, deadlock, and livelock free. The specification SD_SPEC is non-deterministic because the events *ready* and *it are hidden*, i.e. internal. Thus the outside cannot determine whether or not the input by SD_S_SPEC has been transferred to SD_R_SPEC, before the sender has been replaced. To check whether or not this is the case we employed FDR2, Figure 5 shows that FDR supports our expectations. This completes the specification of the protocol.

```
✗• SD_SPEC deterministic [FD]
✓  SD_SPEC deadlock free [F]
✓  SD_SPEC livelock free
```

Figure 5. FDR output after checking determinism, deadlock and livelock properties of SD_SPEC.

3. Formal Protocol Model

The RRABP deals with the situation that the sender, and *only* the sender, can be replaced at any moment. Replacing sender S in the system of Figure 1 (page 205), with a new instance of the sender S, means that receiver R and the unreliable bidirectional communication channel *CHAN* stay operational. Both R and *CHAN* have memory, this means that they can carry on with their normal operation using the messages currently available to them. The synchronisation routine, for the initial synchronisation, must reliably reset R independent of the state R or *CHAN* are in. Furthermore, it needs to deal with the properties of *CHAN*, which can lose or replicate data- and synchronisation-messages.

To overcome the degrading communication channel effects, the protocol applies the idea of the Alternating Bit Protocol. S repeatedly sends out copies of a message until it receives an acknowledgement from R. Because messages as well as acknowledgements can be lost by *CHAN*, R has to acknowledge every message it receives. To filter out replications during the synchronisation process, the protocol uses different messages instead of the tag-bit. The main problem, when designing the synchronisation part of the RRABP, was to find the correct number of synchronisation-messages.

3.1. Determining the Necessary Number of Synchronisation Messages

Initially, we thought one message would be sufficient. This however, is only the case if *CHAN* is replaced together with the sender, *i.e.* if *CHAN* gets replaced together with the sender S^1. The problem that occurs with only a single reset-message is that the reset-acknowledgement may still be waiting for delivery in $C(c, d, 4)$, while the receiver has already processed its first data-message and is about to acknowledge it. If S now gets replaced, the new sender instance assumes, after receiving a reset-acknowledgment from the previous receiver synchronisation, that R is ready for a data-message. Thus the sender will not generate a reset-message, but instead send a data-message with the tag-bit set to 0. The receiver however classifies this data-message as a replication of a previous data-message and therefore discards it. Forcing the sender to always generate at least one reset-message, is not a solution, because this message may be lost by the channel. Finally, this implementation also violates the specification *SD_SPEC* in another respect. The new instance of S can input new data before R has output the data sent by the previous instance of S^2.

It is possible to construct a similar example when using two synchronisation-messages, which does not behave correctly when S gets replaced twice within a very short interval[3]. So we increased the number of synchronisation-messages to three for the RRABP, and we have used FDR to demonstrate that this solved the above problem. The remainder of this section explains how the protocol works. The three synchronisation messages used in the RRABP are: *stop*, *reset*, and *start*. The next section details their meaning and encoding together with the remainder of the protocol vocabulary.

3.2. Protocol Vocabulary

The RRABP uses the following messages:

1. *stop.msg*: Informs the receiver that the sender has just been started and is now synchronising with the receiver.

---

[1] The corresponding CSP_M script is available at:
http://www.abdn.ac.uk/piprg/Papers/CPA2008/RRABP/rrabp_sm_SC_R.csp
[2] The corresponding CSP_M script is available at:
http://www.abdn.ac.uk/piprg/Papers/CPA2008/RRABP/rrabp_sm_S_RC.csp
[3] The corresponding CSP_M script is available at:
http://www.abdn.ac.uk/piprg/Papers/CPA2008/RRABP/rrabp_dm_S_RC.csp

2. *stop.ack*: Acknowledgement from the receiver that it received the stop message.
3. *reset.msg*: Informs the receiver that it should reset itself. This is the second message of the synchronisation sequence.
4. *reset.ack*: Reset acknowledgement from the receiver.
5. *start.msg*: Informs the receiver that the sender wants to start data communications. This is the third and last message of the synchronisation sequence.
6. *start.ack*: Acknowledgement from the receiver that it is ready to start data communication.
7. *data.msg.s.d*: This message transports the data *d* from the sender to the receiver, and *s* represents the tag-bit.
8. *data.ack.s*: This message acknowledges the correct handling of a data message with the tag-bit *s*.

3.2.1. Encoding the Vocabulary

In the description of the protocol we use the following data-types to construct the messages exchanged between sender and receiver:

- *OP* represents the different operations:

$$\textbf{datatype } OP = stop \mid reset \mid start \mid data \tag{12}$$

- *MT* represents the different message types:

$$\textbf{datatype } MT = msg \mid ack \tag{13}$$

- The set *TAG* represents the tag-bit:

$$TAG = \{0, 1\} \tag{14}$$

- The set *DATA* represents the data that can be transferred by the system:

$$DATA = \{0, 1\} \tag{15}$$

The type for all messages is the tuple: *OP.MT.TAG.DATA*. Furthermore, the encoded data-messages are defined as:

- *data.msg.s.d*: This message transports the data *d* from the sender to the receiver, with *s* representing the tag-bit.
- *data.ack.s.d*: This message acknowledges the correct handling of a data message with the tag-bit *s*. The data part *d* of the message must be ignored[4]!

For the synchronisation-messages: *stop.msg.s.d*, *stop.ack.s.d*, *reset.msg.s.d*, *reset.ack.s.d*, *start.msg.s.d*, and *start.ack.s.d*, the tag-bit *s* and the data part *d* are irrelevant.

3.3. Formal Model of Message Exchanges

This section details the communication between sender *S* and receiver *R*. It explains both the internals of sender and receiver.

---

[4]The inclusion of the data part is necessary because FDR insists on channel structure having the same number of components.

3.3.1. Sender

The entry point of the sender is $SENDER(cin, cout)$ (Equation 16), the where cin is the channel from which the process reads the acknowledgements, and $cout$ is the channel to which the process sends synchronisation- and data-messages. The $SENDER(...)$ process transfers stop-messages to the receiver. This is achieved by outputting the stop-message ($stop.msg.0.0$) on channel $cout$ before recursing, until receiving a stop-acknowledgement ($stop.ack.x.y$) in reply. Upon which the process transits to $S\_RESET(cin, cout)$ (Equation 17). Any other synchronisation- ($reset.ack$, $start.ack$) or data-acknowledgement ($data.ack$) pending on channel cin are swallowed by $SENDER(...)$, to avoid deadlocks.

$$\begin{aligned}
SENDER(cin, cout) = \\
cout!stop.msg.0.0 \rightarrow SENDER(cin, cout) \\
\Box\ cin?stop.ack.x.y \rightarrow S\_RESET(cin, cout) \\
\Box\ cin?reset.ack.x.y \rightarrow SENDER(cin, cout) \\
\Box\ cin?start.ack.x.y \rightarrow SENDER(cin, cout) \\
\Box\ cin?data.ack.x.y \rightarrow SENDER(cin, cout)
\end{aligned} \qquad (16)$$

The $S\_RESET(cin, cout)$ process is similar to the $SENDER(...)$ process, except that it sends reset-messages to the receiver and expects a reset-acknowledgement before transiting to the state $S\_START(cin, cout)$ (Equation 17). The process swallows any stop-acknowledgements, which are duplicates of a previously received stop-acknowledgement. However, at no time should the process receive a data-acknowledgement or a start-acknowledgement. Occurrences of these signal that the previously received stop-acknowledgement was not triggered by the sent stop-message. Therefore, the sender must try to reset the receiver once more. To do so the process transits to the state $SENDER(...)$.

$$\begin{aligned}
S\_RESET(cin, cout) = \\
cout!reset.msg.0.0 \rightarrow S\_RESET(cin, cout) \\
\Box\ cin?reset.ack.x.y \rightarrow S\_START(cin, cout) \\
\Box\ cin?stop.ack.x.y \rightarrow S\_RESET(cin, cout) \\
\Box\ cin?start.ack.x.y \rightarrow SENDER(cin, cout) \\
\Box\ cin?data.ack.x.y \rightarrow SENDER(cin, cout)
\end{aligned} \qquad (17)$$

The last process, concerned with resetting the receiver is $S\_START(cin, cout)$ (Equation 18). This process starts the receiver to make it ready to receive data-messages. This is done by outputting the start-message ($start.msg.0.0$) on the channel $cout$ and then recursing, until a start-acknowledgement ($start.ack.x.y$) arrives on channel cin. Upon reception of the start acknowledgement the process transits to $S\_RUN(cin, cout, 0)$, the 0 represents the initial value of the tag-flag. Furthermore, the process filters any reset-acknowledgements. Reception of a data-acknowledgment means that the receiver is completely out of sync, hence it is necessary to restart the reset sequence again. So the process transits to $SENDER(...)$. Receiving of a stop-acknowledgement results in a transit to $S\_RESET(...)$, because this represents the state the receiver is currently in.

$$S\_START(cin, cout) =$$
$$cout!start.msg.0.0 \rightarrow S\_START(cin, cout)$$
$$\square \ cin?start.ack.x.y \rightarrow S\_RUN(cin, cout, 0)$$
$$\square \ cin?reset.ack.x.y \rightarrow S\_START(cin, cout) \quad (18)$$
$$\square \ cin?stop.ack.x.y \rightarrow S\_RESET(cin, cout)$$
$$\square \ cin?data.ack.x.y \rightarrow SENDER(cin, cout)$$

The main task of the process $S\_RUN(\ldots)$ (Equation 19) is to wait for a message (m) to be transmitted from its backend over the channel in. Once it has received m the process transits to $S\_RUN'(\ldots)$ (Equation 20). The process swallows any reset-, start-, and data-acknowledgments, to prevent deadlocks.

$$S\_RUN(cin, cout, s) =$$
$$in?m \rightarrow S\_RUN'(cin, cout, s, m)$$
$$\square \ cin?reset.ack.x.y \rightarrow S\_RUN(cin, cout, s) \quad (19)$$
$$\square \ cin?start.ack.x.y \rightarrow S\_RUN(cin, cout, s)$$
$$\square \ cin?data.ack.x.y \rightarrow S\_RUN(cin, cout, s)$$

$S\_RUN'(cin, cout, s, m)$ (Equation 20) appends the tag-bit to the message m and sends this data-message to the receiver before recursing. The process waits for the corresponding data-acknowledgement upon which it toggles the tag-bit and transits to $S\_RUN(\ldots)$. The process filters data-acknowledgements for a previous data-message as well as any reset- and start-acknowledgements.

$$S\_RUN'(cin, cout, s, m) =$$
$$cout!data.msg.s.m \rightarrow S\_RUN'(cin, cout, s, m)$$
$$\square \ cin?data.ack.s.y \rightarrow S\_RUN(cin, cout, 1-s)$$
$$\square \ cin?data.ack.(1-s).y \rightarrow S\_RUN'(cin, cout, s, m) \quad (20)$$
$$\square \ cin?reset.ack.x.y \rightarrow S\_RUN'(cin, cout, s, m)$$
$$\square \ cin?start.ack.x.y \rightarrow S\_RUN'(cin, cout, s, m)$$

This completes the sender model, now follows the description of the receiver model.

3.3.2. Receiver

The receiver consists of five different processes: $R\_RUN(\ldots)$, $R\_RUN'(\ldots)$, $R\_STOP(\ldots)$, $R\_START(\ldots)$, and $RECEIVER(\ldots)$. During normal operation the receiver toggles between $R\_RUN(\ldots)$ and $R\_RUN'(\ldots)$. The process $RECEIVER(\ldots)$ represents the entry point of the receiver, it determines the current value of the tag-bit. The remaining two processes synchronise the value of the tag-bit when the sender gets replaced.

Initially, the receiver side data handling functionality is in the state $RECEIVER(\ldots)$ (Equation 21). In this state it waits for a data-message to arrive: $cin?data.msg.x.m$, where the variable x holds the value of the tag-bit, and the variable m the received message. After receiving a data-message the process outputs the message (m) to the backend and sends a data-acknowledgment to the sender. After this the receiver transits to the state $R\_RUN(cin, cout, s, m)$ (Equation 22), setting the parameter s to the value of $(1-x)$ (toggling the tag-bit). This initial acceptance of any data-message allows an unstable receiver to synchronise with the sender upon a restart. However, there is the danger of duplicating the last data-message received before the replacement of the receiver. Due to accepting any arriving data-message as valid, there is no need to synchronise the tag-bit. Therefore, the *RECEIVER*(...) acknowledges any arriving stop-, reset-, or start-message and then recurses.

$$
\begin{aligned}
RECEIVER&(cin, cout) = \\
&cin?data.msg.x.m \to cout!data.ack.x.0 \to R\_RUN(cin, cout, 1-x) \\
&\square\ cin?stop.msg.x.y \to cout!stop.ack.0.0 \to R\_RESET(cin, cout) \\
&\square\ cin?reset.msg.x.y \to cout!reset.ack.0.0 \to RECEIVER(cin, cout) \\
&\square\ cin?start.msg.x.y \to cout!start.ack.0.0 \to RECEIVER(cin, cout)
\end{aligned}
\tag{21}
$$

R_RUN(*cin*, *cout*, *s*) (Equation 22) is similar to the previously discussed process *RECEIVER*(...), except that it has the ability to filter out replications of data-messages. The parameter *s* represents value of the tag-bit for new data-messages. Upon inputting a new data-message (*cin*?*data.msg.s.m*) the process outputs the received message (*m*) to the backend. Once this has happened, a data acknowledgement gets sent to the sender and the process toggles the tag-bit while recursing. After inputting a replication of a previous data-message (*cin*?*data.msg*.(1 − *s*).*m*), the process acknowledges it and recurses. Upon receiving a stop message the process outputs a stop-acknowledgement on channel *cout* then it transits to *R_RESET*(...) (Equation 23). The process acknowledges any arriving reset- or start-messages and then recurses.

$$
\begin{aligned}
R\_RUN&(cin, cout, s) = \\
&cin?data.msg.s.m \to cout!data.ack.s.0 \to R\_RUN(cin, cout, 1-s) \\
&\square\ cin?data.msg.(1-s).m \to cout!data.ack.(1-s).0 \to R\_RUN(cin, cout, s, m) \\
&\square\ cin?stop.msg.x.y \to cout!stop.ack.0.0 \to R\_RESET(cin, cout) \\
&\square\ cin?reset.msg.x.y \to cout!reset.ack.0.0 \to R\_RUN(cin, cout, s) \\
&\square\ cin?start.msg.x.y \to cout!start.ack.0.0 \to R\_RUN(cin, cout, s)
\end{aligned}
\tag{22}
$$

R_RESET(*cin*, *cout*) (Equation 23) represents the first tag-bit synchronisation state of the receiver. The receiver enters this state when it receives a stop-message during normal operation conditions. In this state the receiver waits for a reset-message to arrive, which the receiver acknowledges and changes to *R_START*(*cin*, *cout*) (Equation 24). *R_RESET*(...) acknowledges any incoming stop- and start-messages and then recurses.

$$
\begin{aligned}
R\_RESET&(cin, cout) = \\
&cin?reset.msg.x.y \to cout!reset.ack.x.y \to R\_START(cin, cout) \\
&\square\ cin?stop.msg.x.y \to cout!stop.ack.0.0 \to R\_RESET(cin, cout) \\
&\square\ cin?start.msg.x.y \to cout!start.ack.0.0 \to R\_RESET(cin, cout)
\end{aligned}
\tag{23}
$$

The receiver process *R_START*(*cin*, *cout*) (Equation 24) is the second state of tag-bit synchronisation. In this state the receiver expects to receive a start-message, which it acknowledges. The tag-bit synchronisation sequence is now complete and the tag-bit is now set to 0. Now the receiver transits to *R_RUN'*(*cin*, *cout*, 0), with the tag-bit set to 0. In state *R_START*(...) the receiver acknowledges any incoming stop- and reset-messages before recursing.

$$R\_START(cin, cout) =$$
$$cin?start.msg.x.y \rightarrow cout!start.ack.0.0 \rightarrow R\_RUN'(cin, cout, 0)$$
$$\Box \; cin?reset.msg.x.y \rightarrow cout!reset.ack.0.0 \rightarrow R\_START(cin, cout) \quad (24)$$
$$\Box \; cin?start.msg.x.y \rightarrow cout!start.ack.0.0 \rightarrow R\_START(cin, cout)$$

This completes the RRABP message exchange model. The next section details the formal model verification.

4. Formal Model Verification

The formal description of the RRABP does not automatically guarantee that the protocol possesses all desired properties. However, using a formal specification makes it possible to establish the protocol properties. This section shows the compliance of the protocol with the specifications for both: normal operation *COPY* (Section 2.1), and systems with unstable senders *SD_SPEC* (Section 2.2). For model checking the the model checker FDR2 [13] was used.

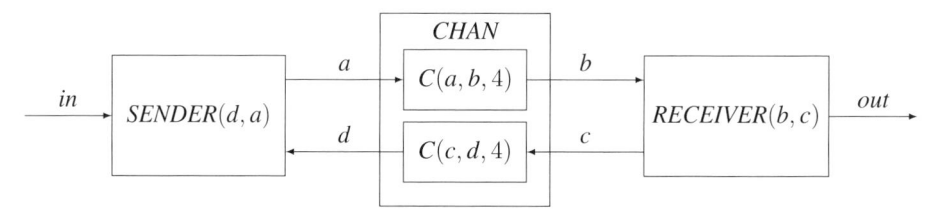

Figure 6. Process Network Layout for *RRABP_NO*.

4.1. Normal Operation

The process *RRABP_NO* (Equation 25) represents a communication system which consists of reliable sender and reliable receiver. They use the RRABP to communicate over an unreliable communication channel. The process, illustrated in Figure 6, is the parallel composition of *SENDER*(...) and *RECEIVER*(...) with *CHAN* (Equation 1), where *CHAN* represents the unreliable bidirectional communication channel.

$$RRABP\_NO = (SENDER(a, d) \underset{\{|a,d|\}}{\parallel} CHAN) \underset{\{|b,c|\}}{\parallel} RECEIVER(a, d) \setminus \{|\,a, b, c, d\,|\} \quad (25)$$

To prove that *RRABP_NO* is equivalent to *COPY* (Equation 3) we perform a failure divergence cross refinement check. Figure 7 shows that *RRABP_NO* has the same traces, failures, and divergences as *COPY*.

Figure 7. FDR screen-shot as proof that *RRABP_NO* is equivalent to *COPY*

This firmly establishes that the RRABP made the unreliable channel *CHAN*, reliable, as long as both sender and receiver stay operational.

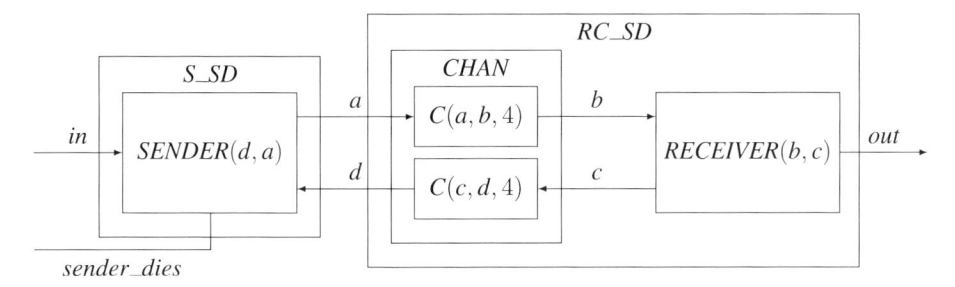

Figure 8. Process Network Layout for *RRABP_SD*.

4.2. Unreliable Sender

The process *RRABP_SD* (Equation 26) represents a communication system with an unreliable sender communicating with a receiver over an unreliable bidirectional communication channel. The sender and receiver of this system communicate using the RRABP. This system must comply with the specification given by *SD_SPEC* (Equation 4). Figure 8 illustrates the process network which represents the *RRABP_SD* process.

$$RRABP\_SD = S\_SD \underset{\{|a,d|\}}{\parallel} RC\_SD \setminus \{|\, a,b,c,d\, |\} \tag{26}$$

SD_RRABP is a parallel composition of two processes:

- *S_SD* (Equation 27) represents the sender of the communication system. The functionality covers unannounced sender replacement at any time. To model sender destruction it can be interrupted at any time by the event *sender_dies*. When this happens *SENDER*(...) passes control back to *S_SD*, which recursively creates a new instance of the sender.

$$S\_SD = SENDER(d,a) \triangle sender\_dies \rightarrow SD\_S \tag{27}$$

- *RC_SD* (Equation 28) represents the receiver and the communication channels of the system.

$$RC\_SD = RECEIVER(b,c) \underset{\{|b,c|\}}{\parallel} CHAN \setminus \{|\, b,c\, |\} \tag{28}$$

To check that the RRABP can deal with an unexpected sender restart, it is necessary to establish that *RRABP_SD* and *SD_SPEC* (Equation 4 on page 207) are equivalent. We establish this equivalence by performing a failure divergence cross refinement check of *SD_SPEC* and *RRABP_SD*. This refinement checks that both processes exhibit the same traces and fail (deadlock) or diverge (livelock) identically. Figure 9 shows that *RRABP_SD* has the same failures and divergences as *SD_SPEC*. This establishes security, stability, and functionality of the protocol even when a sender gets replaced during runtime. The complete model of the RRABP is available as CSP_M script for download at: http://www.abdn.ac.uk/piprg/Papers/CPA2008/RRABP/rrabp_tm_S_RC.csp.

```
✓ SD_SPEC [FD= RRABP_SD
✓ RRABP_SD [FD= SD_SPEC
```

Figure 9. Proof of *RRABP_SD* is equivalent to *SD_SPEC*

5. Conclusion and Further Work

In this paper we establish that the RRABP enables reliable communication over an unreliable channel, which connects an unstable sender with a stable receiver. We underpin this claim by explaining how to model a point-to-point communication system using CSP, including a model of unreliable communication channels. This was followed by a description of the Alternating Bit Protocol, which forms the basis for the Resettable Receiver Alternating Bit Protocol (RRABP). After this, we developed the formal specification of the service the RRABP provides. This represents the specification of the RRABP. The RRABP was then outlined, first verbally and then formally. Finally, this formal model of the RRABP was proven to be equivalent to the previous given specifications.

During the development of the formal protocol description we discovered that a single reset message is not sufficient, if the communication channels stay operational while the sender is exchanged. In fact it was necessary to use a total of three different messages to trigger a reset of the receiver, fewer messages resulted in lost messages.

The RRABP is applicable in any environment where one wants to establish a reliable link with blocking capability over an unreliable (message replication and message loss) communication channel, which does not block. Furthermore, the protocol is applicable in environments with unstable senders. Initially, we targeted the RRABP at the water monitoring system of the WARMER consortium. In this system, unstable in-situ monitoring stations want to transfer data reliably to the data centre. Other environments which require the RRABP features include software defined radio systems which change their signal processing process networks over time, and sensor networks. Besides these systems, we are presently considering using the RRABP to establish a channel between a libCSP2 [14] process network executed by a PC and a libCSP2 process network executing on a soft processor within an FPGA on a PCI card. This requires extension of the RRABP to support bidirectional communication. Another interesting investigation is to determine the effect buffered communication channels have upon the RRABP. A last item, which is worth investigating, is to determine how the protocol copes with unstable senders and receivers.

Acknowledgements

This work has been supported by the EC FP6 project WARMER (**WA**ter **R**isk **M**anagement in **EuR**ope) project (contract no.: FP6-034472).

References

[1] F. Barnes and P. H. Welch. Communicating Mobile Processes. In Ian R. East, David Duce, Mark Green, Jeremy M. R. Martin, and Peter H. Welch, editors, *Communicating Process Architectures 2004*, volume 62 of *WoTUG-27, Concurrent Systems Engineering, ISSN 1383-7575*, pages 201–218, Amsterdam, The Netherlands, September 2004. IOS Press. ISBN: 1-58603-458-8.

[2] Ali E. Abdallah, Cliff B. Jones, and Jeff W. Sanders, editors. *Communicating Sequential Processes: The First 25 Years, Symposium on the Occasion of 25 Years of CSP, London, UK, July 7-8, 2004, Revised Invited Papers*, volume 3525 of *Lecture Notes in Computer Science*. Springer, 2005.

[3] K. Chalmers and J. M. Kerridge. jcsp.mobile: A Package Enabling Mobile Processes and Channels. In Jan Broenink, Herman Roebbers, Johan Sunter, Peter H. Welch, and David Wood, editors, *Communicating Process Architectures 2005*, pages 109–127, sep 2005.

[4] P. H. Welch. Graceful termination – graceful resetting. In Andrè W. P. Bakkers, editor, *OUG-10: Applying Transputer Based Parallel Machines*, pages 310–317, 1989.

[5] B. Sputh and A. R. Allen. JCSP-Poison: Safe Termination of CSP Process Networks. In Jan Broenink, Herman Roebbers, Johan Sunter, Peter H. Welch, and David Wood, editors, *Communicating Process Architectures 2005*, September 2005.

[6] P. H. Welch, N. C. Brown, J. Moores, K. Chalmers, and B. Sputh. Integrating and Extending JCSP. In Alistair A. McEwan, Wilson Ifill, and Peter H. Welch, editors, *Communicating Process Architectures 2007*, pages 349–369, July 2007.

[7] N. C. Brown. C++CSP Networked. In Ian R. East, David Duce, Mark Green, Jeremy M. R. Martin, and Peter H. Welch, editors, *Communicating Process Architectures 2004*, pages 185–200, sep 2004.

[8] G. H. Hilderink. Exception Handling Mechanism in Communicating Threads for Java. In Jan Broenink, Herman Roebbers, Johan Sunter, Peter H. Welch, and David Wood, editors, *Communicating Process Architectures 2005*, pages 317–334, sep 2005.

[9] B. Sputh, O. Faust, and A. R. Allen. A Versatile Hardware-Software Platform for In-Situ Monitoring Systems. In Alistair A. McEwan, Wilson Ifill, and Peter H. Welch, editors, *Communicating Process Architectures 2007*, pages 299–312, July 2007.

[10] B. Sputh, O. Faust, and A. R. Allen. Integration of In-situ and remote sensing Data for Water Risk Management. In *Proceedings of the iEMSs 4th Biennial Meeting: "Integrating Sciences and Information Technology for Environmental Assessment and Decision Making"*, Burlington, USA, July 2008. International Environmental Modelling and Software Society. To be published.

[11] K. A. Bartlett, R. A. Scantlebury, and P. T. Wilkinson. A note on reliable full-duplex transmission over half-duplex links. *Commun. ACM*, 12(5):260–265, 1969.

[12] A. W. Roscoe. *Theory and Practice of Concurrency*. Prentice Hall, Upper Saddle River, New Jersey 07485 United States of America, first edition, 1997. Download: http://web.comlab.ox.ac.uk/oucl/work/bill.roscoe/publications/68b.pdf Last Accessed April 2008.

[13] Formal Systems (Europe) Ltd. *Failures-Divergence Refinement: FDR Manual*. http://www.fsel.com/fdr2_manual.html Last Accessed June 2008.

[14] B. Sputh, O. Faust, and A. R. Allen. Portable CSP Based Design for Embedded Multi-Core Systems. In Frederick R. M. Barnes, Jon M. Kerridge, and Peter H. Welch, editors, *Communicating Process Architectures 2006*, pages 123–134, September 2006.

IC2IC: a Lightweight Serial Interconnect Channel for Multiprocessor Networks

Oliver FAUST, Bernhard H. C. SPUTH, and Alastair R. ALLEN

Department of Engineering, University of Aberdeen, Aberdeen AB24 3UE, UK
{b.sputh, o.faust, a.allen}@abdn.ac.uk

Abstract. IC2IC links introduce blocking functionality to a low latency high performance data link between independent processors. The blocking functionality was achieved with the so-called alternating bit protocol. Furthermore, the protocol hardens the link against message loss and message duplication. The result is a reliable way to transfer bit level information from one IC to another IC. This paper provides a detailed discussion of the link signals and the protocol layer. The practical part shows an example implementation of the IC2IC serial link. This example implementation establishes an IC2IC link between two configurable hardware devices. Each device incorporates a process network which implements the IC2IC transceiver functionality. This example implementation helped us to explore the practical properties of the IC2IC serial interconnect. First, we verified the blocking capability of the link and second we analysed different reset conditions, such as disconnect and bit-error.

Keywords. serial link, hardware channel, embedded systems, system on chip, network on chip, hardware software co-design, multi-core.

Introduction

Parallel processing systems must exchange data in a timely fashion between multiple physically separate processors. According to C. R. Anderson *et al* [1] effective exploitation of multiple processors in a distributed computing environment relies on a low latency, high bandwidth, inter-processor communication network. In the late 1980s the INMOS transputer was a pioneering attempt to built a communication network with these properties for multiprocessor computing [2]. The basic design of the transputer included serial links, that allowed it to communicate with up to four other transputers. The links operated at 5, 10 or 20 Mbit/s — which at the time was faster than existing networks such as Ethernet. The link speed matched the processing speed well, therefore the available resources could be used efficiently. This led to some notable and diverse applications such as neurocomputers [3] and architectures for graphics [4].

While the transputer was simple but powerful compared to many contemporary designs, it never came close to meeting its goal of being used universally in both CPU and microcontroller roles. In the microcontroller realm, the market was dominated by 8-bit machines where cost was the only serious consideration. Here, even the 16-bit transputers were too powerful and expensive for most applications. Furthermore, the concept of communicating processes alienated many users. These are only some of the reasons why the transputer family was not developed further. However, even after the development stopped, the link technology was still used in a limited number of applications. In 1995 the communication links were even standardised as IEEE-1355 [5]. This standard details also "Wormhole Routing" [6], which allows packets of unlimited length to be routed within the network of processors.

The IEEE-1355 link standard continues to generate interest. One reason for this continued interest is the ease of implementation when compared with competing technology such as Ethernet. All twisted pair versions of Ethernet require analogue signal processing of the received signals to extract data – a silicon hungry process [7]. This even led to the claim: "IEEE 1355 data-strobe links: ATM speed at RS232 cost", by Barry M. Cook [8]. On the application side, Greve, O. J. *et al* [9] argue that for heavy mechatronic and robot applications, transputer links can be used to achieve a high-performance and real-time communication between discrete system components. Marcel Boosten *et al* [10] have investigated the construction of a parallel computer using IEEE 1355 high-throughput low-latency DS link networks and high-performance commodity processors running a standard operating system.

Apart from direct applications, the IEEE-1355 technology has influenced a number of other communication standards. The IEEE-1355 encoding scheme has been adopted for the IEEE-1394 standard [11] and the Apple Computer version of IEEE-1394 known as FireWire. Space applications have very high demand on system reliability, because it is difficult, if not impossible, to make changes after the launch of a space craft. These requirements provided the reason why the transputer link technology was adopted as a new communication standard for space applications with relatively minor changes [12]. The European Space Agency plans to use the new SpaceWire standard for most of its future missions. The SpaceWire standard constitutes something like the rebirth of IEEE-1355. Therefore, most of the recent practical projects focus on this standard instead of the original IEEE-1355. In space applications radiation tolerance is very important. B. Fiethe *et al* [13] argue that protection against such effects can be achieved by using parts built of special technology, such as SpaceWire. In a similar argument, Sanjit A. Seshia *et al* [14] state that technology scaling[1] causes reliability problems to become a dominant design challenge. They support their statement with a case study where they analyse the stability of a publicly available implementation of SpaceWire end-nodes. Despite the name, there are attempts to use SpaceWire in earth-bound applications. Sergio Saponara *et al* [15] introduce the SpaceWire standard to the automotive field. This approach is justified, because both space and automotive applications have high demands on reliability of the communication standard.

Apart from these rather high level communication standard considerations, the ideas of transputer links also play a role in the design style for the communication between independent digital circuits. The main reason for this interest is the fact that the communication between independent circuits provides the basis for CSP style hardware design [16]. The two main problems are clock and reset synchronisation between independent circuits. The IEEE-1355 standard provides a solution for both problems. For wire bound communication the clock synchronisation is solved via self-clocking data-strobe signal communication. The reset synchronisation problem is solved with the exchange of silence protocol. However, CSP style communication between independent circuits requires a blocking capability which is not provided by IEEE-1355. The blocking capability ensures that a circuit which is committed to a data transfer can only make progress if the data is exchanged with the communication partner. A minor problem is that many hardware implementations require a higher degree of payload flexibility than the standard can offer. IEEE-1355 defines only data packets with 8 bit payload. This might be too much or too little, depending on the particular application.

This paper discusses the implementation of IC2IC, a lightweight serial interconnect link for multiprocessor networks. This implementation combines IEEE-1355 style synchronisation and encoding with a protocol based blocking functionality. We use an adaptation of the alternating bit protocol (ABP) to ensure the blocking functionality. The protocol guarantees that a particular data packet gets delivered to the receiver. The result is a stiff[2] communica-

---

[1]Decreasing feature size of digital integrated circuits over the years.
[2]The term 'stiff' describes the characteristic of the implementation with respect to the commitment of the

tion channel which is susceptible to external influences on the physical means of data exchange. We discuss the problems which arise from external influences, such as: bit-error, asynchronous reset and disconnect. To solve these problems was a design challenge, because the logic must react to each problem differently and in some cases 'remember' the system state during which the problem occurred. Both, hardening against message loss and formalised synchronisation, are the core concepts which make the IC2IC serial link reliable. Compared with the problems caused by external influences, the introduction of payload flexibility was relatively easy. Nevertheless, payload flexibility is an important feature of the IC2IC link.

The following section provides an in-depth discussion of the IC2IC serial interconnect channel. This includes a short introduction of the alternating bit protocol, which establishes the blocking functionality. Sections 1.1 and 1.2 introduce link signals and packet level of the protocol. After the definition of both signals and packets we define the IC2IC protocol. This protocol handles initialisation, data transfer and error conditions. Section 2 details an example implementation of the IC2IC serial link between two configurable logic devices. Each logic device executes an IC2IC transceiver circuit. With this setup we show blocking and error recovery in an implementation system.

1. IC2IC

This section introduces IC2IC, a lightweight bidirectional channel between two hardware components. The IC2IC functionality extends IEEE-1355 with blocking functionality. From the protocol standard IC2IC inherits the following functionality:

1. Self clocking — This ensures variable transmission speeds.
2. Asynchronous reset — Implemented through exchange of silence.
3. Disconnect detection — Both communication partners can detect and recover from disconnect.

The blocking communication causes problems after an asynchronous reset. The sender needs to know the state of the receiver when the data transfer resumes. The transfer of the last receiver state is a crucial part of the IC2IC protocol, without it the system is prone to data loss and deadlock.

1.1. Alternating Bit Protocol

Figure 1 shows a basic setup of a point to point communication system. In this system, port A (sender) sends data packets to port B (receiver) over a lossy communication channel. The channel can swallow a finite number of packets and duplicate the same packet a finite number of times. The first formalised solution for this problem emerged 1969 as a note on reliable full-duplex transmission over half-duplex links [17]. The solution involves an "alternation bit". This idea was picked up by A. W. Roscoe [18, page 130] who proposed a method called alternating bit protocol to establish a secure communication of the lossy channel. This protocol involves a back channel from port B to port A. This back channel has the same characteristic as the data communication channel. The name alternating bit protocol comes from the fact that each packet is extended by one bit which alternates between 0 and 1 before it is sent along the lossy channel. Multiple copies of the same data packet are sent from port A to port B until port A receives a control packet which acknowledges this data packet. The behaviour of port B depends on the environment. If the environment of port B has picked up

---

communication partners which engage in a message exchange over this link. Once a communication partner has committed there is no way to withdraw.

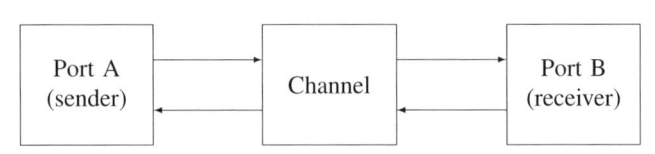

Figure 1. Basic communication system

the data from the previous packet then as soon as port B gets a new packet via the communication channel it sends an acknowledgement packet and offers the data to the environment. Port B acknowledges all resends of this data packet with a copy of the initial acknowledgement packet. In the case that the environment did not pick up the last message port B does not acknowledge the reception of a new data packet. Needless to say that port B does not acknowledge subsequent copies of this data packet until the environment picks up the data of the previous data packet. The two ports can always spot a new message or acknowledgement because of the alternating bit.

Sputh *et al* made relevant adjustments to the alternating bit protocol [19]. They establish safety, stability and functionally of the resettable receiver alternating bit protocol (RRABP). This extended alternating bit protocol creates a reliable and blocking channel between sender and receiver over unreliable non-blocking communication channels. Furthermore, this protocol allows the system to restart the sender at any time: however, not under all conditions without losing a message. The IC2IC serial link protocol is a practical realisation of the formal RRABP.

1.2. Link Signals

Each IC2IC port is connected to four link signals. The two signals tx_data and tx_strobe establish outgoing communication and rx_data and rx_strobe handle incoming communication. Figure 2 shows an IC2IC port as a self-contained entity with the ability to exchange data via the external communication signals. On the signal level the communication happens in two distinct phases, first synchronisation and then data transfer.

Figure 3 shows the timing diagram of the transmission signals in the synchronisation phase of the data exchange. This phase ensures that both communication partners are ready for the data transfer. During this initial phase both data and strobe signals have the same timing. The figure shows two signals, the transmitted data signal tx_data and the received data signal rx_data. At time point 1, the IC2IC transceiver under discussion is ready to transfer data, therefore it assigns low to the tx_data signal. A low timer circuit ensures that the low stays assigned for 100µs before it changes to high for 10µs. At time point 2, the communication partner starts the transmission, by assigning an active low on the rx_data signal. At time point 3, the transceiver generates a falling edge on the tx_data signal. This falling edge causes the communication partner to reset the low timer. This reset synchronises both communication partners. The timer reset is the reason why the 10µs high pulse is not present at time point 4. At time point 5, the falling edges of both communication partners are in sync. This is the start signal for the data transfer. After time point 5, the timing diagram shows the bit pattern for the stop packet.

Figure 4 shows a timing diagram of the IC2IC protocol data transfer signals. The first signal shows the bit-sequence to be transmitted. The second row shows the data (tx_data) signal. The third row shows the strobe (tx_strobe) signal. The data signal encodes the bit sequence directly, i.e. the signal is low when a 0 bit is transmitted and the signal is high when a 1 bit is transmitted. The strobe signal generates an edge whenever the data signal stays constant for more then one bit period T_B. In Figure 4, we assume an initial reset condition. Time point 1 indicates the start of a second consecutive 0 data bit. To communicate the start of the second 0 to the receiver, the strobe signal generates a rising edge. Time point 2 indicates

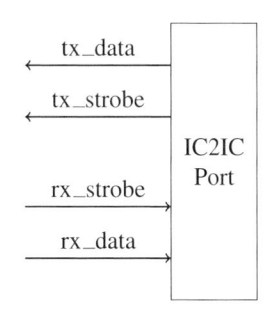

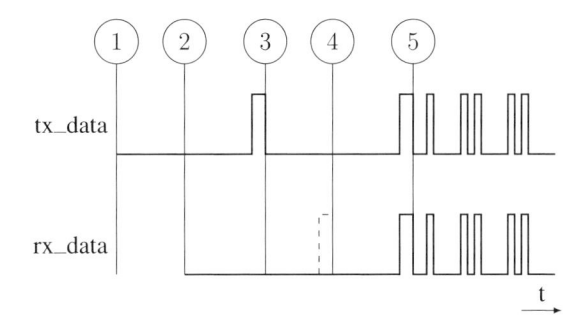

Figure 2. Link signals

Figure 3. IC2IC synchronisation pattern

the start of a second consecutive 1. At this time point the strobe signal transits from high to low thereby creating a falling edge. Similarly, time point 3 indicates the start of a third consecutive 1. At this time point, the strobe signal generates a rising edge.

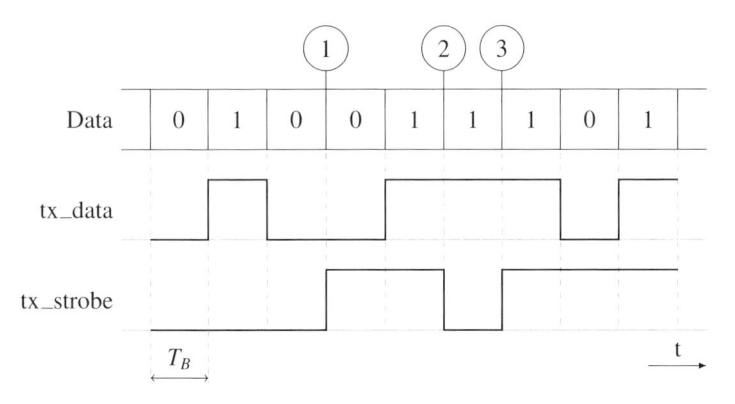

Figure 4. IC2IC data transfer pattern

1.3. Packet Level

The ABP functionality requires data and control packets. Data packets contain header and payload. Control packets are more complex, because they are used for acknowledgement and protocol synchronisation. The following paragraphs detail the composition of data and control packets.

Figure 5 shows the data packet structure. L is the number of payload data bits and $L + 3$ is the total number of bits in one data packet. P is the parity bit, it is set for odd parity in the parity region. $\overline{C}$ indicates not control packet, therefore $\overline{C} = 0$. ABP is the alternating bit protocol bit.

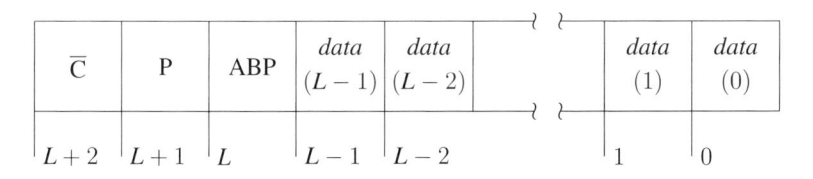

$\overline{C}$	P	ABP	data $(L-1)$	data $(L-2)$	...	data (1)	data (0)
$L+2$	$L+1$	L	$L-1$	$L-2$		1	0

Figure 5. Data packet structure

Figure 6 shows the control packet structure. A control packet always has 6 bits. Setting bit C to 1 identifies this packet as a control packet. P is the parity bit which ensures odd parity in the parity region.

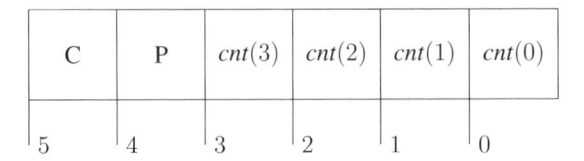

C	P	$cnt(3)$	$cnt(2)$	$cnt(1)$	$cnt(0)$
5	4	3	2	1	0

Figure 6. Control packet structure

The four control packet bits $cnt(3) - cnt(0)$ encode a maximum of $2^4 = 16$ control messages. Table 1 details all messages which are used in the example implementation. The *stop_xy* and *start_xy* messages are used for RRABP protocol synchronisation. *zero_ack* and *one_ack* are the ABP acknowledgement messages sent from port B to port A. The *alive* packet ensures a continuous data stream between the ports.

Table 1. Mapping of the control messages onto the 4 *cnt* bits

Message	cnt bits	Message	cnt bits
stop_msg	0000	stop_ack	1111
start_msg	1110	start_0_ack	1101
start_1_ack	1100	start_rst_ack	1011
zero_ack	1010	one_ack	1000
alive	0111		

The parity bit P of a data packet covers the ABP bit, the L data payload bits and the data / control flag in the next packet. Similarly, the parity bit of a control packet covers the 4 control bits $cnt(3) - cnt(0)$ and the data / control flag in the next packet. Figure 7 shows a parity coverage example. In the example scenario a control packet is followed by a data packet. The parity coverage ensures that an error in any single bit of a packet, including the packet type flag, can be detected even though the packets are not all the same length. The parity bit is set such that the total number of 1s in all the bits covered (including the parity bit) is odd.

	Control packet					Data packet	
C	P	$cnt(3)$	$cnt(2)$	$cnt(1)$	$cnt(0)$	$\overline{C}$	P

Parity coverage

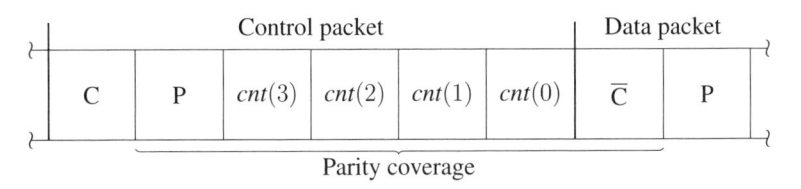

Figure 7. Parity coverage example

1.4. The IC2IC Protocol

The resettable receiver alternating bit protocol, introduced in Section 1.1, handles the protocol synchronisation between port A and port B. However, even the RRABP struggles to handle problems introduced by the physical means of data transmission. These problems include disconnect and bit-errors. In this section we extend the protocol further such that it addresses the practical problems of the IC2IC serial link.

Figure 8 shows the communication diagram for normal operation. Normal operation means there are no errors introduced by the physical means of communication. The figure shows protocol synchronisation, normal data transfer as well as receiver and sender blocking.

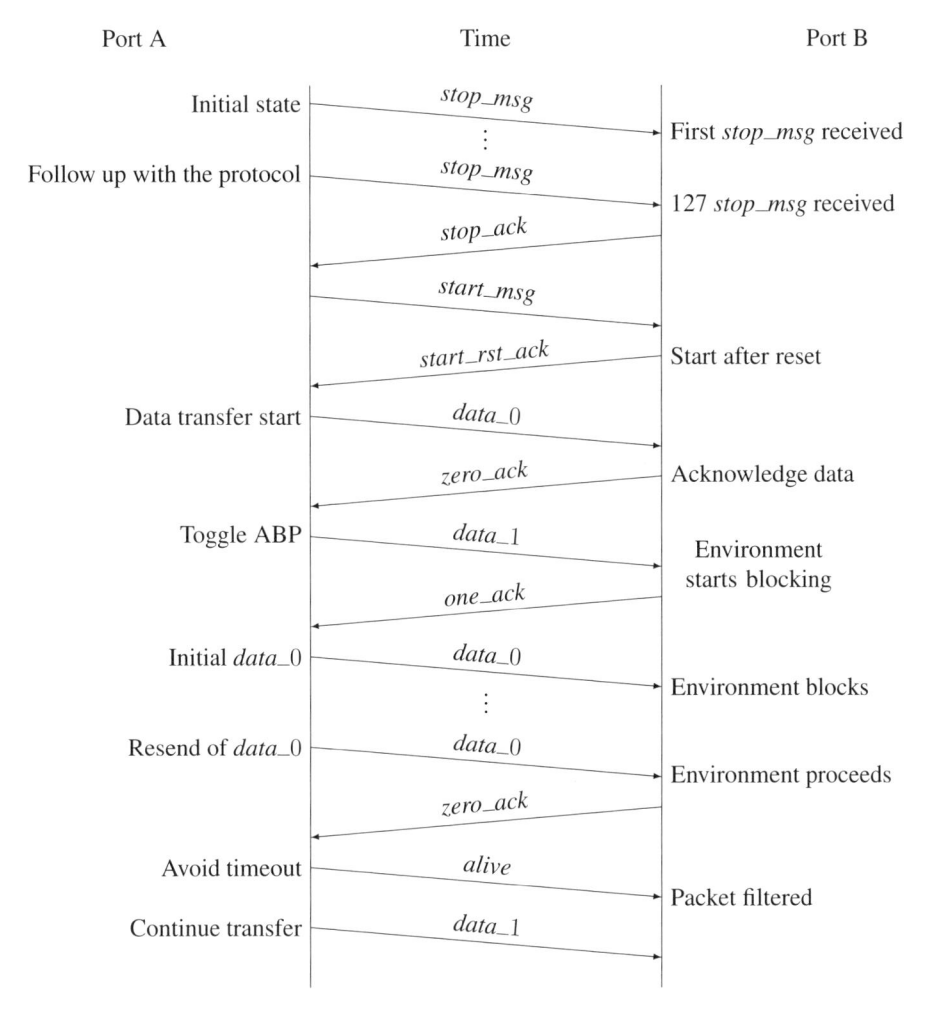

Figure 8. Protocol diagram for normal operation

Port A	Time	Port B
Initial state	stop_msg →	First stop_msg received
Follow up with the protocol	⋮ stop_msg →	127 stop_msg received
	← stop_ack	
	start_msg →	
	← start_rst_ack	Start after reset
Data transfer start	data_0 →	
	← zero_ack	Acknowledge data
Toggle ABP	data_1 →	Environment starts blocking
	← one_ack	
Initial data_0	data_0 →	Environment blocks
	⋮	
Resend of data_0	data_0 →	Environment proceeds
	← zero_ack	
Avoid timeout	alive →	Packet filtered
Continue transfer	data_1 →	

After synchronisation on the physical layer, shown in Figure 3, the IC2IC protocol takes over. The protocol synchronisation starts with a *stop_msg* control packet sent from port A (sender) to port B (receiver). Port B must receive 127 consecutive *stop_msg* messages before the link is considered fit for data transfer. Port B acknowledges the reception of 127 *stop_msg* packets with a *stop_ack* packet. After port A receives the *stop_ack* packet it sends out *start_msg* packets. Upon reception of such a *start_msg* packet, port B responds with *start_rst_ack* packet. This packet indicates that port B came out of a global reset condition. After having received the *start_rst_ack* packet, port A sends out the first data packet, *data_0*, with ABP bit 0. Port B acknowledges the reception of this packet with the *zero_ack* control packet. Upon reception of the *zero_ack* packet port A toggles the ABP bit and sends out the second data packet. Port B receives and acknowledges this data packet, however it is not picked up by the environment of port B. In other words, the environment on the receiver side blocks the communication and port B acts as a buffer (with one place). Upon reception of the acknowledgement, port A sends a *data_0* packet. This packet is not acknowledged by port B, because the buffer is full and the environment still blocks. The absence of acknowledgement forces port A to resend *data_0*. After the receiver environment empties the buffer, port B acknowledges *data_0*. After port A receives this acknowledgement (*zero_ack*) the transmitter recognises that there are no further data vectors to be transmitted. In other words, the transmitter environment blocks. To ensure continuous data traffic between port A and B on the physical layer, port A sends an *alive* control packet. This packet is filtered by port B. At any time port A can resume the data transfer by sending out a *data_1* packet.

Figure 9 shows the communication diagram for the case that a bit error in the *stop_msg* control packet occurs. Upon detection of a bit-error in the *stop_msg* port B forces an asynchronous reset by putting a permanent low on the output signals. This behaviour is detected as a disconnect by port A. In reaction to this disconnect, port A also puts its output signals on low. This silence is observed by both ports for 10ms. Port B remembers that the error occurred during the initial protocol synchronisation. After this silence the communication resumes with the initial synchronisation pattern, described in Section 1.2. After the physical synchronisation, the protocol synchronisation starts. Port B concludes the protocol synchronisation by sending *start_rst_ack*. This indicates that port B came out of a global reset condition and the expected ABP bit value for the first data packet is 0.

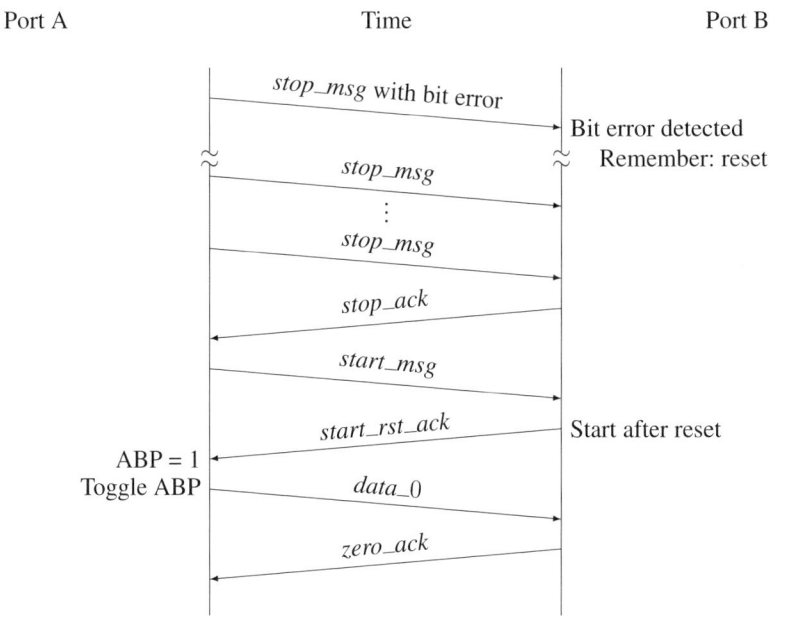

Figure 9. Protocol diagram for *stop_msg* with bit error

Figure 10 shows the communication diagram for the case when a bit error in the *data_1* packet occurs. When port B detects this bit error it sets both output signals to low. This communicates an asynchronous reset to port A. After the silence period has passed, both port A and port B resume the transmission with the physical initialisation sequence. Port B sends out the *start_zero_msg* control packet. This indicates that the value of the last correct received ABP bit was 0. After port A receives this packet it resumes the data transmission with an ABP bit equal to 1, in Figure 10 this is indicated by *data_1*.

Figure 11 shows the communication diagram for the case when a bit error in the control packet *zero_ack* occurs. The error handling is similar to the previous scenario. However, this

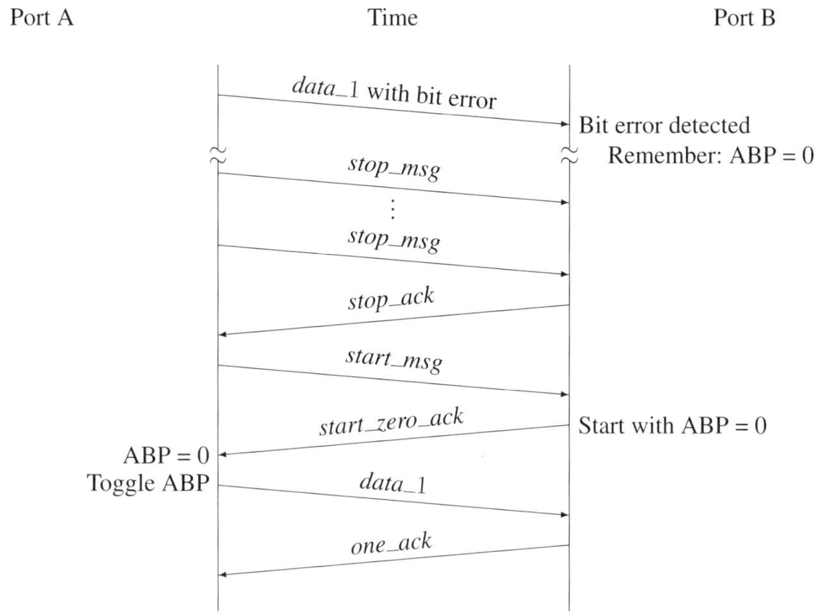

Figure 10. Protocol diagram for *data_1* with bit error

time port A detects the error in the control packet and initiates the exchange of silence. The last correct ABP bit port B has seen was 1, therefore the initialisation protocol is concluded with port B sending *start_zero_ack*.

Figure 12 shows the communication diagram for the case that both ports detect a disconnect during a normal data transfer. Therefore, both initiate the exchange of silence. After the silence period has elapsed, port A and port B are ready to continue the data transfer. The physical synchronisation signals ensure that the link resumes after the ports are connected again. In case of a reconnect, port B concludes the initialisation protocol by sending a *start_one_ack* packet. This indicates that the ABP value of the last correctly received data packet was 1.

2. Example Implementation

The process network of the implementation is organised in a decentralised way. That means, there is no central process which controls the IC2IC functionality. Each process within the network has its own functionality and to a certain extent its own agenda. This leads to a hierarchically flat model.

Figure 13 shows the implementation process network. This process network consists of nine processes and CSP style, i.e. blocking, channels for data transfer. All channels are named *from2to* where *from* indicates the sender process name and *to* indicates the receiver process name. The process network communicates with the local processing environment via *in* and *out* channel. *tx_data* and *tx_strobe* are output signals, i.e. they leave the local processing environment. Similarly, *rx_data* and *rx_strobe* are input signals. The network functionality can roughly be partitioned into two parts. The first part implements the data handling and the second part implements the ABP functionality.

The data handling starts with messages flowing from the environment to the *P* process via the *in* channel. The *P* process appends the ABP protocol bit and sends the extended

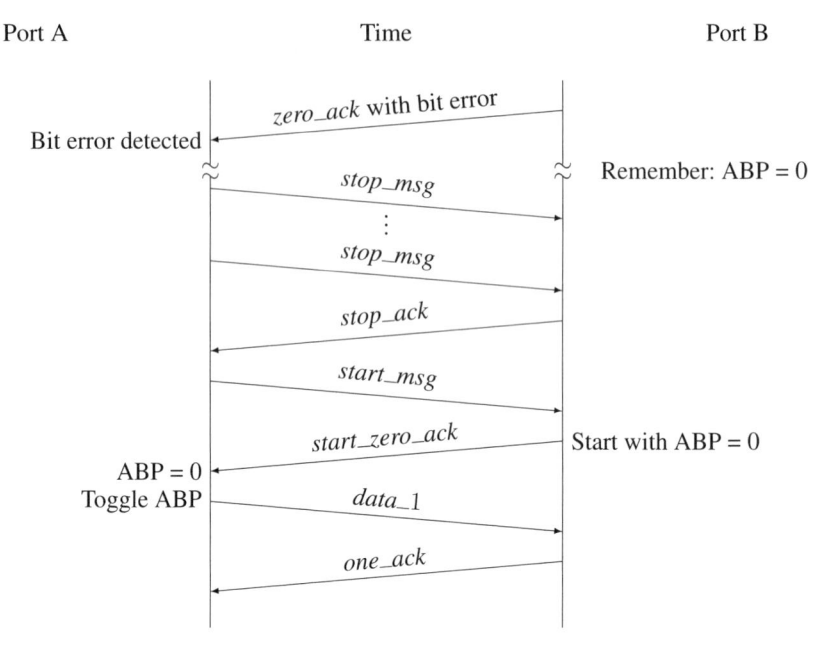

Figure 11. Protocol diagram for *zero_ack* with bit error

message to the *MUXER* process. The *MUXER* extends the message with parity bit and control bit before the message is sent to the *TX* process. The *TX* process translates the message into the self clocking line code for transmission from IC to IC. The message is received with the *RX* process. The *RX* process decodes the transmission signal and sends the received message bits to the *DEMUX* process. The *DEMUX* process assembles and checks the message. The control and parity bit is removed before the message is sent to the *C* process. The *C* process removes the ABP bit before the message is sent to the environment via the *out* channel.

Table 2 shows the packet composition on the TX side and the decomposition on the RX side. The environment provides an L dimensional data vector $Data_L$. The IC2IC[3] process network adds a three bit header which ensures security and robustness of the data transfer. On the RX side (port B) the IC2IC process network removes the header and offers the received L dimensional data vector to the environment.

Table 2. IC2IC layer structure. $Data_L$ is an L dimensional data vector.

TX		RX	
Entity	Packet	Entity	Packet
Environment	$Data_L$	Environment	$Data_L$
P	ABP & $Data_L$	C	ABP & $Data_L$
MUXER	$\overline{C}$ & ABP & $Data_L$	DEMUX	$\overline{C}$ & ABP & $Data_L$
Physical layer			

The three remaining processes, *TX_ABP*, *RX_ABP* and *CNTMSG* implement the ABP protocol functionality. *TX_ABP* controls the ABP protocol on the TX side. Therefore, it is connected to the *P* process via a bidirectional channel. This bidirectional channel enables the *TX_ABP* process to react to input. Furthermore, *TX_ABP* is connected with the *MUXER*

---

[3]Capitalised italic fonts indicate a process name. In this case IC2IC is the name of the process which implements the IC2IC functionality.

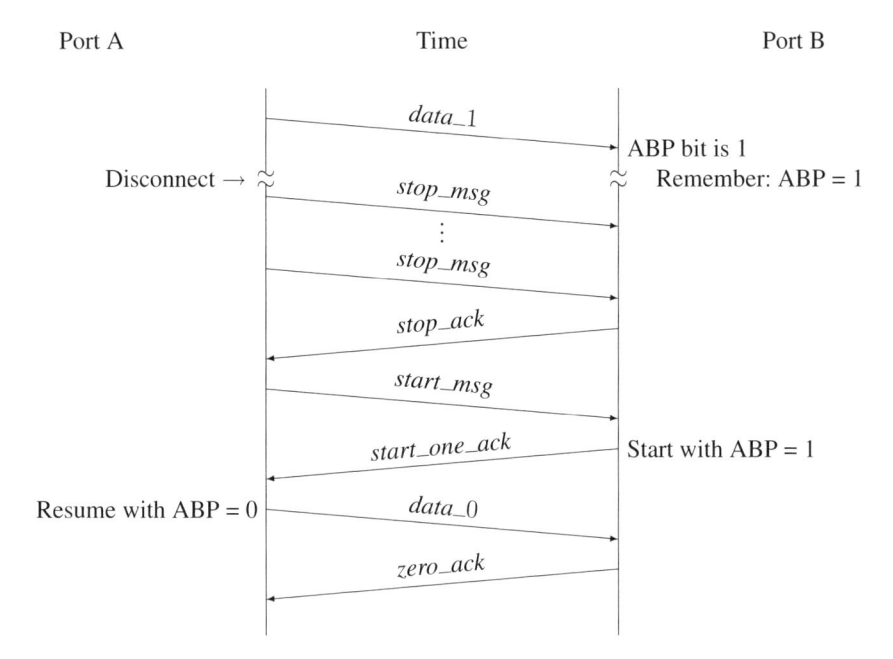

Figure 12. Protocol diagram disconnect error

process. This channel is used to communicate *start* and *stop* messages during the protocol initialisation. The last connection of the *TX_ABP* is with the *CNTMSG* process. Via this channel the *TX_ABP* process receives acknowledgement messages from the communication partner. The *CNTMSG* process routes all other control messages to the *RX_ABP* process. This process is responsible for the protocol on the receiver side. Apart from the control messages it also receives the ABP control bit from the *C* process. Based on this information it generates control messages for the receiver side of the communication partner. In order to send packets to the receiver, the *RX_ABP* process is connected with the *MUXER* process.

Figure 14 shows the reset network which connects the processes of the IC2IC example implementation in parallel with the data and control communication network. The local reset network is used for error handling. A local reset affects the whole IC2IC network implementation, but not the environment. That means after a local reset the IC2IC network is in a predefined state. This state is different from the global reset state, because some processes 'remember' particular information such as the value of the ABP bit. The *DEMUX* process informs the *RX* process that a protocol error, such as a parity error, happened during transmission. The *RX* process is the central point in the reset network, because it controls the *local_rst* signal. This signal is asserted after an *rx_rst* event is received or a disconnect is detected. This *local_rst* enables the process network to react to protocol error or disconnect.

The following sections provide an in-depth discussion of the individual processes. This discussion introduces the functionality of the processes during normal operation and their behaviour during local reset.

2.1. P Process

In its initial or reset state process *P* requests the ABP bit value from the *TX_ABP* process. Once *P* has received this value it is ready to receive a message from the environment via the *in* channel. After having received a message the *P* process appends the ABP token and sends the resulting packets to the *MUXER* process. This is the last step before *P* recurses. A local

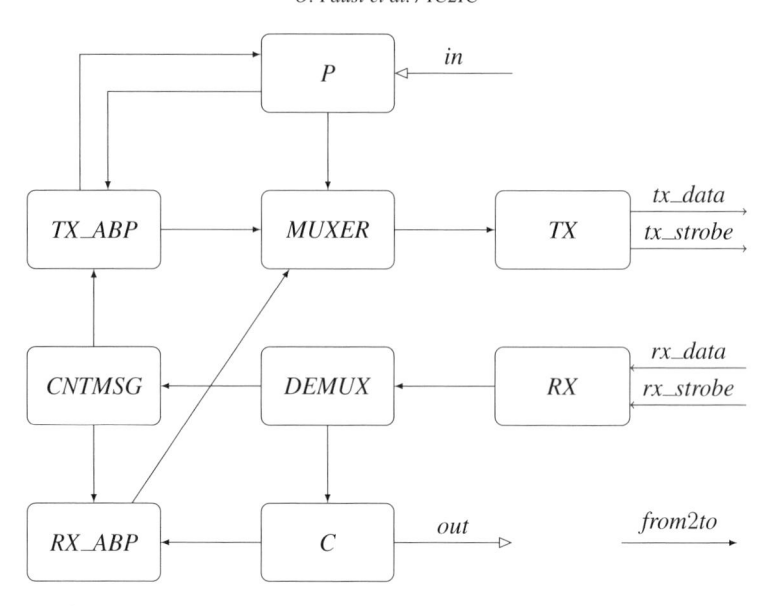

Figure 13. IC2IC process network with all the communication channels between the processes

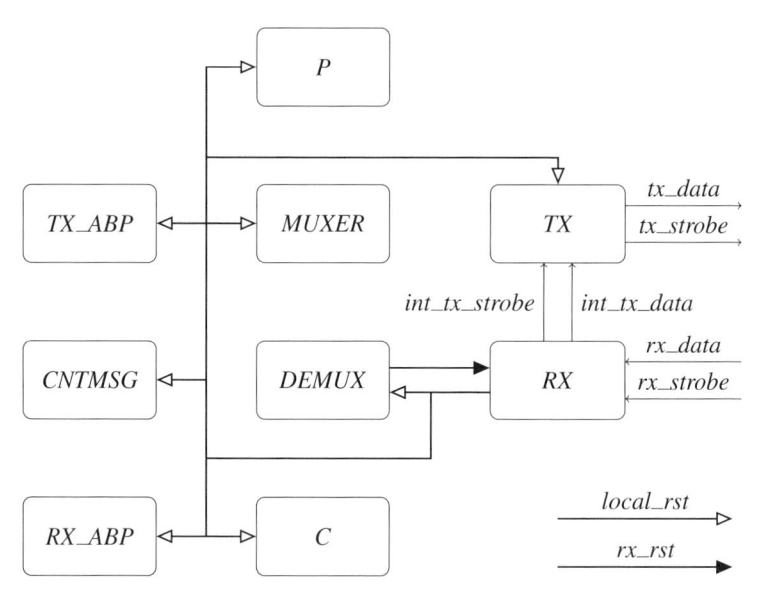

Figure 14. IC2IC reset network

reset forces this process into the initial state. However, it remembers the last data vector from the environment.

2.2. TX_ABP Process

TX_ABP handles the transmitter part of the ABP protocol. In the initial state, it establishes the protocol synchronisation with the receiver part of the communication partner. This is achieved by sending out *stop_msg* control packet to the *MUXER* process. The *stop_msg* are sent until the first *stop_ack* control packet is received via the channel which connects it to the

CNTMSG process. After having received the *stop_ack* message the *TX_ABP* process sends out the *start_msg*. The receiver can respond to the start message with one of three different control packets:

1. *start_rst_ack* – This control packet is used to indicate that port B (receiver) came out of a global reset.
2. *start_zero_ack* – This control packet indicates that the ABP bit of the last accepted data packet had ABP = 0.
3. *start_one_ack* – This control packet indicates that the ABP bit of the last accepted data packet had ABP = 1.

After initialisation, the *TX_ABP* process establishes the ABP functionality on the transmitter side. It provides the *P* process with the correct value of the ABP bit. The value of the ABP bit is negated when the acknowledgement packet (*one_ack* and *zero_ack*) indicates the same value as the current ABP bit. For *TX_ABP* local and global resets have the same effect. During the initialisation the communication partner provides the ABP bit information via one of the three possible responses to *start_msg*.

2.3. C Process

The *C* process consumes the messages from the *DEMUX* process and transfers them to the environment. The process acts as a message buffer, which means that initially it is able to receive a packet before the environment can block the acknowledgement of subsequent packets. The ABP bit is stripped from this packet and internally stored as well as sent to the *RX_ABP* process. After the ABP bit is stripped from the packet, the process is ready to communicate the message to the environment. In case the environment blocks this communication, the *C* process consumes all subsequent packets from the *MUXER* process but it does not send any *ABP* bits to the *RX_ABP* process. This effectively implements the blocking functionality, because the sender will continuously resend the same packet. In effect, this prevents the sender side from making progress, i.e. the sender is blocked. This block is released when the receiver environment picks up the packet from the *C* process. A local reset forces this process into the initial state, but it retains all information local to the process. Furthermore, the ability to deliver buffered data to the environment is not affected.

2.4. RX_ABP Process

The *RX_ABP* process handles the receiver side of the protocol. In the initial state the process is poised to receive 127 *stop_msg* control packets from the *CNTMSG*. The 128th and all subsequent *stop_msg* are acknowledged by sending *stop_ack* packets to the *MUXER*. Subsequently, a *start_msg* is acknowledged with *start_rst_ack*. This concludes the protocol synchronisation and *RX_ABP* handles the receiver (port B) functionality of the ABP protocol. In this state the *RX_ABP* process waits for input from *C* or *CNTMSG*. The *C* process can request the acknowledgement of new data packets. In case a *stop_msg* is received from the *CNTMSG* process or a local reset occurs the *RX_ABP* process returns to the initial protocol handling state. However, this time, the initial protocol is concluded with either *start_zero_ack* or *start_one_ack* depending on the last data packet which was acknowledged.

2.5. MUXER Process

The *MUXER* process relays or multiplexes the packets from *P*, *TX_ABP* and *RX_ABP* to the *TX* process. Furthermore, the *MUXER* process inserts an *alive* message if none of the connected processes supplies a message. This ensures a constant data transmission which allows the *RX* process in the receiver (port B) to detect a disconnect. The message transfer to the *TX* process is a serial bit-stream. For the *MUXER* process local and global reset are identical.

2.6. DEMUX Process

The *DEMUX* process receives a serial bit-stream from the *RX* process. This bit-stream is decoded and the individual messages are assembled. During this assembly the parity is checked. If the parity check is successful, control packets are passed to the *CNTMSG* process and data packets are passed to the *C* process. All packets are passed on without the parity bit. In case the parity check fails the *DEMUX* process sends out a reset signal to the *RX* process via the *rx_rst* channel. For the *DEMUX* process local and global reset are identical.

2.7. RX Process

The *RX* process performs either control or data extraction tasks. In normal operation, *RX* extracts the data bits from the received *rx_data* and *rx_strobe* signals and sends them on to the *DEMUX* process. During control tasks the *RX* process asserts the *local_rst* signal and manages the output signals: *tx_data* and *tx_strobe*. *RX* has two distinct control tasks, exchange of silence and physical synchronisation. There are two conditions which cause *RX* to start an exchange of silence:

1. No signal transition (edge) was observed during a 1ms time window on either *rx_data* or *rx_strobe*. The absence of signal transitions indicates a disconnect from the communication partner.
2. An event is received via the *rx_rst* line. This indicates the *DEMUX* process has detected a bit error.

During exchange of silence *tx_data* and *tx_strobe* are set to low. After the silence period has elapsed, *RX* performs initial synchronisation. Initial synchronisation is also triggered by a global reset. During initial synchronisation *RX* behaves according to the specification given in Section 1.2.

2.8. TX Process

The *TX* process abstracts the layer-0 functionality of the transmitter. This abstraction involves synchronisation and data transfer functionality. Section 1.2 describes the physical synchronisation and data transfer which is carried out by the *TX* process. When a local reset signal is received the *TX* process hands over the control of the output signals to the *RX* process.

2.9. IC2IC test setup

Figure 15 shows the block diagram of the IC2IC test setup. Each FPGA executes the IC2IC process network. The test setup provides the means to introduce bit errors and disconnect the link at arbitrary time points. To have more testing flexibility, in addition to the IC2IC process network the ML403 hosts also a soft processor. This soft-processor was used to generate test sequences and monitor the received and transmitted messages. The SRSv02 hosts an additional process which mirrors all the messages. In other words, the test setup constitutes a loop, where data source and sink are located in the ML403.

The following list details the tests conducted with this setup:

- The system starts up and transfers data according to the protocol diagram for normal operation, shown in Figure 8. The data transfer starts regardless of the sequence in which the communication partners come online.

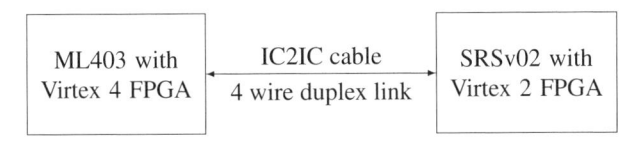

```
┌─────────────┐                      ┌─────────────┐
│  ML403 with │      IC2IC cable     │  SRSv02 with│
│ Virtex 4 FPGA│◄──────────────────►│ Virtex 2 FPGA│
│             │    4 wire duplex link│             │
└─────────────┘                      └─────────────┘
```

Figure 15. Block diagram of the IC2IC test setup

- The data stream was interrupted by disconnect at various time points. The stream recovered according to the protocol diagram for disconnect error shown in Figure 12.
- The data stream was interrupted by bit errors at various time points. The stream recovered according to the protocol diagrams for bit errors in various packets, see Figures 9, 10 and 11.

All these tests were conducted with a payload of $DATA_L = 8, 16$ and 32 bits. For each payload setup the test duration was at least 10 hours, i.e. the stream was running for this time. According to the test software, executed by the soft-processor, during all these tests no messages were lost or duplicated. This establishes confidence in the implementation stability.

The tests validate all IC2IC protocol features which were discussed in Section 1.4. Furthermore, the IC2IC process network was implemented in two different flexible logic devices[4]. This shows that the implementation is portable to a wide range of flexible logic devices. Finally, the link was tested with the help of a soft-processor. The results show that the blocking functionality is pervasive. That means, if the link is interrupted the software, executed by the soft-processor, is not able to send or receive a message, i.e. it is blocked. This shows blocking communication over two borders, software / hardware and IC (Virtex 4) / IC (Virtex 2).

3. Conclusion

In this paper we introduced IC2IC a lightweight serial interconnect channel for multiprocessor networks. This serial link establishes a low latency high performance data link between independent processors. The result is a reliable way to transfer bit level information from one IC to another IC. The blocking functionality and ability to recover from various error conditions distinguishes the IC2IC link from other serial interconnect links.

The practical part of the paper introduces the alternating bit protocol. This protocol constitutes the cornerstone of the IC2IC implementation, because it provides blocking functionality and hardens the link against message loss and message duplication. It was not possible to implement the alternating bit protocol directly, because in the basic form the protocol does not have the flexibility to cope with various practical reset conditions. Therefore, we extended the alternating bit protocol in order to cope with error conditions such as disconnect, bit error and receiver reset. The extended version of the protocol is called IC2IC. On the physical layer the IC2IC link uses a data strobe signal setup and the initial synchronisation is done with a specific synchronisation pattern.

The example implementation constitutes an IC2IC transceiver system. We designed the system as a network of independent processes which communicate via blocking channels. With this approach we carried on the idea of blocking channels. Apart from the communication channels the processes are also connected to a local reset signal. A local reset is asserted in case of an error on the physical link. The local reset signal interrupts all processes and forces them into a predefined state. The processes have the ability to 'remember' certain information after local reset. We use this memory functionality to avoid deadlock states and possible data loss after resynchronisation.

---

[4]Two different generations of the Xilinx Virtex family.

Process networks for hardware logic are a very elegant way to utilise the inherent parallelism of these devices. Key to these process networks are the blocking communication channels between them. It is relatively easy to establish blocking channels between processes within one chip. However, due to degrading effects on the physical means of communication and the unpredictability of the communication partner, it is difficult to establish blocking communication between logic circuits located in different ICs. The properties of the IC2IC serial link were validated with various tests on a hardware setup. These tests built up confidence in the claim that IC2IC serial links release process networks from the prison of only one IC. The IC2IC link enables blocking communication between two ICs. This blocking point to point communication is the corner stone for CSP style process networks which span over multiple ICs.

With the IC2IC protocol each individual payload message is acknowledged. Therefore, the IC2IC serial link is best described as buffered channel. The buffer has room for two massages, one in the producer (P) and one in the consumer (C). These buffers are necessary to expose a uniform interface to the environment. So, to their environment the IC2IC input and output looks like an ordinary hardware channel – even though the messages are transferred or received from another chip. If this interface is not enforced, back to back synchronisation on individual messages is possible. The possibility of back to back synchronisation is one of the strongest points in favour of the ABP protocol and indeed the IC2IC serial link. With a buffered channel there is always the uncertainty about whether or not a message has reached the destination. This uncertainty is potentially dangerous especially when important control messages are transferred. This predictability of the IC2IC serial link leads to simple abstract models. In CSP the IC2IC serial link with uniform interfaces is modelled as a buffered channel with two places. In case these interfaces are not enforced, the IC2IC serial link can be abstracted as a CSP style channel.

3.1. Future Work

The development of the IC2IC protocol is the first step towards process networks implemented in multiple ICs. The next development step is concerned with particular examples. These examples help to establish the merits of the protocol. The translation step from the theoretical model to the implementation is not proven, therefore examples are necessary to built up trust in the protocol. Based on particular examples, speed and robustness properties can be tested. The speed property is concerned with the data rate. There is a theoretical maximum data rate and specific data rates for individual applications. To compare the individual with the maximum data rates reveals some insights into the IC2IC protocol. Robustness considerations are concerned with particulars of the channel. The communication channel between two ICs might introduce bit errors. Future work will be concerned with analysing the effects of bit errors on the protocol. Furthermore, the communication channel might not be stable. That means, there are random disconnects due to bouncing connections. Some work will focus on simulating such bouncing conditions and analysing the performance of the IC2IC protocol.

After having established trust in the protocol implementation we can start looking towards more involved systems. Process networks are the most widely used mechanism to model complex systems with non-linear behaviour. The size of the process networks, i.e. the number of processes involved, depends on the complexity of the task. For many practical problems the task is ill defined, therefore the complexity of the system is not defined. However, more complex systems can react to a wider range of environments. That means, systems which model artificial intelligence tend to work better with larger process networks. The IC2IC protocol might be one way to extend such networks beyond the borders of a single IC without functional side effects. The only negative side effect might be insufficient communication speed. But, this is a field of further research.

Acknowledgements

This work was supported by the European FP6 project "WARMER" (contract no.: FP6-034472).

References

[1] C. R. Anderson, Marcel Boosten, R. W. Dobinson, S. Haas, R. Heeley, N. A. H. Madsen, B. Martin, J. Pech, D. A. Thornley, and C. L. Ullod. IEEE 1355 DS-Links: Present Status and Future Prospects. In Peter H. Welch and Andrè W. P. Bakkers, editors, *Proceedings of WoTUG-21: Architectures, Languages and Patterns for Parallel and Distributed Applications*, pages 69–80, mar 1998.
[2] Colin Whitby-Strevens. The transputer. *SIGARCH Comput. Archit. News*, 13(3):292–300, 1985.
[3] A. Johannet, G. Loheac, L. Personnaz, I. Guyon, and G. Dreyfus. A transputer based neurocomputer. In Traian Muntean, editor, *OUG-7: Parallel Programming of Transputer Based Machines*, pages 120–127, sep 1987.
[4] M. Meriaux, A. Atamenia, and E. Lepretre. A transputer-based architecture for graphics. In Traian Muntean, editor, *OUG-7: Parallel Programming of Transputer Based Machines*, pages 297–306, sep 1987.
[5] Standard for Heterogeneous Inter-Connect (HIC). Low Cost Low Latency Scalable Serial Interconnect for Parallel System Construction, 1995. 445 Hoes Lane, P.O. Box 1331, Piscataway, NJ 08855-1331, USA.
[6] W. J. Dally and C. L. Seitz. Deadlock-Free Message Routing in Multiprocessor Interconnection Networks. *IEEE Trans. Comput.*, 36(5):547–553, 1987.
[7] Barry M. Cook and C. P. H. Walker. SpaceWire - DS-Links Reborn. In Frederick R. M. Barnes, Jon M. Kerridge, and Peter H. Welch, editors, *Communicating Process Architectures 2006*, pages –, sep 2006.
[8] Barry M. Cook. IEEE 1355 data-strobe links: ATM speed at RS232 cost. *Microprocessors and Microsystems*, 21(7-8):421–428, March 1998.
[9] O. J. Greve, M. H. Schwirtz, Gerald H. Hilderink, Jan F. Broenink, and Andrè W. P. Bakkers. An A/D D/A board using IEEE-1355 DS-Links for Heterogeneous Multiprocessor Environment. In Peter H. Welch and Andrè W. P. Bakkers, editors, *Proceedings of WoTUG-21: Architectures, Languages and Patterns for Parallel and Distributed Applications*, pages 27–38, mar 1998.
[10] Marcel Boosten, R. W. Dobinson, B. Martin, and P. D. V. van der Stok. A PCI-based Network Interface Controller for IEEE 1355 DS-Links. In Peter H. Welch and Andrè W. P. Bakkers, editors, *Proceedings of WoTUG-21: Architectures, Languages and Patterns for Parallel and Distributed Applications*, pages 49–68, mar 1998.
[11] IEEE-1394-1996 High Speed Serial Bus, 1996. 445 Hoes Lane, P.O. Box 1331, Piscataway, NJ 08855-1331, USA.
[12] ECSS-E-50-12A SpaceWire – Links, nodes, routers and networks. (24 January 2003).
[13] B. Fiethe, H. Michalik, C. Dierker, B. Osterloh, and G. Zhou. Reconfigurable system-on-chip data processing units for space imaging instruments. In *DATE '07: Proceedings of the conference on Design, automation and test in Europe*, pages 977–982, San Jose, CA, USA, 2007. EDA Consortium.
[14] Sanjit A. Seshia, Wenchao Li, and Subhasish Mitra. Verification-guided soft error resilience. In *DATE '07: Proceedings of the conference on Design, automation and test in Europe*, pages 1442–1447, San Jose, CA, USA, 2007. EDA Consortium.
[15] Sergio Saponara, Esa Petri, Marco Tonarelli, Iacopo Del Corona, and Luca Fanucci. FPGA-based networking systems for high data-rate and reliable in-vehicle communications. In *DATE '07: Proceedings of the conference on Design, automation and test in Europe*, pages 480–485, San Jose, CA, USA, 2007. EDA Consortium.
[16] Ad M. G. Peeters. Implementation of Handshake Components. In Ali E. Abdallah, Cliff B. Jones, and Jeff W. Sanders, editors, *25 Years Communicating Sequential Processes*, volume 3525 of *Lecture Notes in Computer Science*, pages 98–132. Springer, 2004.
[17] K. A. Bartlett, R. A. Scantlebury, and P. T. Wilkinson. A note on reliable full-duplex transmission over half-duplex links. *Commun. ACM*, 12(5):260–265, 1969.
[18] A. W. Roscoe. *Theory and Practice of Concurrency*. Prentice Hall, Upper Saddle River, New Jersey 07485 United States of America, first edition, 1997.
[19] Bernhard Sputh, Oliver Faust, and Alastair R. Allen. RRABP: Point-to-Point Communication over Unreliable Components. In P.H. Welch et al., editor, *Communicating Process Architectures 2008*, pages 203–217, 2008.

Asynchronous Active Objects in Java

George OPREAN, Jan B. PEDERSEN

School of Computer Science, University of Nevada
opreang@unlv.nevada.edu, matt@cs.unlv.edu

Abstract. Object Oriented languages have increased in popularity over the last two decades. The OO paradigm claims to model the way objects interact in the real world. All objects in the OO model are passive and all methods are executed synchronously in the thread of the caller. Active objects execute their methods in their own threads. The active object queues method invocations and executes them one at a time. Method invocations do not overlap, thus the object cannot be put into or seen to be in an inconsistent state. We propose an active object system implemented by extending the Java language with four new keywords: active, async, on and waitfor. We have modified Sun's open-source compiler to accept the new keywords and to translate them to regular Java code during desugaring phase. We achieve this through the use of RMI, which as a side effect, allows us to utilise a cluster of work stations to perform distributed computing.

Keywords. Active Objects, Asynchronous Active Objects, Java, distributed computing.

Introduction

Within the last two decades object oriented programming (OOP) has become increasingly popular, not least because of the introduction of the Java programming language by Sun in 1995. One of the basic ideas of object oriented programming is encapsulation of method and data; each object holds its own data as well as methods operating on the data.

However, objects are *passive*, that is, when a method is invoked on an object, the executing thread is that of the caller. The method is not executed by the object itself. Since many different threads of control can hold a reference to the same object, many different threads can invoke methods at the same time, possibly leaving the state of the object inconsistent. Using the *synchronized* keyword only provides a fake sense of security [1] as a method called from within a synchronized method can call methods in other objects that can cause a call back into the original object and modify the internal data.

The heart of the problem is simply that an object is a passive structure that is being violated by various threads. If the object itself were in charge of executing its own methods in its own thread of control, then such unfortunate action can be avoided. In order to invoke a method on an *active object* the object will have to accept the method invocation in the form of the parameters as well as the name of the method to be executed, and subsequently execute the code of the requested method in its own thread of control, thus excluding other threads from manipulating its data and executing its methods.

Active Objects and the Real World

The argument that "objects are considered harmful" has been borrowed from the process oriented design community [1]. It is also argued that object oriented design is not a good reflection of how interaction between objects take place in the real world (which we model when creating software); the problem again lies with the passiveness of the objects: If you are standing in front of your friend and want to borrow $20 you ask him, and he digs into his pocket and hands it to you. You do not ask and then reach into his pocket to retrieve the money. If that is the case, then why would we model systems this way? In standard OO, if you hold a reference to *friend*, and you wish to invoke the *borrowMoney()* method, then the call *friend.borrowMoney()* is executed in your own thread of control, not in a separate thread (or in *friend*'s thread), thus breaking all similarity to the way the real world works. Interestingly enough, in OO, invoking a method is often referred to as "sending a message" (as in message passing, which in distributed computing means transferring data from one process to another through communication). This is *actually* what we *want*, but certainly not what we do in a system with passive objects.

An active object receives messages (representing the requested method invocation) from the caller, performs the computation, and returns the result. An active object is comparable to the well known technique of Remote Procedure Call (RPC) [2] or in OO terms Remote Method Invocation (RMI) [3], which both involve transferring method parameters and results back and forth between the calling process (client) and the remote object (server).

An active object system can be mimicked, and ultimately ours is indeed implemented using RMI; as a matter of fact, a synchronous active object system can be easily used to implement a RMI system without stub generation.

The active object model has been the subject of intensive research since late 80s and now the active objects are used in diverse areas: query processing [4], building concurrent compiler [5], implementing services in Telecommunication Management Services [6], developing applications for smartphones [7,8] or developing collaborating games [9].

Asynchronous Active Objects

Our active object system can easily be used to implement any RMI system; no stub generation is needed as the system uses reflection and a general server manager to accept requests to create remote objects on remote machines. Though RMI does provide a way to improve execution time by executing code on machines that are perhaps better suited for a specific part of the computation task, it is not considered a typical technique for parallel computation; recall, a remote method invocation is, as a local method invocation, synchronous. This means that the caller blocks until the remote method invocation returns.

A well known technique for parallelising computations is using message passing; the program is decomposed into a number of processes whose only way to synchronise is through communication using message passing. Messages must be explicitly sent and received. This can be done both synchronously or asynchronously.

The idea behind our asynchronous active objects is based on a merger between RMI and asynchronous message passing. An active object is placed on a remote machine when it is created. Any method invocation is done (automatically) though RMI (using a general client/server system and reflection), but such calls can be declared to be asynchronous. This means that the caller does not block and wait until the result of the method invocation is returned, but can continue executing immediately thereafter. When the return value is needed, a new **waitfor** statement is executed. If the value arrives before this statement is executed, the effect of the **waitfor** statement is a binding of the returned value to the variable without blocking. If the return value is not yet present, the statement will block the execution until the value is available. Another way to implement asynchronous active objects is with the use of a future object, that acts as a placeholder for the returned result [10,11].

Moreover, we modified the object creation expression syntax to allow the programmer to specify the computer on which the active object will reside. This is similar to spawning a process with MPI_spawn. Thus we have obtained a system that can be used for developing parallel and distributed applications using asynchronous active objects.

As a desirable side effect, the model allows us to write parallel code that uses asynchronous active objects. In section 6.1 we describe three different distributed computations that we implemented using asynchronous active objects.

1. Related Work

The first time the concept of active objects (actors) was presented was Hewitt's actor model in 1977 [12]. In his model, the actor was the central entity that executed its actions: communicates with other actors, creates actors and changes its behaviour. The location of the actors can be distributed and they can execute actions in parallel. Later a mathematical definition for the behaviour of an actor system was presented [13]. This actor (active object) model is suited perfectly for creating parallel and distributed applications. Unfortunately the actor model was not as popular as procedural, functional or OO programming. The parallelism model offered by the procedural or OO model is not as powerful as the actor model, so a number of ways to integrate active objects into the OO paradigm have been proposed.

Java RMI proposed a solution for accessing remote (distributed) objects in the same manner as accessing local objects. The location of the object would be transparent to the programmer, using the same '.' operator to invoke methods on local or remote objects. While making the process of developing distributed applications easier, the RMI (and RPC) model is a synchronous client/server model. The client has to wait for the response from the server before it can resume its execution. Better performance can be obtained if the client can use this "waiting time" to do other operations that do not involve using the return value from the server. If we consider a real world example, when someone wants to prepare breakfast for her family and realizes that she does not have milk for the cereal, she can ask somebody (an active object) to get some cereal (the result of the method call) from the store. During the time she gets the cereal (is waiting for the result), she does not just wait and do nothing (this is what happens with synchronous communication). It is much more efficient to do some other activities related to preparing breakfast that has nothing to do with the cereal: get the milk from the refrigerator, get the cereal bowls from the cabinet, set the spoons on the table and maybe pour orange juice in some glasses. Once she gets the cereal, breakfast can be served.

Active objects have been the subject of intensive research since mid to late 80s. Issues like garbage collection of active objects [14,15,16,17,18], exception handling [19,11], type theory of active objects [20,21,22], transition from active objects to autonomous agents [23,24] and many more have been addressed. As a result of this research, a number of different ways to integrate concurrency into object-oriented languages, using active objects, exist. These approaches are categorised in [25]: the library approach (create new libraries for concurrent and distributed programming), the integrative approach (extend the language, rather than the library), and the reflective approach (that is a "bridge" between the two previous approaches).

We have opted for the integrative approach, that consists of merging concepts of object and process/activity, thus creating the new concept of active object.

An Active Object pattern can be used to implement active objects. This pattern decouples method execution from method invocation to enhance concurrency. A number of components like Proxy, Scheduler, Servant, Activation Queue, Method Request and Future have to be implemented for this pattern to work [10]. The programmer does not only have to concentrate on implementing the actual active object, but also understand how these components interact and then implement them. Java Dynamic Proxies offers another solution to implementing the active objects [26].

Claude Petitpierre integrated active objects in the Java language by adding new keywords [27]. His approach uses the MPI model of sending and receiving messages: one object calls a method on a second object (ready to send the message) and the second object has to accept that method (ready to receive the message). The keywords added are: *active*, *accept*, *select* and *waituntil*. His approach models only synchronous active objects. The C++ language has also been extended, with *active* and *passive* keywords, to integrate active objects. One example by Chen et al. is [28]. The implementation of active objects has a transaction service to maintain the atomicity of a client's invoking sequence. If a client declares a transaction for an active object, all the calls to that active object will be blocked until the transaction is over.

Creating an external library that can be integrated with the Java language (similar to the MPI library [29] for C and Fortran, and MPJ (A message passing library) for Java [30]) is another possibility of having active objects in the OO model. The French National Institute for Research in Computer Science and Control has created such a library called ProActive [11]. Regular passive objects can be turned into active objects, asynchronous method invocation is supported and a future object is used as a placeholder for the actual result. Restrictions are imposed when using this library: final methods cannot be used in active objects, final classes cannot be used to instantiate active objects, and the programmer cannot override functions like *hashCode()* or *equals()*. Moreover, the use of the *toString()* method with future object can lead to deadlock.

Another Java library that can be used for developing concurrent systems is *Communicating Sequential Processes for Java* (JCSP) [31,32,33]. It uses Hoare's algebra of Communicating Sequential Processes (CSP), a mathematical model for concurrency that can guarantee (and prove) that the multithreaded application developed with it does not have deadlock, livelocks, race hazards, or resource starvation. JCSP views the world as layered networks of communicating processes. The processes interact via channels (or other CSP synchronisation objects like barriers or shared-memory CREW locks) and not via method invocation (our active objects communicate with each other by sending messages through RMI). Similar to our active object system, JSCP offers the programmer an alternative to developing concurrent application without the use of synchronized, wait, notify or notifyAll primitives (the JCSP actually uses internally these primitives and so does our active object system). The JCSP.net [34] (an extension of JCSP for networking) allows processes to be distributed on the network. Objects sent across JCSP.net channels are serialised. These channels can be set to allow automatically download the class if it does not exist on the receiving side (we have used RMI for our system for this purpose).

Java is not the only object-oriented language that has libraries that can be utilised when developing applications that involve active objects. DisC++ [35] and ACT++2.0 [36], are two libraries for concurrent programming in C++.

A number of other, less mainstream, but nevertheless exciting, languages that support active objects exist: Cω, a research language based on the join calculus [37], which is an extension of C# for asynchronous wide-area concurrency [38]. Active objects in Cω are supported through the techniques of using chords (A chord is a method with multiple headers, all providing parameters for the body, and all but one being called asynchronously); ACTALK is a framework integrated with Smalltalk programming environment [39]; Hybrid is an object-oriented language where objects are active entities [40]; Correlate is a concurrent object-oriented language that supports active objects as a unit of concurrency [41]; TCOZ is an extension of Object-Z with timed CSP and timing constructs [42].

2. Implementation Choices

A number of important choices must be made when embarking on language development. Some of the most important ones are as follows:

- Should we chose to develop an brand new language or provide an extension to an existing language? Thousands of very exciting languages exist, all developed with one thing in mind: solving particular problems that the developer feels are not handled well by existing languages. However, convincing people to adopt new languages seems to be a daunting task. We decided to extend Java, a language that has been accepted in almost all areas of computing.
- How should the new functionality be provided? Since we are using Java, two distinct choices are available. First: extend than language with new keywords, and amend the compiler or provide a pre-processor. (This was the original approach taken by the developers of C++). Second: provide a library of (precompiled) classes that implements the functionality (JCSP uses this approach). We chose the first approach; providing new keywords and amending the compiler; A down side of this approach is of course the difficulty in maintaining the compiler when a new version of the language comes out, but we believe that the learning curve of learning to write concurrent programs with new keywords/syntax compared to the library approach, is less steep, and the code looks cleaner. Also, we believe that this enables novice programmers to more easily pick up the concept of active objects.
- Should the implementation allow for remote active objects or only local ones? If we only allow local active objects, data sharing can be realized through the use of the standard Java primitives like the `protected` keyword and the use of locks and monitors. This is not possible if we wish to allow for remote active objects.
- To support remote active objects, we need to decide how networked data transfer should be handled. Again, two different approaches exist; first: implement data exchange through TCP/UDP sockets; second: use a library like mpiJava [43], which implements a messaging system though TCP, or three: use the Java provided RMI functionality. Implementing messaging directly using TCP is the most speed efficient method, but we decided to use RMI for a number of reasons: It is already provided in Java, but more importantly, the RMI system provided by Java automatically allows for classes not already on the remote system to be downloaded through a web service.

This implementation of asynchronous active objects in Java is a prototype that in time might reveal points in its implementation that need to be optimised, especially if serious parallel computing is to be done using an asynchronous active object system. However, we believe that the choices we have made serve as a good basis for an asynchronous active object system integrated in into Java.

3. Asynchronous Active Objects in Java

We have implemented our asynchronous active object system in Java, by extending the language to integrate both synchronous and asynchronous active objects; we chose Java for a number of reasons:

- The language is already object oriented.
- It supports reflection.
- It has RMI 'built in'.
- The Java compiler 1.6 is available as open-source.
- Auto-boxing is done automatically (since Java 1.5).

- It is platform independent.

We define an *asynchronous active Java object* as an object that has the following characteristics:

- It must be *active*, that is, it executes the methods in its own thread.
- It must be possible to place an active object on any machine reachable on the network that supports Secure Shell scripting (ssh) [44] and the Java Runtime Environment (JRE) [45].
- Method invocation can be synchronous or asynchronous.
- A way to obtain the result of an asynchronous invocation must exist.

3.1. New Java Keywords

We decided to add new keywords to Java 1.6, and then utilise the Java compiler to parse and compile them. This addition consists of 4 new keywords/constructs:

- A new **active** modifier, which can only be placed on a class declaration.
- An extended object creation expression:

    new <active class>(...) **on** "machine name";

    which creates an active object instance of <active class> on machine "machine name". The last part (**on** "machine name" is optional, and if left out the object will be created on the local machine.
- An extended method invocation expression:

    <active object>.<method>(...) **async**;

    The **async** keyword makes the method invocation asynchronous, that is, the control returns immediately.
- A new blocking **waitfor** statement:

    waitfor <active object> <variable>;

    This causes the execution of the thread to be temporarily suspended until the asynchronous method invocation has returned a value into the variable.

3.2. Restrictions on Using the New Keywords

The design of our active object system restricts the usage of the new keywords or constructs:

- The new creation expression can only be used to create active objects. Passive objects use the regular syntax.
- The **async** keyword can only be used after method invocations. If multiple method calls are chained in the same expression, **async** only applies to the last method invocation:

    activeObj.foo().bar() **async**

    the *bar* method will be called asynchronously.
- Logically, asynchronous method invocation may only used as an expression *statement* (where its return value, if not *void*, is not immediately needed). So, it cannot be used on the right RHS of an assignment or as an argument in another method invocation.
- Passing *null* as an argument of a method of an active object is not allowed. Reflection is used to determine which method needs to be invoked. Because of method overloading, the name of the method is not enough and the parameter types are used to select the method that needs to be invoked. *null* has no type, so the decision can not be made.

- Waiting for the results of an asynchronous invocation on the same active instance has to be done in the same order as the invocation. If *foo* and *bar*, in this order, are invoked on the same active object, then first the result of *foo* has to be waited for and then the result of *bar*.

4. Implementation

In the OO model, an object sends a message to another object (invoking a method is sometimes erroneously described as sending a message) synchronously. We have extended this model of interaction by allowing objects to "communicate" both synchronously and asynchronously. The active object executes the method invocations in its own thread and has a queue of pending messages that must be processed. An active object can be created on any reachable machine from the network that supports Secure Shell (ssh). The communication with active objects is done by exchanging real messages and is realized through RMI.

Let us assume that the machine creating the active object is called *client* and the machine where the active object resides is called *server*. The *client* can send the *server* two types of messages:

- **create message** asking the *server* to create a new active object.
- **invoke message** asking the *server* to execute a method on one of the active objects that it hosts.

4.1. Creating an Active Object

The syntax for creating an active object on *server* is:

activeObj = new ActiveClass(args) **on** *"server"*

where again, the **on** *"machine"* part is optional, and if omitted, the object will reside one the local machine. The creation of an active object happens synchronously and the *server* replies with another message sending the client a handle to the active object, as shown in Figure 1.

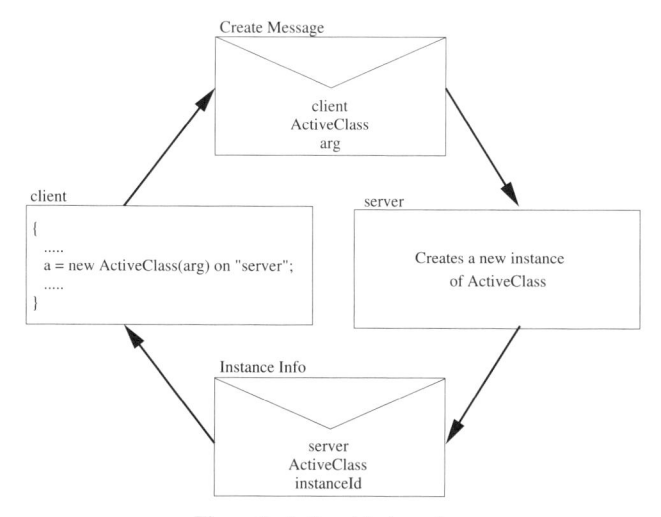

Figure 1. Active object creation.

The **create message** contains: the name of the machine that initiated the call (used internally by the server to keep track of invocations from each client since active object references can be passed from machine to machine), the type of the class, and the arguments of the constructor. The *server* sends the *client* a 'remote reference' represented by an *InstanceInfo* object that contains the following information: the name of the machine where the object resides, the type of the object and a unique identifier to distinguish this instance from other instances on the same *server*. The *client* then uses this remote reference every time it wants to send a message to the active object.

4.2. Invoking Methods on an Active Object

Unlike passive objects, the methods of active objects can be invoked asynchronously:

activeObj.foo(a,b,c) **async**

The *client* sends an **invoke message** asking the *server* to execute the *foo* method on the *activeObj* and return the result asynchronously. The *client* uses the 'remote reference' sent by the server, as shown in Figure 2.

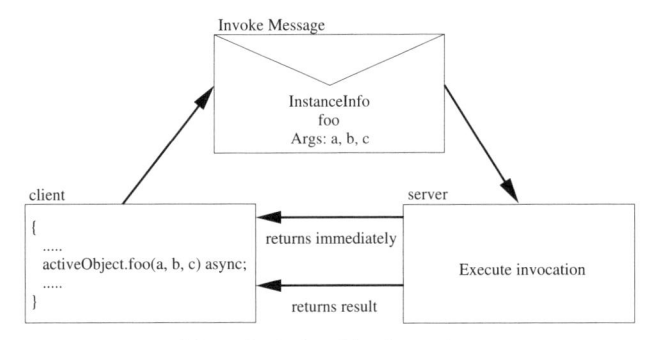

Figure 2. Active object invocation.

The call returns immediately since the **async** keyword is used and the client continues executing the rest of the code. We have implemented the asynchronous invocation by creating a thread that performs the actual call. This thread uses RMI to invoke the method on the active object. This RMI call is synchronous, but the main thread continues with the rest of the client code, thus simulating asynchronous behaviour. When the *server* finishes executing the method, the thread that carried out the execution signals the main thread that the result is available. If a method is invoked synchronously on an active object, then the main thread blocks and waits for the result, unless the method returns void. In this case the invocation is sent to the server and the main thread continues its execution.

4.3. Waiting for the Result of Asynchronous Invocation

When asynchronous interaction between the *client* and the *server* is used, the *client* has to get the result of the method invocation at some point. The **waitfor** statement is used for that:

waitfor *activeObj var ;*

From the programmer's point of view, the **waitfor** statement can be interpreted as: "I'm waiting for the result of an asynchronous invocation on the active object, *activeObj*, and save the return value in the variable *var*". The **waitfor** statement has to be used only for asynchronous methods that return a value. If two methods that return a value, *foo* and *bar* in this order, are invoked on the same active object, then the return values have to be waited in the order of invocation, first the result of *foo* and then the result of *bar*. The compiler will not complain if the the order of waiting for the result is reversed, but a *ClassCastException* will be thrown at runtime if the types of the two return values are not compatible.

The programmer may forget (or choose not to) to wait for the result of an asynchronous call. In this case, when a method finishes executing its body, all the results of the asynchronous invocations that were not waited for are discarded/ignored. This implies that all the results of asynchronous calls have to be waited for in the same method in which they were made.

4.4. Message Ordering

An active object accepts requests from any machine that has a reference to it. Active objects can be passed around, so not only the machine that created the object can have a reference to it. The active object will execute the requests in the order of arrival, on a first come, first served basis. No ordering can be imposed on invocations from different machines on the same active objects, in accordance with Lamport's happen-before relation [46]. Though, if two invocations, *foo* and *bar* in this order, are invoked on the same active object, then the *foo* will be executed before *bar*.

4.5. ClientManager and ServerManager

The core components of our active object system comprises two classes: *ClientManager* and *ServerManager*. The communication between the machine that creates the active object and the machine that hosts the active object is realized through these classes. The creation of an active object or a method invocation of an active object is translated at compile time to an invocation to one of the following *ClientManager* methods:

- *invokeConstructor(...)* creates an active object and returns a 'remote reference' of type *InstanceInfo*. This remote reference is used by the *client* every time it wants to invoke a method on the active object.
- *invokeMethod(...)* invokes a method on an active object.

The *ClientManager* keeps track of all the active object invocations that are initiated from the machine on which it runs. As some of these invocations are executed asynchronously, the return values of these calls are also managed by the *ClientManager*. Besides the two core functions, the *ClientManager* also has additional helper methods:

- *getMethodId(...)* returns a unique identifier that is associated with every method that has an active invocation in its body. It is inserted by the compiler at the beginning of each method body that has at least one active object invocation.
- *removeUnwaitedCalls(...)* removes all the un-waited calls within a method body. It is inserted at the end of all methods that have at least one active object invocation.

The *ServerManager* plays a similar role on the server side (a server is any machine that hosts an active object). It accepts **create** and **invoke** messages and sends these requests to the corresponding active object. There is only one *ServerManager* on each machine that participates in the execution and it has to be started before the program is executed. Starting the managers is done by executing an ssh script.

4.6. Compiler Modifications

The programmer does not know about the *ClientManager* and *ServerManager* classes, but instead he writes Java code using the added keywords. At compile time, during desugaring, the new syntax is replaced by calls to *ClientManager* methods.

The modified compiler performs two important tasks: first it checks if the syntax is correct and that the new keywords are used properly and secondly, during the desugaring phase, it replaces the new syntax with regular Java code.

The **active** keyword is removed from the class declaration. During the attribution phase additional information is saved in the node that represents a class to identify an active class.

The generation of the *methodId*, the translation of the new creation expression, asynchronous invocation and **waitfor** statement to regular Java code will be shown in the Mandelbrot example (section 6.1).

5. Example

To demonstrate the use of the asynchronous active object system we have implemented a simple publisher/subscriber system; to be exact, we have implemented the example given in the Cω tutorial [47]. The Cω implementation implements 3 classes: *ActiveClass*, *Subscriber*, and *Distributor*; in addition it implements an interface called *eventSink*. Since we have the notion of active objects built into the language and the compiler, we do not need to define a class for an active object, and thus need only classes for the *Subscriber* and the *Distributor*, as well as the driver with the main method (Figure 3).

The main method creates a distributor and 3 subscribers, and adds the subscribers to the distributor, who in time invokes the `post` method of these; subsequently the subscribers print out the message received.

In this example we do not make use of the ability to create *remote* active objects; if the optional **on** *"machine"* part of the new object creation is left out, then the new object will be created on the local host. See section 4.1.

The output obtained (one of the many possible) when executing the code is:

```
a got the message: First message
b got the message: Second message
b got the message: Third message
c got the message: Third message
a got the message: Second message
a got the message: Third message
```

Note, there exist no total ordering for all the messages, but a total ordering within the each subscriber does exist. This is a sided effect of the message ordering constraints that we implemented (See section 4.4), something which the Cω implementation does not guarantee.

In the next section we describe another desirable side effect of our asynchronous active object system; namely the ability to implement parallel computation using active objects.

6. Active Objects for Distributed Computing

By utilising asynchronous method invocation, our system allows parallel computation in a very simple manner. Moreover, the active objects do not have to be located on the same machine, but can be distributed over the network. The distributed support offered by Java is RMI. By calling methods on active objects synchronously we obtain the same effect as RMI without searching the registry for the remote object. By calling methods asynchronously (add the **async** keyword at the end of the invocation) and by allowing the caller to continue

```
public active class Distributer {
    private ArrayList<Subscriber> subscribers = new ArrayList<Subscriber>();

    public void subscribe(Subscriber s){
        subscribers.add(s);
    }
    public void post(String message){
        for (Subscriber s:subscribers){
            s.post(message) async;
        }
    }
}

public active class Subscriber {
    private String name;

    public Subscriber(String name) {
        this.name = name;
    }
    public void post(String message){
        System.out.println(name+" got the message: "+message);
    }
}

public class Demo {
    public static void main(String args[]){
        Distributer d = new Distributer();
        Subscriber a = new Subscriber("a");
        d.subscribe(a) async;
        d.post("First message");
        Subscriber b = new Subscriber("b");
        d.subscribe(b) async;
        d.post ("Second message");
        Subscriber c = new Subscriber("c");
        d.subscribe(c) async;
        d.post("Third message");
    }
}
```

Figure 3. The active *Distributor* and *Subscriber* and passive driver class.

his execution without waiting for the result at the invocation time, we obtained a parallel computing model.

6.1. Mandelbrot Example

One often used pattern in distributed systems is the master/slave architecture. The master divides the work into tasks and sends the tasks to the slaves. The slaves execute the task and send the result back to the master. We will show how this master/slave model can be designed with our system by implementing the Mandelbrot Set computation. The active code for this implementation can be seen in Figure 4 (see lines 1, 11, 12, and 16 for the use of these new keywords). Executing our compiler with the code from Figure 4 will generate the Java code shown in Figure 5.

The first and the last statements of the *main* method of the *MandelbrotClient* class were added by our compiler. The *methodId* (line 9) is used by the *ClientManager* to handle method invocations and the last statement (line 26) discards all the results of any asynchronous invo-

```
public active class Mandelbrot {
    public MandelBrotSet compute(...) {
        ... compute the mandelbrot set for the area given by the parameters
    }
}

public class MandelbrotClient {
    public static void main(String[] args) {
        Mandelbrot[] mandelbrot = new Mandelbrot[noProc];
        for(int i=0; i<noProc; i++) {                                                       10
            mandelbrot[i] = new Mandelbrot() on computerNames[i];
            mandelbrot[i].compute(...) async;
        }
        MandebrotSet mandelSet;
        for(int i=0; i<noProc; i++) {
            waitfor mandelbrot[i] mandelSet;
            ... process the return value
        }
    }
}                                                                                           20
```

Figure 4. Mandelbrot fractal computation using asynchronous active objects.

```
public class Mandelbrot {
    public MandelBrotSet compute(...) {
        ... compute the mandebrot set for the area given by the parameters
    }
}

public class MandelbrotClient {
    public static void main(String args) {
        Long methodId = ClientManager.getMethodId();
        InstanceInfo[] mandelbrot = new InstanceInfo[noProc];                               10

        for(int i=0; i<noProc; i++) {
            mandelbrot[i] = ClientManager.invokeConstructor(
                "Mandelbrot", new Object[]{}, computerNames[i]);
            ClientManager.invokeMethod(
                methodId, mandelbrot[i], "compute", new Object[]{....}, true, false);
        }

        MandebrotSet mandelSet;
        for(int i=0; i<noProc; i++) {                                                       20
            ReturnObject returnObject0 = ClientManager
                .waitForThread(methodId, mandelbrot[i], false);
            mandelSet = (MandelbrotSet)returnObject0.getReturnValue();
            ... process the return value
        }
        ClientManager.removeUnwaitedCalls(methodId);
    }
}
```

Figure 5. Mandelbrot code generated by our compiler.

cations from this method body that were not waited for. The construction of an active object (line 11 in Figure 4) is translated to a call to *invokeConstructor* (line 13-14 in Figure 5) and the active object method invocation (line 13 in Figure 4) is translated to a *invokeMethod* call (lines 15-16 in Figure 5). Finally the **waitfor** (line 16 in Figure 4) statement is translated into two statements: first a call to *waitForThread* to wait for the result to be available and secondly code for getting the actual result.

6.2. Matrix Multiplication Example

We have also implemented Fox's Pipe-and-Roll matrix multiplication algorithm [48] with active objects. The algorithm divides the matrices, A and B, into sub matrices in a grid manner and then the following steps are repeated: broadcast A (the first matrix) to the right, do local multiplication of incoming A and local B (second matrix), and shift B up with wraparound. An active object (slave) receives an A sub matrix and a B sub matrix and performs the three steps from the algorithm. A master object controls the number of iteration for the algorithm and on each iteration it sends the slaves the task they have to perform. The results obtained after running our example is presented in section 7.2.

6.3. Pipeline Computation Example

Another example that we have implemented with our active object system is a pipelined computation. Our example contains 5 processes (or 5 steps) that take 2, 6, 6, 8, and 2 seconds respectively to execute (to produce a result at the end of the pipe). The throughput is no better than one result ever 8 seconds, and the latency is no better than 24 seconds. This pipeline is illustrated in Figure 6.

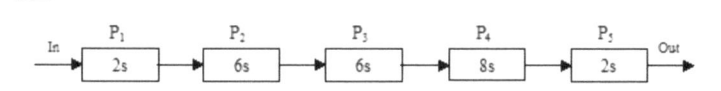

Figure 6. Original Pipeline.

As stated, we cannot get a better throughput than one result every 8 seconds. This can be optimised to get a better throughput by adding dispersers and collectors, and by replicating the processes that take more time to execute. One possible arrangement of these processes is shown in Figure 7.

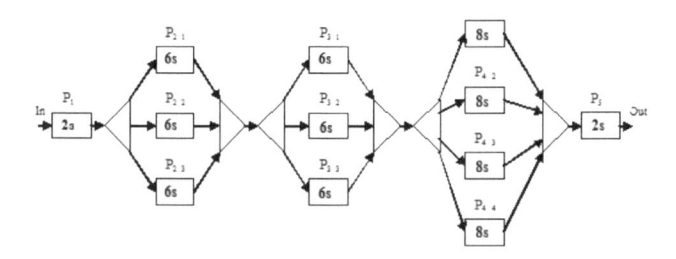

Figure 7. Optimised Pipeline with Dispersers and Collectors.

The dispersers, the collectors and the replicated processes are all active objects. The disperser creates the necessary number of processes and the collector. Once a process finishes its work, it forwards the result to the collector, which in turn forwards it to the next process or disperser. The results of the pipeline computation are presented in section 7.3.

With this new setup, we should be able to improve the throughput to one result every (no better than) 2 seconds. Naturally there might be a little overhead from passing data through dispersers and collectors.

7. Results

Although the main focus of this paper is the implementation of an active object system in Java, we do wish to present the initial results of the experiments of *parallel programming with active objects*.

7.1. Mandelbrot Set

We have tested the Mandelbrot Set computation with active objects against the sequential Mandelbrot. We started with a complex plane of 5,000 x 5,000 points and increased each dimension of the complex plane by 5,000 points each time, up to 20,000 x 20,000 points. Additionally, the parallel versions were tested on 4, 8, 16, 32 and 64 slaves (plus a master) for each complex plane size. The speedups obtained are presented in Figure 8.

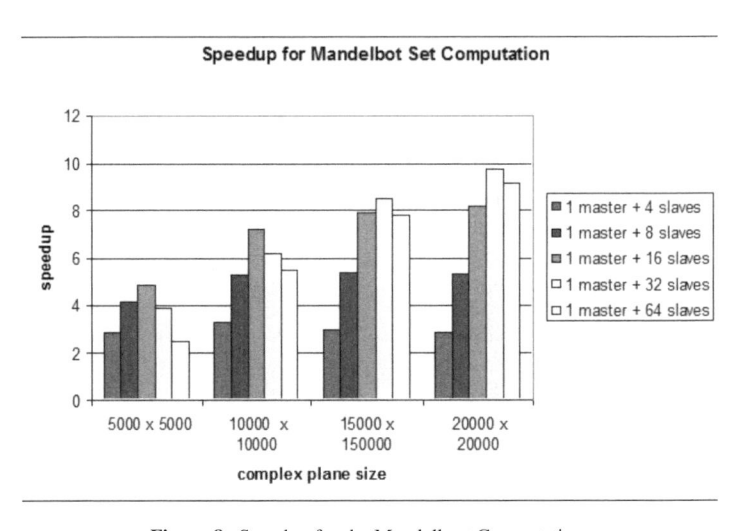

Figure 8. Speedup for the Mandelbrot Computation.

7.2. Matrix Multiplication Results

We have tested the Matrix Multiplication for matrix sizes of 512 by 512, 1,024 by 1,024, 2,048 by 2,048 and 4,086 by 4,086 on 4, 16 and 64 processors (slaves). The benefits of distributing the workload on multiple computers and doing the computation in parallel starts paying off for the matrix size 2,048 by 2,048 and 16 processors. There is some overhead of starting each slave (active object) and sending the parameters across the network (in each step, each active object "pipes" its A sub matrix and "rolls" is B matrix). The speedups obtained are presented in Figure 9.

7.3. Pipeline Computation Results

The regular pipeline example presented in section 6.3 creates a result each 24 seconds. By interspersing dispersers and collectors, and at the same time replicate the processes in the stages that take longer, the optimised version improves the throughput to one result every 2.008 seconds on average (2 is the minimum we could ever hope to achieve).

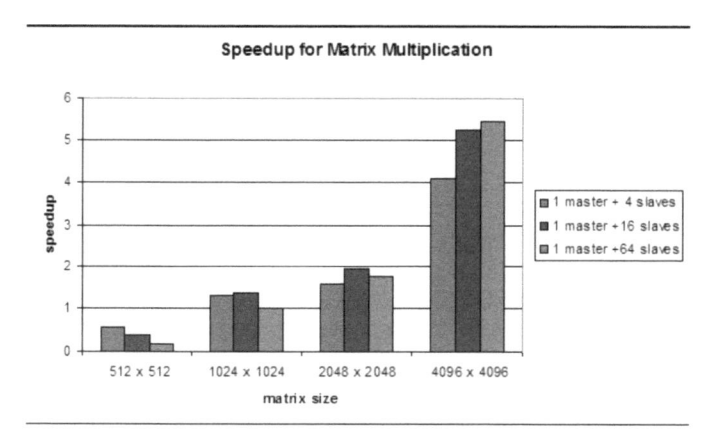

Figure 9. Speedup for the Matrix Multiplication algorithm.

8. Conclusion

In the standard Object Oriented (OO) model everything is considered an object. Objects communicate with each other by sending messages. Some argue that this model is a good reflection of the real world and of how objects interact. However, all the objects are passive and all the method invocations are executed in the caller's thread. A passive object reuses the thread of the object that created it.

The active object model represents a better reflection of how objects in the real world interact. Each object executes the method invocations in its own thread and only sends the result to the caller. Moreover, the communication can be done both synchronously and asynchronously (in the OO model the communication is always synchronous).

We have modified the Java language to integrate active objects by adding four new keywords: **active**, **async**, **on** and **waitfor**. Because methods can be invoked asynchronously using the **async** keyword, active objects can be used for developing parallel applications (the caller does not wait for the result at the moment of the invocation but continues its execution in parallel with the active object invocation). Active objects do not have to reside on the same machine, but they can be distributed over the network. The location of an active object is specified when the object is created and it can not be changed afterward, even though active object references can be passed around. Our system allows the existence of both passive and active objects, just like the real world. We used the active object model to implement a number of applications: Mandelbrot set computation, pipe-and-roll matrix multiplication, and pipeline computation. Even though linear speedup was not achieved, the results are encouraging and demonstrate the feasibility of using active objects in Java for certain types of distributed computation.

In general we we have shown one way of implementing asynchronous active objects in Java, using RMI and compiler modifications, which provides an easily approachable introduction to active objects for Java programmers.

Finally, since current versions of the Java Virtual Machine (JVM) are implemented to take advantage of multi-core processors, our active object system provides a straight forward way to write code for such architectures. As long as the work load associated with executing a method (invoked asynchronously) is significant, the advantage of utilising multiple cores will outweigh the overhead.

9. Future Work

Our system demonstrates one way to implement active objects in Java by extending the language. The system that we have created enforces some restrictions of how the new keywords should be used, but can be further extended and improved to eliminate some of these restrictions. Some improvements are:

- starting/stopping the *ServerManager* from code instead of ssh script
- warning the user if not all asynchronous calls with return value have a matching **waitfor** statement
- creating an exception handling mechanism (*ClientManager* catches all the exceptions sent by the active objects and does not forward them to the caller)
- keeping the active objects once they are created and allowing them to be accessed by other applications (creating a remote directory)
- Rather than creating a new thread for each asynchronous method invocation and then discarding it after the RMI call returns, a pool of threads can be kept, these can be reused.
- As illustrated in the Mandelbrot example, the **waitfor**s are handled in the same order as the invocations, this could be a potential slowdown. A possible solution to this could be a method for doing a number of **waitfor**s in parallel.
- If an active object resides on the local machine, the use of RMI will be a significant overhead, and totally unnecessary; an obvious optimisation would be to simply exchange messages within the shared memory space using locks and monitors.

One interesting, and important issue of using active objects is that of deadlocks. It is not hard to write a program that causes a deadlock; recursive calls to active objects, mutually recursive calls between active objects are simple examples that will certainly cause a deadlock. Like many other programming models, there are certain difficulties that cannot automatically be avoided. However, it it possible to observe and report such deadlocked situations; by logging all calls and keeping an up-to-date call graph in a central manager process, deadlocks can be reported to the programmer at run-time. This is an extension that we would like to add to the active object system in the future.

References

[1] Peter H. Welch. Communicating processes, components and scaleable systems.
http://www.cs.kent.ac.uk/projects/ofa/jcsp/components-6up.pdf, May 2001.
[2] James E. White. RFC 707: A high-level framework for network-based resource sharing, December 1975.
[3] Sun Microsystems. Java remote method invocation.
http://java.sun.com/javase/technologies/core/basic/rmi/index.jsp.
[4] R. Jungclaus, G. Saake, and C. Sernadas. Using Active Objects for Query Processing. In *Object-Oriented Databases: Analysis, Design and Construction (Proc. of the 4th IFIP WG 2.6 Working Conf. DS-4, Windermere, UK, 1990)*, pages 285–304, 1991.
[5] Patrik Reali. Structuring a compiler with active objects. In *roceedings of the Joint Modular Languages Conference on Modular Programming Languages (LNCS 1897)*, pages 250–262. Springer Verlag, 2000.
[6] K. X. S. Souza and I. S. Bonatti. Using distributed active object model to implement tmn services. In M. E. Cohen and D. L. Hudson, editors, *Proceedings of the ISCA 11th International Conference*, San Francisco, California, 1996.
[7] Active Objects in Symbian OS.
http://wiki.forum.nokia.com/index.php/ Active_Objects_in_Symbian_OS.
[8] An Introduction to Active Objects for UIQ 3-based Phones.
http://developer.sonyericsson.com/site/global/techsupport/tipstrickscode/symbian/p_active_objects_uiq3.jsp.
[9] T.A. Busby and L.T. Chen. Developing collaborative games using active objects.
http://jdj.sys-con.com/read/35903.htm.

[10] R. G. Lavender and D. C. Schmidt. Active object: an object behavioral pattern for concurrent programming. In *Pattern languages of program design 2*, pages 483–499. Addison-Wesley Professional, 1996.
[11] The French National Institute for Research in Computer Science and Control. ProActive: A comprehensive solution for multithreaded, parallel, distributed, and concurrent computing. http://proactive.inria.fr/release-doc/html/index.html.
[12] C. Hewitt. Viewing control structures as patterns of passing messages. *Artificial Intelligence*, 8(3):323–364, June 1977.
[13] Gul Agha. *Actors: a model of concurrent computation in distributed systems*. MIT Press, Cambridge, MA, USA, 1986.
[14] D. Kafura, M. Mukherji, and D.M. Washabaugh. Concurrent and distributed garbage collection of active objects. *IEEE Transactions on Parallel and Distributed Systems*, 6(4):337–350, 1995.
[15] D. Kafura, D. Washabaugh, and J. Nelson. Garbage collection of actors. *Conference on Object Oriented Programming Systems Languages and Applications*, pages 126–134, 1990.
[16] D. Kafura, D. Washabaugh, and J. Nelson. Progress in the garbage collection of active objects. *ACM SIGPLAN OOPS Messenger*, 2(2):59–63, 1991.
[17] I. Puaut. A distributed garbage collection of active objects. *ACM SIGPLAN Notices*, 29(10):113–128, 1994.
[18] A. Vardhan and G. Agha. Using passive object garbage collection algorithms for garbage collection of active objects. *International Symposium on Memory Management*, pages 106–113, 2002.
[19] C. Dony, C. Urtado, and S. Vauttier. Exception handling and asynchronous active object: Issues and proposal. *Lecture Notes in Computer Science. Advanced Topics in Exception Handling Techniques*, 4119:81–100, 2006.
[20] O. Nierstrasz. Regular types for active objects. *SIGPLAN: Conference on Object Oriented Programming Systems Languages and Applications*, pages 1–15, 1993.
[21] O. Nierstrasz and M. Papathomas. Towards a type theory for active objects. In *SIGPLAN: Proceedings of the workshop on Object-based concurrent programming at the Conference on Object Oriented Programming Systems Languages and Applications*, pages 89–93, Ottawa, Canada, 1991.
[22] F. Puntigam and C. Peter. Types for active objects with static deadlock prevention. *Fundamenta Informaticae*, 48(4):315–341, December 2001.
[23] Z. Guessoum and J. Briot. From active objects to autonomous agents. *IEEE Concurrency*, 7(3):68–76, 1999.
[24] O. Nierstrasz and M. Papathomas. Viewing objects as patterns of communicating agents. In Norman Meyrowitz, editor, *Proceedings of the Conference on Object-Oriented Programming Systems, Languages, and Applications European Conference on Object-Oriented Programming (OOPSLA) (ECOOP)*, pages 38–43, Ottawa, ON CDN, [10] 1990. ACM Press, New York, NY, USA.
[25] J. Briot, R. Guerraoui, and K. Lohr. Concurrency and distribution in object-oriented programming. *ACM Computing Surveys*, 30(3):291–329, 1998.
[26] B. Pryor. Implementing active objects with java dynamic proxies. http://benpryor.com/blog/2006/03/08/implementing-active-objects-with-java-dynamic-proxies/.
[27] C. Petitpierre. Synchronous active objects introduce CSP's primitives in Java. In James Pascoe, Peter H. Welch, Roger Loader, and Vaidy Sunderam, editors, *Proceedings of Communicating Process Architecture 2002*, pages 109–122, Reading, United Kingdom, September 2002. IOS Press.
[28] W. Chen, Z. Ying, and Z. Defu. An efficient method for expressing active object in c++. *SIGSOFT Software Engineering Notes*, 25(3):32–35, 2000.
[29] J. Dongarra. MPI: A message passing interface standard. *The International Journal of Supercomputers and High Performance Computing*, 8:165–184, 1994.
[30] B. Carpenter, V. Getov, G. Judd, T. Skjellum, and G. Fox. MPJ: MPI-like message passing for Java. *Concurrency: Practise and Experience*, 12(11), September 2000.
[31] Peter H. Welch and Jeremy M. R. Martin. A csp model for java multithreading. In *PDSE '00: Proceedings of the International Symposium on Software Engineering for Parallel and Distributed Systems*, page 114. IEEE Computer Society, 2000.
[32] Peter H. Welch. Communicating Sequential Processes for Java (JCSP) http://www.cs.kent.ac.uk/projects/ofa/jcsp/.
[33] Peter H. Welch. Process oriented design for Java: Concurrency for all. In H. R. Arabnia, editor, *International Conference on Parallel and Distributed Processing Techniques and Applications (PDPTA 2000)*, volume 1, pages 51–57. CSREA, 2000.
[34] Peter H. Welch, Jo R. Aldous, and Jon Foster. Csp networking for java (jcsp.net). In *ICCS '02: Proceedings of the International Conference on Computational Science-Part II*, pages 695–708. Springer-Verlag, 2002.
[35] G. Rimassa F. Bergenti, A. Poggi and M. Somacher. DisC++: A software library for object oriented

concurrent and distributed programming.
[36] D. Kafura, M. Mukherji, and G.R. Lavender. Act++ 2.0: A class library for concurrent programming in C++ using actors. *Journal of Object Oriented Programming*, pages 47–55, October 1993.
[37] Cédric Fournet, Georges Gonthier, Jean-Jacques Lévy, Luc Maranget, and Didier Rémy. A calculus of mobile agents. In *Proceedings of the 7th International Conference on Concurrency Theory (CONCUR'96)*, pages 406–421. Springer-Verlag, 1996.
[38] N. Benton, L. Cardelli, and Cédric Fournet. Modern concurrency abstractions for C#. *ACM Trans. Program. Lang. Syst.*, 26(5):769–804, 2004.
[39] J. Briot. Actalk : A framework for object oriented concurrent programming - design and experience. In *Proceedings of the 2nd France-Japan Workshop (OBPDC'97)*. Herms Science, 1999.
[40] O.M. Nierstrasz. Active objects in hybrid. In Norman Meyrowitz, editor, *Proceedings of the Conference on Object-Oriented Programming Systems, Languages, and Applications (OOPSLA)*, volume 22(12), pages 243–253, New York, NY, 1987. ACM Press.
[41] F. Matthijs, W. Joosen, J. Van Oeyen, B. Robben, and P. Verbaeten. Mobile active objects in Correlate. In *ECOOP'95 Workshop on Mobility and Replication*, Aarhus, Denmark, 1995.
[42] J. S. Dong and B. Mahony. Active objects in tcoz. In *ICFEM '98: Proceedings of the Second IEEE International Conference on Formal Engineering Methods*, page 16, Washington, DC, USA, 1998. IEEE Computer Society.
[43] mpiJava. http://www.hpjava.org/mpiJava.html.
[44] T. Ylonen. RFC 4254: The secure shell (SSH) connection protocol, January 2006.
[45] Sun Microsystems. Java runtime environment. http://java.sun.com/j2se/desktopjava/ jre/index.jsp.
[46] L. Lamport. Time, clocks and the orderings of events in a distributed system. *Communications of the ACM*, 21:558–565, 1978.
[47] Active Objects Tutorial. http://research.microsoft.com/COmega/doc/comega_tutorial_active_objects.htm.
[48] G. Fox, S. Otto, and A. Hey. Matrix algorithms on a hypercube i: Matrix multiplication. In *Parallel Computing*, pages 17–31, 1987.

JCSPre: the Robot Edition to Control LEGO NXT Robots

Jon KERRIDGE, Alex PANAYOTOPOULOS and Patrick LISMORE

School of Computing, Napier University, Edinburgh UK, EH10 5DT
`j.kerridge@napier.ac.uk`, `alex@flocci.org`, `plismore@gmail.com`

> **Abstract.** JCSPre is a highly reduced version of the JCSP (*Communicating Sequential Processes* for Java) parallel programming environment. JCSPre has been implemented on a LEGO Mindstorms NXT brick using the LeJOS Java runtime environment. The LeJOS environment provides an abstraction for the NXT Robot in terms of Sensors, Sensor Ports and Motors, amongst others. In the implementation described these abstractions have been converted into the equivalent active component that is much easier to incorporate into a parallel robot controller. Their use in a simple line following robot is described, thereby demonstrating the ease with which robot controllers can be built using parallel programming principles. As a further demonstration we show how the line following robot controls a slave robot by means of Bluetooth communications.
>
> **Keywords.** parallel, robot controllers, JCSP, LEGO NXT, LeJOS, Bluetooth.

Introduction

The LEGO[1] NXT construction kit is often used for the construction of simple robots, in research and teaching and learning applications. Students and researchers of the CSP model would benefit from having a programming platform specifically for the programming of LEGO robots using CSP techniques. Researchers [1, 2] at The University of Kent have created their own parallel robot programming system which has many aspects in common with the environment described in this paper. However, their system is built on the Transterpreter platform, using the occam programming language, rather than the more widely used Java language. We therefore set out to create and develop an environment, for the LEGO NXT, based on a Java environment using a highly reduced version of the Communicating Sequential Processes for Java (JCSP) parallel programming environment, using an equally small Java virtual machine, called LeJOS. This means that developers of robot control systems can utilise parallel programming techniques based on a Java environment using a common development tool such as Eclipse.

The aim was to find out if a port of a reduced JCSP environment and LeJOS to the LEGO NXT was possible and feasible. Feasible is used in the sense that, even though a port was achieved, the size of control system that could be implemented would be so small as to render the system unusable. If the implementation proved feasible, then it would form the basis of a teaching module that introduces the concepts of process-based parallel programming using robotics as the underlying motivation. Furthermore, in using an environment based on Java and Eclipse we would be building on the students' prior experience and also showing them a concurrent programming environment that does not

---

[1] LEGO is a registered trade mark of the LEGO Group of Companies.

explicitly expose the underlying Java thread model. This may mean that we can wean students away from the widely-held belief that concurrent programming is hard.

The next section places the work in context. Section 2 describes the architectural framework and shows how LeJOS NXT is integrated with the cut-down JCSP and how this is then used to create active processes that can be used to build robot controllers. Section 3 explains how such robot controllers can be designed and section 4 shows an example of the design process for a simple line following robot. Section 4 describes the Bluetooth communication aspects that allow a slave robot to follow the movement of a master line following robot. The paper concludes with a summary of the achievements and the identification of future work.

1. Background

The LEGO NXT comes with a disc containing the proprietary LEGO programming language, called NXT-G. This is a visual language based on LabVIEW, another proprietary language created by National Instruments [3]. LabVIEW, and by extension NXT-G, has a control structure based upon that of dataflow diagrams [4]. While this means that there is an inherent parallelism in any LabVIEW program, it does not conform to the CSP model that this paper is most concerned with.

In addition, the LabVIEW language has two other shortcomings. Firstly, as a graphical programming language, it can be difficult for a more traditional programmer to learn [4]. Secondly, it is a closed source proprietary language, whose continued support is fully dependent on National Instruments. Any proprietary product raises concerns about the longevity of the format, since support for that format is dependent on the continued existence of the vendor [5].

The programming language occam (and its extension, occam-pi) is specifically designed around the CSP model of parallel programming [6], and it has been widely used in research [7]; but its use in the wider community is very limited in comparison to the world-wide use of Java.

One approach to the implementation of a CSP model on various devices is that of the Transterpreter project. This is a native occam runtime written in C, and developed at The University of Kent – originally for the LEGO Mindstorms RCX brick (the predecessor to the NXT brick) [1]. The Transterpreter is designed to be portable to a wide range of devices, and is currently being ported to the NXT brick by the Transterpreter team [2].

As well as running the Transterpreter as an interpreter, another approach is to use it to compile occam programs to native machine code, either directly or via an intermediary language such as C. This can be shown to give a measurable performance increase over interpreted occam [8]. The Transterpreter is not the only project aimed at creating a native occam runtime environment for embedded devices, [9] describes a similar project aimed at the Cell BE microarchitecture.

Although occam itself may be the most appropriate language for CSP programming, the language is still unfamiliar or unavailable to the majority of programmers. This has prompted the development of various occam-like libraries for more prevalent languages, such as C++ [10], Python [11], and notably Java.

The Java Communicating Sequential Processes (JCSP) framework is a library that allows a complete occam-like CSP programming package for Java, including all basic occam features [12], and some of the ideas of the pi-calculus/occam-pi [13], and whose channel classes have been formally proved to be equivalent to the CSP concepts upon which they are based [14]. JCSP was developed at the University of Kent, in consultation with the CSP and occam/transputer communities [15].

The JCSP framework is subdivided into four top-level packages, of which only parts of two of these are required for the JCSPre environment.

- `org.jcsp.lang`: This package provides the core of the CSP / occam core concepts. These are provided by classes such as CSProcess, Guard, Alternative and Parallel, each of which are named after the CSP concept which they implement. Most of the complexity of the `lang` package comes in the implementation of the channel classes, which are subdivided by source (One or Any), destination (One or Any) and type (integer or object). There are also "Alting" versions of each one, which can be included in an Alternative. To hide the complexity of the implementation from the user, a convenience class (Channel) is provided. The `lang` package also provides a few other classes not in the original CSP paper, but which have shown to be frequently used concepts.

- `org.jcsp.util`: This package provides some extensions to the simple channel behaviour of the CSP model. Various styles of buffering are provided to extend the simple semantics of the CSP channel model. A sub-class, `org.jcsp.util.filter`, allows objects to be transformed as they are transmitted over a channel – either at the input end, the output end, or both [16]. This package also contains some `exception` definitions that are required by the previous package.

LeJOS is a project to port a Java virtual machine to the various LEGO robotic bricks. The port for the NXT brick is known as LeJOS NXJ [17]. The most recent version, at the time of development, was LeJOS NXJ 0.3.0 alpha. Although the LeJOS virtual machine is based upon that of Java, there are several limitations which prevent it from being a fully compatible JVM. These include:

- No garbage collection is performed on the generated code, meaning that objects should be reused whenever possible,
- The Java `switch … case` statements are not implemented, meaning that a series of `if … then` statements need to be used instead,
- No arithmetic operations can be performed on variables of type `long`.
- Arrays cannot be longer than 511 items in length.
- The `instanceof` and `checkcast` operations are not properly implemented for interfaces and arrays.
- The core Java class `Class` is not implemented, and the `.class` construct is non-functional. This means that programming techniques involving reflection will not work.
- As well as the above design limitations, there are numerous bugs in the 0.3.0alpha release, as should be expected from any alpha release. One known bug is that multidimensional arrays of objects do not work correctly [18], (removed in LeJOS NXJ 0.6.0 beta).
- Inspection of the LeJOS API specification shows that the `java.lang` package, whose classes are described as "fundamental to the design of the Java programming language" by the core Java API, is missing a large portion of its classes [19], although the collection of classes implemented are similar to those implemented by other "micro edition" virtual machines, such as the CLDC [20].

The first two restrictions have been removed in the latest beta release of LeJOS but have yet to be incorporated into our implementation as the JCSPre as implemented had been tested and shown to be functional. In particular, reliance on garbage collection in small, resource constrained devices should be avoided whenever possible. The control processes described later all use `int` channels for data communication to avoid the creation of objects.

2. Architectural Framework

The system requires a multi-layer approach, with each layer adding more functionality and being more tailored to the specific outcome of a CSP-based LEGO robot.

2.1 Architecture: Description and Diagram

The basis of the architecture is the LEGO NXT brick itself, whose firmware supports the installation of a LeJOS kernel. The nature of the layered architecture is shown in Figure 1. The LeJOS NXJ abstraction layer provides a set of Java classes and interfaces that allow the creation of control systems using normal Java programming methods. The JCSPre is implemented using the LeJOS kernel abstraction layer. The Active Sensors then provide a process based realisation for the LEGO NXT sensors, see 2.5. The only part the final developer of a robot control system is concerned with is the high level Control aspect.

	Control	
		Active Sensors, Motors and other processes
LeJOS NXJ abstraction		JCSPre
	LeJOS kernel	
	LEGO NXT Firmware	

Figure 1: the layered architecture of the JCSPre environment.

The complete architecture has been integrated into the Eclipse development environment. In particular, a control system can be developed and downloaded into a LEGO robot from within the Eclipse environment.

2.2 LeJOS Kernel

The LeJOS virtual machine and core packages were described previously. During the development of the system, several bugs and shortcomings were uncovered. This prompted the development of workarounds, in the form of Java classes created in the package `org.jcsp.nxt.util`. These in the main dealt with `exceptions`, which in the LeJOS implementation did not have a `String` parameter.

2.3 LeJOS NXJ Abstractions

The LeJOS project includes a set of classes specific to the functioning of the LEGO NXT robot, concerning input, output and communication. The packages provided for these functions are:

- `LeJOS.navigation`: A set of high-level classes which use the NXT internal compass and tachometer to move an NXT robot around at specified angles and distances.

- `LeJOS.nxt`: The majority of the robot input/output classes. Classes are provided to read the state of various input sources (battery level, push buttons, light and colour sensors, compass and tilt sensors, sound sensors, touch sensors and ultrasonic sensors), various output sources (motors, LCD screen, and speakers), as well as access to the flash storage facilities of the robot, and a "polling" method to wait for input from various sources.

- `LeJOS.nxt.comm`: Provides classes for communication with other robots and with a computer. These include interfaces for USB communication and Bluetooth wireless communication, and a class implementing the LEGO Communication Protocol used by all NXT devices.

- `LeJOS.subsumption`: This package contains classes which implement the LeJOS project's interpretation of the subsumption architecture.

2.4 JCSPre : Design and Evaluation

The core of the JCSPre takes the form of a ported set of classes from the JCSP framework; the essential subset. To identify this subset, a core set of features was identified as either necessary to the CSP model, or vital to the implementation of the JCSPre. The required classes were determined using the class dependency diagramming tools available in Eclipse. These features include, together with the JCSP required classes:

- The ability to declare an object to be a CSP process, and implement a run method for that object – the JCSP interface `CSProcess`.

- The ability to run several processes in parallel: the JCSP classes `Parallel` and `ParThread`.

- The ability to create channels, with a single input end and a single output end, thus allowing one process to pass an object to a second object. The object-passing channel variant is necessary in order to be able to pass `String` objects to the display. The integer-passing channel variant is necessary in order to provide a memory-efficient method of passing around large volumes of data, such as sensor levels. These are implementations of the JCSP channel interfaces `One2OneChannel` and `One2OneChannelInt` and their associated implementation classes. The Factory design pattern was widely used within JCSP, especially for the creation of Channels. These Factory classes were removed and the channel classes re-implemented appropriately to reduce the size of the final code footprint on the LEGO NXT. In addition none of the `Any` versions of the channels was implemented.

- The ability to choose between multiple input sources based upon the order in which they become ready, which requires the incorporation of the classes `Alternative`, `Skip` and `Barrier` and the abstract class `Guard`. In addition, the abstract classes `AltingChannelInput` and `AltingChannelInputInt` are required.

- The ability to set a timer; used for creating robot behaviours that last a specific length of time, or for creating "ticks" at which to schedule events. This requires the implementation of the class `CSTimer`.

The above comprises the complete set of classes and interfaces required from the core JCSP package `org.jcsp.lang`, which are included in the JCSPre package `org.jcsp.nxt.lang`.

2.5 Active Components

The main JCSP AWT libraries have a set of input/output components which it calls "Active Components". They include such common components such as text boxes, sliders and software buttons, but each is built to conform to the specification of a CSP process. Instead of accessing an active component through its listener, as would happen if using the standard Java libraries, an active component is accessed through its input and output channels.

Changes in an active component's state are represented by signals on the output channel. Changes in its configuration are effected by writing a signal to its input channel. So, for example, one might create an active "slider" process; configure its start point, end point, and interval via its input channel; and then connect another process to its output channel to read in a signal every time the user changes the value on the slider. This provides a much tidier interface to such active components because all the code associated with a particular component is contained within one class and the developer is unaware of the existence of the listener. In well written Java code the component instance and its listeners should be in different classes, which make understanding the code much more difficult because the reader has to refer to two class definitions.

The JCSPre mimics this active component style of implementation when implementing robot input and output components, for two main reasons. Firstly, as the JCSPre is meant as a port of the JCSP, a consistent approach should be used between the two projects. Secondly, the JCSPre, as a teaching tool, should seek to use a simple and consistent approach throughout its own library of classes. This approach is, necessarily, one which complies very closely with the CSP model. One possible disadvantage to implementing components in an active way could be the performance implications of having so many processes running in parallel on what is essentially a limited resource environment.

2.5.1 Transforming Listeners to Channels

The LeJOS implementation of the LEGO sensors (light, sound, ultrasound etc.) works in a typical Java fashion – a Listener object is created for various sensors, and when a sensor undergoes a state change, a method of the Listener object is called. The combination of the component object and its listener object can thus form the basis of an active component: the listener state change method, when called, writes a new token to the output channel of the active component. This style of implementation is required for any LEGO sensor that can be used to input data into the controller. This includes not only the input sensors, but motors which are able to input rotational information.

For the purposes of explanation we describe the implementation of the Active Touch Sensor process, simply because it is the least complicated but contains the essence of any of the active components.

The enclosing package `org.jcsp.nxt.io` is specified {1}[2] and then the required imports are defined from either the LeJOS NXT environment {2-4} or the JCSPre environment {5-7}. The specific classes are defined simply because the LeJOS environment only adds those required classes to the resulting downloadable classes, rather than the whole package. The class `ActiveTouchSensor` {8} implements the interfaces

---

[2] The notation {n} is used to indicate a line number in a code listing.

CSProcess, so it can be executed in a Parallel and SensorPortListener so that is can be notified of any changes in the sensor's state.

```
01   package org.jcsp.nxt.io;
02   import Lejos.nxt.TouchSensor;
03   import Lejos.nxt.SensorPort;
04   import Lejos.nxt.SensorPortListener;
05   import org.jcsp.nxt.lang.CSProcess;
06   import org.jcsp.nxt.lang.ChannelOutputInt;
07   import org.jcsp.nxt.lang.AltingChannelInputInt;

08   public class ActiveTouchSensor implements CSProcess, SensorPortListener {
09     private ChannelOutputInt outChannel;
10     private AltingChannelInputInt config;
11     private TouchSensor sensor;
12     private int lastValueTransmitted = -9999;
13     private int pressed;
14     private boolean booleanMode = false;
15     private boolean running = false;

16     public static final int ACTIVATE = -1;
17     public static final int BOOLEAN = -2;

18     public ActiveTouchSensor( SensorPort p,
19                               ChannelOutputInt outChannel,
20                               AltingChannelInputInt config ) {
21        this.sensor = new TouchSensor( p );
22        p.addSensorPortListener( this );
23        this.outChannel = outChannel;
24        this.config = config;
25     }
26     public void stateChanged( SensorPort p, int oldVal, int newVal ) {
27        if (booleanMode) {
28              if ( running && (newVal > 512) )
29                    pressed = 0;
30              else
31                    pressed = 1;
32              if (pressed != lastValueTransmitted) {
33                    outChannel.write( pressed );
34                    lastValueTransmitted = pressed;
35              }
36        }
37        else {
38              if (running && (newVal != lastValueTransmitted)) {
39                    outChannel.write( newVal );
40                    lastValueTransmitted = newVal;
41              }
42        }
43     }
44
45     public void run() {
46           int signal = config.read();
47           while( signal != Integer.MAX_VALUE ) {
48                if( signal == ACTIVATE )
49                      running = true;
50                else if (signal == BOOLEAN)
51                      booleanMode = true;
52                signal = config.read();
53           }
54     }
55   }
```

The private variables of the class are then defined {9-15} of which outChannel {9} and config {10} provide the channel interface to this process. Two publicly available constants are available, ACTIVATE {16} and BOOLEAN {17} that can be used to configure the component.

The constructor for the process {18-25} requires three parameters: the identity of the `SensorPort` (S1, S2, S3 or S4) {18} and the processes' output {19} and configuration {20} channels. The `SensorPort` identity is then used to create {21} an instance of a `TouchSensor` called `sensor`, which is private to this process. The process is then added as a sensor port listener {22}.

The interface `SensorPortListener` requires the implementation of a method called `stateChanged` {26-44}. The method is only active if the sensor process has received the `ACTIVATE` message on its configuration channel represented by `running` having the value `true`. The behaviour of the method depends whether the touch sensor is being operated in a 'pressed' mode or is required to output a value that is proportional to the amount of pressure applied to the sensor. In the 'pressed' mode the sensor simply outputs a 0 if the sensor is subject to no pressure and 1 if any pressure has been applied {28-31}. All sensors return a value in the range 0 to 1023 and we use `512` {28} simply as a means of ensuring that the sensor button has been pressed. A change of value is only written to the output channel provided the new value is different from that previously output {32-36}. In proportional mode the method outputs any change of value in the sensor provided it is different from the value that was last transmitted {38-40}.

The `run` method {45-54} is required by the interface `CSProcess`. Initially the sensor is inactive and so must receive at least one message on its configuration channel. The process thus waits until it receives at least one message on its configuration channel {46}, which is read into its `signal` variable. We have chosen not to use the poison mechanism [13] to cause a network of processes to stop processing but instead circulate `Integer.MAX_VALUE` as the universal system halt message. This choice was made on the grounds of reducing the size of the code required to support channels within the resulting system. The incoming configuration value is used to set the value of the local variables `running` {49} or `booleanMode` {51}. Values other than those expected on the configuration channel are ignored. The design of the run method is such that once the process has been set running and possibly been placed in the Boolean mode it waits for another input {52} on its configuration channel. The only sensible valid message is to read the halt value. Thus the process is, for the most part, waiting for a communication on its configuration channel, which consumes no processor resource. The underlying LeJOS kernel will only cause the invocation of the `stateChanged` method when it detects a change in the hardware state associated with this Touch Sensor's Sensor Port.

An additional complexity associated with the Active Light Sensor is that the sensor can be configured to recognise different input values: such as when detecting the input from either a black or white surface. It can also be configured to use, or not use, the sensor's floodlights. The `stateChanged` method can also be configured so that it only outputs changes in state that are larger than some configuration defined "delta" value.

Each sensor implemented in the LeJOS NXT environment has a corresponding implementation in the Active layer of the layered architecture shown in Figure 1.

3. Designing Robot Control Systems

The JCSPre environment comprises a number of elements contained within the `org.jcsp.nxt` structure which contains the packages `filters`, `io`, `lang`, `plugnplay`, `bluetooth` (see section 5) and `util`. The designer then utilizes processes contained within these packages to build their control system in conjunction with any control process(es) and user interface components that need to be coded specially. The designer has to construct a process network diagram of their controller, which will comprise mostly processes contained within the Active Sensor, Motors and other Processes layer (Figure 1). They then have to implement their control and user interface processes so the control system can be tested.

3.1 Filter Processes

In designing systems using this style of process construction it became apparent that a set of processes that could take an input on one channel and then output the value on a specific channel depending upon its value relative to some configuration parameters was very useful. Such processes are contained in the `filters` package Thus a Binary Filter process is provided that outputs an input value on one of two output channels depending on the input value relative to a single level value initially input on the process' configuration channel. Similarly, a Ternary Filter is provided that outputs on one of three output channels an input value depending on its value relative to two configuration values. Thus we can determine whether an input value is below a low value, between a low and high value or above the high value. These filter processes are generic in that all sensors return integer values typically between 0 and 1023 and thus the configuration values simply have to be set according to the requirements of the control application. Each filter process outputs the input value on whichever of its output channels is used. Figure 2 shows a 6-way filter constructed from one `BinaryFilter` and two `TernaryFilter`s. Each filter process has a configuration channel (c) which is used to initialise the filter with its boundary value(s). The output channels are labelled h – high, m – mid and l – low.

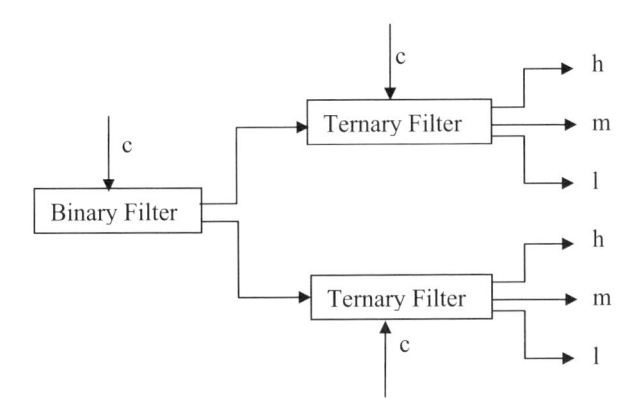

Figure 2: construction of a 6-way filter.

3.2 Input – Output Processes

This set of processes comprises the Active Sensor Processes and the Motor control processes. A variety of these motor control processes is provided depending on whether they are simply output processes or require some form of input capability because they are used as a sensor as well.

3.3 The lang and util Packages

These packages simply implement the JCSPre system and, as such, can be treated as black boxes.

3.4 The plugnplay Package

This package contains 'useful' processes that can be used to create common parallel network design idioms such as processes that copy an input from a single channel to a number of output channels as a sequence of channel outputs. In systems designed for large systems that use the full JCSP capability this is achieved using Delta and DynamicDelta processes. In the LEGO NXT environment this is not feasible because these processes use a large number of internal processes. This would tend to use scarce resources in an uneconomic manner.

3.5 Feasibility Testing

Initially, a program was created that used the underlying LeJOS thread model to discover how many child threads could be spawned. Each child process never terminates and it was discovered that 160 child processes could be spawned. A variant was created in which each child process held a simple piece of data and this resulted in the creation of 90 threads. This demonstrated that the number of threads was limited simply by the amount of memory used in each thread. A JCSPre program was created which comprised a pipeline of AddN processes, which add a constant value to an input value that is then output. An initial producer process sends an integer through the pipe of AddN processes to a consumer process that output a message on the robot LCD display. The system broke when the number of AddN processes reached 78. This was deemed sufficient to continue development because each AddN process has internal state and an input and an output channel. The point at which the system failed was indicated by an exception produced by the LeJOS virtual machine.

4. A Simple Line Following Robot

The photograph in Figure 3 shows a simple two motored robot with a rear trolley wheel (not visible) where each motor has an associated light sensor placed sufficiently far apart and to the front of the robot to be able to detect a line 15mm wide.

Figure 3: a simple two-motored robot with rear trolley wheel.

The control system is very simple in that if a sensor detects black, the colour of the line to be followed, the associated motor stops rotating, otherwise the motor rotates at a predetermined speed. Due to varying lighting conditions of the space in which the robot can operate it is necessary to configure the sensors so that it knows the sensor values for black and white. The speed at which the motors rotate can be set through a user interface, which also guides the user through the sensor configuration process.

4.1 The Control Process Network

Figure 4 shows the process network diagram for the line following robot. The left hand side shows the channels (solid lines) that are used during configuration of the system. The right hand side shows the channels (solid lines) used when the robot is running. In reality, all channels are required for the robot to function correctly.

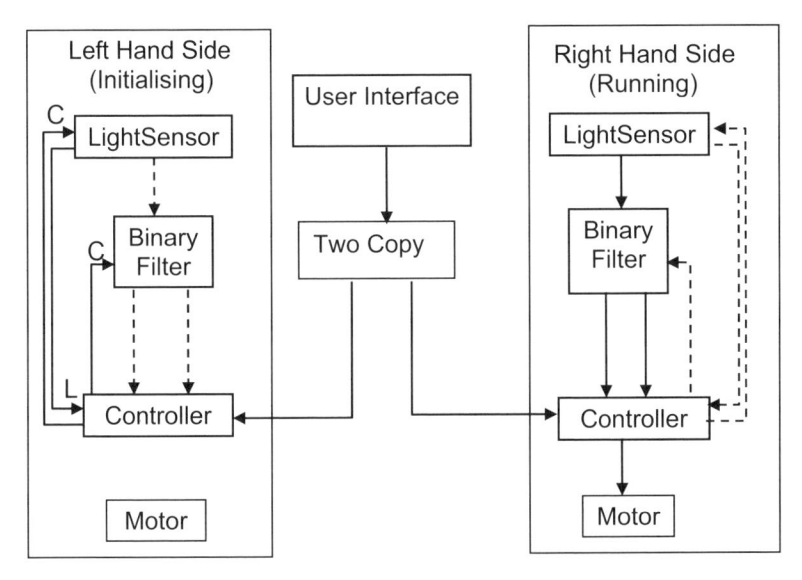

Figure 4: process network diagram for the simple line following robot.

The User Interface process guides the user through the configuration process. Messages are displayed on the LEGO NXT screen telling the user the stage reached. The user then responds by pressing buttons as required. Thus to configure the sensor black level the robot is placed over a black line such that both sensors are observing a black input. A button is then pressed, which is communicated to both sides of the control system to their Controller processes. This causes a message to be sent to the LightSensor process on its configuration channel (C) to determine the input level, which is returned to the Controller process using another channel, light level (L). The display then shows a message asking for the configuration of the white level, at the completion of which the Controller process has both the black and white levels. The point half way between these values is then sent to the BinaryFilter process on its configuration channel (C). The UserInterface process now requests the user to specify the speed at which the robot's wheels are to rotate. This is achieved using the left and right arrow buttons. The user then places the robot at the start of the line to be followed. Another button is then pressed to signify that the robot should start to follow the line which is sent to both Controller processes.

During movement the robot simply uses its LightSensor process (see the Right Hand Side Structure, solid lines) to detect a change in light value, which it outputs to the BinaryFilter process. Depending on the input value the BinaryFilter process outputs the input value on one of its output channels. The Controller process then waits for an input on either of its input channels and depending upon which channel the message is received it either outputs a stop or rotate message on the channel connecting it to the Motor process. This portion of the control loop is shown in the following code snippet.

An Alternative a is defined {55-56} which allows the Controller to choose amongst inputs from the Filter process on filterHigh and filterLow and the buttonPress channel from the user interface. The Controller comprises an unending loop {57-78}. The speed of wheel rotation is read from the buttonPress channel {58} followed by a simple button press that is interpreted as the go signal {59}. A further loop is now entered which terminates when the robot stops and a stop signal is received on the buttonPress channel {74}. A selection is made on the alternative a {62} which returns the index of the enabled input channel that has been selected. The following if statements {63-68, 69-72, 73-76} determine the action that should be undertaken depending upon the input that has been read. Normally, A switch .. case structure would be used but the LeJOS kernel does not contain such a capability as indicated earlier.

```
56    Alternative a = new Alternative(new Guard[]{ filterHigh, filterLow,
57                                                 buttonPress});
58    while (true) {
59      rotate = buttonPress.read(); // read speed of wheel rotation
60      buttonPress.read(); // the go signal
61      boolean going = true;
62      while (going) {
63            altIndex = a.select();
64            if (altIndex == 0) {
65                    filterHigh.read();
66                    motorSpeed.write(ActiveMotor.SPEED);
67                    motorSpeed.write(rotate);
68                    motorSpeed.write(ActiveMotor.FORWARD);
69            }
70            else if (altIndex == 1) {
71                    filterLow.read();
72                    motorSpeed.write(ActiveMotor.STOP);
73            }
74            else (altIndex == 2) {
75                    buttonPress.read(); // the stop signal
76                    going = false;
77            }
78     }
79  }
```

In the case of an input read from filterHigh {64} then a sequence of messages are sent to the motor process{65-67} telling it to move forward at the indicated speed of rotation, which can be changed for each iteration of the main loop. In the case of an input from filterLow {70}, which represents the sensing of the black line, the motor process is sent the stop signal {71}. Finally, a button press can be read {74} which has the effect of terminating the internal loop by setting going false {75}. It should be noted that the robot comes to a stop if both its sensors sense black as the same time, which then enables the pressing of a button. Recall that each side of the robot has its own Controller process and thus these are both running concurrently.

In total, the system comprises 12 processes: a process for each side of the robot plus the processes shown in Figure 2.

During a series of Christmas Lectures [21], aimed at 14-17 school students, participants were asked to control the same robot, with no change to its `Controller` process to guide it through a slalom course using a black lollipop. The black lollipop was placed under the required sensor, by the participant, in order to control the movement of the robot through the slalom course. It was then shown that by placing a black line down the middle of the course the same effect could be achieved.

5. Slave Robots using Bluetooth

To evaluate the Bluetooth capability of the LEGO NXT robot an experiment was proposed in which one, master, robot followed a line and a second, slave, robot, without sensors, was sent messages telling it how to move. The slave robot was furnished with a pen so that it could draw the path it had taken on a sheet of paper. The path the master robot took could then be compared with that followed by the slave robot. The nature of the messages sent from the master to the slave is very simple. They were a single integer that indicated whether a motor was moving or rotating. It was presumed that a motor remained in the same state until the controller changed it.

In order to set up a Bluetooth connection an interaction between the two communicating robots is required. Each robot has to be made aware of the possible other devices with which it can communicate. Once this is established then subsequent connection at run-time is relatively easy. A Java `DataInputSteam, dis` {79} and a `DataOutputStream, dos` {80} are defined. A LeJOS Bluetooth connection `btc` is then defined {81}. The connection is made {82} after which the data streams can be opened {83, 86} and an initial interaction undertaken {84, 85}. The code snippet ignores exception handling and shows the output side of the initial interaction. The receiving process simply opens an input data stream, reads from it and then opens a data output stream..

```
80    DataInputStream dis = null;
81    DataOutputStream dos = null;
82    BTConnection btc;
83    btc = Bluetooth.waitForConnection();
84    dos = btc.openDataOutputStream();
85    dos.writeInt(1);
86    dos.flush();
87    dis = btc.openDataInputStream();
```

Taking the architecture shown in Figure 4, a further process, Bluetooth Transmit, was added to the master robot network, which had two input channels, one from each Motor process. A corresponding output channel was added to each Motor process. Whenever the state of a motor changed, it output a corresponding value to the Bluetooth Transmit process, which then caused this value to be transmitted to the slave robot over the Bluetooth connection.

The slave robot had a much simpler process architecture comprising a Bluetooth Receive process together with two Motor processes for each wheel of the robot as shown in Figure 5. The BTReceive process inputs data values from the Bluetooth data input stream connection that has been created. Depending on the value a message is sent to one of the Motor processes. A Motor process is expecting to receive a zero (0), indicating it should stop or a one (1) indicating it should rotate. The motor stays in that state until otherwise changed.

The code snippet {88-94} captures this behaviour. The channels outChannelLeft and outChannelRight provide the connection between the BTReceive process and the Motor processes respectively.

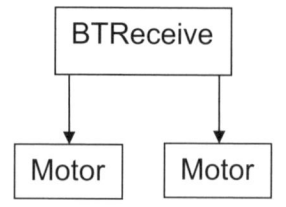

Figure 5: slave robot process network.

The Bluetooth Receive process receives an input {89}, which depending on the value {90, 92} causes a message to be sent to the appropriate Motor process {91, 93}, which then effects the required change in motor behaviour. The values cause a motor to switch on or switch off.

```
88    while(true){
89      value = dis.readInt();
90      if(value == 0 || value == 1){
91          outChannelLeft.write(value);
92      }else if(value == 2 || value == 3){
93          outChannelRight.write(value - 2);
94      }
```

A video showing this basic operation is available [22]. It demonstrates that the basic operation of the slave robot is a reasonable mimic of the master robot. However because the robot simply moves its motors for the same time as indicated by the master robot it is possible for the slave to wander from the desired path. In particular, the orientation of the slave robot and in particular its trolley wheel alignment is crucial if the robot is not to wander from the path from the outset. This behaviour can be observed in the video, which also shows that once the trolley wheel effect has been removed the remainder of the movement is in fact a reasonable copy of the path taken by the master robot.

6. Conclusions and Further Work

The achievement of the work reported in this paper was that it demonstrated the feasibility of building robot controllers for small factor robot systems such as the LEGO NXT robot using parallel programming techniques in a Java environment. This was achieved by building a highly reduced version of an extant parallel environment (JCSP) and implementing this on top of a publicly available small Java kernel and virtual machine (LeJOS). A process based abstraction of each of the components that are available in the LeJOS NXJ packages were developed so that a truly parallel control network could be implemented. The basic functionality was demonstrated by means of a simple line following robot. Finally, it was shown that a master line following robot could control the operation of a slave robot such that the slave robot could mimic the path followed by the master using a very simple communication message sequence that utilized the Bluetooth communication capability of the LEGO NXT robot.

The LeJOS system contains an implementation of a subsumption architecture. This work is currently ongoing in terms of a process based implementation of such an architecture. The latest release of LeJOS is referred to as a Beta release and overcomes many of the shortcomings that caused much work, for example, the lack of `switch` statements. The release also contains a means of sending messages from the robot to the PC used to download the code in the form of a remote console capability. This capability needs to be incorporated into the Eclipse environment to ensure that users, and in particular students can develop and test their designs in a single development environment.

The major challenge though is to develop the system to a point where sufficient basic processes have been created to enable the easier design of parallel robot control systems. It is anticipated that this will be achieved as part of the development of an undergraduate module that will teach parallel programming techniques through the use of robot controllers.

Acknowledgments

Both Alex Panayotopoulos and Patrick Lismore acknowledge the support they received from the Student Awards Agency Scotland for payment of their fees for the programmes of study they were undertaking while they developed parts of JCSPre and the demonstrations. The work was undertaken as part of the project element of a Masters and Bachelors qualification respectively.

References

[1] J. Simpson, C. L. Jacobsen, and M. C. Jadud, "A Native Transterpreter for the LEGO Mindstorms RCX," in A. McEwan, S. Schneider, W. Ifill, and P. H. Welch (Eds.), Communicating Process Architectures 2007, pp. 339-348, IOS Press, Amsterdam, 2007.
[2] Transterpreter.org, "LEGO Mindstorms RCX and NXT," 2008. Retrieved 25 March 2008 from http://www.transterpreter.org/docs/LEGO/index.html.
[3] LEGO.com, "Mindstorms NXT Software," 2008. Retrieved 7 March 2008 from http://mindstorms.LEGO.com/Overview/NXT_Software.aspx.
[4] J. Kodosky, "Is LabVIEW a General-purpose Programming Language?," 2008. Retrieved 7 March 2008 from http://zone.ni.com/devzone/cda/tut/p/id/5313.
[5] A. Tan, "Call for Asia to Adopt ODF," 2006. Retrieved 12 March 2008 from http://www.zdnetasia.com/news/software/0,39044164,39380446,00.htm.
[6] G. Theodoropoulos, "Modelling and Distributed Simulation of Asynchronous Hardware," Simulation Practice and Theory, 7(6), pp. 741-767, 2000.
[7] C. G. Ritson and P. H. Welch, "A Process-Oriented Architecture for Complex System Modelling," in A. McEwan, S. Schneider, W. Ifill, and P. H. Welch (Eds.), Communicating Process Architectures 2007, pp. 249-266, IOS Press, Amsterdam, 2007.
[8] C. L. Jacobsen, D. J. Dimmich, and M. C. Jadud, "Native Code Generation Using the Transterpreter," in P. H. Welch, J. Kerridge, and F. R. M. Barnes (Eds.), Communicating Process Architectures 2006, pp. 269-280, IOS Press, Amsterdam, 2006.
[9] S. Jørgensen and E. Suenson, "trancell - an Experimental ETC to Cell CE Translator," in A. McEwan, S. Schneider, W. Ifill, and P. H. Welch (Eds.), Communicating Process Architectures 2007, pp. 287-297, IOS Press, Amsterdam, 2007.
[10] N. Brown, "C++CSP2: A Many-to-Many Threading Model for Multicore Architectures," in A. McEwan, S. Schneider, W. Ifill, and P. H. Welch (Eds.), Communicating Process Architectures 2007, pp. 183-205, IOS Press, Amsterdam, 2007.
[11] J. M. Bjørndalen, B. Vinter, and O. Anshus, "PyCSP - Communicating Sequential Processes for Python," in A. McEwan, S. Schneider, W. Ifill, and P. H. Welch (Eds.), Communicating Process Architectures 2007, pp. 229-248, IOS Press, Amsterdam, 2007.
[12] L. Yang and M. R. Poppleton, "JCSProB: Implementing Integrated Formal Specifications in Concurrent Java," in A. McEwan, S. Schneider, W. Ifill, and P. H. Welch (Eds.), Communicating Process Architectures 2007, pp. 67-88, IOS Press, Amsterdam, 2007.
[13] P. H. Welch, N. Brown, J. Moores, K. Chalmers, and B. H. C. Sputh, "Integrating and Extending JCSP," in A. McEwan, S. Schneider, W. Ifill, and P. H. Welch (Eds.), Communicating Process Architectures 2007, pp. 349-370, IOS Press, Amsterdam, 2007.
[14] P. H. Welch and J. Martin, "Formal Analysis of Concurrent Java Systems," in P. H. Welch and A. W. P. Bakkers (Eds.), Communicating Process Architectures 2000, pp. 275-301, IOS Press, Amsterdam, 2000.

[15] JCSP Project, "CSP for Java (JCSP)," 2008. Retrieved 3 April 2008 from http://www.cs.kent.ac.uk/projects/ofa/jcsp/explain.html.
[16] JCSP Project, "CSP for Java (JCSP) 1.1-rc3 API Specification," 2008. Retrieved 3 April 2008 from http://www.cs.kent.ac.uk/projects/ofa/jcsp/jcsp-1.1-rc3-doc/.
[17] LeJOS Project, "NXJ Technology," 2008. Retrieved 16 April 2008 from http://LeJOS.sourceforge.net/p_technologies/nxt/nxj/nxj.php.
[18] LeJOS Project, "README," 2008. Retrieved 16 April 2008 from software download of LeJOS-NXJ-win32, version 0.3.0alpha, available at http://sourceforge.net/project/showfiles.php?group_id=9339.
[19] LeJOS Project, "LeJOS NXT API Documentation," 2008. Retrieved 16 April 2008 from http://LeJOS.sourceforge.net/p_technologies/nxt/nxj/api/index.html.
[20] Sun Microsystems, "CLDC Library API Specification 1.0," 2008. Retrieved 16 April 2008 from http://java.sun.com/javame/reference/apis/jsr030/.
[21] Institution of Engineering and Technology, "Christmas Lecture 2007," 2007. Retrieved 18 June 2008 from http://tv.theiet.org/technology/infopro/898.cfm.
[22] P. Lismore, "2 NXT Robots With Parallel Java Brains Test 2," 2008. Retrieved 22 April 2008 from http://www.youtube.com/user/plismore.

A Critique of JCSP Networking

Kevin CHALMERS, Jon KERRIDGE and Imed ROMDHANI

School of Computing, Napier University, Edinburgh, EH10 5DT
{k.chalmers, j.kerridge, i.romdhani}@napier.ac.uk

Abstract. We present a critical investigation of the current implementation of JCSP Networking, examining in detail the structure and behavior of the current architecture. Information is presented detailing the current architecture and how it operates, and weaknesses in the implementation are raised, particularly when considering resource constrained devices. Experimental work is presented that illustrate memory and computational demand problems and an outline on how to overcome these weaknesses in a new implementation is described. The new implementation is designed to be lighter weight and thus provide a framework more suited for resource constrained devices which are a necessity in the field of ubiquitous computing.

Keywords. JCSP, JCSP.net, distributed systems, ubiquitous computing.

Introduction

JCSP (Communicating Sequential Processes for Java) [1, 2] is a Java implementation of Hoare's CSP [3] model of concurrency. Previous work was presented [4] that brought about the integration of the differing versions of JCSP, as well as augmenting and extending some of the underlying mechanisms inside the core package. In this paper, we present information on the current implementation of the network package and raise some issues with the current approach. This allows us to address these limitations in a new implementation of JCSP networking, while also attempting to bring the networking package closer to the same level of functionality as the core package.

The rest of this paper is broken up as follows. In Section 1, we present some background information on JCSP and network based approaches for CSP. In Section 2, motivation for this work is given. Section 3 provides a description of the current architecture of JCSP Networking and in Section 4 an analysis and criticism of the current implementation is given. Finally, in Section 5, conclusions are drawn and future work considered.

1. Background

In this article we are concerned with the networking capability of JCSP [5] as opposed to the core functionality. The main library for JCSP is aimed at concurrency internal to a single JVM, whereas the network library was designed to permit transparent distributed parallelism using the same basic interfaces as present in core JCSP. Unlike core JCSP, there is no channel object that spans the network. Instead, JCSP networking operates using a channel end concept, with a node (a single JVM within the network) declaring an input end, and other nodes connecting to this input end via an output end. The input end and output end make up a virtual networked channel between two nodes. The input end and output end have the same interfaces as `ChannelInput` and `ChannelOutput` in the core JCSP package. Beyond this, JCSP networking adds little to the functionality of the main packages, and due to the recent improvements in JCSP [4] can now be considered lacking certain constructs that would make it on par with core JCSP.

Of particular importance in this case are the lack of a basic network synchronisation primitive (the JCSP `Barrier`) and the multi-way synchronisation primitive (JCSP `AltingBarrier`) which leads to a lack of transparency between locally synchronizing processes and distributed synchronizing processes. These constructs and other, Java 1.5 specific, considerations were given as future work in [4]. This paper brings these constructs closer to implementation by illustrating the need to address some of the underlying architectural decisions made for JCSP networking. First we shall look at other implementations of CSP inspired networking architectures.

1.1 Network Based CSP Implementations

There is now a wealth of CSP inspired distributed concurrency libraries available, ranging from **occam-π** and the pony architecture [6] through C++CSP [7] and Python [8]. Most base their architecture on the T9000 model [9] for creating virtual channels across a communication mechanism.

JCSP Networking [5] enables the virtual connections to be created via `NetChannelLocation` structures sent between nodes to allow virtual connections to be created. A `NetChannelLocation` signifies the location of a `NetChannelInput` end which a `NetChannelOutput` end can connect to; the input end acts as a server connection to an output end. The location structures are fairly trivial, being formed by the address of the node on which the channel is declared, and a Virtual Channel Number (VCN) uniquely identifying the channel on said node, although other methods involving channel names may be used. Initial interaction between nodes is usually managed by a Channel Name Server (CNS) which allows registration of channel names by the server end of a connection, and the resolution of these names – thus providing the location – by a client end of a connection. After initial interaction, locations can be exchanged directly using networked channels or all channels may be declared with the CNS. `NetConnections` are also available, and methods for permitting the mobility of channels and processes (via code mobility [10]) are also available [11].

C++CSP networking [7] enhances the original C++CSP library [12] by adding the capability for two C++CSP programs to communicate via TCP/IP based sockets. Unlike JCSP, there is no CNS – channels connect using unique names on the node. VCNs exist in the underlying exchanges between nodes. C++CSP networking is limited by how it sends data, due to differing machine architectures that may be in operation, and a lack of an object serialisation approach in C++ similar to that found in Java. Recently C++CSP was updated to version 2 [13], a move that concentrated more on utilizing multi-core systems than implementation of a networking architecture.

pony [6] is the **occam-π** approach to networking, and shares a number of similarities with JCSP. Instead of a CNS, pony uses an Application Name Server (ANS) and controls the system of nodes via a main node. Channel mobility is also controlled and there is no current implementation of process mobility. The architecture of pony has been inspired by the need to implement transparent concurrency and parallelism in a cluster computing environment, which are more controlled than standard distributed systems architectures, towards which JCSP is more aimed.

CSP.NET [14] is an implementation of CSP for Microsoft's .NET Framework, inspired a great deal by JCSP. Developed in C#, the main advantage of this library over

JCSP is the number of languages in .NET that can now utilize the library. CSP.NET does rely on the remoting capabilities of .NET, and is therefore technology restricted – remoting being the RPC system built explicitly into .NET, requiring .NET to operate. JCSP operates in a manner that is decoupled from the communication mechanism, and is thus independent of it. Initial performance analysis of CSP.NET has shown little difference in performance in comparison to JCSP. The most recent version of CSP.NET is now available commercially (www.axon7.com) and no longer relies on specific features of .NET, but this has yet to be formally reported.

Finally, PyCSP [8] is implemented in the Python programming language. Although at last reporting only having basic networking capabilities, the current approach uses remote objects as opposed to an underlying communication mechanism. The aim of PyCSP appears to be geared towards cluster computing, making a solid networking infrastructure essential in the future.

1.2 Performance Evaluations of Network Based CSP Implementations

Some work towards measuring performance of network enabled CSP implementations has been conducted in previous research. Many of these approaches have focused only briefly on the performance of the communication mechanism, and instead examine the performance of parallelized tasks within the architecture. Brown [7] has examined latency and performance overheads in C++CSP, but no extensive testing of performance of the communication mechanism using different data packet sizes was made. Instead, work was allocated and different packet sizes used to measure performance. Although this can lead to some information about the communication performance, it does not analyze it in great enough detail to find the difference in performance between C++CSP networking and standard communication mechanisms.

Schweigler [6] has done extensive tests examining the CPU overhead and throughput of pony, as well as some comparisons with JCSP and work allocation. Again, little analysis as to the actual costs of sending messages between nodes is given, and most of the conclusions on such overheads are interpreted from the throughput and comparison when parallelizing a task.

Lehmberg's analysis of CSP.NET [14] provides only simple analysis of performance without comparison to other approaches, although a brief comparison to JCSP has been made. The authors note that the tests performed are by no means thorough enough to constitute a benchmark.

For JCSP, little real analysis of the performance of the communication mechanisms has been made. Schaller [15] assessed the different speed ups of Java parallel computing libraries when performing certain tasks across multiple nodes. Vinter [16] has examined similar properties with other packages in Java, performing different tasks but forgoing any analysis of the communication mechanism. Kumar [17] has examined JCSP performance in the context of multiplayer computer games, and although providing some interesting results on the scalability of JCSP, little analysis of the communication was made.

To fully understand how suitable JCSP is in comparison with other approaches to communication, analysis of the current mechanisms has to be made. In Section 4, we provide some of the required parameters. First, a brief description of Java serialisation and object migration is presented to help understand these parameters more fully.

1.3 Serialisation and Object Migration in Java

It is important to understand serialisation in Java as it puts into context some of the performance values we shall be discussing in Section 4. Java serialisation is the process of converting a Java object into a sequence of bytes for storage or transfer via a communication mechanism. This is usually performed by an `ObjectOutputStream` which acts as a filter on another output stream type to serialize objects. Reflection (the ability to interrogate an object to discover its properties and methods) is used to gather the values within the object and the recreation of the object at the destination from its name (i.e. `java.lang.Integer`).

Control signals are sent with serialized object data to allow recreation of the object at the destination. As a case study, we shall examine the bytes representing an `Integer` object, highlighting some of these control signals as necessary. Further information on serialisation is found in the Java 2 SDK documentation (http://java.sun.com/j2se/1.3/docs/guide/). All values are single bytes unless stated otherwise.

When a new `ObjectOutputStream` is instantiated, a handshake message is sent to allow correct behavior at the destination. Normally four bytes are sent which represent two 16-bit integers: `STREAM_MAGIC` (-21267) and `PROTOCOL_VERSION` (5). These are sent once between an output stream and an input stream. We will not consider the handshake as normal data for this reason.

The next value represents the type of message being sent. For `Integer` this is `TC_OBJECT` (115), which signifies a standard object. Other control signals for base data and arrays also exist. Next is a control signal for the class description, `TC_CLASSDESC` (114), followed by the name of the class as a string (with a 16-bit length header). For `Integer` this is `java.lang.Integer`. A 64-bit integer representing the unique serialisation identifier follows and is used to ensure that the class of the given name is the same at the destination.

A single byte representing control flags to determine how the object was serialized (for example, custom methods can be used within an object) is next and then a 16-bit value representing the number of object attributes. With `Integer` there is only the wrapped primitive integer. The attribute types are given which may be other objects, thus invoking the previous steps for describing the object. For `Integer` the attribute is a primitive integer represented by 'I' (74). The attribute names are given as strings with length headers – `Integer`'s attribute being `value`. A final control signal for the end of the class description is then written – `TC_ENDBLOCKDATA` (120).

If the class of the sent object is a subclass the description of the parent class must also be sent. The parent class may declare attributes not visible to the subclass, but are used by inherited methods. `Integer` extends `Number` (`java.lang.Number`), which has no internal attributes.

When all classes have been described, a byte to signal the attribute values of the object, `TC_BASE` (112) is written. The values of the attributes are written to the stream in the order they were declared. For `Integer`, the 4 bytes representing the 32-bit integer `value` are written.

Taking into account the control signals and descriptions sent for an `Integer` object, we can calculate a total of 77 bytes sent to represent a 4 byte value. This is a significant overhead although `Integer` is a special case with a direct primitive representation of the sent object. However, it does point to the need to avoid serialisation if possible. We have not discussed the recreation of the object at the destination which involves using reflection on the sent name to get the specific class, creating a new instance of the class, and assigning the values of the properties using reflection. This is a costly process in comparison to sending primitives.

Java serialisation tries to overcome overhead problems by using references to previously sent objects and classes within the object stream. For example, if an object is sent twice over a stream, instead of a new complete message, a `TC_REFERENCE` (113) byte is sent, followed by a 32-bit value representing the place in the stream of the object. Class descriptions may also be referenced if an object of the same type is sent more than once. The former cuts the 77 bytes of `Integer` to 5 bytes and the latter cuts the message to 10 bytes. This requires lookup tables within the stream object to accomplish. Over time, these lookup tables can become large, and may cause a memory exception. To combat this, the object streams can be explicitly reset, which causes the output stream to clear its lookup table and send a signal to the input stream to do likewise. This does mean that complete descriptions of classes need to be sent again. Also, the serialiser has no method to determine if an object has been modified between sends, therefore any modification will not be seen at the destination if the object stream is not reset. The serialiser has no method to distinguish between mutable and immutable objects.

The reason to examine serialisation is that JCSP networking relies heavily upon it. To avoid referencing problems, the underlying streams are reset after each message, thus every object is sent as a new object. The first instinct for doing this is the possibility that a user may be optimizing their own application to avoid garbage collection, thus using a pool of messages. JCSP uses this mechanism internally within the networked channels. Each output channel is allocated two message packets that are swapped between sending. If the underlying object streams were not reset after every communication the channel would appear to only send two objects, and then only ever send those two objects. Unfortunately, resetting the streams in this manner leads to other problems, which we shall discuss in Section 4. It should be possible to replace the serialiser with a more efficient implementation, but this is currently left for future work.

2. Motivation

We are examining JCSP in the context of ubiquitous computing (UC) [18], which is the idea of having an environment populated by numerous computationally able elements, that interact together to provide new and unique services. To accomplish UC, dynamic architectures and movement is envisioned; for example to enable agents to move between elements to accomplish goals. An architecture the size and complexity envisioned by UC requires abstractions that enable simpler design and reasoning. The π-Calculus [19] and similar mobile abstractions have been put forward as a possible model for this understanding [20], and we are interested in the practical examination of such abstractions, using JCSP as a case study. The scalability of mobile channels and processes has been shown by the work of Ritson [21], with an architecture involving millions of interacting mobile processes operating simply with no great problems for design or analysis.

The reason for examining JCSP is that it is more mature than similar frameworks when considering distributed mobility [11]. Java is also a more commonly available framework, particularly for mobile devices which enable a close approximation of the capabilities of the elements available in a UC environment. Work on the Transterpreter [22] may lead to **occam-π** being available on more devices, but there is currently no network capability.

We are also hoping to develop a universal mechanism to allow the abstractions that we require within multiple frameworks, a discussion of which is given in Section 5. Although we have taken JCSP as a case study, it cannot be considered that Java will be available on all the elements in a UC environment. As an outcome of our work we also address some of the future work given in [4].

3. Current JCSP Networking

Figure 1 illustrates the current architecture of JCSP networking. Within the diagram, ovals represent processes and rectangles objects. Channels are represented by arrows, and dashed lines between components represent object based connections (uses, contains). Channels that have potential infinite buffering are indicated with an infinity symbol. This diagram appears different from those previously presented for JCSP networking as there is no `NetChannelOutputProcess`. To reduce the number of Java threads supporting the JCSP infrastructure, this has been integrated into `NetChannelOutput` object.

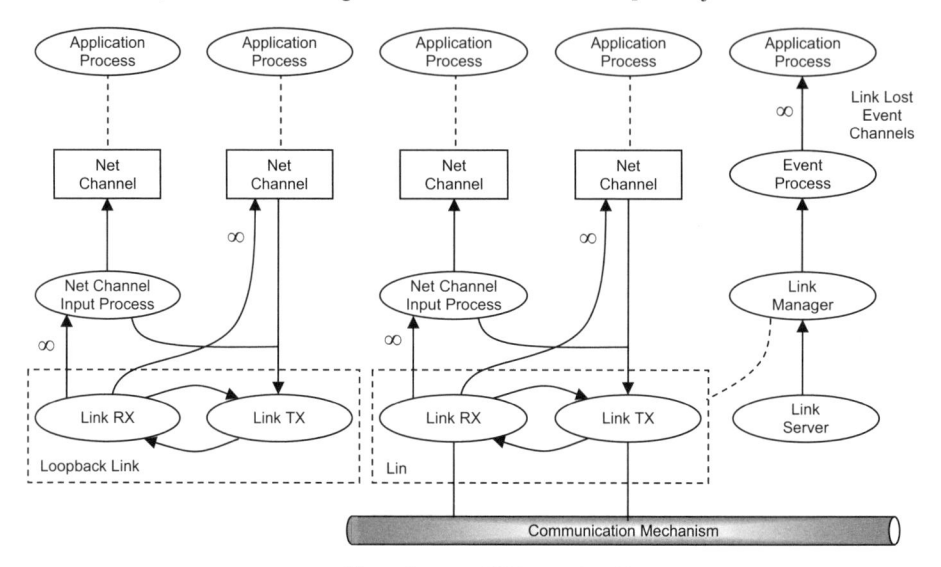

Figure 1: current JCSP network architecture.

The `Link` encapsulates the connection to another node within the system by sending and receiving messages via the underlying stream mechanism, which is dependent on the inter-node communication fabric. The `Link` process has two sub-processes: `LinkTX` which is responsible for sending messages, and `LinkRX` which is responsible for receiving messages. As a node may be connected to multiple remote nodes, there may be multiple pairs of these processes.

The other form of connection is the `LoopbackLink` which simulates a local connection. The `LoopbackLink` allows a channel output to connect to a channel input on the same node. Because JCSP allows for this eventuality, the `LoopbackLink` must always be in operation. There is only ever one `LoopbackLink` and corresponding TX and RX processes active on a node.

The `LinkServer` processes listen for incoming connections from other nodes, and start up a new `Link` process accordingly. With TCP/IP the `LinkServer` process encapsulates a normal server socket. In most cases, only one `LinkServer` is created; but if different communication mechanisms are in use or the node must listen on multiple interfaces, then multiple `LinkServer` processes may be active.

The `Link` processes are managed by a `LinkManager` process. `LinkServer` processes connect to the `LinkManager` process to communicate new connections. `Link` processes are also connected to the `LinkManager` to allow notification of a connection failure. An `EventProcess` is spawned by the `LinkManager`, which is used to communicate link failures to the application level. An application process must create a new `LinkLostEventChannel` to allow this message to be received. The `EventProcess` is a sequential delta outputting upon the `LinkLostEventChannels` any `LinkLost` message it receives.

Channels similarly have a manger called `IndexManager`. This is not shown in Figure 1, but it contains connections to all networked channel constructs. Unlike `LinkManager`, this is a shared data object controlled via Java synchronized methods. Whenever a new channel is created, it is registered with the `IndexManager`. The `Link` processes use the `IndexManager` to access the channel ends during operation.

As mentioned, networked channels come in two forms: `NetChannelInput` and `NetChannelOutput`. An output end is connected directly to its corresponding `LinkTX` process. As network channels are Any2One, an input end may receive messages from any `LinkRX` process. The messages are not sent directly to the channel end, but are sent to a `NetChannelInputProcess` which then forwards the message onto the channel end. The channels from the `LinkRX` to the channel end / process are buffered with an infinite buffer, meaning that there is no risk of deadlock on the `LinkRX` process when it sends the message to a channel. The amount of buffering needed is bounded by the number of external processes trying to communicate to that particular network channel. It holds received, but not yet accepted, messages. That number cannot be pre-calculated, since it may, of course, change during run-time. However, it will always be finite!

3.1 Current Functionality

The basic operation during a send / receive operation occurs as follows:

1. An application calls `write` on the `NetChannelOutput`.
2. The `NetChannelOutput` wraps the sent object inside a message object, and writes this to the `LinkTX` process. The message object contains details on destination and source to allow delivery and subsequent acknowledgement. The `NetChannelOutput` then waits for an acknowledgement message from the `LinkRX`, blocking the writer (so as to maintain synchronisation semantics of CSP channels).
3. The `LinkTX` serializes the message onto the connection stream to the remote node.
4. The `LinkRX` on the remote node deserializes the message from the stream, retrieves the destination index, and requests the channel from the `IndexManager`.
5. To allow quick acknowledgement, the `LinkRX` attaches the channel to its partner `LinkTX` to the message. The message is then written to the `NetChannelInputProcess`.
6. The `NetChannelInputProcess` reads the message and writes the data part to the channel end. This is a blocking write, so until the receiving process is ready to read the message the `NetChannelInputProcess` waits. Once the write has completed, an acknowledgement message is written to the channel attached to the message during the previous step.
7. The `LinkTX` serializes the message onto the connection stream to the remote node.
8. The `LinkRX` on the remote node deserializes the message from the stream, retrieves the destination index, and requests the corresponding channel from the `IndexManager`.

9. As the message is an acknowledgement, the message is written directly to the channel end.
10. On receiving the acknowledgment message, the `NetChannelOutput` message can complete the write operation and release the writer.

These ten steps capture most of the functionality underlying the network architecture. Other conditions, such as link failure, message rejection, etc, are not covered here. With these steps in mind, we can move forward and critique the current implementation.

4. Critique of JCSP Networking

Our test bed consists of a PC communicating to a PDA device via a wireless network. The PC is a Pentium IV 3 GHz machine with 512 MB of RAM running Ubuntu Linux. The PDA is an HP iPaq 2210 with a 400 MHz processor and 64 MB of memory, shared between storage and applications. The operating system on the device is Windows Mobile 4.2, and thus provides a similar API to a standard Windows based desktop. The wireless network is an 802.11b network running at 11 Mbps. The PC is connected via a standard Ethernet interface to the router, and the PDA is connected via a wireless interface. Considering how small and resource restrictive components in ubiquitous computing may be, this test bed is fairly powerful. However, this setup allows us to discover limitations. Of particular note is the JVM running on the PDA. This is an IBM J9 JVM, and due to resource limitations can only create just under 400 simple threads with little internal stack. As every process in JCSP is handled by a thread, this allows us to examine JCSP networking in a very limited environment, not envisioned during original development.

We are interested in analyzing the resource usage and general performance of JCSP, and have therefore sent objects of various sizes and complexities via normal networked streams, buffered network streams, normal JCSP network channels and unacknowledged JCSP network channels. The buffered streams are required as JCSP buffers its own streams when used within a TCP/IP environment. The unacknowledged channels are a feature of JCSP networking and it was hoped that examination of these would permit understanding of the overheads of message sending. As we shall see, it has helped us discover another problem instead. As JCSP sets Nagle[1] 'off' for its TCP/IP connections, all the results presented also have Nagle deactivated.

As mentioned, different complexities and sizes of objects have been examined. By complexity, we refer to the number of aliased objects that exist within the sent object itself. Here we will be presenting `TestObject4` to demonstrate properties. `TestObject4` is the largest object we have used, in byte size, and is complex. It inherits from `TestObject`. The class definitions are:

```
public class TestObject implements Serializable {
  protected Integer[] ints;
  protected Double[] dbls;
  ...}

public class TestObject4 extends TestObject {
  private TestObject testObject;
  private Integer[] localInts;
  private Double[] localDbls;
```

---

[1] Nagle increases performance for small packet sizes by condensing numerous small messages into single packets.

```
public TestObject4(int size) {
      ints = new Integer[size];
      dbls = new Double[size];
      localInts = new Integer[size];
      localDbls = new Double[size];
      for (int i = 0; i < size; i++) {
            ints[i] = localInts[i] = new Integer(i);
            dbls[i] = localDbls[i] = new Double(i);
      }
}

public static TestObject create(int size) {
      TestObject4 tObj1 = new TestObject4(size);
      TestObject4 tObj2 = new TestObject4(size);
      tObj1.setTest(tObj2);
      tObj2.setTest(tObj1);
      return tObj1;
}
}
```

To create an instance of the object, `create (int size)` is used. A single `TestObject4` has four internal arrays (two for `Integer` and two for `Double`), with the internal objects within these arrays being aliased. `TestObject4` has a reference to another `TestObject4`, which in turn references the original object. Therefore, there are numerous aliases within the objects being sent. The tests use internal array sizes from 0 to 100.

To understand the complexity and size of `TestObject4`, we can use the following formulae. For the number of unique objects sent, relative to n (the size of the internal arrays) we have:

$$2 \cdot (TestObject4\ (1)) + Inherited\ Array\ Objects\ (2) + Own\ Array\ Objects\ (2) + 2 \cdot n$$

The number of object references sent is greater than this value as the objects in the arrays declared in `TestObject` are sent as reference data. The total number of object references can be gained by multiplying n by 2 again:

$$2 \cdot (TestObject4\ (1)) + Inherited\ Array\ Objects\ (2) + Own\ Array\ Objects\ (2) + 2 \cdot 2 \cdot n$$

Calculating the amount of data in bytes is more difficult, due to the message headers as described in Section 1.3. The simplest method is:

$(n = 0) \rightarrow 326\ bytes$
$(n = 1) \rightarrow 500\ bytes$
$(n > 1) \rightarrow 500 + ((n - 1) * 68)\ bytes$

The increase of 68 bytes per increment in size of the message is due to the size of the object being sent. An `Integer` wraps 4 bytes, and `Double` 8 bytes. Two of each object type is created in total – one for each `TestObject4` – for a total of 32 bytes. Each object also takes up 4 bytes of reference information, and eight object references are created in total – four for the new objects and each new object is aliased once. This requires 36 bytes, which gives us 68 bytes in total.

The reason to use complex objects that are however small in size is threefold. Firstly, the platform used is restricted in performance, and thus small message sizes are the most likely to be sent. As we are concerned with UC, it is the abstraction we are more concerned with, and it may be that numerous communications are occurring between these devices. This is unlike parallel computing approaches, where large blocks of data are processed to try and increase performance by having a processor spend most of its time processing as opposed to communicating. Tests have also been conducted using large byte arrays of data, but this does not allow capturing of the serialisation and message overhead we present here.

Secondly, we are trying to discover the cost of sending messages via JCSP, taking into account serialisation. Sending large objects is not the norm within Java if we consider other remote communication mechanisms such as Java Remote Method Invocation (RMI). Therefore, we hope to analyze the overheads of sending messages via JCSP in comparison to the underlying stream mechanism. A comparison with RMI is left for future work as RMI is not a standard feature on mobile devices.

Thirdly, the maximum size of the object we present is smaller than the buffer underlying the network streams within JCSP. With these experiments, we are hoping to avoid the operation of the buffer being automatically flushed due to filling. Any overhead associated with this operation can be captured during large block sending.

4.1 Resource Usage

Our first criticism of JCSP networking is the required resources to start a networked node. If we examine Figure 1, we can see that each connection to another node requires two processes; each input channel requires a process; and a `LinkServer`, a `LinkManager`, an `EventProcess`, and two processes for loopback are created at startup. For an initial unconnected node, with no input channels declared we have 6 threads created (including the main thread). Of these processes, the `LinkServer` is required to accept incoming connections.

When the node connects to a remote Channel Name Server (CNS) during initialisation, the number of threads required is 11 (including main). The connection to the CNS involves the two `Link` processes, a service process for connection to the CNS, a `NetChannelInputProcess` for the connection from the CNS to the service process, and when the first `NetChannelOutput` is created, a small process to handle link failures is spawned also. Only the two `Link` processes are required.

During a connection operation, five processes are created and subsequently destroyed to handle handshaking between the two nodes. The `Parallel` in standard JCSP is robust, and will try its best to manage all the used process threads in the system. However, many of these processes are spawned using the `ProcessManager` object; therefore the threads are not taken from the standard pool but are recreated every time, although it would be possible to modify `ProcessManager` to utilize the `Parallel` pool. The starting and subsequent destruction of threads may not be considered a serious strain on the system per se, but it may increase the active thread count beyond the system limit. `Links` may also be created to connect to a remote node already connected to, and the handshake process takes place to determine if the new `Link` should be kept. This requires the creation and subsequent destruction of temporary threads, which again may cause the thread count to increase beyond system capabilities.

The number of processes created as operations continue is also substantial. Each connection to a new node requires two further processes, and each new `NetChannelInput` requires a further process. It can be seen that it is not hard to reach the 400 thread limit within the PDA without inclusion of the application processes. As an example, being connected to five nodes, with ten input channels (two for each node) and the initial CNS connection will require a total of 31 processes.

The main reason for the heavy resource usage would appear to stem from the common CSP / **occam** philosophy of when in doubt, use a process. In Java this is sometimes an expensive approach to take, particularly when considering resource constrained devices. One of the main problems is the use of extra processes as managers, and processes for controlling the `NetChannelInput`. Channels should be lightweight, but a thread in Java is not lightweight so using a thread for a channel is wasteful. This approach is also dangerous in a dynamic architecture when, for example, the application process forgets to destroy the channel when it is finished with it, resulting in the process being lost and the resources not reclaimed. The garbage collector will not recover these as there is still an active thread holding the resources. As one of the goals of JCSP is to transparently allow channels to be either local or distributed, we cannot rely on a process actively destroying an unused channel if it has no knowledge of whether it is networked or not. Modifying the existing input channel so that it uses fewer resources is therefore a necessary goal.

4.2 Complexity

The next criticism we level at JCSP networking may be considered subjective. It concerns the internal complexity of the implementation. The basic premise of JCSP networking is trivial; there are two arrays of Any2One channels creating a crossbar between the channel ends and `Link` processes. One of the supposed properties of JCSP networking is the fact that the architecture is removed from the underlying communication mechanism, meaning that JCSP can be implemented over any guaranteed delivery mechanism. The argument is that if the correct addressing mechanism is used, then JCSP can operate around it. Although this statement is true, it is difficult to achieve, requiring understanding JCSP networking internals. Without the source code, it would be incredibly difficult for a custom communication mechanism to be used. If JCSP truly sat above the communication medium, then all that should be required are the necessary input and output streams, and an addressing mechanism.

As an example, the TCP/IP version of the `Link` process must implement numerous abstract methods from the `Link` class, including writing and reading of test objects, handshaking, waiting for `LinkLost` replies, and reading and writing of `Link` decisions (whether this `Link` should be used). There are many methods that need implementation for addressing, protocols, and a `NodeFactory` (used to initialize a JCSP `Node` during startup). Some of these require knowledge of objects such as `Profile` and `Specification` which are undocumented. There is also a reliance on the `org.jcsp.net.security` and `org.jcsp.net.settings` sub-packages. This is to name a few of the hurdles that must be overcome to allow JCSP to operate upon a new communication mechanism.

4.3 Message Cost

Object messages are an expensive method of transferring data. Any sent object must be wrapped within the object message before being sent to the other node, and acknowledgement messages are themselves objects. If we consider the amount of extra data sent using serialisation and as reflection is used during recreation, this leads to an overhead. The message types are defined within an object hierarchy, with specialized messages extending simpler ones. As inheritance information is also sent within a serialized object, this adds a further overhead.

For instance, a send message to another node requires a source and a destination value, which are two 64-bit values, but the size of the message object is 249 bytes without any data inside it. An acknowledgment message is 205 bytes. This is a significant overhead considering that the information required is only 16 bytes (the source and destination).

There is also extra information sent within the message, such as a channel name for resolution by the receiving node if the destination index is not known, and a flag indicating if the message should be acknowledged. Name lookup puts an extra strain on the node as it must find the name in a table prior to message delivery. For named channels the CNS should really be used.

Having the messages wrapped up in objects also restricts JCSP interacting with other process based frameworks. Under the current implementation it would be impossible for JCSP to send a message to pony for example. This reflects badly on the use of JCSP in a ubiquitous computing environment, as we cannot expect all devices to be able to use a JVM. The nature of distributed systems also requires a great deal more platform heterogeneity, and currently JCSP does not offer this.

4.4 Objects Only

Following from the previous point is the inability of JCSP to send anything but objects between nodes. In principle this is not a problem if we consider JCSP in only the Java context, but it does again make it difficult to communicate with other platforms. It would also be useful to send raw data between nodes as required, which can be done in principle as a byte array in Java is considered an object, but there is again an overhead involved due to serialisation.

This limitation also means that primitive data must be wrapped in an object prior to sending. This brings a further overhead. The core JCSP packages implement a primitive integer channel to allow a slight increase in performance, and it should be reasonable that JCSP networking do this as well. To achieve this in the current implementation would require the channel to wrap the primitive in an object for sending, as the message object only uses an object as data to transfer.

4.5 Performance

We now present experimental data regarding the general performance of JCSP networking when sending messages. The results presented are the mean of a roundtrip of 60 objects of a given size, taken from the PDA. The results from the point of view of the PC are the same. These values are gathered by sending and receiving back an object of the given size, acquiring the time for performing this action 10 times, and this in turn is performed 10 times. Thus in total 100 objects are sent and received, but this is split into batches of 10 to allow a finer grained analysis. Finally, of these 10 batches, the largest and smallest two results are removed to smooth the data. Initially, the average of 100 send-receive operations was taken, but due to unexpected peaks and valleys in the results, a closer examination was taken. Although it has shown the peaks and valleys are not the result of individual runs, the real reason has yet to be determined.

Our first set of data highlights the efficiency gained by placing an 8 KB buffer on the stream when compared to an unbuffered stream. These are Java `BufferedInputStream` and `BufferedOutputStream` objects surrounding the streams provided by Java `Socket` objects. Java serialisation causes individual bytes to be written directly to the stream, which results in numerous small packets being sent in a TCP/IP environment. As JCSP networking places these 8 KB buffers on its streams, this allows us to see why initially JCSP performs better than standard networking on the mobile devices we are investigating. Some preliminary experiments using standard desktop machines appears to show that this advantage is lost as normal buffering and processor speed on desktop architectures compensates for the lack of buffering. It can be deduced that the PDAs have little or no network buffering to increase performance, leading to the assumption that JCSP is more efficient than it actually is. The comparison of buffered and unbuffered streams is presented in Figure 2.

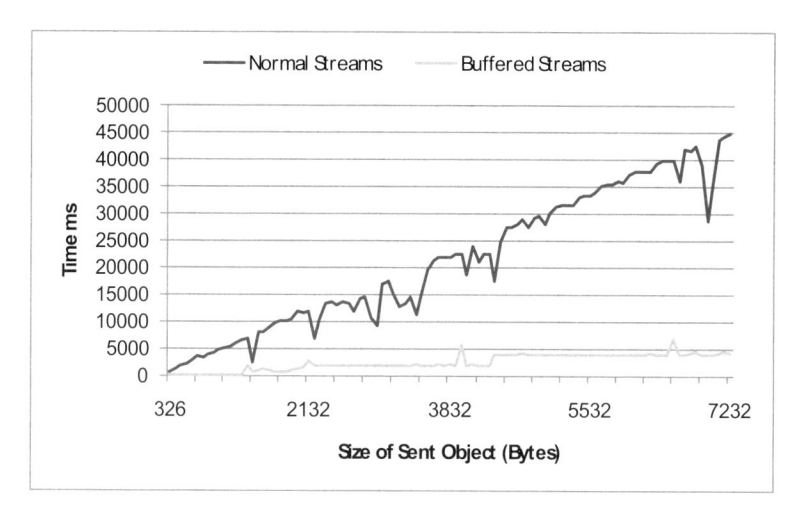

Figure 2: normal streams versus buffered btreams.

There are some interesting points to consider. As previously mentioned, there are valleys and peaks evident which are currently under investigation, initial signs pointing to the internal workings of the JVM or the PDA itself when dealing with certain sizes of data being the cause. The other interesting phenomenon is the steps in the buffered stream results. These steps reflect the extra packets of data sent as the Maximum Transmission Unit (MTU) of the network is reached. The MTU is the maximum single packet size that can be sent in one operation. The MTU for the wirelesos network is 2272 bytes and the MTU for the Ethernet connection is 1492 bytes as set by Linux. The buffer is greater than this value and therefore the data is split into separate packets during transmission.

Our next set of results illustrates the difference between networked channel operations and buffered stream operations. These are presented in Figure 3. The same step increase is apparent, although the networked channels are distinctly pushed left. This is due to the extra overhead of sending objects via JCSP as described in Section 4.3. It also appears that the first step in the results occurs at the MTU size for the Ethernet (1492) and the second occurs around twice the MTU for wireless packets (2272). The reason for this has not yet been ascertained. A packet is fragmented by the router when being transmitted from the PDA to the PC to allow the larger wireless packet to be sent as smaller packets on the Ethernet. There is no such constraint from the PC to the PDA as the wireless network can manage packets from the Ethernet. The PC should be capable of handling the reconstruction of fragmented packets without significant delays. Therefore, the steps should occur at twice the MTU of the Ethernet, especially as the PDA is capable of sending data far quicker than it can receive. This is shown in Figure 5.

We can also see the same peaks appearing in the JCSP networked channel results as th buffered stream results, strengthening our belief that it is not the implementation causing a problem.

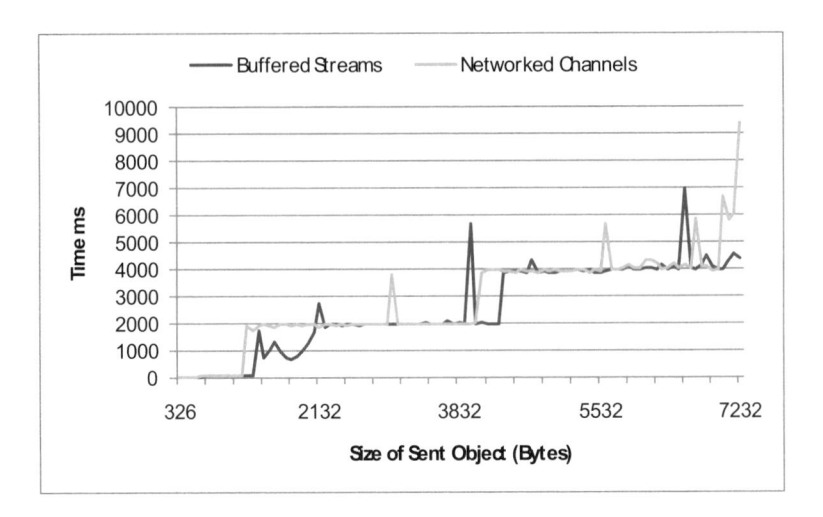

Figure 3: buffered streams versus networked channels.

These results allow us to show that JCSP does have an overhead within its communication mechanism. This overhead can have consequences, especially for resource constrained devices. As Figure 3 illustrates, an object that is sent via a channel may take two seconds on a roundtrip when the buffered stream equivalent has minimal time. This is due to sending extra data packets and if the overhead was reduced, the difference in performance could be compensated. The extra bytes sent for the message (249) and the Ethernet packet size (1492) imply that one sixth of a single packet is taken up by information beyond the sent data. Therefore there is a one in six chance that a message sent via JCSP will require an extra packet in a normal network. For a send-receive operation (as presented in Figure 3), this increases the time by two seconds within our test bed.

4.6 High Priority Links

Our next set of results illustrates a danger in the implementation of JCSP Links. These processes are given highest priority in JCSP to decrease latency by having the TX/RX processes start as soon as possible. The argument is that these processes may be blocked while trying to send or receive data if they are not given high enough priority. There is a converse to this argument. To illustrate the danger, we present the results of the PDA only receiving (and not sending back) messages from normal networked channels and unacknowledged network channels. These are given in Figure 4.

As this chart illustrates, sending unacknowledged messages takes more time than acknowledged ones. This should not be the case. First, there are fewer messages sent (no acknowledgements), and second the PC should be able to send messages faster than the PDA can read them, due to performance differences between the two machines and their network interfaces. What is happening here is that the PC is sending messages too fast for the PDA to cope. Data is appearing on the network before the PDA has time to process the previous message. The readings are taken from the application level process, and it is being superseded by the LinkRX process as it receives new messages. his results in the application process taking longer to receive messages as it must wait for the LinkRX. Underlying every network channel is an infinite buffer to receive messages upon to avoid a LinkRX process deadlocking. This leads to the LinkRX being capable of constantly writing to the channel if it has data ready, without the channel having time to respond.

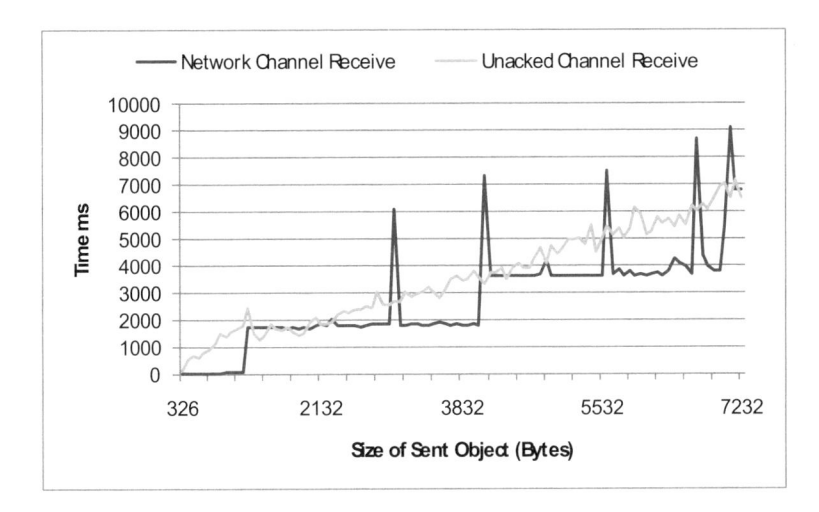

Figure 4: normal versus unacknowledged network channels receiving.

This may appear an unfair comment since unacknowledged channels should not be used in such a manner (they are used to avoid the Channel Name Server and networked connections from blocking).

The problem can also lie in channels which are buffered, have multiple incoming connections, and are accepting large packet sizes. This can lead to an application process waiting until data is received by the `LinkRX` processes. As the user has no control over the priority of the `Link` processes, they have no method to decide whether distributed I/O or application processes should be given highest priority.

A simple analogy of this is a producer-consumer program that operates with an infinitely buffered channel. If the sender is given higher priority than the receiver, then the receiver is theoretically starved as it cannot continue until the sending process has completed sending. Over time, the buffer in the channel grows and we are in danger of running out of memory. In practice this is not strictly true, as the receiver will be allowed to consume some of the messages. This may occur in JCSP over time using a standard infinitely buffered channel.

Buffered networked channels are also present in the current JCSP implementation. These are implemented by buffering the channel between the `NetChannelInputProcess` and `NetChannelInput`. Therefore the `NetChannelInputProcess` may not be blocked while writing a message to the `NetChannelInput`, depending on the buffer used within the channel. The standard JCSP buffers may be used within these channels, and thus there may in fact be two infinite buffers filling into one another.

The other interesting point that this graph illustrates is the repetition of the step function for the receiving results on the PDA. When comparing the results for communication from PC to PDA and from PDA to PC (Figure 5) we see that the greatest time is taken when the PDA is receiving data. Reducing this time would increase performance in our test bed considerably and is worth future investigation. The communication synchronizes, so the results should be the same if it were network capabilities restricting performance.

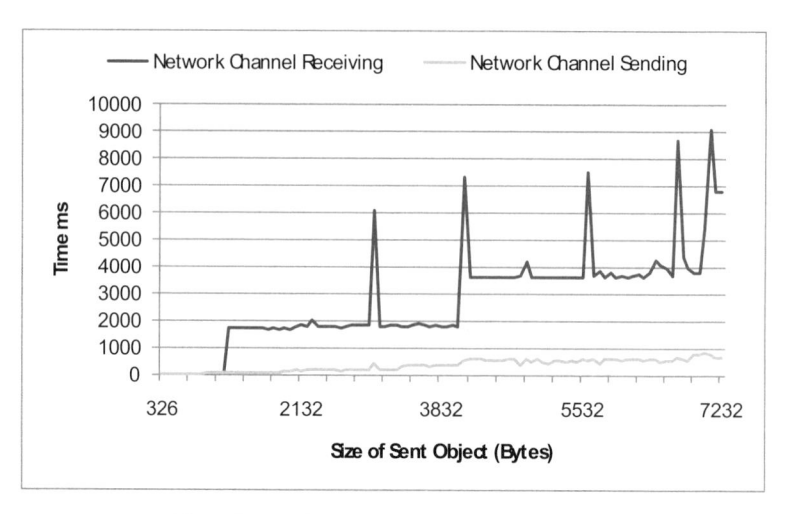

Figure 5: network channel sending and receiving on PDA.

4.7 Lack of Exceptions

The next limitation of JCSP is the lack of well documented exceptions being passed up to the application level processes. JCSP networking relies on I/O operations, and these can fail for reasons outside the control of JCSP. To combat this, exceptions such as LinkLost may be thrown, or a LinkLostEventChannel can be checked for any lost connections. These are not always caught. An example of such an operation is when an output end sends a message to the corresponding input end. If, prior to the acknowledgement being sent, the connection between the two nodes fails, the output end is not informed, and is left to hang waiting for the acknowledgement that will never come. As we do not have guarded output in JCSP networking we cannot recover from such an eventuality and must restart the system. A simple solution to overcome this problem is given in Section 5.

We may get an exception when something bad happens in the form of a ChannelDataRejectedException. This is a strange exception to be thrown for failed I/O operations. RejectableChannels are a deprecated construct within the core library and the reliance of JCSP networking on them should be removed. It would appear that the reason for having RejectableChannels was originally to handle I/O exceptions so that exceptions could be passed to the application level. As these I/O exceptions may still occur, a mechanism must be put in place to pass the exceptions onto the application process in a manner that allows a networked channel end to still appear as a normal channel end if required. This is described in Section 5.

4.8 Lack of Universal Protocol

Our final concern reflects on the issues raised in sections 4.3 and 4.4. JCSP utilizes objects as messages between distributed nodes. This is not a concern if we only wish JCSP to communicate with itself. However, we now have a numerous implementations of networked CSP inspired architectures across a great number of platforms. It is impossible for JCSP to communicate with pony or PyCSP in its current form without some form of Java object interpreter built into the respective frameworks. This is a limitation. When concerned with distributed systems, we should strive to allow inter-system communication whenever possible. It is reasonable for JCSP to communicate with pony, and vice-versa, but this is not possible. A universal communication mechanism and protocol should be developed to allow these separate frameworks to communicate as much as possible. This is covered in Section 5.

5. Conclusions and Ongoing Work

We have shown that JCSP networking currently has a number of problems – especially when considering small power/memory devices – and our hope is to address these with a new implementation of JCSP networking. In summary, we have argued that:

- The architectural implementation leads to high resource overhead
- The architectural implementation is complex when compared to the basic premise of JCSP networking, making extension difficult
- Message packets are large in comparison to the amount of information actually sent within them
- The current implementation by default only allows serializable objects to be sent
- Performance of the basic communication mechanism is almost on par with the underlying stream, but message overheads have an effect
- The default high priority link is restrictive as some applications may require lower priority I/O
- Exception raising is not guaranteed
- There is no interoperability between frameworks due to JCSP relying on objects as transmitted messages

Our new implementation of JCSP networking aims to overcome these problems while also trying to bring the new architecture to the same level as the core for functionality.

5.1 A New Architecture for JCSP Networking

Our new architecture is based on the existing JCSP networking implementation, as well as taking inspiration from C++CSP Networked and pony. We have aimed to retain the existing interfaces whenever possible to allow existing users the same familiarity with the library.

The new architecture has the initial aim of reducing the resource overheads discovered in JCSP networking by removing unnecessary processes. To support this, we have removed `NetChannelInputProcess` and `LoopbackLink`, as well as converting the management processes into shared data objects.

Our new approach is based on a layered model. This allows functional separation and allows simple extension / modification. The underlying process model is still in effect, and the new architecture model is almost exactly the same. Figure 6 illustrates the new architectural model.

The key difference is how the components communicate together. Each layer only understands certain message types, thus promoting separation. As a layered approach is taken, messages only travel as far up or down the layered model as required, providing a further degree of separation. This allows simple additions and modifications in specific components to allow extension of the architecture. For example, networked barriers have been implemented by providing the same mechanisms in the Event Layer as there are for channels.

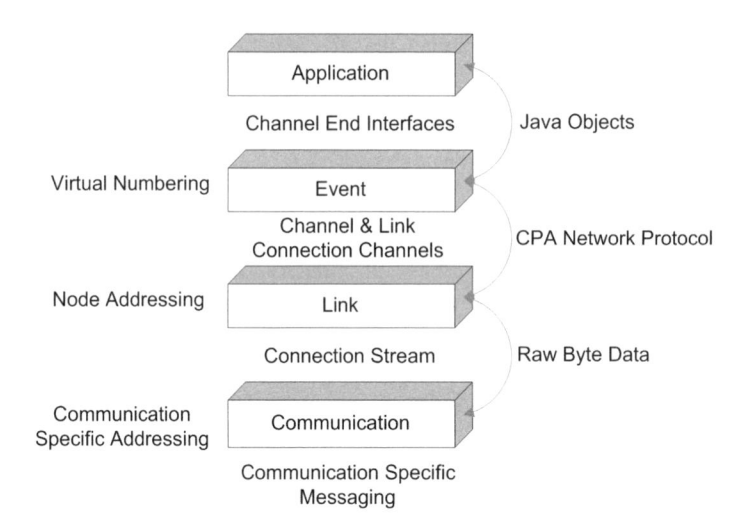

```
                    ┌─────────────────┐
                    │   Application   │
                    └─────────────────┘
                  Channel End Interfaces        Java Objects
                    ┌─────────────────┐
Virtual Numbering   │      Event      │
                    └─────────────────┘
                     Channel & Link             CPA Network Protocol
                    Connection Channels
                    ┌─────────────────┐
Node Addressing     │      Link       │
                    └─────────────────┘
                    Connection Stream           Raw Byte Data
Communication       ┌─────────────────┐
Specific Addressing │  Communication  │
                    └─────────────────┘
                   Communication Specific
                        Messaging
```

Figure 6: new JCSP architectural model.

5.1.1 Networked Barrier and AltingBarrier

The barrier operates on a client-server basis, with one JCSP node acting as a server for n client barriers. Each barrier end may have a number of processes synchronizing upon it, and for optimisation purposes it is only when all locally enrolled processes have synchronized does a client barrier end communicate with the server end. Once all client ends have synchronized, the server end releases them. Thus, the networked barrier operates in two phases; local synchronisation and then distributed synchronisation

We are also considering how to implement a networked multi-way synchronisation within the architecture, but this is far more difficult. The two phase approach used in the standard barrier cannot be reused for a direct networked implementation of `AltingBarrier` due to the implementation within the JCSP core. Here, a central control object is used, which ensures that only one process is actively alting on any `AltingBarrier` at any one time. This is irrespective of the number of `AltingBarriers` in the system. This controls the multi-way synchronisation in a manner that allows fast resolution of choice between multiple multi-way synchronizing events.

The problem with this controlled model is that it does not scale to multiple processors. If each process must wait to access this coordination object, then only one process is in operation at any time. This is fine in a single processor, concurrent system as only one process can only ever be in operation. With a multi-processor environment the problem is that all processors must wait while one accesses the coordination object. This is a worst case scenario, but does highlight the problem faced. A distributed event based system faces the same problem. However, we currently believe the two phase approach used for networked barriers is the most likely approach for efficiency reasons when dealing with a distributed multi-way synchronisation. A possible approach is to use a process within the networked `AltingBarrier` client end to control the synchronisation and communicate with the declaring `AltingBarrier` server end. This approach should allow a single networked `AltingBarrier` to exist on a single node. However, the goal would be to allow multiple networked `AltingBarriers` on a node.

Another key feature is the use of a communication protocol independent of Java or any other platform. Instead of relying on Java objects, data primitives are used.

5.1.2 A Universal Protocol for CPA Based Networking

The aim of our new protocol is to promote communication independent from the data being sent. Through this we hope to achieve a standard mechanism that can be exploited across the various CSP based network architectures. It is perfectly reasonable to expect JCSP to communicate with KRoC, and thus we have aimed at a simple protocol that is easily portable to other platforms. The standard message types can be well defined, and thus far we have encountered only message types that require three values: a *type*, and two *attributes*. These can be expressed using a byte and two 32 bit signed integers. We are also removing the need for object messages which reduces the message overhead. At present, this reduces the 249 byte message header to 9 or 13 bytes.

This does not take into account data passed within the message itself, and this is considered a special case. We define message types using values, therefore the message type can also be used to determine whether or not the message has a data segment. If it does, the number of bytes can be sent as a 32 bit integer, and then the data itself can be transferred. The new key feature is that channels are now responsible for converting objects to and from bytes, and the messages themselves must only contain byte data. The conversion method can be specified by the JCSP user, thus providing data independence as there is no longer a reliance on Java serialisation. For example, a JCSP system could send a message to a pony system by utilising a converter that implemented strict **occam** rules for data structures. Schweigler's [6] work in this area provides a strong basis to build upon.

5.1.3 Channel Mobility

We are also hoping to build channel mobility directly into the protocol to allow mobility of channels between platforms. Unfortunately, at present this cannot be fully accomplished due to conflicting approaches for mobility proposed for JCSP [11] and pony [6]. Both of these approaches have advantages and disadvantages, and a coming together is required for mobility to be implemented directly into the protocol.

5.1.4 Other Features

One shortcoming overcome in the new model is the exception handling mechanism, as specified in Section 4.7. There is now a unique exception that can be thrown by a networked JCSP system; the `JCSPNetworkException`. This is a silent exception, in that it does not need to be caught explicitly by the JCSP user, thus permitting channel operations to throw the exception but not break the existing core channel interfaces.

We have also implemented a solution to the deadlock caused by an output channel waiting indefinitely for an acknowledgement from a broken connection. `NetChannelOutput` objects now register with their respective `Link` components on creation, and unregister on destruction / failure. This allows a `Link` to send a message to the `NetChannelOutput` on failure. As this message is written to the same channel as an acknowledgement would be expected the `NetChannelOutput` can receive this message and act accordingly. As this message may also be sent prior to an initial write, the channel can be checked at the start of a write operation, avoiding unnecessary attempts to write to a dead `Link`.

5.1.5 Verified Model

Our design has been approached with model checking in mind. We have performed some preliminary investigation with Spin [23], with promising results thus far. We do have a number of properties we still wish to examine. The reason to use Spin as opposed to FDR is due to the usage of mobile channels within the JCSP architecture, and Spin allows channel mobility explicitly within its models. The model built within Spin can be directly composed into Java due to the core of JCSP being in place.

5.1.6 Ongoing Work

There is still work to do on the new implementation of JCSP networking. In particular, if the defined protocol is to be considered as a method to allow intercommunication between different platforms, then further investigation must be undertaken for other common / expected messages within these frameworks. As an example, the protocol currently does not implement any notion of claiming a channel end, although this is used within pony for shared channel ends.

Connections are currently not implemented in the new architecture. It is possible to implement connections using normal networked channels, but this requires building a connection message protocol that will be sent via the communication protocol. A more practical approach is to implement these message types directly into the communication protocol, and then develop management and component ends within the Event Layer.

5.2 Future Work

We have highlighted a number of other pieces of future work beyond the new implementation of JCSP networking. Work on enhancing the serialisation capabilities of Java to accommodate JCSP will likely lead to an increase in performance for both small factor and desktop applications. We also hope to perform comparisons with RMI, taking into account simplicity, code mobility and performance. With networked connections, it should be possible to create remote interfaces that are externally similar to RMI. Finally, for our own experimental test bed, the possibility of increasing performance for the PDA receiving data is interesting.

5.3 Conclusions

We have shown that JCSP networking has a significant number of problems which lead to certain impracticalities when considering JCSP in small factor devices. Particularly, we have shown that that there are overheads due to excess process creation and destruction, as well as overheads for message transfer. We have also illustrated some dangers and argued on the complexity of the implementation.

Finally, we have discussed a new approach for JCSP networking which will lead to a more ubiquitous approach for CSP networking as a whole. We hope that this approach can be replicated across the various CSP based frameworks to allow stronger integration, allowing simpler exploitation of multiple platforms.

References

[1] P. H. Welch, "Process Oriented Design for Java: Concurrency for All," in H. R. Arabnia (Ed.), *International Conference on Parallel and Distributed Processing Techniques and Applications (PDPTA '2000)Volume 1*, pp. 51-57, CSREA Press, 2000.

[2] D. Lea, "Section 4.5: Active Objects," in *Concurrent Programming in Java: Second Edition*, Boston: Addison-Wesley, 2000, pp. 367-376.
[3] C. A. R. Hoare, *Communicating Sequential Processes*. Prentice Hall, Inc., 1985.
[4] P. H. Welch, N. Brown, J. Moores, K. Chalmers, and B. H. C. Sputh, "Integrating and Extending JCSP," in A. McEwan, S. Schneider, W. Ifill, and P. H. Welch (Eds.), *Communicating Process Architectures 2007*, pp. 349-370, IOS Press, Amsterdam, 2007.
[5] P. H. Welch, J. R. Aldous, and J. Foster, "CSP Networking for Java (JCSP.net)," in P. M. A. Sloot, C. J. K. Tan, J. J. Dongarra, and A. G. Hoekstra (Eds.), *International Conference Computational Science — ICCS 2002, Lecture Notes in Computer Science 2330*, pp. 695-708, Springer Berlin / Heidelberg, 2002.
[6] M. Schweigler and A. T. Sampson, "pony - The occam-π Network Environment," in P. H. Welch, J. Kerridge, and F. R. M. Barnes (Eds.), *Communicating Process Architectures 2006*, pp. 77-108, IOS Press, Amsterdam, 2006.
[7] N. Brown, "C++CSP Networked," in I. East, J. Martin, P. H. Welch, D. Duce, and M. Green (Eds.), *Communicating Process Architectures 2004*, pp. 185-200, IOS Press, Amsterdam, 2004.
[8] J. M. Bjørndalen, B. Vinter, and O. Anshus, "PyCSP - Communicating Sequential Processes for Python," in A. McEwan, S. Schneider, W. Ifill, and P. H. Welch (Eds.), *Communicating Process Architectures 2007*, pp. 229-248, IOS Press, Amsterdam, 2007.
[9] Inmos Limited, "The T9000 Transputer Instruction Set Manual," SGS-Thompson Microelectronics 1993.
[10] A. Fuggetta, G. P. Picco, and G. Vigna, "Understanding Code Mobility," *IEEE Transactions on Software Engineering,* 24(5), pp. 342-361, 1998.
[11] K. Chalmers, J. Kerridge, and I. Romdhani, "Mobility in JCSP: New Mobile Channel and Mobile Process Models," in A. McEwan, S. Schneider, W. Ifill, and P. H. Welch (Eds.), *Communicating Process Architectures 2007*, pp. 163-182, IOS Press, Amsterdam, 2007.
[12] N. Brown and P. H. Welch, "An Introduction to the Kent C++CSP Library," in J. F. Broenink and G. H. Hilderink (Eds.), *Communicating Process Architectures 2003*, pp. 139-156, IOS Press, Amsterdam, 2003.
[13] N. Brown, "C++CSP2: A Many-to-Many Threading Model for Multicore Architectures," in A. McEwan, S. Schneider, W. Ifill, and P. H. Welch (Eds.), *Communicating Process Architectures 2007*, pp. 183-205, IOS Press, Amsterdam, 2007.
[14] A. A. Lehmberg and M. Olsen, "An Introduction to CSP.NET," in P. H. Welch, J. Kerridge, and F. R. M. Barnes (Eds.), *Communicating Process Architectures 2006*, pp. 13-30, IOS Press, Amsterdam, 2006.
[15] N. C. Schaller, S. W. Marshall, and Y.-F. Cho, "A Comparison of High Performance, Parallel Computing Java Packages," in J. F. Broenink and G. H. Hilderink (Eds.), *Communicating Process Architectures 2003*, pp. 1-16, IOS Press, Amsterdam, 2003.
[16] B. Vinter and P. H. Welch, "Cluster Computing and JCSP Networking," in J. Pascoe, P. H. Welch, R. Loader, and V. Sunderam (Eds.), *Communicating Process Architectures 2002*, pp. 203-222, IOS Press, Amsterdam, 2002.
[17] S. Kumar and G. S. Stiles, "A JCSP.net Implementation of a Massively Multiplayer Online Game," in P. H. Welch, J. Kerridge, and F. R. M. Barnes (Eds.), *Communicating Process Architectures 2006*, pp. 135-149, IOS Press, Amsterdam, 2006.
[18] M. Weiser, "The Computer for the 21st Century," *Scientific American,* September, 1991.
[19] R. Milner, J. Parrow, and D. Walker, "A Calculus of Mobile Processes, I," *Information and Computation,* 100(1), pp. 1-40, 1992.
[20] R. Milner, "Ubiquitous Computing: Shall we Understand It?," *The Computer Journal,* 49(4), pp. 383-389, 2006.
[21] C. G. Ritson and P. H. Welch, "A Process-Oriented Architecture for Complex System Modelling," in A. McEwan, S. Schneider, W. Ifill, and P. H. Welch (Eds.), *Communicating Process Architectures 2007*, pp. 249-266, IOS Press, Amsterdam, 2007.
[22] C. L. Jacobsen and M. C. Jadud, "The Transterpreter: A Transputer Interpreter," in I. East, D. Duce, M. Green, J. Martin, and P. H. Welch (Eds.), *Communicating Process Architectures 2004*, pp. 99-107, IOS Press, Amsterdam, 2004.
[23] G. J. Holzmann, *The Spin Model Checker: Primer and Reference Manual*. Addison-Wesley, Boston, 2003.

Virtual Machine Based Debugging for occam-π

Carl G. RITSON and Jonathan SIMPSON

Computing Laboratory, University of Kent, Canterbury, Kent, CT2 7NF, England.

{c.g.ritson, j.simpson}@kent.ac.uk

Abstract. While we strive to create robust language constructs and design patterns which prevent the introduction of faults during software development, an inevitable element of human error still remains. We must therefore endeavor to ease and accelerate the process of diagnosing and fixing software faults, commonly known as *debugging*. Current support for debugging occam-π programs is fairly limited. At best the developer is presented with a reference to the last known code line executed before their program abnormally terminated. This assumes the program does in fact terminate, and does not instead live-lock. In cases where this support is not sufficient, developers must instrument their own tracing support, "printf style". An exercise which typically enlightens one as to the true meaning of concurrency... In this paper we explore previous work in the field of debugging occam programs and introduce a new method for run-time monitoring of occam-π applications, based on the Transterpreter virtual machine interpreter. By adding a set of extensions to the Transterpreter, we give occam-π processes the ability to interact with their execution environment. Use of a virtual machine allows us to expose program execution state which would otherwise require non-portable or specialised hardware support. Using a model which bears similarities to that applied when debugging embedded systems with a JTAG connection, we describe debugging occam-π by mediating the execution of one execution process from another.

Keywords. concurrency, CSP, debugging, occam-pi, virtual machine.

Introduction

While a program may be correct by design, as long as there remains a gap between that design and its implementation then an opportunity exists for errors to be introduced. In the ideal world our implementations would be automatically verified against higher-level specifications of our designs, but despite much work in the area, such checking is not yet supported for all cases in occam-π. Hence, support for finding of and fixing errors is essential to the rapid development of large-scale and complex applications. Debugging is the process of locating, analyzing, and correcting suspected errors [1].

The Kent Retargetable occam-π Compiler (KRoC) supports basic post-mortem debugging [2], which gives the developer access to the last executed line and process call at the time a fatal error such as integer overflow or array bounds is detected. This provides the developer with a starting point for debugging and assists with the correction of many trivial errors; however, it does not elucidate the, often complex, interactions which lead up to application failure, nor does it provide any assistance for non-fatal errors. The work presented in this paper is an extension of earlier work and is aimed at providing the developer with uniform and accessible tracing and replay facilities for occam-π applications.

Without any explicit debugging support developers must implement their own tracing support. This typically takes the form of a shared messaging channel along which component processes emit their status. The programmer writes "print" commands into critical sections of the application and then views the trace output. Concurrency, however, creates many issues for this debugging approach. Buffering on the messaging channel or the presence of multiple threads of execution will cause the trace output to deviate from the actual execution flow of the application. This hinders the detection of errors in the interaction between processes. In turn, the introduction of the messaging channel itself, a new shared resource on which component processes synchronise, subtly changes the scheduling behaviour of the program obscuring race condition related faults. This is often call the *probe effect* and we describe it further in section 5.1.

In this paper we briefly review previous work in the field of parallel application debugging, and more specifically the debugging of occam programs (section 1). Following on we introduce a new method for run-time monitoring of occam-π applications, based on the Transterpreter [3] virtual machine interpreter. Section 3 describes how virtual machine instances can interact with the interpreter to intercede on each other's execution. This allows an occam-π process to mediate the execution of another occam-π program, a model which bears similarities to that applied in debugging embedded systems with a JTAG connection [4]. Our new extensible bytecode format which provides access to debugging information is detailed in section 4. In section 5 we describe additional run-time debugging support we have added to the Transterpreter virtual machine, and then in section 6 we apply it and virtual machine introspection to the task of tracing occam-π programs. Conclusions are presented in section 7, and pointers to future work in section 8.

1. Related Work

Viewing the state of an executing program is one method of obviating its behaviour and debugging program code. Past work in the area of debugging occam has concentrated heavily on networks of Transputers, as these were the primary target for large scale occam application development. As such, tracing and debugging of program execution across networks of processors increased the complexity of past solutions. In addition to this observation, past solutions can be divided into those that traced and interacted with running programs and those that acted on stored post-mortem traces.

Stepney's GRAIL [5] extracted run-time data from a running network of Transputers, instrumenting the target occam program at compile-time with a modified version of the INMOS compiler and extracting the data over a custom bus external to the Transputer network [6], so as not to interrupt the communications occurring on the communications links. Maintaining the original timings and communication semantics as a program not being debugged is critical.

Cai and Turner identified in [7] the danger of added debugging hook processes altering the timing of a parallel program and propose a model under which a *monitor* has execution control over all processes and a logical clock is maintained for the execution of each process. The logical clock provides a way to measure and balance the interaction added to the network by the monitor hooks, and is similar to the approach we propose in section 5.1. The paper also identifies the problem of maintaining timings for real-time systems whilst monitoring the system, suggesting the use of dummy test-sources or buffering of real-time data.

Debugging methodologies that require annotations or changes to the code have the potential to introduce additional complexity, changing the run-time dynamics of the program. May's Panorama [8], designed for the debugging of message-passing systems, identifies an approach for post-mortem debugging which records a log of communications. Replaying these communications offers a look at the particular execution of the program whilst introducing only a minimal overhead at run-time. The Panorama system had the downside of requiring all communications to be modified to use a different call and recompiled against a custom library. This need for program modification, instrumentation or annotation is common to many approaches that do not have the benefit of an interpreted run-time environment.

The INMOS Transputer Debugging System (TDS) [9] allowed the user to inspect the occam source and display run-time values of variables. Programs for use with TDS were compiled in a debug mode which embedded extra information: workspace offsets, types and values of constants, protocol definitions, and workspace requirements for procedures and functions. The TDS provides access to running processes in the network by "jumping through" channels to the process on the other end of the channel. Deadlock detection requires source modification, as a deadlocked process can not be jumped to using the channel jump system. Processes suspected of causing deadlock must be modified to include a secondary process and channel which will not deadlock and allows a jump into the process for state inspection.

Finally, Zhang and Marwaha's Visputer [10] provided a highly developed visual tool for editing and analysing occam process networks. An editing suite allowed the assisted building of programs using toolkit symbols. A network editor facilitated the configuration of networks of Transputer processors. Pre-processing of source files inserted instrumentation library calls to support post-mortem debugging and performance profiling. A static analyser also predicted performance and detected deadlocks. Importantly Zhang and Marwaha pointed out that occam "has a language structure that is highly suitable for graphical representation". The work presented in this paper is intended to provide a means of extracting the information required to exploit this feature of occam.

2. The Transterpreter Virtual Machine

The Transterpreter, or TVM (Transterpreter Virtual Machine), is a virtual machine interpreter for running occam-π applications compiled to a modified form of Transputer bytecode [3]. Written in platform independent ANSI C, the TVM emulates a hybrid T8 Transputer processor. Most T8 instructions are supported, with additional instructions added to support dynamics and mobility added by occam-π.

The Transputer was a three-place stack machine, and executed a bytecode where the most common instructions and their immediates required only a single byte to represent. Large instructions were composed from sequences of smaller instructions prior to execution. Hence the Transputer bytecode is compact, and simple to interpret. A modified version of the INMOS compiler with occam-π support is used to generated Extended Transputer Code (ETC), which is then converted by a linker into bytecode for the TVM.

The TVM has a very small memory footprint making it suitable for use on small embedded systems, such as the LEGO Mindstorms RCX [11] and Surveyor SRV-1 [12] mobile robotics platform. The work presented in this paper focuses on execution on desktop class machines; however, the portability of the bytecode representation used by the TVM allows emulation of embedded systems on a desktop machine. We envisage that the techniques detailed here be used in such a manner as to debug embedded application code via emulation when the target platform has insufficient resources to support debugging in-place.

3. Introspection

Recently, we have transitioned the TVM from a static library design to a fully re-entrant implementation. This shift allows us to run multiple virtual Transputers concurrently, or at least with the appearance of concurrency. The virtual Transputers can communicate with each other, or more specifically processes in one TVM instance can communicate with processes in another, efficiently and transparently. Being software defined, our Transputers are not restricted in the number of communications links they support, and hence one virtual link is provided per-shared channel, freeing the programmer from multiplexing a fixed number of links. Mobile data [13] and mobile channels [14] are also supported, allowing the construction of arbitrary process networks which in turn span multiple virtual machines.

Each TVM instance has its own registers and run-queues. Instances can be scheduled cooperatively (only switched when they relinquish control), or pre-emptively (time-sliced). The specific algorithm used to schedule between TVM instances is defined by the virtual machine wrapper, for most purposes a round-robin algorithm is used (the same as occam processes); however, priority is also supported.

On a small robotics platform, the Surveyor Corporation SRV-1 [12], we have used multiple TVM instances to execute a cooperatively scheduled firmware and a time-sliced user application. The user application is loaded at run-time, and may be swapped out, terminate or even crash, without interfering with the firmware. The firmware executes as required, mediating access to the hardware on behalf of the user application.

Our virtual Transputers, TVM instances, give us the isolation and encapsulation we need to start allowing one occam-π program to mediate and intercede on the execution of another. This concept, of multiple concurrent virtual machine interpreters, underpins the debugging framework presented in the rest of this paper.

3.1. Interface

The application running in each virtual machine instance can access the virtual machine runtime via a special `PLACED` channel. Channel read requests to the `PLACED` channel return a *channel bundle* (channel type [14]) which provides request and response channels for the manipulation of the current virtual machine instances and the creation of new sub-instances. For each sub-instance created a further channel bundle is returned that can be used to control the instruction-by-instruction execution of the sub-instance.

The virtual machine interface channel also provides access to the interpreter's bytecode decoder. Using this interface, a virtual machine can load bytecode into new sub-instances, and access additional information stored in the bytecode. Details of our bytecode format and decoder interface can be found in section 4.

Having decoded bytecode we created a VM instance associated with it. Once *top level process* parameters are supplied, and the instance started, the control interface can be used to mediate its execution in a number of ways. The following subsections detail requests which can be used to control execution.

3.1.1. run

Execute bytecode for a number of instruction dispatches, or until a breakpoint or error is encountered. With the exception of breakpoints this causes the sub-instance to act as a normal virtual machine instance. At present this operation is implemented synchronously and is uninterruptable, blocking execution of the parent virtual machine. However, there is no reason that this could not be enhanced to permit interleaved execution. Where processing facilities exist the sub-instance could also be executed concurrently on a separate processor.

3.1.2. step

Decode and dispatch a single bytecode instruction. Feedback is given when the instruction is decoded and then after it is dispatched, allowing the supervising process to keep track of execution state without querying the entire virtual machine state.

3.1.3. dispatch

Execute, *dispatch*, an instruction not contained in the program bytecode. This can be used to alter program flow, for example to execute a *stop process* instruction to pause the running process and scheduling the next. Alternatively this request can be used to inject a completely new stream of instructions into the virtual machine instance.

3.1.4. get.state / set.state

Get and set requests for the state provide access to the virtual machine registers, instruction pointer, operand stack and clock. By combining these requests with the *dispatch* request a debugger can save virtual machine state, change processes and later restore state.

3.1.5. read / write

Read and write requests for all basic types provide access to the virtual machine's memory. As these can only be executed when the virtual machine is stopped, they have predictable results outside their influence on program behaviour. If virtual memory is in use then address translation occurs transparently.

4. Bytecode

To support our debugging enhancements we have developed a new extensible bytecode format for the TVM. Until the introduction of this new format, a number of different fixed formats were used for the different platforms supported by the TVM. By creating a new format we have unified bytecode decoding support code, and have removed the need to rewrite the decoder when the bytecode is extended.

4.1. Encoding

We call our new encoding *TEncode*. TEncode is a simple binary markup language, a modified version of IFF.

TEncode streams operate in either 16-bit or 32-bit mode, which affects the size of integers used. A stream is made up of *elements*. Each element consists of a 4-byte identifier, followed by an integer, then zero or more bytes of data. Elements always begin on integer aligned boundaries, the preceding element is padded with null bytes ("\0") to maintain this. All integers are big-endian encoded, and of consistent size (2-bytes or 4-bytes) throughout a stream. Integers are unsigned unless otherwise stated.

```
Identifier  := 4 * BYTE
Integer     := INT16 / INT32
Data        := { BYTE }
Padding     := { "\0" }
Element     := Identifier, Integer, Data, Padding
```

Identifiers are made up of four ASCII encoded characters stored in 8-bit bytes, and are case-sensitive. The last byte indicates the type of data held in the element. The following types are defined:

- **B**yte string . Integer field encodes number of bytes in Data field, excluding padding null-bytes.
- **I**nteger . Integer field encodes signed numeric value, no Data bytes follow.
- **L**ist of elements . Integer field encodes number of bytes in Data field which will be a multiple of the integer size, Data field contains nested elements, and may be parsed as a new or sub-Stream.

String (UTF8 null-terminated) . Integer field encodes number of bytes in Data including null terminator, but not padding.

Unsigned integer . Integer field encodes unsigned numeric value, no Data bytes follow.

With the exception of signed and unsigned integer types, the Integer of all elements defines the unpadded size of the Data which follows. Decoders may use this relationship to skip unknown element types, therefore this relationship must be preserved when adding new types.

A TEncode stream begins with the special `tenc` or `TEnc` element, the integer of which indicates the number of bytes which follow in the rest of the stream. A lower-case `tenc` indicates that integers are small (16-bit), whereas an upper-case `TEnc` indicates integers are large (32-bits). The `TEnc` element contains all other elements in the stream.

```
ByteString    := 3 * BYTE, "B", Integer, Data [, Padding ]
SignedInt     := 3 * BYTE, "I", Integer
ElementList   := 3 * BYTE, "L", Integer, Stream
UTF8String    := 3 * BYTE, "S", Integer, {<character byte>},
                 "\0" [, Padding ]
UnsignedInt   := 3 * BYTE, "U", Integer

Element       := ByteString / SignedInt / ElementList / UTF8String /
                 UnsignedInt

Header        := ("tenc", INT16) / ("TEnc", INT32)
Stream        := { Element }
TEncode       := Header, Stream
```

Decoders ignore elements they do not understand or care about. If multiple elements with the same identifier exist in the same stream and the decoder does not expect multiple instances then the decoder uses the first encountered. When defining a stream, new elements may be freely added to its definition across versions; however, the order of elements must be maintained in order to keep parsing simple.

4.2. Structure

The base Transterpreter bytecode is a TEncode stream defined as follows:

```
TEnc <stream length>
  tbcL <length>
    endU <endian (0=little, 1=big)>
    ws U <workspace size (words)>
    vs U <vectorspace size (words)>
    padB <length> <padding>
    bc B <length> <bytecode>
```

As previously stated, the `TEnc` element marks the beginning of a TEncode stream. The stream contains a number of `tbcL` elements, each defines a bytecode chunk. A stream may contain multiple bytecode chunks in order to support alternative compilations of the same code, for example with different endian encodings. The `endU` element specifies the endian type of the bytecode. The `ws U` and `vs U` elements specify the workspace and vectorspace memory requirements in words. The `padB` element ensures there is enough space to in-place decode the stream. Finally, the `bc B` element contains the bytecode which the virtual machine interpreter executes.

Following the mandatory elements a number of optional elements specify additional properties of a chunk of bytecode. By placing these elements after the mandatory elements a stream decoder on a memory constrained embedded system can discard all unnecessary elements, such as debugging information, once the mandatory elements have been received and decoded.

A tlpL element defines the arguments for the *top level process* (entry point). The foreign function interface table, and associated external library symbols are provided in a ffiL element. A symbol table, defining the offsets, memory requirements and type headers of processes within the bytecode is provided by a stbL element. Finally a dbgL element specifies debugging information.

The debugging information takes the following form:

```
dbgL <length>
  // File names
  fn L <length>
    fn S <length> <file name>
  // Line Numbering Data
  lndB <length>
    <bytecode offset> <file index> <line number>
    ... further entries ...
```

A table of source file names is defined by fn S elements. A table of integer triples is then specified by the lndB element. The integers in each triple correspond to a bytecode offset, the index of the source file (in the source file name table), and the line number in the specified source file. The table is arranged such that bytecode offsets are ascending and offsets that fall between entries in the table belong to the last entry before the offset. For example if entries exist for offsets 0 and 10, then offset 5 belongs to offset 0.

4.3. In-place Decoding

On memory constrained embedded platforms it is important to minimise and preferably remove dynamic memory requirements. For this reason we designed our new bytecode format so as not to require dynamic memory for decoding. We achieve this by providing for rewriting of the bytecode in memory as it is decoded. The C structure tbc_t is placed over the memory of the tbcL element of the TEncode stream and the component fields written as their TEncode elements are decoded. The bytecode field is a pointer to the memory address of the data of the bc B element. tlp, ffi, symbols and debug fields are pointers to the in-place decodes of their associated elements. The pointers are NULL if the stream does not contain the associated element.

```
struct tbc_t {
  unsigned int  endian;
  unsigned int  ws;
  unsigned int  vs;

  unsigned int  bytecode_len;
  BYTE          *bytecode;

  tbc_tlp_t     *tlp;
  tbc_ffi_t     *ffi;
  tbc_sym_t     *symbols;
  tbc_dbg_t     *debug;
};
```

4.4. Interface

As with the introspection interface discussed in section 3, the bytecode decoder also has a channel interface accessible from within a virtual machine instance. When passing an encoded bytecode array to the virtual machine, a channel bundle is returned which can be used to access the decoded bytecode. The following subsections detail some of the available requests.

4.4.1. create.vm

Create a new virtual machine instance based on this bytecode. A control channel bundle is returned for the new instance.

4.4.2. get.file

Translate a source file index to its corresponding file name. The file name is returned as a mobile array of bytes.

4.4.3. get.line.info

Look up the line numbering information for a bytecode address. The source file index and line number are returned if the information exists.

4.4.4. get.symbol / get.symbol.at

Look up a process symbol by name or bytecode address. The symbols bytecode offset, name, type description and memory requirements are returned if the symbol exists.

4.4.5. get.tlp

Access information on the top-level-process (entry point), if one is defined in the bytecode. If defined, the format mask and symbol name of the process are returned. This information is required to setup the entry point stack.

5. Debugging Support

We have added additional features to the TVM in order to support debugging. The following section details some of the more significant changes we have made and how they may be applied to the problem of debugging occam-π programs.

5.1. The Probe Effect

The *probe effect* is a term used to describe the consequence that observing a parallel or distributed system may influence its behaviour. This is a reminiscent of the Heisenberg uncertainty principle as applied in quantum physics. If the decisions made in non-deterministic elements of a program differ between runs under and not under observation, then the observed behaviour will not be the actual operation behaviour of the program. The impact of this on debugging is that, on observation of a known faulty program, errors do not occur or different errors appear.

To prevent the occurrence of the probe effect we must ensure the non-deterministic behaviour of the program is not altered by observation. In occam-π there are two sources of programmed non-determinism: alternation constructs and timers. A program free of these constructs is completely deterministic; in fact a program is still deterministic in the presence of simple timer delays [15].

Previous work has shown that by maintaining the sequence of events in a program and recording the decisions made by non-deterministic elements, it is possible to replay parallel programs [16]. In the work presented here we do not attempt to constrain the non-determinism which occurs from external inputs to the program, only to control the internal non-determinism. Given that the implementation of virtual machine constructs such as alternation is unaffected by changes made to monitor program execution, we need only manage the non-determinism caused by timers.

Our present approach to constraining timer based non-determinism is to use a logical time clock, as opposed to a real time clock [7]. This logical clock ticks based on the count of instructions executed. By applying a suitable multiplier the instruction count is scaled to give a suitable representation of time.

The accuracy of the logical clock's representation of time depends on the scaling value used. At program start-up the virtual machine (or firmware components of it), calculate the average execution time of a spread of instructions, and use this to derive a scaling factor. This scaling factor is then periodically adjusted and offset, such that the clock time matches the actual time required to execute the program so far. If virtual machine execution pauses, for example waiting on the timer queue, then an adjustment need only be stored to the new time when execution resumes.

In order to replay a program's execution we need only store the initial scaling factor and the instruction offset and value of subsequent adjustments. If adjustments are stored in three 32-bit integers, then any single adjustment will be 12 bytes in size. Assuming adjustments are made approximately once a second then around 40KiB of storage is required for every hour of execution. As the program is replayed, adjustments are applied at their associated instruction offsets irrespective of the actual execution speed of the program.

5.2. Memory Shadowing

We have added support for shadowing each location of the *workspace* (process stack) with type information of the value it holds. This information assists in the debugging of programs in the absence of additional compiler information about the memory layout. The type mapping is maintained at run-time for the virtual machine registers and copied to and from the shadow map as instructions are interpreted. For example if a pointer to a workspace memory location is loaded and then stored to another memory location, then the type shadow will indicate the memory is a workspace pointer. The shadow types presently supported are:

Data: The memory location holds data of unknown type.
Workspace: Pointer to a workspace memory location.
Bytecode: Pointer to a location in bytecode memory, e.g. an instruction or constant pointer.
Mobile: Pointer into mobile memory.
Channel: Memory is being used as a channel.
Mobile type: Mobile type pointer or handle.

By iterating over the shadow of the entire memory, a debugger can build an image of a paused or crashed program. To assist in this, the memory shadow holds flags about whether the memory is utilised or not. Workspace words are marked as in-use when they write to, and marked as free whenever a process releases them via the *adjust workspace* instruction. As memory known to be unused need not be examined, false positives and data ghosts are reduced.

In addition to a utilisation flag, call stack and channel flags are also recorded. Whenever a *call* instruction is executed, the workspace word used to store the return address is marked. This allows later reconstruction of the call stack of processes. Any word used as a channel in a channel input or output instruction is also marked to facilitate the building of process network graphs. If a process's workspace holds a channel word, or a pointer to a channel word, then we can assume it is a communication partner on that channel. When more than two such relationships exist then pointers can be assumed to represent the most recent partners on the channel, for example a channel declared in a parent process being used by two processes running in parallel beneath it.

6. A Tracing Debugger

The virtual machine extensions and interfaces so far detailed can be used to implement various development and debugging tools. In this section we will detail a tracing debugger.

6.1. Instancing Bytecode

First we must decode the bytecode we intend to run, this is done by making a `decode.bytecode` request on the introspection channel. Having successfully decoded bytecode we can request a virtual machine instance to execute it. The following code example demostrates decoding bytecode and creating a virtual machine instance. A virtual machine introspection bundle is assumed to be defined as vm.

```
MOBILE []BYTE data:
INT errno:
SEQ
  ... load bytecode into data array ...
  vm[request]  ! decode.bytecode; data
  vm[response] ? CASE
    CT.BYTECODE! bytecode:
    bytecode; bytecode
      -- bytecode decoded
      SEQ
        bytecode[request]  ! create.vm
        bytecode[response] ? CASE
          CT.VM.CTL! vm.ctl:
          vm; vm.ctl
            -- VM instance created
          error; errno
            ... handle VM creation error, e.g. out of memory ...
    error; errno
      ... handle decoding error, e.g. invalid bytecode ...
```

6.2. Executing Bytecode

Before we can execute bytecode in our newly created virtual machine instance we must supply the parameters for the top-level-process. In a standard occam-π program these provide access to the keyboard input, and standard output and error channels. In our example tracing code we will assume the program being processed has a standard top-level-process; however, to handle alternate parameter combinations the `get.tlp` (section 4.4.5) request would be used to query the bytecode. In the following code example we supply the top level parameters as channel bundles, preparing the bytecode for execution:

```
CT.CHANNEL? kyb.scr, scr.svr, err.svr: -- server ends
CT.CHANNEL! kyb.cli, scr.cli, err.cli: -- client ends
SEQ
  ... allocate channels ...
  -- fork off handling processes
  FORK keyboard.handler (kyb.cli, ...)
  FORK screen.handler (scr.svr, ...)
  FORK error.handler (err.svr, ...)
  -- set top level parameters
  vm.ctl[request]  ! set.param.chan; 0; kyb.scr
  vm.ctl[response] ? CASE ok
  vm.ctl[request]  ! set.param.chan; 1; scr.cli
  vm.ctl[response] ? CASE ok
```

```
vm.ctl[request]   ! set.param.chan; 2; err.cli
vm.ctl[response]  ? CASE ok
```

Having provided the top level parameters we can begin executing instructions from the bytecode. We want to trace the entire execution of the program thus we use the `step` request on the VM control channel bundle. If we did not require information about the execution of every instruction then we could use the `run` request. The following code snippet demonstrates the `step` request, which returns two responses, one on instruction decode and one on dispatch:

```
IPTR iptr:
ADDR wptr:
INT op, arg:
SEQ
  vm.ctl[request]   ! step
  vm.ctl[response]  ? CASE decoded; iptr; op; arg
  -- decoded instruction "op", with argument "arg"
  --   new instruction pointer: iptr
  vm.ctl[response]  ? CASE dispatched; iptr; wptr
  -- dispatched instruction
  --   new instruction pointer: iptr
  --   new workspace pointer: wptr
```

6.3. Tracking Processes

We want to create a higher level view of the program execution than the raw instruction stream as it executes. Ideally we want to know the present line of execution of all the processes. To do this we need to track the executing processes and attribute instruction execution to them. We need to monitor *start process* and *end process* instructions, additionally we must track any instruction that alters the workspace pointer (stack) such as *call* and *adjust workspace pointer*.

The following code snippet converts the decoding and dispatching of instructions into a stream of `executing`, `start.process`, `end.process` and `rename.process` tagged protocol messages. Each message carries the process workspace, which acts as the process identifier. `rename.process` messages indicate a change of process identifier, or a shift in workspace pointer.

```
ADDR id:
IPTR iptr:
WHILE TRUE
  ADDR new.id:
  SEQ
    -- report present process position
    out ! executing; id; iptr
    -- start process and end process
    vm.ctl[request]   ! step
    vm.ctl[response]  ? CASE decoded; iptr; op; arg
    IF
      op = INS.OPR
        VM.STATE state:
        SEQ
          vm.ctl[request]   ! get.state
          vm.ctl[response]  ? CASE state; state
          CASE arg
            INS.STARTP
```

```
                    -- process workspace is on the operand stack
                    out ! start.process; state[stack][0]
              INS.ENDP
                    out ! end.process; id
              ELSE
                    SKIP
        TRUE
              SKIP
      -- workspace pointer altering operations
      vm.ctl[response] ? CASE dispatched; iptr; new.id
      IF
        (op = INS.ADJW) OR (op = INS.CALL)
              out ! rename.process; id; new.id
        (op = INS.OPR) AND (arg = INS.RET)
              out ! rename.process; id; new.id
        TRUE
              SKIP
      -- update workspace pointer
      id := new.id
```

6.4. Visualisation

The bytecode decoder interface detailed in section 4.4 is used to look up the present source position of each process. This allows us to generate output on the current source file and line being executed. Visualising this information as a graph of active processes we can see the executing program, although the specifics of this visualisation is an area for future work (see section 8). Call graphs can be generated by monitoring *call* and *return* instructions, and by recording the process which executes a *start process* instruction as the parent of the process started, we can see the program structure.

Adding to the graph of active processes, we also build a graph of relationships between processes. Whenever a channel communication instruction is executed we record the channel pointer in a channel table. Processes which access the same channel become related; edges can be drawn between them in the graph of process nodes. On *adjust workspace pointer* instructions, and other memory deallocation instructions channel words which fall out of scope are deleted from the channel table. This information can be used to visualise the relationships between the active processes, either in real-time as communication occurs, or statically as a cumulative image recorded over time.

Figure 1 applies these techniques to generate an idealised visualisation of the *commstime.occ* example from the KRoC compiler distribution.

7. Conclusions

We have created a method for occam-π programs to inspect and intercede on the execution of other occam-π programs. This access is driven through a channel based interface; which permits reasoning about programs which use it, in-line with that applicable to a standard occam-π program. Using this new interface it is possible to develop debugging tools for occam-π using occam-π, which allows us to use a full parallel programming environment in order to tackle the challenges we face developing such tools.

By constraining the probe effect caused by non-determinism related to the use of time in programs, we in turn permit cyclic debugging: the process of repeated execution of a program with slight modifications to its composition of inputs in an attempt to locate and fix an error. Additionally, we expose as much memory typing information as we can in order to help build complete debugging tools.

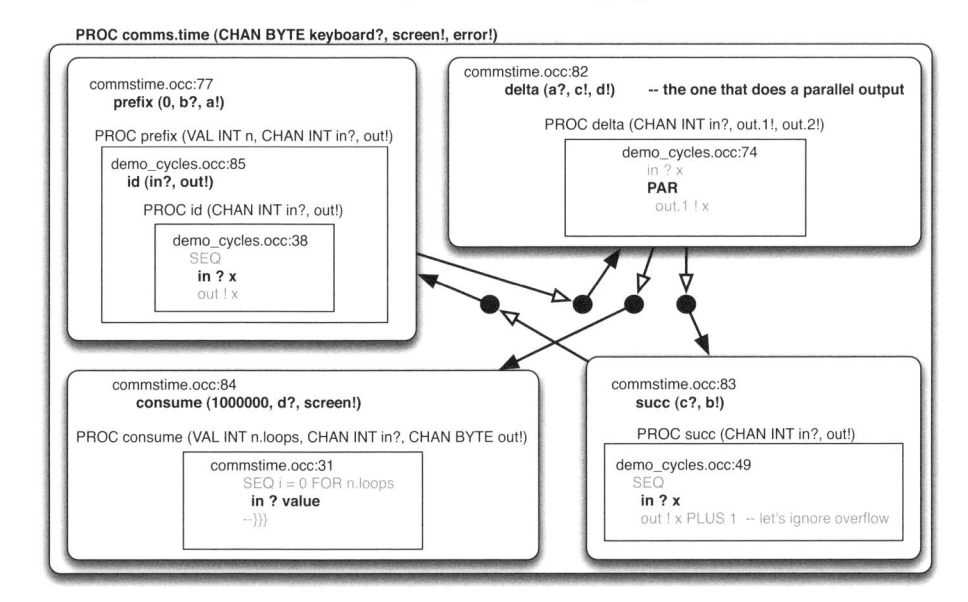

```
PROC comms.time (CHAN BYTE keyboard?, screen!, error!)
┌─────────────────────────────────────────────────────────────────────────────────────┐
│  ╭──────────────────────────────────────╮   ╭──────────────────────────────────────╮ │
│  │ commstime.occ:77                     │   │ commstime.occ:82                     │ │
│  │    prefix (0, b?, a!)                │   │    delta (a?, c!, d!)  -- the one that does a parallel output │
│  │ ┌────────────────────────────────┐   │   │ PROC delta (CHAN INT in?, out.1!, out.2!) │
│  │ │ PROC prefix (VAL INT n, CHAN INT in?, out!) │   │ ┌────────────────────────────────┐ │
│  │ │ demo_cycles.occ:85             │   │   │ │ demo_cycles.occ:74             │ │
│  │ │    id (in?, out!)              │   │   │ │    in ? x                      │ │
│  │ │ PROC id (CHAN INT in?, out!)   │   │   │ │    PAR                         │ │
│  │ │ ┌──────────────────────────┐   │   │   │ │    out.1 ! x                   │ │
│  │ │ │ demo_cycles.occ:38       │   │   │   │ └────────────────────────────────┘ │
│  │ │ │    SEQ                   │   │   │   │                                    │ │
│  │ │ │    in ? x                │   │   │   │                                    │ │
│  │ │ │    out ! x               │   │   │   │                                    │ │
│  │ │ └──────────────────────────┘   │   │   │                                    │ │
│  ╰──────────────────────────────────────╯   ╰──────────────────────────────────────╯ │
│                                                                                     │
│  ╭──────────────────────────────────────╮   ╭──────────────────────────────────────╮ │
│  │ commstime.occ:84                     │   │ commstime.occ:83                     │ │
│  │    consume (1000000, d?, screen!)    │   │    succ (c?, b!)                     │ │
│  │ PROC consume (VAL INT n.loops, CHAN INT in?, CHAN BYTE out!) │ PROC succ (CHAN INT in?, out!) │
│  │ ┌────────────────────────────────┐   │   │ ┌────────────────────────────────┐ │
│  │ │ commstime.occ:31               │   │   │ │ demo_cycles.occ:49             │ │
│  │ │    SEQ i = 0 FOR n.loops       │   │   │ │    SEQ                         │ │
│  │ │    in ? value                  │   │   │ │    in ? x                      │ │
│  │ │    --}}}                       │   │   │ │    out ! x PLUS 1  -- let's ignore overflow │
│  │ └────────────────────────────────┘   │   │ └────────────────────────────────┘ │
│  ╰──────────────────────────────────────╯   ╰──────────────────────────────────────╯ │
└─────────────────────────────────────────────────────────────────────────────────────┘
```

Figure 1. Idealised visualisation of commstime.occ. Curved boxes represent processes. Squared boxes represent calls in the call stack. Bold text is the present line being executed, as taken from the source file. Dots are channels, with arrowed lines representing communication requests. Hollow arrows are previously executed input or output, and filled arrows are active blocked requests.

Overall, we hope that the work presented in this paper will form the basis for developing visualising and debugging facilities for the occam-π tool-chain.

8. Future Work

Virtual machine sub-instances presently run synchronously to their parents, a logical enhancement to our work is the multiplex execution of virtual machine instances. Further to this, by applying algorithms developed for the KRoC runtime support library, CCSP, we expect to be able to extend the TVM to execute multiple instances concurrently on multi-core shared-memory hardware. These concurrently executing instances will be able to communicate, wait and synchronise on each other safely while maximising use of any underlying hardware parallelism.

Given the powerful underlying tool for inspecting the behaviour of running parallel programs we have developed, we would ideally like to make this power available in an easy to interpret graphical form. Exton identified four main areas of importance in the visualisation of program execution: abstraction, emphasis, representation and navigation [17]. Being able to provide a good abstraction level for observation of run-time activity, and the ability to adjust the level in tune with the user's expertise is critical to sensibly observing executing networks of processes. The design of a suitable visual model to represent the state of a running occam-π program is an as yet unexplored area, and may potentially provide useful insight to inform related work in visual design of parallel programs.

Dynamicism of the network topology and processes provides additional complexity to the debugging process beyond that encountered with occam 2 running on the Transputer. Dijkstra [18] states that "our powers to visualise processes evolving in time are relatively poorly developed", a statement that encapsulates some of the difficulty encountered when trying to reason about systems using dynamic process creation and mobile channels. Working to allow end-users to explore the execution of their program and interact with its run-time behaviour to diagnose problems and mistakes would begin to provide some semblance of modern, user-friendly debugging to the occam-π tool-chain.

Acknowledgements

This work was funded by EPSRC grant EP/D061822/1. This work would not be possible if not for the past work and input of other researchers at the University of Kent and elsewhere, in particular, Christian Jacobsen, Matt Jadud and Damian Dimmich. We also thank the anonymous reviewers for comments which helped us improve the presentation of this paper.

References

[1] Charles E. McDowell and David P. Helmbold. Debugging concurrent programs. *ACM Comput. Surv.*, 21(4):593–622, 1989.
[2] David C. Wood and Frederick R. M. Barnes. Post-Mortem Debugging in KRoC. In Peter H. Welch and Andrè W. P. Bakkers, editors, *Communicating Process Architectures 2000*, pages 179–192, sep 2000.
[3] Christian L. Jacobsen and Matthew C. Jadud. The Transterpreter: A Transputer Interpreter. In Dr. Ian R. East, Prof David Duce, Dr Mark Green, Jeremy M. R. Martin, and Prof. Peter H. Welch, editors, *Communicating Process Architectures 2004*, volume 62 of *Concurrent Systems Engineering Series*, pages 99–106. IOS Press, Amsterdam, September 2004.
[4] *IEEE Standard 1149.1-1990 Test Access Port and Boundary-Scan Architecture-Description*. IEEE, 1990.
[5] Susan Stepney. GRAIL: Graphical representation of activity, interconnection and loading. In Traian Muntean, editor, *7th Technical meeting of the occam User Group,Grenoble, France*. IOS Amsterdam, 1987.
[6] Susan Stepney. Understanding multi-transputer execution. In *IT UK 88, University College Swansea, UK*, 1988.
[7] Wentong Cai and Stephen J. Turner. An approach to the run-time monitoring of parallel programs. *The Computer Journal*, 37(4):333–345, March 1994.
[8] John May and Francine Berman. Panorama: a portable, extensible parallel debugger. In *PADD '93: Proceedings of the 1993 ACM/ONR workshop on Parallel and distributed debugging*, pages 96–106, New York, NY, USA, 1993. ACM Press.
[9] C. O'Neil. The TDS occam 2 debugging system. In Traian Muntean, editor, *OUG-7: Parallel Programming of Transputer Based Machines*, pages 9–14, sep 1987.
[10] K. Zhang and G. Marwaha. Visputer–A Graphical Visualization Tool for Parallel Programming. *The Computer Journal*, 38(8):658–669, 1995.
[11] Jonathan Simpson, Christian L. Jacobsen, and Matthew C. Jadud. Mobile Robot Control: The Subsumption Architecture and occam-pi. In Frederick R. M. Barnes, Jon M. Kerridge, and Peter H. Welch, editors, *Communicating Process Architectures 2006*, pages 225–236, Amsterdam, The Netherlands, September 2006. IOS Press.
[12] Surveyor Corporation SRV-1: http://www.surveyor.com/blackfin/.
[13] F.R.M. Barnes and P.H. Welch. Mobile Data, Dynamic Allocation and Zero Aliasing: an occam Experiment. In Alan Chalmers, Majid Mirmehdi, and Henk Muller, editors, *Communicating Process Architectures 2001*, volume 59 of *Concurrent Systems Engineering*, pages 243–264, Amsterdam, The Netherlands, September 2001. WoTUG, IOS Press. ISBN: 1-58603-202-X.
[14] F.R.M. Barnes and P.H. Welch. Prioritised Dynamic Communicating Processes: Part I. In James Pascoe, Peter Welch, Roger Loader, and Vaidy Sunderam, editors, *Communicating Process Architectures 2002*, WoTUG-25, Concurrent Systems Engineering, pages 331–361, IOS Press, Amsterdam, The Netherlands, September 2002. ISBN: 1-58603-268-2.
[15] A. Cimitile, U. De Carlini, and U. Villano. Replay-based debugging of occam programs. *Software Testings, Verfication and Reliability*, 3(2):83–100, 1993.
[16] T.J. Leblanc and J.M. Mellor-Crummey. Debugging parallel programs with instant replay. *Computers, IEEE Transactions on*, C-36(4):471–482, April 1987.

[17] Chris Exton and Michael Kölling. Concurrency, objects and visualisation. In *ACSE '00: Proceedings of the Australasian conference on Computing education*, pages 109–115, New York, NY, USA, 2000. ACM Press.

[18] Edsger W. Dijkstra. Letters to the editor: go to statement considered harmful. *Commun. ACM*, 11(3):147–148, 1968.

Process-Oriented Collective Operations

John Markus BJØRNDALEN [a,*] and Adam T. SAMPSON [b]
[a] *Department of Computer Science, University of Tromsø*
[b] *Computing Laboratory, University of Kent*

Abstract. Distributing process-oriented programs across a cluster of machines requires careful attention to the effects of network latency. The MPI standard, widely used for cluster computation, defines a number of *collective operations*: efficient, reusable algorithms for performing operations among a group of machines in the cluster. In this paper, we describe our techniques for implementing MPI communication patterns in process-oriented languages, and how we have used them to implement collective operations in PyCSP and occam-π on top of an asynchronous messaging framework. We show how to make use of collective operations in distributed process-oriented applications. We also show how the process-oriented model can be used to increase concurrency in existing collective operation algorithms.

Keywords. CSP, clusters, concurrency, MPI, occam-π, parallel, PyCSP, Python.

Introduction

Distributing process-oriented programs across a cluster of machines requires careful attention to the effects of network latency. This is especially true for operations that involve a group of processes across the network, such as barrier synchronisation. Centralised client-server implementations are straightforward to implement and often scale acceptably for multicore machines, but have the potential to cause bottlenecks when communication latency is introduced.

The MPI standard, widely used for cluster computation, provides a number of *collective operations*: predefined operations that involve groups of processes in a cluster. Operations such as barriers, broadcasting and reductions are already used in distributed process-oriented programs.

Implementations of MPI are responsible for providing efficient implementations of each operation using algorithms and communication mechanisms suited to the particular cluster technology for which they are designed. We examine one of these implementations, OpenMPI [1,2], to study how these algorithms can be expressed using process-oriented systems such as PyCSP [3,4] and occam-π [5].

We are studying how these algorithms can be reused in distributed process-oriented programs. We show how process-oriented implementations of these algorithms can express a higher level of parallelism than is present in OpenMPI. The Python language makes it easy for us to express these concurrent algorithms in a reusable and concise way.

For CSP-based programs executing on clusters, we believe that effective group operations can be implemented efficiently using a hybrid approach: existing process-oriented communication techniques within each node, and MPI-inspired collective operations between nodes in the cluster. Expressing collective operations using process-oriented techniques allows effective integration of the two approaches.

---

[*]Corresponding Author: *John Markus Bjørndalen, Department of Computer Science, University of Tromsø, N-9037 Tromsø, Norway.* Tel.: +47 7764 5252; Fax: +47 7764 4580; E-mail: `jmb@cs.uit.no`.

To improve the efficiency of communication between nodes, we have created a lightweight asynchronous messaging framework for occam-π and PyCSP.

This paper is organised as follows: In section 1 we provide a short introduction to PyCSP. In section 2 we describe our implementations of several of the OpenMPI collective operations using Python and PyCSP, and show some of our improvements to them. In section 3, we describe our implementations of lightweight asynchronous messaging. In section 4, we describe how our work can be applied in a CSP-based framework for configurable group operations. This work is a consequence of our experiences with CoMPI [6,7], a runtime configuration system for LAM-MPI [8], and the PATHS [9,10,11] system, used with a tuple-space system to achieve better performance than LAM-MPI for collective operations on clusters of multiprocessor machines. A configuration system like this can help us tune both an application and the collective operations it uses for improved performance on a given cluster. Part of the motivation for this paper is to explore initial possibilities for the configuration project.

1. A Quick Introduction to PyCSP

This section provides an overview of the facilities offered by PyCSP, a library to support process-oriented programming in Python. A more detailed description can be found in "PyCSP – Communicating Sequential Processes for Python" [3], which describes an earlier version of PyCSP; the examples in this paper make use of features introduced in PyCSP 0.3, such as network channels. The source code for PyCSP can be found on the PyCSP homepage [4].

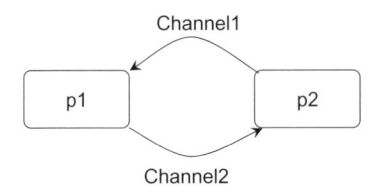

Figure 1 Basic PyCSP process network, with two processes: P1 and P2. The processes are connected and communicate over two channels: Channel1 and Channel2.

As in all process-oriented systems, the central abstractions offered by PyCSP are the *process* and the *channel*. A PyCSP program is typically composed of several processes communicating by sending messages over channels. Figure 1 shows an example with two processes, P1 and P2, communicating using channels Channel1 and Channel2.

Listing 1 shows a complete PyCSP implementation of the process network in Figure 1. A PyCSP process is defined by writing a function prefixed with the `@process` decorator. This ensures that a PyCSP `Process` object is created with the provided parameters when the function is called.

PyCSP processes are executed using the `Sequence` and `Parallel` constructs, as shown in the listing. `Parallel` starts all the provided processes in parallel, then waits for them all to finish; `Sequence` executes them in sequence, waiting for each process to finish before starting the next.

Channels are created by instantiating one of the channel classes. This example uses `One2OneChannel`, the simplest type of channel, which connects exactly two processes without buffering. The methods `write` and `read` are used to send and receive messages across the channel. Other channel classes provide sharing, buffering, and similar features.

PyCSP provides support for network communication using *named channels*. A channel is registered with the networking library by calling `registerNamedChannel(chan, name)`,

```
import time
from pycsp import *

@process
def P1(cin, cout):
    while True:
        v = cin()
        print "P1, read from input channel:", v
        time.sleep(1)
        cout(v)

@process
def P2(cin, cout):
    i = 0
    while True:
        cout(i)
        v = cin()
        print "P2, read from input channel:", v
        i += 1

chan1 = One2OneChannel()
chan2 = One2OneChannel()

Parallel(P1(chan1.read, chan2.write),
         P2(chan2.read, chan1.write))
```

Listing 1: Complete PyCSP program with two processes

which creates an entry in a nameserver. A remote process can then get a proxy to the channel by calling `getNamedChannel(name)`, which may be used in the same ways as a local channel object.

2. Collective Operations

The MPI collective operations are used to execute operations across a group of processes, either to synchronise the processes, exchange data, or execute a common operation on data provided by one or more of the participating processes. Examples of the provided operations are Barrier, Broadcast and Reduce.

OpenMPI uses several algorithms to implement each collective operation. When a program calls any of the operations, OpenMPI chooses which implementation to use based on parameters such as the size of the group (the number of MPI processes involved in that call), and the sizes of the messages involved. Additionally, parameters may be specified when starting MPI programs to influence the choices and select alternative implementations.

Note that for this paper, we are restricting ourselves to discussing the *basic* component of OpenMPI to limit the complexity of the discussions and code examples. The basic component includes well-known algorithms for implementing reasonably efficient group operations in clusters. We are also restricting ourselves to a selection of the group operations, focusing on those where we identify patterns that can be reused and the most common operations.

2.1. MPI Terminology

In MPI, the *size* of a group of processes is the number of processes within that group. Within an MPI program, a group is represented using a *communicator*. The communicator can be thought of as an object that keeps track of information such as the size of the group, which algorithms to use for various operations, and how to reach each of the other processes within the group.

Within a given group, each process has a *rank*, which is a unique ID identifying that process in the group. When a process wants to send a message to another process in the group, the sender invokes the *send* operation on the communicator representing the group, and specifies the rank of the receiving process.

The MPI standard defines a number of operations for sending and receiving messages, including variations such as asynchronous send and receive operations. These low-level operations are used to construct the higher-level collective operations.

Group operations, or collective operations, are called with the same parameters for all members of the group. The ranks of the processes are used to identify which processes is responsible for what part of the algorithm. For example, in the Broadcast operation, one process broadcasts a value to all the other processes. The sender is called the *root* process, and is identified by all participating processes by specifying which rank is the root process in the parameters to the operation.

2.2. Broadcast

In the MPI Broadcast operation, a *root* process distributes the same piece of data to all the other processes. For simplicity, we will assume that the root process is rank 0. OpenMPI handles the cases where root is not 0 by re-mapping the ranks of each participating process (to a virtual rank). Adding this to the implementations would not significantly alter the algorithms.

Listing 18 (in the Appendix) shows the *linear* implementation of Broadcast from OpenMPI 1.2.5. The linear algorithm is usually used for very small groups, where having the other processes communicating directly with the root node will be more efficient than more complex algorithms. In the algorithm, the root process calls `isend_init` (corresponding to `MPI_Isend`) to queue a message for each of the other processes in the group. It then blocks, waiting for all the send operations to complete, using the `wait_all` call.

Starting a number of requests then waiting for them all to complete is a common technique used in MPI programs to perform multiple send or receive operations in parallel. This approach is equivalent to the CSP `Parallel` construct, which starts a set of parallel processes then waits for them to complete; the MPI pattern can be directly translated into a `Parallel`.

Listing 2 shows our PyCSP implementation of the same algorithm. The code uses Python list expressions to create a list of `pass_on` processes which are responsible for passing on the message to processes with rank 1 to `groupsize - 1`. The list of processes are then run in parallel using the `Parallel` construct[1].

```
@process
def pass_on(comm, targ_rank, data):
    comm.send_to(targ_rank, data)

def Broadcast_linear(comm, data):
    ''linear broadcast''
    if comm.rank == 0:
        Parallel(*[pass_on(comm, targ, data) \
                   for targ in range(1, comm.groupsize)])
    else:
        data = comm.recv()
    return data
```

Listing 2: Broadcast with linear algorithm

The code to call `pass_on` in parallel for several destinations will be used in several collective operations; we have separated it out into a helper function, `pass_on_multiple`, as shown in listing 3.

---

[1] The * prefix in Python allows a list to be used directly as parameters to a function.

```
@process
def pass_on(comm, targ_rank, data):
    comm.send_to(targ_rank, data)

def pass_on_multiple(comm, targ_ranks, data):
    Parallel(*[pass_on(comm, targ, data) for targ in targ_ranks])

def Broadcast_linear(comm, data):
    ''linear broadcast''
    if comm.rank == 0:
        pass_on_multiple(comm, range(1, comm.groupsize), data)
    else:
        data = comm.recv()
    return data
```

Listing 3: Broadcast with linear algorithm. Separating out the Parallel construct.

One of the interesting lessons from this is that we can get some of the same benefits as asynchronous operations (communicating with multiple peers in parallel), without complicating the API by adding separate asynchronous versions of the operations.

2.2.1. Nonlinear Broadcast

For larger groups, the Broadcast operation uses a hierarchical approach with a binomial broadcast algorithm. The general idea is to arrange the ranks into a tree structure, where each internal node is responsible for forwarding a message to its children. The root process is the root of the tree.

```
def Broadcast_binomial(comm, data):
    ''linear broadcast''
    # binomial broadcast
    if comm.rank != 0:
        data = comm.recv()  # read from parent in tree
    pass_on_multiple(comm, binom_get_children(comm.rank, comm.groupsize),
                     data)
    return data
```

Listing 4: Broadcast with non-linear algorithm

To perform a binomial broadcast, we use the helper function `binom_get_children` to get the ranks of our children, and then call `pass_on_multiple` to pass on the data to all our children in parallel. The OpenMPI implementation (Listing 19, in the appendix) interleaves the code to compute the tree structure with the communication code; the Python implementation is cleaner and more succinct. In addition, separating the tree-structuring code out into a function makes it possible to reuse it elsewhere, or to replace it with a different function to structure the tree in a different way, specified using the configuration system described in section 4.

2.3. Reduce

The Reduce operation combines the data produced by all the participating processes using a provided function on two operands; in functional terms, it performs a fold across all the results. Depending on how the reduction function is defined, the order in which the reductions occur can make a difference to the result; even a simple operation such as addition can produce different results for different orderings if it is acting on floating-point data. The simplest approach is for the root process to receive the results in order, performing a reduction step after each one; this is shown in listing 5. This approach will always produce a consistent result regardless of the function used.

```
def Reduce_linear(comm, data, func):
    if comm.rank == 0:
        # Start with the highest ranks, reducing towards the root.
        v = comm.recv_from(comm.groupsize - 1)
        for i in range(comm.groupsize - 2, 0, -1):
            t = comm.recv_from(i)
            v = func(t, v)
        v = func(data, v)
        return v
    else:
        # Non-root processes pass the values directly to the root
        comm.send_to(0, data)
        return data
```

Listing 5: Linear reduce

If the reduction function permits it, we can significantly increase the efficiency of the algorithm for larger groups by using a binomial tree. An equivalent of the OpenMPI binomial reduction is shown in listing 6, using the helper function `get_reduce_peers_binom` to compute the tree structure.

Again, the Python implementation is much more concise than OpenMPI's version; the OpenMPI implementation, which manages buffers at this level in the code, is 199 lines of C. The PyCSP collective operations avoid this duplication by managing buffers in the message transport layer.

```
def Reduce_binom1(comm, data, func):
    recv_from, send_to = get_reduce_peers_binom(comm)
    v = data
    for peer in recv_from:
        t = comm.recv_from(peer)
        v = func(v, t)
    for peer in send_to:
        comm.send_to(peer, v)
    return v
```

Listing 6: Logarithmic reduce similar to OpenMPI

The code in listing 6 allows processes executing on different nodes to communicate and reduce data concurrently, but expresses only a limited degree of parallelism.

If a process early in the list is late sending its response message, it will block reception of messages from processes later in the list. To reduce the effects of this, the message transport layer may buffer incoming messages. However, if the messages are large, buffer size limitations may cause messages to be blocked anyway.

A better solution is to express the desired parallelism explicitly, receiving from all processes in the list in parallel. With this approach, shown in listing 7, the message transport layer has enough information to schedule message reception in an appropriate order.

```
def Reduce_binom2(comm, data, func):
    # If communication overhead and timing is the issue, this one might
    # help at the potential cost of delaying the reduction until all
    # messages have arrived.
    recv_from, send_to = get_reduce_peers_binom(comm)
    v = data
    res = recv_from_multiple(comm, recv_from)
    for r in res:
        v = func(v, r)
    for peer in send_to:
        comm.send_to(peer, v)
    return v
```

Listing 7: Reduce using parallel receive

The drawback of that method is that we delay execution of the reduce function until we have all the messages; we cannot use the time that we spend waiting for messages to perform reductions on the messages we have received so far. We must also keep all the received messages in memory until the reduction is performed.

Listing 8 shows an alternative implementation that creates a chain of `_reduce_1` processes using the "recursive parallel" process-oriented pattern. Each process receives a message from its designated peer rank and another from the previous process in the chain, reduces the two, then passes the result on to the next process in the chain. This allows messages from all the peers to be received in parallel, while the reductions happen in as efficient a manner as possible. The code demonstrates another feature of the PyCSP `Parallel` construct: it returns a list of the return values from the PyCSP processes that were executing within that construct.

```
def Reduce_binom3(comm, data, func):
    """Recursive processes, calling 'func' on the current and next in the
    chain. Runs the recv operations in parallel, but 'func' will naturally
    be sequential since it depends on the result from the next stage in
    the chain.
    """
    @process
    def _reduce_1(comm, func, peers):
        if len(peers) > 1:
            res = Parallel(recv_from_other(comm, peers[0]),
                           _reduce_1(comm, func, peers[1:]))
            return func(res[0], res[1])
        else:
            return comm.recv_from(peers[0])

    recv_from, send_to = get_reduce_peers_binom(comm)
    v = data
    if len(recv_from) > 0:
        res = Sequence(_reduce_1(comm, func, recv_from))
        v = func(v, res[0])
    for peer in send_to:
        comm.send_to(peer, v)
    return v
```

Listing 8: Reduce using recursion to create a chain of reduce processes in each node

A further optimisation would be to execute the reduce operation out-of-order, as the messages arrive. We have not implemented this in time for this paper.

2.4. Barrier

```
def Barrier_non_par(comm):
    # Non-parallel version
    if comm.rank > 0:
        # non-root processes send to root and receive ack when all
        # have entered
        comm.send_to(0, None)
        comm.recv_from(0)
    else:
        # Root process collects from all children and broadcasts back
        # response when all have entered
        for i in range(1, comm.groupsize):
            comm.recv_from(i)
        for i in range(1, comm.groupsize):
            comm.send_to(i, None)
```

Listing 9: Linear sequential barrier

The Barrier operation stops processes from leaving the barrier until all processes in the group have entered. When the final process has entered, it releases all the other processes which can then leave the barrier.

Listing 9 shows an implementation of barrier that is equivalent to the OpenMPI linear barrier. Non-root processes send to the root process, and block waiting for a return message from the root process. The root process first receives from all the other processes (the enter phase), and then sends back a reply (the release phase).

An improvement of the above is to do all the send and receive operations in parallel, as in listing 10, but the advantage will probably be limited as the root process is still likely to be a communication bottleneck.

```
def Barrier_linear(comm):
    if comm.rank > 0:
        # non-root processes send and receive a token from the root process
        comm.send_to(0, None)
        comm.recv_from(0)
    else:
        # root process recvs from all other processes, then sends reply
        recv_from_multiple(comm, range(1, comm.groupsize))
        pass_on_multiple(comm, range(1, comm.groupsize), None)
```

Listing 10: Linear parallel barrier

OpenMPI provides a logarithmic barrier for larger groups. A PyCSP implementation based on the OpenMPI implementation is shown in listing 11.

The processes in the group are arranged in a tree. Internal nodes in the tree first receive messages from all of their children telling them that the children have all entered the barrier. The `binom_get_children` function returns the list of children for any node in the tree. For leaf nodes, the list will be empty.

After receiving all the messages, the internal non-root node sends and then receives from its parent. When it receives the "leave" message from the parent, it knows that the algorithm is in the release phase and can pass this on to the children. The root node only needs to receive from all its children in the tree to know that all processes have entered the barrier, and can subsequently send the release messages allowing the child processes to leave the barrier.

Compared to OpenMPI, the main difference is that we are doing parallel receives and sends when communicating with the children in the tree while OpenMPI receives and sends the messages in sequence. Given the message sizes and number of child processes for a typical barrier operation, this may not make as much of a difference as for other operations.

```
def Barrier_binom(comm):
    # binomial/log barrier
    # NB: OpenMPI uses root = rank0 and a similar tree to broadcast
    children = binom_get_children(comm.rank, comm.groupsize)

    recv_from_multiple(comm, children)
    if comm.rank != 0:
        # non-root need to communicate with their parents
        # (pass up, recv down)
        parent = binom_get_parent(comm.rank, comm.groupsize)
        comm.send_to(parent, None)
        comm.recv_from(parent)
    # send to children
    pass_on_multiple(comm, children, None)
```

Listing 11: Logarithmic barrier

2.5. Allreduce

```
def Allreduce(comm, data, func=lambda x,y: x+y):
    data = Reduce(comm, data, func)
    data = Broadcast(comm, data)
    return data
```

Listing 12: Allreduce

The Allreduce operation is a combination of the Reduce and Broadcast operations: it first calls Reduce to collect a value in the root process, and then calls Broadcast to distribute the collected value from the root process back out to all the other processes.

Listing 12 shows a PyCSP implementation that is equivalent to the OpenMPI implementation. The main weakness with this implementation is that it has two distinct phases, and a process that is finished with the Reduce phase is blocked waiting for the broadcasted value.

It is possible to merge the two phases by essentially running a number of concurrent Reduce operations rooted in each participating process or in subgroups within the larger group; this can decrease the time taken for the operation at the cost of increasing the number of messages sent. A potential problem is that for some functions and data types, we might end up with different results in the different processes, as they are performing Reduce operations with the intermediate results in different orders; this does not occur in the existing implementation, because the result is only computed by the root process.

2.6. Gather

The Gather operation gathers values from each process in the group and collects them in the root process. The PyCSP version returns a list of all the received values to the root process, and None to the non-root processes. OpenMPI uses an algorithm similar to listing 13, calling receive for each of the other processes in sequence.

```
def Gather_seq(comm, data):
    # No log-version of this one, as in OpenMPI
    if comm.rank != 0:
        # Non-root processes send to root
        comm.send_to(0, data)
        return None
    else:
        vals = [data]
        for i in range(1, comm.groupsize):
            vals.append(comm.recv_from(i))
        return vals
```

Listing 13: Sequential gather

For messages that are large enough, an implementation that throttles communication may stall unexpected messages from a sender that is ready while the reader is waiting for data from a sender that is not ready.

As with `pass_on_multiple`, we provide a `recv_from_multiple` helper function to receive data from several other ranks in parallel. This is similar to doing a sequence of `MPI_Irecv` operations followed by an `MPI_Waitall` to wait for them to complete.

The function returns the received messages in a list, and optionally inserts them into a provided dictionary, with the rank of the sending process used as a key. The dictionary representation is convenient for functions that want to reorder the results; see the final line of listing 14 for an example.

```
def Gather_par(comm, data):
    # Using a PAR to receive messages from the other processes
    if comm.rank != 0:
        # Non-root processes send to root
        comm.send_to(0, data)
        return None
    else:
        # Uses a parallel recv. This ensures that we don't stall a send
        # while waiting for a send from a slower process earlier in
        # the list.
        # Using a dict to return the values ensures that we can return a
        # correctly sorted list of the return values (sort on keys/sender
        # ranks even if we use a different order on the ranks to recv from).
        res = { 0 : data}
        recv_from_multiple(comm, range(1, comm.groupsize), res_dict = res)
        return [res[k] for k in range(comm.groupsize)]
```

Listing 14: Parallel gather

2.7. Allgather

```
def Allgather(comm, data):
    vals = Gather(comm, data)          # rank 0 gets all of them
    vals = Broadcast(comm, vals)       # rank 0 will forward all values
```

Listing 15: Allgather

The Allgather operation is a simple combination of the Gather and Broadcast operations. The root process first gathers data from all the other processes, then distributes all the gathered data to all the other processes.

A potential problem with this implementation is that some of the nodes may finish the gather phase early and sit idle waiting for the broadcast phase. To increase parallelism, we may try to use a an implementation that resembles the Alltoall operation below. Whether that will improve the performance of the operation will have to be experimentally validated.

2.8. Scatter

```
def Scatter_seq(comm, data):
    # Sequential implementation, as in OpenMPI basic
    if comm.rank != 0:
        return comm.recv_from(0)
    else:
        for i in range(1, comm.groupsize):
            comm.send_to(i, data[i])
        return data[0]

def Scatter_par(comm, data):
    if comm.rank != 0:
        return comm.recv_from(0)
    else:
        Parallel(*[pass_on(comm, i, data[i])
                   for i in range(1, comm.groupsize)])
        return data[0]
```

Listing 16: Scatter

The scatter operation is similar to the broadcast operation in that the root process sends data to each of the other processes. The difference is that scatter takes a sequence of values, partitions that up into `size` segments and sends one part to each of the processes. Upon completion, each process will have an entry corresponding to its own rank.

The Python implementation of scatter takes a list and sends entry `i` to the process with rank `i`. `Scatter_seq` in listing 16 is equivalent to OpenMPI's scatter implementation, sending a message to each of the other ranks in sequence.

The problem with this approach is that the runtime system has no idea about the next message in the sequence until the current send operation completes. Sending them in parallel allows the message transport layer to consider all of the messages for scheduling and sending when possible. `Scatter_par` is an alternative implementation which explicitly sends all messages in parallel.

2.9. Alltoall

```
def scatter_send(comm, data, targets):
    Parallel(*[pass_on(comm, i, data[i]) for i in targets])

@process
def scatter_send_p(comm, data, targets):
    scatter_send(comm, data, targets)

def Alltoall(comm, data):
    other = [i for i in range(comm.groupsize) if i != comm.rank]

    # OpenMPI version posts all send/recvs and then waits on all of them
    # This can be trivially implemented using one or more PARs
    res = { comm.rank : data[comm.rank]}
    Parallel(scatter_send_p(comm, data, other),
             recv_from_multiple_p(comm, other, res_dict=res))
    return [res[k] for k in sorted(res.keys())]
```

Listing 17: Alltoall

The Alltoall operation is similar to all the processes in the group taking turns being the root process in a Scatter operation. When the Alltoall operation completes, each process in the group has received a value from each of the other processes. The process with rank `i` will have a list of the `i`th entry that each process contributed. One way of viewing this operation is that it transposes the matrix formed by combining the lists contributed by each process.

The implementation in listing 17 works similarly to the OpenMPI implementation. OpenMPI uses non-blocking sends and receives, posting the receives first and then the sends, then waiting for all the operations to complete.

The PyCSP implementation uses a technique similar to the Scatter function for sending entries, but factors that out as a `scatter_send` function and a `scatter_send_p` process. This simplifies sending in parallel with receiving, and allows for reuse: `scatter_send` is also used in the Scatter operation.

PyCSP supports an alternative method of creating processes, using the Process class directly with a regular function. Using this, we could have written:

`Parallel(Process(scatter_send, comm, data, other),...)`

This avoids needing to define a separate `scatter_send_p` process. However, explicitly creating a process simplifies reuse, and we believe that the syntax in listing 17 is more readable.

3. Implementations

We have implemented collective operations using two process-oriented systems: PyCSP [3] and occam-π [5]. It is straightforward to implement the algorithms we have described using channels on a single host, but collective operations are most useful across clusters of machines. Both systems provide transparent networking facilities – PyCSP using the Pyro [12] remote objects framework, and occam-π using pony [13], an implementation of networked channel bundles.

However, with both Pyro and pony, the communication latency over a CSP channel that has been extended over a network is very high compared to local communication. This is because a channel communication must be acknowledged in order to provide the correct CSP blocking semantics, requiring one or more round trips across the network. However, we do not need full CSP channels for the collective operations we have described; asynchronous delivery of messages is sufficient. We can take advantage of this to remove the need for acknowledgements in our implementations of collective operations.

We have implemented an asynchronous network messaging system for both PyCSP and occam-π. The two implementations are broadly similar but have some important differences in their low-level implementation, since PyCSP maps processes to kernel-level threads, whereas occam-π has its own lightweight process scheduler. MPI implementations are typically single-threaded, so we incur some extra latency from context-switching in both implementations, but this is insignificant compared to the latency inherent in network communication; the smarter communication scheduling permitted by process-oriented implementations should more than make up for the difference, similar to the work we describe in section 4.

3.1. Interface

Our messaging system provides facilities similar to Erlang [14] or MPI's low-level messaging system: a system consists of several *nodes* (equivalent to MPI ranks), each of which has zero or more receiving *ports*. Messages are delivered asynchronously from a node to any port in the system. In process-oriented terms, ports can be considered to be infinitely-buffered shared channels. A receiving process may choose to receive from one port or several; this makes it possible to selectively wait for messages from a particular set of other processes.

We aim to provide a reasonably natural interface for the process-oriented programmer, allowing them to mix asynchronous communications with regular local communications and timeouts in alternation. The programmer uses a client-server interface to interact with a *communicator* process on each node, which supports *send* and *receive* operations. In PyCSP, *receive* simply blocks until an appropriate message is received; in occam-π, *receive* immediately returns a mobile channel down which received messages will be delivered.

3.2. Network Communications

Network communications are implemented using TCP sockets, with one socket used between each pair of communicating nodes, carrying messages for multiple ports in both directions. At the network level, messages are just sequences of bytes with a simple header giving their length and addressing information. It is up to the application to serialise messages for network communication; this can be done simply using Python's `pickle` module, or occam-π's `RETYPES` mechanism.

To discover the network addresses of other nodes, we need a name server mechanism. Nameserver lookups are not performance-critical, so this can most easily be implemented on top of an existing network mechanism such as Pyro (as has been done for PyCSP) or pony; alternatively, a second TCP protocol can be used (as in the occam-π implementation). The nameserver is generally a long-running service that handles multiple applications, so it must cope gracefully with nodes connecting and disconnecting at any time.

Since the protocol is simple, it would be straightforward to make our two implementations interoperate, provided they agree on the contents of the messages being sent. The choice of TCP rather than UDP for asynchronous messaging may seem surprising, but we wanted to be able to use our system over unreliable networks without needing to worry about retransmitting lost packets. In a busy system, TCP piggyback acknowledgements will largely negate the need for TCP-level acknowledgement packets anyway.

3.3. IO Scheduling

Each node has several TCP sockets open to other nodes. In a large cluster, this could potentially be a very large number of sockets; it is necessary to manage communications across the collection of sockets in an efficient manner.

At the lowest level, communication is performed on TCP sockets using system calls to read and write data, which may block if the socket's buffer is full. This is a problem for a system with lightweight processes, since blocking one kernel-level thread could result in multiple lightweight processes being blocked. Sockets may be placed into non-blocking mode, which causes the system calls to return immediately with an error code rather than blocking, but system calls are relatively expensive; we cannot afford to repeat a call until it succeeds.

The approach currently used in occam-π is to use a separate kernel-level thread for each operation that may block [15]. This is rather inefficient, in that we must have one thread running for each blocked operation, and each blocking call requires a context switch to a worker thread, followed by another back again when the call is complete. Furthermore, while PyCSP does not yet make use of lightweight processes, it is still inefficient to have very large numbers of threads performing parallel system calls.

Programs that communicate using large numbers of sockets are generally better written in an event-driven style, using the POSIX `poll` or `select` system calls to efficiently wait for any of a set of non-blocking sockets to become ready [16]. When a socket is ready, the next `read` or `write` call on it will not block. (`poll` is poorly-named; it is simply a more efficient version of `select`, and blocks in the kernel rather than actually polling.)

As with the Haskell I/O multiplexing system [17], we provide an *IO scheduler* process. This process maintains a set of open sockets, and repeatedly calls `poll` across the complete set on the user's behalf. When a socket becomes ready, the IO scheduler wakes up an appropriate user process to perform the IO, by communicating down a channel. It then waits for confirmation that the IO has been completed, and returns to the `poll` loop. The user process may opt to keep the socket in the set (if it expects to perform more IO in the future), or to remove it.

The user may add new sockets to the set at any time. This is achieved using a standard trick [18]: in addition to the sockets, the `poll` also waits on the reading end of a pipe. Writing a character to the pipe will interrupt the `poll`, allowing it to handle the user's request.

All processes on a node share the same IO scheduler. The result is that processes using sockets may be written in a conventional process-oriented manner, with a process to handle each individual socket. Processes may block upon socket IO without needing to worry about accidentally blocking other processes or incurring the overheads of using a separate thread. They may even use alternation to wait for any of a set of sockets to become ready. However, the underlying mechanism used to implement communications is the more efficient event-based model supported by poll, which means that very large numbers of sockets can be handled with minimal performance overheads. We have provided these facilities with no modification of the runtime system or operating system kernel.

Again, we have implemented this for both PyCSP and occam-π. The two implementations are very similar; the only difference is that the PyCSP IO scheduler can perform IO itself on behalf of user processes, in order to reduce the number of kernel-level thread context switches necessary.

3.4. Messaging System

Communications for each node are coordinated by a communicator process, which is itself a group of parallel processes.

For each TCP connection to another node, there is an *inbound* process that handles received messages, and an *outbound* process that handles messages to be sent. Both maintain an internal buffer to avoid blocking communications. The inbound process uses the IO scheduler to wait for the socket to become readable, and the outbound process uses the IO scheduler to wait for it to become writable (when it has messages to send).

For the occam-π implementation, the buffers are simple FIFO arrays of bytes, and the inbound messages are delivered to the switch process in strict order of reception. For PyCSP, the inbound buffer is slightly more complex: it is a queue from which messages may be removed at arbitrary points, so that the receive operation may pull out only the messages it is interested in.

A central *switch* process handles user requests, and messages received from inbound processes. It routes incoming and outgoing messages to the correct process. New TCP connections (and their corresponding inbound and outbound processes) are created when the first message is sent, or when an incoming TCP connection is received, the latter being handled by a separate *listener* process.

4. Configuration and Mapping of Operations and Algorithms

Previous experience with the PATHS [9,10,11] and CoMPI [6] systems has shown that changing the algorithms, mapping and communication patterns can significantly improve the performance of operations compared to the standard implementations in LAM-MPI [19]. Similar experiences have led to efforts such as MagPIe [20] and Compiled Communication [21].

The approach taken by PATHS and CoMPI differs in that the configurations are built at runtime, using a specification which can be retained to help analysis of the system afterwards. This simplifies analysis and comparison of several different configurations as both the specifications and the trace data can be compared and studied. Tools and systems for doing this include STEPS [22], EventSpace [23] and the PATHS visualiser and animation tool for display walls [24].

One of the drawbacks of the current implementations of the PATHS system is that the concurrency facilities available are very low-level: threads, mutexes and condition variables. This is one area where CSP-based languages and systems may provide an improvement compared to the current PATHS implementations.

4.1. PATHS

The PATHS system is based around *wrappers*; a wrapper encapsulates an existing object. A thread can call methods in the original object by having a reference to the wrapper. The wrapper usually forwards the call, but can optionally perform extra computations or modify the behaviour of the object's methods. For example, a wrapper could be used to provide tracing or caching of method calls on the object.

Several wrappers can be stacked on top of each other to create a *path* to the object (similar to a call path or stack trace). This means that several different types of wrappers can be combined to provide extra functionality. The call path is built using a *path specification*, which is a list of the wrappers that a call goes through along with the parameters used to instantiate each wrapper.

The system allows several different call paths to be specified to reach the same object, allowing a thread to use multiple paths to the same object, or more commonly, to allow multiple threads to access the same object through separate paths[2]. Paths can be combined, allowing the system to aggregate calls (especially useful for caching).

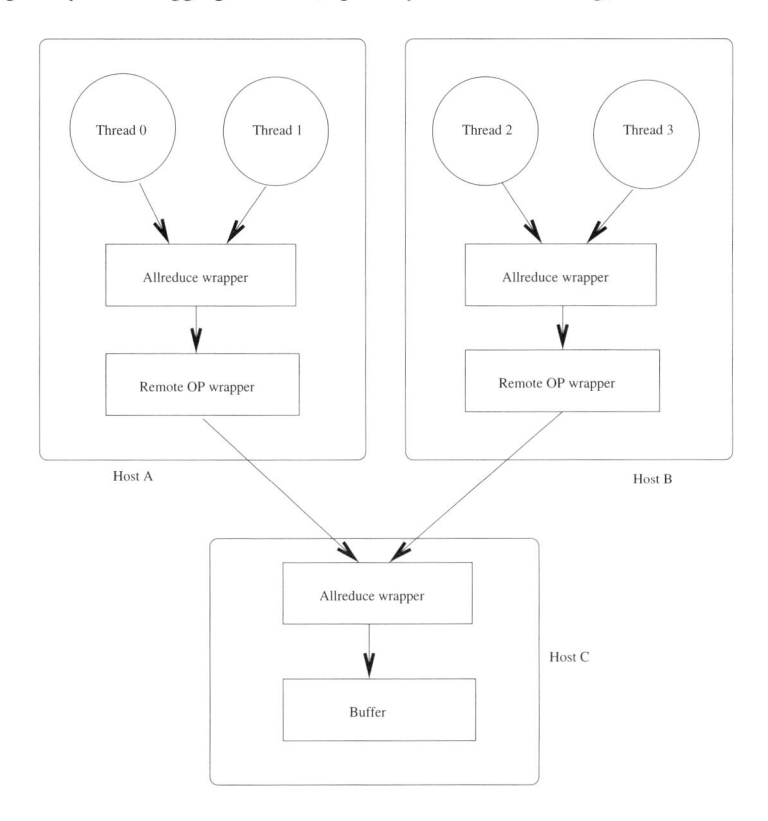

Figure 2 Example configuration of an Allreduce operation using the PATHS system

Figure 2 shows an example configuration of an Allreduce operation, using *Allreduce* wrappers. When several threads share an Allreduce wrapper, the first $n-1$ threads contribute their values and block, waiting for the result. Thread number n contributes the last value at that stage and forwards the partial result to the next wrapper in the path. When all results have been combined in the root object, the result is propagated back up the tree, waking up blocked threads.

Remote operation wrappers are used to implement remote method calls, allowing a path to jump to another computer. The parameters to the remote operation wrapper include information such as the host name and protocol to use when connecting to the other host. This information is also useful when analysing the running system, as it provides the necessary information to determine exactly where each wrapper was located.

The combination of all path specifications is called a *path map*, and describes how a system is mapped onto a cluster as well as how an algorithm is implemented by combining wrappers. *Trace wrappers*, which can be added to the path descriptions, can log timestamps and optionally data going through the wrapper. By combining the traces with the path map, we can analyse the system's performance and identify problems such as bottlenecks.

The CoMPI [6] system is an implementation of the PATHS concepts in LAM-MPI.

---
[2]Multiple threads can also share the same path

4.2. Towards CSP-Based Wrappers

One of the motivations for this paper is to provide background and tools for a CSP-based version of the PATHS system. As this part of the project is still in the early phases, we will just give a quick overview of a few important points.

A PATHS wrapper is an object that encapsulates some behaviour, can contain some state, and that may forward or initiate communication with other wrappers and objects. This corresponds well with the notion of a process in CSP-based systems.

Setting up a path and binding the wrappers together is similar to activity common in CSP-based programs. In PATHS, a process asks for a path to a given object by calling a `get_path` function. `get_path` retrieves a path specification, either from a library that can be loaded at run-time (usually Python modules or Lisp code), or by querying a central path specification server.

The specification includes all the necessary information to instantiate the wrappers on the hosts in the system (or locate wrappers that have already been instantiated) and bind the wrappers together. This corresponds well with spawning CSP processes and plugging them together using channels.

5. Conclusion

We have described how a number of the MPI collective operations can be implemented using PyCSP, showing how MPI implementation idioms can be translated into process-oriented equivalents. The same techniques can be used to implement collective operations using other process-oriented systems such as occam-π. We find that implementing collective operations using a system with high-level concurrency facilities typically results in more concise, clearer code.

We have shown how process-oriented techniques can be used to express a greater level of concurrency in our implementations of the collective operations than in OpenMPI's implementations. In particular, our implementations generally allow more communications to proceed in parallel, potentially reducing communication bottlenecks and thus making more effective use of the cluster communication fabric.

The basis of process-oriented systems in process calculi makes it possible to formally reason about the behaviour of collective operations. For example, we will be able to ensure that new cluster communication strategies are free from deadlock.

In order to support future experiments with collective operations, we have built an asynchronous network messaging framework for both occam-π and PyCSP. We have demonstrated how an IO scheduler process can be used to efficiently handle large numbers of network connections in a process-oriented program without the latency problems inherent in existing approaches.

This framework will have other uses as well; we have already used it for a distributed complex systems simulation, using a two-tiered strategy with channel communication between process on a single and asynchronous network messaging between hosts. In the future, we would like to make its use more transparent, supporting a direct channel interface and channel end mobility; we would also like to increase the interoperability of the occam-π and PyCSP implementations, and extend this to other process-oriented systems such as JCSP, in order to support distributed multi-language systems.

We have described how systems using process-oriented collective operations may be dynamically reconfigured at runtime using an approach based on the PATHS system. We have shown a straightforward mapping between PATHS' object-oriented wrappers and processes in a process-oriented system.

We have not yet extensively benchmarked our implementations of collective operations. While the expressiveness and correctness of the process-oriented implementations are clear advantages, we still need to measure their performance in comparison with the existing Open-MPI implementations in medium- and large-scale cluster applications. We are optimistic about the potential performance gains from the application of process-oriented techniques to cluster computing based on earlier experiences with the PATHS system.

Acknowledgements

This work was partially funded by EPSRC grant EP/E053505/1 and a sabbatical stipend from the University of Tromsø.

References

[1] Edgar Gabriel, Graham E. Fagg, George Bosilca, Thara Angskun, Jack J. Dongarra, Jeffrey M. Squyres, Vishal Sahay, Prabhanjan Kambadur, Brian Barrett, Andrew Lumsdaine, Ralph H. Castain, David J. Daniel, Richard L. Graham, and Timothy S. Woodall. Open MPI: Goals, concept, and design of a next generation MPI implementation. In *Proceedings, 11th European PVM/MPI Users' Group Meeting*, Budapest, Hungary, September 2004.
[2] OpenMPI homepage. http://www.open-mpi.org/.
[3] John Markus Bjørndalen, Brian Vinter, and Otto Anshus. PyCSP - Communicating Sequential Processes for Python. In A.A.McEwan, S.Schneider, W.Ifill, and P.Welch, editors, *Communicating Process Architectures 2007*, pages 229–248, jul 2007.
[4] PyCSP distribution. http://www.cs.uit.no/~johnm/code/PyCSP/.
[5] The occam-π programming language. http://occam-pi.org/.
[6] Espen Skjelnes Johnsen, John Markus Bjørndalen, and Otto Anshus. CoMPI - Configuration of Collective Operations in LAM/MPI using the Scheme Programming Language. In *Applied Parallel Computing. State of the Art in Scientific Computing*, volume 4699 of *Lecture Notes in Computer Science*, pages 189–197. Springer, 2007.
[7] John Markus Bjørndalen, Otto Anshus, Brian Vinter, and Tore Larsen. Configurable Collective Communication in LAM-MPI. *Proceedings of Communicating Process Architectures 2002, Reading, UK*, September 2002.
[8] G. Burns, R. Daoud, and J. Vaigl. LAM: An Open Cluster Environment for MPI. http://www.lam-mpi.org/, 1994.
[9] John Markus Bjørndalen, Otto Anshus, Tore Larsen, and Brian Vinter. PATHS - Integrating the Principles of Method-Combination and Remote Procedure Calls for Run-Time Configuration and Tuning of High-Performance Distributed Application. In *Norsk Informatikk Konferanse*, pages 164–175, November 2001.
[10] John Markus Bjørndalen, Otto Anshus, Tore Larsen, Lars Ailo Bongo, and Brian Vinter. Scalable Processing and Communication Performance in a Multi-Media Related Context. *Euromicro 2002, Dortmund, Germany*, September 2002.
[11] John Markus Bjørndalen. *Improving the Speedup of Parallel and Distributed Applications on Clusters and Multi-Clusters*. PhD thesis, Department of Computer Science, University of Tromsø, 2003.
[12] Irmen de Jong. Pyro: Python remote objects. http://pyro.sourceforge.net/.
[13] M. Schweigler. *A Unified Model for Inter- and Intra-processor Concurrency*. PhD thesis, Computing Laboratory, University of Kent, Canterbury, UK, August 2006.
[14] Jonas Barklund and Robert Virding. Erlang 4.7.3 Reference Manual, February 1999. http://www.erlang.org/download/erl_spec47.ps.gz.
[15] F.R.M. Barnes. Blocking System Calls in KRoC/Linux. In P.H. Welch and A.W.P. Bakkers, editors, *Communicating Process Architectures*, volume 58 of *Concurrent Systems Engineering*, pages 155–178, Amsterdam, the Netherlands, September 2000. WoTUG, IOS Press. ISBN: 1-58603-077-9.
[16] Dan Kegel. The C10K problem, Sep 2006. http://www.kegel.com/c10k.html.
[17] Simon Marlow, Simon Peyton Jones, and Wolfgang Thaller. Extending the Haskell foreign function interface with concurrency. In *Haskell '04: Proceedings of the 2004 ACM SIGPLAN workshop on Haskell*, pages 22–32, New York, NY, USA, 2004. ACM.
[18] D.J. Bernstein. The self-pipe trick. http://cr.yp.to/docs/selfpipe.html.
[19] LAM-MPI homepage. http://www.lam-mpi.org/.

[20] Thilo Kielmann, Rutger F. H. Hofman, Henri E. Bal, Aske Plaat, and Raoul A. F. Bhoedjang. MagPIe: MPI's collective communication operations for clustered wide area systems. In *Proceedings of the seventh ACM SIGPLAN symposium on Principles and practice of parallel programming*, pages 131–140. ACM Press, 1999.
[21] Amit Karwande, Xin Yuan, and David K. Lowenthal. CC-MPI: a compiled communication capable MPI prototype for Ethernet switched clusters. In *Proc. of the ninth ACM SIGPLAN symposium on Principles and practice of parallel programming*, pages 95–106. ACM Press, 2003.
[22] Lars Ailo Bongo. Steps: A Performance Monitoring and Visualization Tool for Multicluster Parallel Programs, June 2002. Large term project, Department of Computer Science, University of Tromsø.
[23] Lars Ailo Bongo, Otto Anshus, and John Markus Bjørndalen. EventSpace - Exposing and observing communication behavior of parallel cluster applications. In *Euro-Par*, volume 2790 of *Lecture Notes in Computer Science*, pages 47–56. Springer, 2003.
[24] Tor-Magne Stien Hagen. The PATHS Visualizer. Master's thesis, Department of Computer Science, University of Tromsø, 2006.

Appendix: OpenMPI Source Code

```
/*
 *      bcast_lin_intra
 *
 *      Function:       - broadcast using O(N) algorithm
 *      Accepts:        - same arguments as MPI_Bcast()
 *      Returns:        - MPI_SUCCESS or error code
 */
int
mca_coll_basic_bcast_lin_intra(void *buff, int count,
                               struct ompi_datatype_t *datatype, int root,
                               struct ompi_communicator_t *comm)
{
    int i;
    int size;
    int rank;
    int err;
    ompi_request_t **preq;
    ompi_request_t **reqs = comm->c_coll_basic_data->mccb_reqs;

    size = ompi_comm_size(comm);
    rank = ompi_comm_rank(comm);

    /* Non-root receive the data. */

    if (rank != root) {
        return MCA_PML_CALL(recv(buff, count, datatype, root,
                                 MCA_COLL_BASE_TAG_BCAST, comm,
                                 MPI_STATUS_IGNORE));
    }

    /* Root sends data to all others. */

    for (i = 0, preq = reqs; i < size; ++i) {
        if (i == rank) {
            continue;
        }

        err = MCA_PML_CALL(isend_init(buff, count, datatype, i,
                                      MCA_COLL_BASE_TAG_BCAST,
                                      MCA_PML_BASE_SEND_STANDARD,
                                      comm, preq++));
        if (MPI_SUCCESS != err) {
            return err;
        }
    }
    --i;

    /* Start your engines.  This will never return an error. */
```

```
    MCA_PML_CALL(start(i, reqs));

    /* Wait for them all.  If there's an error, note that we don't
     * care what the error was -- just that there *was* an error.  The
     * PML will finish all requests, even if one or more of them fail.
     * i.e., by the end of this call, all the requests are free-able.
     * So free them anyway -- even if there was an error, and return
     * the error after we free everything. */

    err = ompi_request_wait_all(i, reqs, MPI_STATUSES_IGNORE);

    /* Free the reqs */

    mca_coll_basic_free_reqs(reqs, i);

    /* All done */

    return err;
}
```

Listing 18: Linear Broadcast in OpenMPI 1.2.5

```
int
mca_coll_basic_bcast_log_intra(void *buff, int count,
                               struct ompi_datatype_t *datatype, int root,
                               struct ompi_communicator_t *comm)
{
    int i;
    int size;
    int rank;
    int vrank;
    int peer;
    int dim;
    int hibit;
    int mask;
    int err;
    int nreqs;
    ompi_request_t **preq;
    ompi_request_t **reqs = comm->c_coll_basic_data->mccb_reqs;

    size = ompi_comm_size(comm);
    rank = ompi_comm_rank(comm);
    vrank = (rank + size - root) % size;

    dim = comm->c_cube_dim;
    hibit = opal_hibit(vrank, dim);
    --dim;

    /* Receive data from parent in the tree. */

    if (vrank > 0) {
        peer = ((vrank & ~(1 << hibit)) + root) % size;

        err = MCA_PML_CALL(recv(buff, count, datatype, peer,
                                MCA_COLL_BASE_TAG_BCAST,
                                comm, MPI_STATUS_IGNORE));
        if (MPI_SUCCESS != err) {
            return err;
        }
    }

    /* Send data to the children. */

    err = MPI_SUCCESS;
    preq = reqs;
    nreqs = 0;
    for (i = hibit + 1, mask = 1 << i; i <= dim; ++i, mask <<= 1) {
        peer = vrank | mask;
```

```
            if (peer < size) {
                peer = (peer + root) % size;
                ++nreqs;

                err = MCA_PML_CALL(isend_init(buff, count, datatype, peer,
                                              MCA_COLL_BASE_TAG_BCAST,
                                              MCA_PML_BASE_SEND_STANDARD,
                                              comm, preq++));
                if (MPI_SUCCESS != err) {
                    mca_coll_basic_free_reqs(reqs, preq - reqs);
                    return err;
                }
            }
        }
    }

    /* Start and wait on all requests. */

    if (nreqs > 0) {

        /* Start your engines.  This will never return an error. */

        MCA_PML_CALL(start(nreqs, reqs));

        /* Wait for them all.  If there's an error, note that we don't
         * care what the error was -- just that there *was* an error.
         * The PML will finish all requests, even if one or more of them
         * fail.  i.e., by the end of this call, all the requests are
         * free-able.  So free them anyway -- even if there was an
         * error, and return the error after we free everything. */

        err = ompi_request_wait_all(nreqs, reqs, MPI_STATUSES_IGNORE);

        /* Free the reqs */

        mca_coll_basic_free_reqs(reqs, nreqs);
    }

    /* All done */

    return err;
}
```

Listing 19: Non-linear Broadcast in OpenMPI 1.2.5.

Representation and Implementation of CSP and VCR Traces

Neil C.C. BROWN [a] and Marc L. SMITH [b]

[a] *Computing Laboratory, University of Kent,*
Canterbury, Kent, CT2 7NZ, UK, `neil@twistedsquare.com`
[b] *Computer Science Department, Vassar College,*
Poughkeepsie, New York 12604, USA, `mlsmith@cs.vassar.edu`

Abstract. Communicating Sequential Processes (CSP) was developed around a formal algebra of processes and a semantics based on traces (and failures and divergences). A trace is a record of the events engaged in by a process. Several programming languages use, or have libraries to use, CSP mechanisms to manage their concurrency. Most of these lack the facility to record the trace of a program. A standard trace is a flat list of events but structured trace models are possible that can provide more information such as the independent or concurrent engagement of the process in some of its events. One such trace model is View-Centric Reasoning (VCR), which offers an additional model of tracing, taking into account the multiple, possibly imperfect views of a concurrent computation. This paper also introduces "structural" traces, a new type of trace that reflects the nested parallelism in a CSP system. The paper describes the automated generation of these three trace types in the Communicating Haskell Processes (CHP) library, using techniques which could easily be applied in other libraries such as JCSP and C++CSP2. The ability to present such traces of a concurrent program assists in understanding the behaviour of real CHP programs and for debugging when the trace behaviours are wrong. These ideas and tools promote a deeper understanding of the association between practicalities of real systems software and the underlying CSP formalism.

Keywords. Communicating Sequential Processes, CSP, View-Centric Reasoning, VCR, traces.

Introduction

Communicating Sequential Processes (CSP) [1,2] is a model of concurrency based on processes synchronising on shared events. To support this idea, Hoare developed a process algebra to permit the specification of concurrency, and defined a semantics based on traces, failures (deadlocks) and divergences (livelocks). This paper is only concerned with traces: records of a program's event behaviour.

The recent growth in multicore processors has led to the need for programming models that can exploit concurrency. In contrast to the popular locks and threading models, CSP and process-oriented design promise an elegant and powerful alternative when used properly. The occam-π programming language [3] and libraries for several other mainstream programming languages use CSP as the basis for their concurrency. CSP and formal methods have a reputation as being challenging and divorced from practice. However, applying CSP without proper training can diminish its advantages.

When programmers encounter problems in programming concurrent systems, they naturally turn to familiar methods of debugging – for example, adding "print" statements at points of interest in their programs. Programmers new to concurrency soon discover that due to non-deterministic scheduling their programs can behave differently from run to run, and that adding these debugging statements can also change the behaviour of the program.

Programmers are effectively seeking to understand their program's behaviour. Debugging statements primarily reveal the behaviour of a single component in the system, not synchronisations between components. In contrast to print statements, traces record these interactions between components – such as channel communications, barriers, etc.

An additional frustration in using disciplined process-oriented programming (such as in occam-π) can be the addition of the necessary wiring that will enable debugging messages to be printed. Programmers must redesign their process networks to route output channels in order to print messages from processes. This further complicates mastering the new programming model.

We have augmented a CSP implementation for Haskell (Communicating Haskell Processes, CHP) [4] to provide a convenient tracing facility. When enabled, it records channel communications and barrier synchronisations that have completed. This is akin to a CSP trace of the program.

When a problem arises, this trace facility allows the programmer to see what events actually occurred, and their order. They can compare this behaviour with their intentions. While this is no substitute for formal reasoning and/or model checking, this is a practical aid for programmers, and introduces them to the notion of tracing. We believe tracing is a viable entry point to the CSP theory for programmers unfamiliar with formal reasoning and model checking tools such as FDR2[5], which can check various safety, refinement and liveness properties of CSP processes.

Our approach is intended to aid programmers in debugging real programs, which may not have been formally modelled. This is distinct from tools such as FDR2 and ProBE, which allow a programmer to investigate the behaviour of CSP models in a language-independent manner.

Recording traces of a computation brings with it a set of questions and challenges. Specifically, how shall traces be represented in a meaningful way, and how shall we implement these traces? View-Centric Reasoning (VCR) [6,7] was developed to address some of challenges posed by the CSP observer, including the bookkeeping practice of reducing observed concurrency to sequential interleavings of events. VCR offers an additional model of tracing, taking into account the multiple, possibly imperfect views of a concurrent computation.

One aspect of traces that programmers may find challenging is the flattening of their parallel-composed process network into a single sequential trace. For this reason, we also introduce structural traces. Structural traces contain notions of sequential and parallel composition that enable these traces to reflect to some degree the sequential and parallel logic of original program.

In this paper we will introduce all three trace types fully (section 1) before explaining how we have implemented the recording of each of these trace types (section 2). We will then present examples of each of the traces (section 3) before looking at related and future work (sections 4 and 5).

1. Background

Tracing is the recording of a program's behaviour. Specifically, a trace is a record of all the events that a process has engaged in. A trace of an entire program is typically an interleaving (in proper time order) of all the events that occurred during the run-time of the program. In this section we briefly recap CSP and VCR tracing, and introduce a new form of trace that will be used in this paper: structural tracing.

For our trace examples in this section, we will use the following CSP system:

$$(a \to b \to c \to \text{STOP}) \underset{\{a\}}{\|} (a \to d \to e \to \text{STOP})$$

1.1. CSP Tracing

A CSP trace is a sequential record of all the events that occurred during the course of the program. Hoare originally described the trace as being recorded by a perfect observer who saw all events. If the observer saw two (or more) events happening simultaneously, they were permitted to write down the events in an arbitrary order.

For example, these are the possible maximal traces of our system:

$$\langle a, b, c, d, e \rangle \quad \langle a, b, d, c, e \rangle \quad \langle a, b, d, e, c \rangle$$
$$\langle a, d, b, c, e \rangle \quad \langle a, d, b, e, c \rangle \quad \langle a, d, e, b, c \rangle$$

1.2. VCR Tracing

A VCR trace is similar to a CSP trace, but instead of recording a sequence of single events in a trace, the observer records a sequence of multisets. In the original model of VCR traces, each multiset represented a collection of simultaneous events, preserving information that the CSP observer had lost. In practice, simultaneity is a difficult concept to define, reason about or observe.

We have therefore adapted the meaning of event multiset traces in VCR. In our implementation in CHP each multiset is a collection of *independent* events. We term a and b to be independent events if a did not observably require b to occur first, and vice versa. The implication is that it was possible for events a and b to occur in either order without altering the meaning of the program.

By convention in this paper, we write multisets of size one without set notation. These are the possible maximal traces of our system:

$$\langle a, b, c, d, e \rangle \quad \langle a, b, d, c, e \rangle \quad \langle a, b, d, e, c \rangle$$
$$\langle a, d, b, c, e \rangle \quad \langle a, d, b, e, c \rangle \quad \langle a, d, e, b, c \rangle$$
$$\langle a, b, \{c, d\}, e \rangle \quad \langle a, b, d, \{c, e\} \rangle$$
$$\langle a, d, b, \{c, e\} \rangle \quad \langle a, d, \{b, e\}, c \rangle$$
$$\langle a, \{b, d\}, \{c, e\} \rangle$$

VCR also permits multiple observers and imperfect observation (some events may not be recorded), but unless specifically stated, in this paper we will be using a single observer that records all events.

1.3. Structural Tracing

To explore other forms of trace recording, we have created structural tracing. CSP and VCR tracing are both based on recording the events in a single, fairly flat trace. Structural tracing is a constrasting approach where a structure is built up that reflects the parallel and sequential composition of the processes being traced. Traces can be composed in parallel with the $\|$ operator, and these parallel-composed traces may appear inside normal sequential traces.

There is only one possible maximal trace of our system:

$$\langle a,b,c \rangle \parallel \langle a,d,e \rangle$$

Note that a single occurrence of an event is recorded multiple times (once by each process engaging it), which is distinct from both CSP and VCR tracing. This makes structural traces quite different from the classic notion of tracing.

1.4. Communicating Haskell Processes

Communicating Haskell Processes (CHP) is a new concurrency library in the tradition of JCSP et al for the Haskell programming language [4]. We have chosen to implement our traces in CHP because it was the easiest framework to add tracing to, given the differing implementations of the various CSP libraries and the authors' expertise. The techniques presented here should be immediately portable to any other CSP framework.

CHP is primarily built on top of Software Transactional Memory (STM), a library for implementing concurrency in Haskell [8]. STM is built on the idea of atomic transactions that modify a set of shared transactional variables. Event recording (at least for CSP and VCR traces) takes place in the same transaction as actually completing the event itself, and thus the recording is indivisible from the event.

2. Implementation

2.1. General

The Communicating Haskell Processes library has two types of synchronisations: channel communications (that transmit data) and barrier synchronisations (that have no data). This is also true for most CSP-derived languages and libraries. We have altered CHP so that after the completion of every successful event, the event is recorded. For CSP and VCR this is done by the process that completes the synchronisation. For structural traces, every process that engages in the event records it.

This highlights the fundamental difference between structural tracing and the other two forms; under structural tracing, events are recorded multiple times, but under CSP and VCR tracing events are only recorded once.

Tracing can be enabled or disabled at the start of the program's execution. Switching on tracing is as simple as changing a single line in the user's program. This makes the tracing facility very easy to use.

2.2. CSP Traces

CSP tracing is the most straightforward to implement. Hoare's perfect observer can be implemented by modifying a single sequence shared between all processes (with appropriate mechanisms, such as a mutex, to handle the concurrent updates). Due to Haskell's lists having $O(1)$ time complexity for prepending but not for appending, we add all new events at the head of the list, effectively recording the trace backwards.

2.2.1. Recording Bottleneck

The main problem with CSP tracing is that it serialises the whole program. All processes end up contending for access to the central trace, and thus with multiple threads this can slow down the program. An alternative approach would be to have each process record its own trace with events timestamped (as in notions of timed CSP), and merge these sorted traces by timestamp from sub-processes into their parent process when the sub-processes complete.

The problem with such timestamps is that modern computers (from the point of view of a portable high-level language) often have timers with relatively coarse frequency, such as once per millisecond. CHP is able to complete around a hundred events in that time, and thus events that occur within a millisecond of each other would be recorded in arbitrary order. Therefore we still record CSP traces using a single shared sequence.

2.3. VCR Traces

As described in section 1.2, we group independent events into multisets for VCR traces. If we can deduce a definite ordering of two events, we take the latter event to be dependent on the former. Otherwise, we take the events to be independent. For example, we know that two events must be sequential if they are both performed by the same process-thread. To be able to reason about this and other details, we record events with associated process identifiers.

2.3.1. Process Identifiers

We use process identifiers to deduce information about sequencing. Consider these two CSP systems:

$$(a \rightarrow (b \rightarrow \text{SKIP} \mathbin{|||} c \rightarrow \text{SKIP})) \mathbin{;} (d \rightarrow e \rightarrow \text{STOP})$$
$$(f \rightarrow \text{SKIP} \mathbin{|||} g \rightarrow \text{SKIP}) \mathbin{;} (h \rightarrow \text{STOP} \mathbin{|||} i \rightarrow \text{STOP})$$

We wish to be able to deduce the following facts, where $<$ is the strict partial ordering of the events in time:

$$a < b, a < c$$
$$b < d, c < d$$
$$d < e$$
$$f < h, g < h$$
$$f < i, g < i$$

Process identifiers are one or more integers (sequence numbers) joined together by a sub-process operator, $\triangleright_x$ where x is itself an integer (a parallel branch identifier). The top-level process of the program starts with the identifier 0. Process identifiers only change when the process runs a parallel composition. Before and after the composition, the last (rightmost) sequence number is incremented. The identifiers for the sub-processes are formed by appending $\triangleright_x 0$ to the current identifier, where x is distinct for each sub-process.

Our example programs, with process identifiers, are shown in figures 1 and 2. Each event is given in a circle. The open double-headed arrows indicate sequencing, and sub-processes are shown in a vertical relation, with an empty (dot) event as a parent. The events are grouped into dashed boxes, which represent the actual processes that will be created when the program is run. Next to each event is the associated process identifier that will be recorded with it.

The sequence numbers in the top row of each figure are not in error. Events d and e are associated with the same sequence number. Event e will definitely be recorded after event d (since the same process is doing them in sequence) so we do not need to use the process identifier to tell them apart. It is only in the case of parallel composition that we must change the sequence numbers, in order to deduce an ordering between the parent's events (such as a and d) and the sub-processes' events (b and c). The incrementing before and after both parallel compositions is what leads to the sequence numbers being 1 and 3 for the second figure.

The following Haskell pseudo-code explains the algorithm for comparing process identifiers:

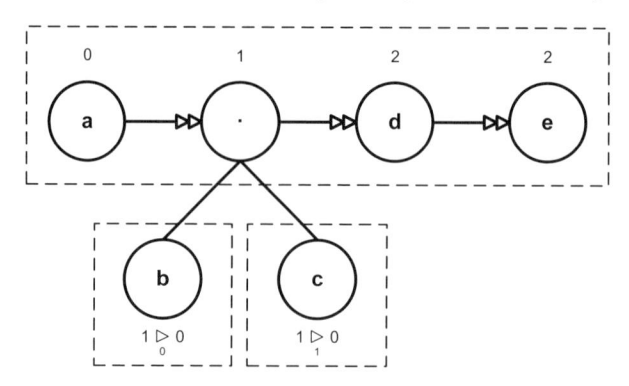

Figure 1. The program $(a \to (b \to \text{SKIP} \ ||| \ c \to \text{SKIP})) ; (d \to e \to \text{STOP})$, with the identifiers for each event

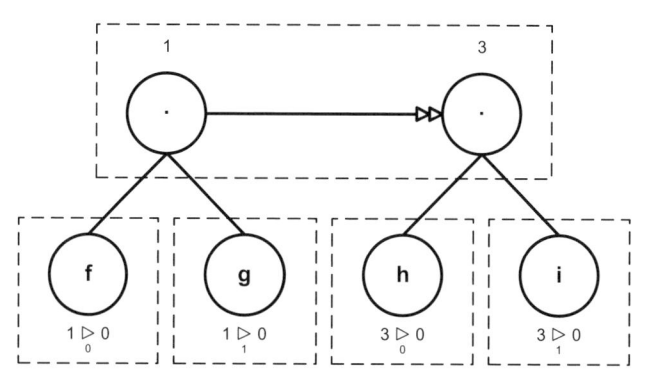

Figure 2. The program $(f \to \text{SKIP} \ ||| \ g \to \text{SKIP}) ; (h \to \text{STOP} \ ||| \ i \to \text{STOP})$, with the identifiers for each event

data *Maybe a* = *Nothing* | *Just a*

type *ProcessId* = (*Integer* , *Maybe* (*Integer* , *ProcessId*))

lessThan : : *ProcessId* -> *ProcessId* -> *Bool*
lessThan (*x*, *Nothing*) (*y*, *Nothing*) = *x* < *y*
lessThan (*x*, *Just* (*xpar*, *xpid*)) (*y*, *Just* (*ypar*, *ypid*))
  = **if** *x* == *y*
     then
       (**if** *xpar* == *ypar*
          then *lessThan xpid ypid*
          else *False*) −− Independent
     else *x* < *y*

The data type indicates that a process identifier is a pair of an integer, and an optional (indicated by the *Maybe* type) parallel branch identifier with the further part of the process identifier. So the identifier $2 \triangleright_1 4$ would be represented as (2, *Just* (1, (4, *Nothing*))).

If the process identifiers are both single sequence numbers, we simply compare these sequence numbers. If they are a compound identifier, we again start by comparing the sequence numbers. If they are not equal, we return the value of $x < y$. If however they are equal, we compare their parallel branch identifiers. If these identifiers are not equal, we know the identifiers come from parallel siblings, and we return false. If they are equal, we proceed with comparing the remainder of the process identifiers.

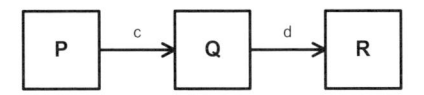

Figure 3. A standard pipeline of three processes

Due to the incrementing of the sequence number before and after running sub-processes, it can be seen that no recorded identifier will ever be an exact prefix of another. In our diagrams (figures 1 and 2), the parent with the same prefix is represented as a dot, and engages in no events itself. Thus, pairs of recorded identifiers will always differ within the length of the shorter identifier, or both identifiers will be equal.

2.3.2. Recording Rules

In the previous section we explained our process identifiers and a partial order over them. We will now show how we use this to record VCR traces, for the moment considering events that only involve one process.

When recording an event in a VCR trace, a process must look at the most recent set of parallel events, here termed Θ. Θ is a set[1] of pairs of an event (for which we use α, β, etc) and a process identifier (for which we use p, q, etc). We must determine, for a given (β, q) whether to add the new event to Θ or whether to start a new set of parallel events. Our rule is straightforward: if $\exists (\alpha, p) \in \Theta : p < q \vee p = q$, we must start a new set of parallel events because an event in the most recent set provably occurred before the event we are adding. Otherwise, add the new event to Θ (the events are independent).

Consider the following CSP traces of our example processes from figures 1 and 2, annotated with process identifiers:

$$\langle a[0], c[1 \triangleright 0]_1, b[1 \triangleright 0]_0, d[2], e[2] \rangle$$

$$\langle f[1 \triangleright 0]_0, g[1 \triangleright 0]_1, i[3 \triangleright 0]_1, h[3 \triangleright 0]_0 \rangle$$

Following our rules, the same traces recorded in VCR form would be:

$$\langle a, \{b, c\}, d, e \rangle$$

$$\langle \{f, g\}, \{h, i\} \rangle$$

2.3.3. Multiple Processes

We have so far considered events with just a single process. In reality, events involve multiple processes synchronising. Consider a process pipeline, where three processes are connected with channels c and d, as shown in figure 3. We will assume here that the reader happens to record the events (recall that events are recorded by the last process to engage in the synchronisation). Process Q will read from channel c, and record this event. Then it will write to channel d and the process R will read and record the event. Because the process identifiers for Q and R have no ordering, the events will be recorded as being independent. However, the communication on channel d was clearly dependent on the communication on from channel c, since process Q communicated on channel c before communicating on channel d.

In order to combat this problem, we modify our synchronisation events so that the party that records the event can know the process identifier of all the other processes that engaged in the synchronisation. The event is recorded with a set of process identifiers (in barriers, there may be more than two participants) rather than a single identifier.

---

[1] Technically in VCR, it is a multiset, but due to our recording strategy, no pair of event and process identifier will occur multiple times in the same set.

Thus we can adjust our previous rule, now that Θ is a set of pairs of single event identifier and a *set* of process identifiers (for which we will use capital P, Q, etc). For adding a new pair of event and identifier set (β, Q): if $\exists (\alpha, P) \in \Theta, \exists p \in P, \exists q \in Q : p < q \vee p = q$, start a new set of parallel events. Otherwise, add the new event to Θ.

In our previous example, the communication on channel c will have been recorded with the process identifiers for processes P and Q. When process R then records the communication on channel d with the identifiers for Q and R, the events will be seen to be dependent.

The reader may wonder what happens should it also be the case that $\exists p \in P, \exists q \in Q : q < p$ in the above rule. This is not possible in the CHP library because the recording of an event is bound into the synchronisation, so events will always be recorded in order. It cannot be the case that an event a that provably occurred *before* b will be recorded *after* b.

2.3.4. Relevance of Event Identifiers

The event identifier currently plays no role in our rules – it is usually subsumed by considering the identifiers of processes engaging in the events. Therefore the only cases where it would be of use are where an entirely different (unrelated) set of processes may engage on two successive occurrences of the same event.

Communication on unshared channels, and channels with only one shared end, are taken care of by considering the processes involved. In these cases, one process will always be involved[2] in the subsequent communications, and we can use this to deduce sequence information. Therefore we need only consider here channels that are shared at both ends.

With channels that are shared at both ends, it is possible for P and Q to communicate once on the channel, then release the ends, after which R and S claim them and perform another communication. In such a situation, we do not consider the second communication to be dependent on the first, and so we do not use the event identifier to deduce any sequence information.

2.3.5. Independence and Inference

The VCR theory describes events in the same multiset as being observed simultaneously, but we have altered this and implemented our traces to record independent events in the same multiset. This means that some of the theoretically possible VCR traces can never actually be recorded in our implementation, and also that some sequence information can be retroactively inferred.

Recall our CSP system from the example in section 1:

$$(a \to b \to c \to \text{STOP}) \underset{\{a\}}{\|} (a \to d \to e \to \text{STOP})$$

Imagine that the first two events to occur are a and b, giving us a partial trace: $\langle a, b \rangle$. According to the VCR theory, the possible maximal traces that may follow from this are:

$$\langle a, b, c, d, e \rangle \quad \langle a, b, d, c, e \rangle \quad \langle a, b, d, e, c \rangle$$
$$\langle a, b, \{c, d\}, e \rangle \quad \langle a, b, d, \{c, e\} \rangle \quad \langle a, \{b, d\}, \{c, e\} \rangle$$

Which trace is recorded will depend on which event occurs next. If the next event is c, the theoretical traces are:

$$\langle a, b, c, d, e \rangle \quad \langle a, b, \{c, d\}, e \rangle$$

---
[2]Or a process and its sub-processes, that we can deduce sequence information about.

In fact, the trace recorded in our implementation will always be the second trace. Because c and d are independent events, they will always be recorded in the same multiset. If c happens first, the partial trace will be $\langle a, b, c \rangle$. But then when d is recorded, because it is independent of all events in the most recent multiset (c, in a singleton multiset), it will be added to that multiset, forming $\langle a, b, \{c, d\} \rangle$, then become the second trace shown above. The trace $\langle a, b, c, d, e \rangle$ can never actually be recorded using our implementation.

If instead the next (third) event is d, the possible theoretical traces are:

$$\langle a, b, d, c, e \rangle \quad \langle a, b, d, e, c \rangle \quad \langle a, b, d, \{c, e\} \rangle$$
$$\langle a, b, \{c, d\}, e \rangle \quad \langle a, \{b, d\}, \{c, e\} \rangle$$

For similar reasons to the previous example, it is not possible to record any of the top three traces. The pairs of events (b, d), (c, d) and (c, e) are each independent, so the traces would always be recorded using the latter two traces (with appropriate multisets) rather than the earlier three traces with only singleton multisets.

It is even possible to infer sequence from a trace with multisets. Consider the trace $\langle a, b, \{c, d\}, e \rangle$. If event d had occurred before c, then b and d would have been grouped into the same multiset (since they are independent). Therefore to produce this trace, c must have happened before d.

These issues reflect the difference between the theory of VCR and our actual implementation of its tracing. Our trace may seem to add obfuscation over and above the CSP trace. But it also abstracts away from some of the unnecessary detail. Grouping two events into a multiset implies that they *could* have happened in either order. For trace compression (see section 2.5), comprehension and visualisation, a more regular trace that abstracts away from some of the scheduling-dependent behaviour of the program may well be appealing in some circumstances.

2.4. Structural Traces

A structural trace is the most natural – and fastest – to record. Each running process records events it has engaged in using a local trace. Since the trace is unshared, there is no contention or need for locks, and hence it is faster than any other method of recording. Sequential events are recorded by adding them to a list. As with the other traces, it adds to the front of the list (reversing the order) to be faster.

When a process runs several sub-processes in parallel, they all also separately record their own trace. Upon completion they all send back their traces to the parent process. These traces are received by the parent and form a hierarchy in the parent's trace. This process of joining traces means that the end result is one single trace that represents the behaviour of the entire program.

The drawback with structural traces is that information about the ordering of events between non-descendant processes is lost. Consider the trace:

$$\langle a, a, a, a, a, a \rangle \parallel \langle b, b, b, b, b, b \rangle$$

We cannot tell from this trace whether all the events a happened before all the events b, vice versa, a strict interleaving or any other possible pattern.

2.5. Compression

One problem of recording the trace of a system is that a large amount of data is generated. IF the system may generate an event every microsecond[3], that is a million events a second.

---
[3]This figure is a very rough average of channel communication times on modern machines from various CSP implementations.

Assuming on a 32-bit machine that an event could be reduced to eight bytes per event (four bytes for an identifier, four for a pointer in a linked list or similar), in the worst case nearly a gigabyte of data would be generated approximately every two minutes. Generally programs are likely to fall far below this upper bound, but ideally we would like to reduce the space required.

Processes usually iterate (by looping or recursion), and thus display repetitive behaviour. This repetition is often diluted by observing the behaviour of several processes, but even large process networks can display regular behaviour.

Repeated patterns appearing in the trace means that it should be possible to compress the trace by removing this redundancy. To gain much benefit from this, we need to compress the trace on-line, while the program is still running, rather than off-line after the program has finished.

The obvious compression approaches are forms of run-length encoding that can reduce consecutive repeated behaviour, and dictionary-based compression methods that can spot common patterns and reduce them to an index in a lookup table of frequently occurring patterns.

We also note that viewing the compressed version is often more comprehensible to the programmer than its raw, uncompressed form. Most of the structural traces in this paper have been left in their compressed form for that reason.

2.6. Rolling Trace

A primary use of traces is for post-mortem debugging, especially in the case of deadlock or livelock. In this case the programmer is interested to see what the program was doing leading up to the failure.

This failure could occur after the program has been running for hours, or days. The earlier behaviour of the program will probably not be of interest. Therefore a good policy in these cases would be to keep a rolling trace which records, say, the most recent thousand events. This would require a constant amount of memory rather than accumulating the trace as normal (recording using an array, rather than a linked list).

This would allow tracing to place some time overhead on the program, but only a small amount of memory overhead, and would still be useful in the case of program failure.

3. Example Traces

In this section we present examples of traces of several different small programs. Each trace is recorded from a different run of the program, so the traces will not (necessarily) be direct transformations of each other.

3.1. CommsTime

Commstime is a classic benchmark used in process-oriented systems. Its configuration is shown in figure 4.

For this benchmark, we show approximately six iterations of the commstime loop.

3.1.1. CSP

⟨ prefix-delta, delta-recorder, delta-succ, succ-prefix,
prefix-delta, delta-succ, delta-recorder, succ-prefix,
prefix-delta, delta-succ, delta-recorder, succ-prefix,
prefix-delta, delta-succ, delta-recorder, succ-prefix,
prefix-delta, delta-succ, delta-recorder, succ-prefix,
prefix-delta, delta-succ, succ-prefix, delta-recorder,
delta-succ ⟩

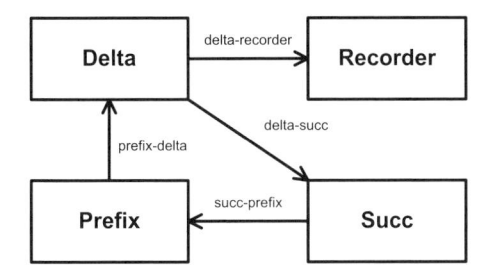

Figure 4. The CommsTime Network

3.1.2. VCR

⟨ {prefix-delta}, {delta-succ}, {delta-recorder, succ-prefix},
{prefix-delta}, {delta-succ}, {delta-recorder, succ-prefix},
{prefix-delta}, {delta-recorder, delta-succ}, {succ-prefix},
{prefix-delta}, {delta-recorder, delta-succ}, {succ-prefix},
{prefix-delta}, {delta-recorder, delta-succ}, {succ-prefix},
{prefix-delta}, {delta-recorder, delta-succ}, {succ-prefix}, {delta-succ} ⟩

3.1.3. Structural

⟨⟨⟨7*⟨delta-succ? , succ-prefix!⟩, delta-succ?⟩
|| ⟨7*⟨prefix-delta! , succ-prefix?⟩ , prefix-delta!⟩
|| ⟨6*⟨delta-recorder?⟩⟩
|| ⟨7*⟨prefix-delta? , ⟨delta-recorder! || delta-succ!⟩⟩⟩⟩⟩

3.1.4. Summary

The CSP trace reflects the interleaving that occurred between all the parallel processes, and shows how the order of events changes slightly at the beginning and end of the trace.

The lines in the middle of the VCR trace are probably what we would expect; singleton sets of prefix-delta and succ-prefix, and a set of the two parallel events from the delta process: delta-recorder and delta-succ. However, the early traces show that there is a different recording that can occur, with the delta-succ event happening first, and the delta-recorder happening in parallel with succ-prefix. This is a valid behaviour of our process network.

It is easy to see the four different processes in the structural trace, each with distinct behaviour. This would be visible even if the channels were not labelled. The regular repetition of all of the processes is also clear.

3.2. Dining Philosophers

The dining philosophers is a classic concurrency problem described by Hoare in his original book on CSP [1]. We use the deadlocking version for our benchmark. To keep the traces simpler and to provoke deadlock more easily, we use only three philosophers. The fork-claiming channels are named according to the philosopher they are connected to – the names of all of the channels are shown in figure 5. We show only the tail-end of the traces (leading up to the deadlock) as the full traces are too long to display here.

3.2.1. CSP

⟨ ..., fork.right.phil1, fork.left.phil0, fork.left.phil2, fork.right.phil2, fork.right.phil0, fork.left.phil1, fork.right.phil0, fork.left.phil0, fork.right.phil1, fork.left.phil2, fork.left.phil1, fork.right.phil2, fork.right.phil1, fork.left.phil0, fork.left.phil2, fork.right.phil0, fork.right.phil2, fork.left.phil1, fork.left.phil0, fork.right.phil0, fork.right.phil1, fork.left.phil2, fork.left.phil1, fork.right.phil1, fork.right.phil2, fork.left.phil0, fork.left.phil2, fork.right.phil2, fork.right.phil0, fork.left.phil1, fork.right.phil0, fork.left.phil0, fork.left.phil2, fork.left.phil0 ⟩

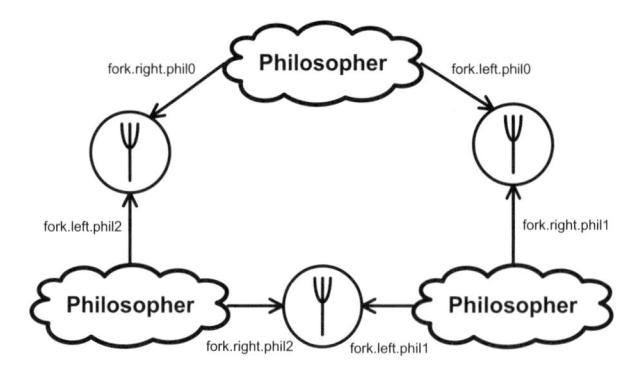

Figure 5. Dining Philosophers with Three Philosophers

3.2.2. VCR

⟨ ..., {fork.left.phil2}, {fork.right.phil0, fork.right.phil2}, {fork.left.phil0, fork.right.phil0, fork.left.phil1},
{fork.right.phil1, fork.left.phil2}, {fork.left.phil1, fork.right.phil1}, {fork.left.phil0, fork.right.phil2},
{fork.left.phil2, fork.right.phil2}, {fork.right.phil0, fork.left.phil1}, {fork.left.phil0, fork.right.phil0},
{fork.right.phil1, fork.left.phil2}, {fork.left.phil1, fork.right.phil1}, {fork.left.phil0, fork.right.phil2},
{fork.left.phil2, fork.right.phil2}, {fork.right.phil0, fork.left.phil1}, {fork.left.phil0}, {fork.right.phil0,
fork.right.phil1}, {fork.left.phil1, fork.right.phil1}, {fork.left.phil0, fork.left.phil1, fork.left.phil2} ⟩

3.2.3. Structural

Unfortunately our current implementation of structural tracing in CHP cannot show its traces in the presence of deadlock, due to the fast way the traces are recorded (in local per-process storage) and the way that the error manifests (being caught outside the scope of this storage). We hope to remedy this situation in the near future.

3.2.4. Summary

The dining philosophers problem is an example of a larger process network with less regular behaviour. Another problem is that it is hard to identify which messages are "pick up" messages and which are "put down" messages. An automated tool for dealing with traces could easily fix this by labelling alternating messages on the same channel differently. At the end of each trace it can be seen (most visibly in the VCR trace) that each philosopher communicated to its left-hand fork but not its right-hand fork, immediately before the deadlock.

3.3. Token Cell Ring

This example is a token-passing ring using alting barriers (multiway synchronisations that support choice). The process network is a ring of cells, where each cell is enrolled on two alting barriers (here termed "before" and "after") and has a channel-end from a "tick" process. A cell can have two states: full or empty.

If the cell is full, it offers to either engage on its "after" barrier (to pass the token on) or read from its "tick" channel. If the barrier is the event chosen, the cell then commits to read from its "tick" channel. If the cell is empty, it offers to either engage on its "before" barrier (to receive a token) or read from its "tick" channel. Again, if the barrier is chosen, it then commits to reading from the "tick" channel.

The writing ends of all the tick channels are connected to a "ticker" process that writes in parallel to all its channels, then repeats. Six cells are wired in a ring, and initially the first (index: 0) is full, with the rest empty. All the barriers are named according to the cell

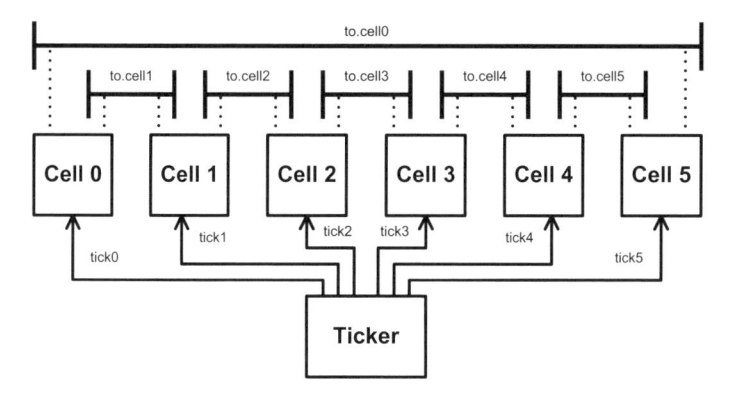

Figure 6. Token Cell Ring Network

following it – that is, the barrier that is "after" for cell 2 and "before" for cell 3 is named "to.cell3".

This network is portrayed in figure 6. The idea is a simple version of the model used for blood clotting in the TUNA project [9]. Eight iterations (i.e. eight ticks) of the network were run.

3.3.1. CSP

⟨ to.cell1, tick0, tick1, tick3, tick4, to.cell2, tick2, tick5,
tick2, tick0, tick1, tick3, tick5, to.cell3, to.cell4, tick4,
to.cell5, tick0, tick2, tick5, tick1, tick3, to.cell0, tick4,
tick0, tick2, tick3, tick5, to.cell1, tick1, to.cell2, tick4,
tick1, tick2, to.cell3, tick4, tick0, tick3, to.cell4, tick5,
tick2, tick3, tick4, tick5, tick0, to.cell5, tick1,
tick0, tick2, tick3, tick5, to.cell0, tick1, tick4,
tick1, tick2, tick4, tick0, tick3, to.cell1, tick5,
tick0, tick1, tick2, tick3, tick4, tick5 ⟩

3.3.2. VCR

⟨ {tick1, tick2, tick4, tick5, to.cell1}, {tick0, tick3, to.cell2},
{tick0, tick1, tick2, tick3, tick4, tick5},
{tick1, tick2, to.cell3}, {tick0, tick3, tick4},
{tick2, tick3, tick5, to.cell4}, {tick0, tick1, tick4, tick5},
{tick2, tick3, tick4, to.cell5}, {tick0, tick1, tick5},
{tick0, tick1, tick2, to.cell0}, {tick3, tick4, tick5, to.cell1},
{tick0, tick1}, {tick2, tick3, tick4, tick5, to.cell2},
{tick0, tick1, tick3, to.cell3}, {tick4, to.cell4}, {tick2, tick5, to.cell5},
{to.cell0}, {tick0}, {tick1}, {tick2}, {tick3}, {tick5}, {tick4} ⟩

3.3.3. Structural

⟨⟨⟨to.cell1* , 4*⟨tick0?⟩ , to.cell0* , tick0? , to.cell1* , 2*⟨tick0?⟩ , to.cell0* , tick0? , to.cell1* , tick0?⟩
 || ⟨to.cell1* , tick1? , to.cell2* , 4*⟨tick1?⟩ , to.cell1* , tick1? , to.cell2* , tick1? , to.cell1* , tick1? , to.cell2* , tick1?⟩
 || ⟨to.cell2* , tick2? , to.cell3* , 4*⟨tick2?⟩ , to.cell2* , tick2? , to.cell3* , 2*⟨tick2?⟩ , to.cell2* , tick2?⟩
 || ⟨8*⟨tick0! || tick1! || tick2! || tick3! || tick4! || tick5!⟩⟩
 || ⟨tick3? , to.cell3* , tick3? , to.cell4* , 4*⟨tick3?⟩ , to.cell3* , tick3? , to.cell4* , 2*⟨tick3?⟩⟩
 || ⟨2*⟨tick4?⟩ , to.cell4* , tick4? , to.cell5* , 3*⟨tick4?⟩ , to.cell4* , tick4? , to.cell5* , 2*⟨tick4?⟩⟩
 || ⟨3*⟨tick5?⟩ , to.cell5* , tick5? , to.cell0* , 2*⟨tick5?⟩ , to.cell5* , tick5? , to.cell0* , 2*⟨tick5?⟩⟩⟩⟩

3.3.4. Summary

We have separated the CSP and VCR traces approximately into the eight iterations of the process network – one per line. It can be seen that each iteration interleaves its events slightly differently as the token passes along the process pipeline. It is possible to see the movement of the token in each of the different styles of tracing.

3.4. I/O-PAR Example

An I/O-PAR process is one that behaves deterministically and cyclically, by first engaging in all the events of its alphabet, then repeating this behavior. I/O-PAR (and I/O-SEQ) processes have been studied extensively by Welch, Martin, Roscoe and others in [10,11,12,13,14]. Roscoe and Welch separately proved I/O-PAR processes to be deadlock-free, and further that I/O-PAR processes are closed under composition. This proof is not simple, and reasoning about I/O-PAR processes from their traces is not straight-forward.

For an example of a simple IO-PAR network, we use the example originally presented in [15]. One process repeats a in parallel with b ten times. The other process repeats b in parallel with c ten times. The two processes are composed in parallel, synchronising on b together.

3.4.1. CSP

$\langle$ a, b, c, c, a, b, a, b, c, a, b, c, a, b, c, a, b, c, a, b, c, a, b, c, a, b, c, a, b, c $\rangle$

3.4.2. VCR

$\langle$ {a, b, c}, {a, b, c}, {a, b, c}, {a, b, c}, {a, b, c}, {a, b, c}, {a, b, c}, {a, b, c}, {a, b, c}, {a, b, c} $\rangle$

3.4.3. Structural

$\langle\langle 10*\langle a \parallel b\rangle\rangle \parallel \langle 10*\langle b \parallel c\rangle\rangle$

3.4.4. Summary

Given a simple process network, the structural trace again has a direct mapping to the original program. The VCR trace shows the regular parallelism in the system, whereas the CSP trace reveals a slight mis-ordering in the second triple of events. This was predicted in the original paper; two c events happen in-between two b events in the CSP trace, but this slush is ironed out in the VCR trace.

4. Related Work

While VCR is a model of true concurrency, and the basis for implementing a tracing facility in CHP, it is by no means the only such model. A model of true concurrency is one which does not reduce concurrency to a nondeterministic sequential interleaving of events. A comprehensive survey paper by Cleaveland, et al. [16] discusses models of true concurrency versus interleaving models (such as CSP). Of the models discussed in [16], the earliest example of a model of true concurrency is Petri nets [17]. An introduction to Petri nets can be found in [18]. Kahn nets [19] provide a fixed-point semantics for the concurrency found in dataflow systems. While both are models of true concurrency, neither Petri nets nor Kahn nets are trace-based models. Three trace-based models of true concurrency are discussed in [16], and they are Mazurkiewicz traces [20], pomsets (partially-ordered multisets) [21], and event structures, by Winskel [22]. Mazurkiewicz traces define an independence relation on events to identify potential concurrency in traces of execution, which is in the same spirit of events contained in CHP's event multisets. From a purely model-theoretic standpoint, pomsets and event structures are similar in spirit to VCR's parallel event multisets and ROPEs (randomly ordered parallel events).

The independence relation discussed here is also similar to Lamport's seminal work on the happened-before relation [23]. Lamport defined a partial ordering in time of events in a system based on causality, inspired by the notion from special relativity that there is no single definitive ordering, and that different observers can disagree on ordering. This is the same idea that inspired VCR's tracing model. Lamport's work was in distributed systems with asynchronous communications. In this paper we have adapted the relation to a hierarchy of parallel processes that communicate synchronously. Our techniques do not inherently prohibit use in a distributed system, but are primarily suited to non-distributed systems.

Unlike much work on clocks (such as vector clocks [24]) in distributed systems, we do not attempt to synchronise time between different processes. Each process has its own process identifier (akin to a local clock), but it is never changed by, or synchronised with, other process identifiers. Our process identifiers reflect the process hierarchy, which bears some resemblance to work on hierarchical vector clocks [25] that tries to have a vector clock per level of the process hierarchy.

5. Future Work

There are several interesting avenues for future work, one of them inspired by two other tools. PRoBE is a formal CSP tool that allows step-by-step interactive exploration of the state space for a CSP program. At each step, one of the next possible events is chosen, and the program proceeds to the next step. The Concurrent Haskell Debugger [26] is a tool designed to visualise and step through Concurrent Haskell programs. It includes the capability to speculatively search ahead in the space of possible execution orderings to try to locate potential deadlocks [27].

We believe that the ideas behind these two systems could be combined, using the approach of the Concurrent Haskell Debugger with the trace recording facilities presented here, to provide programmers with a debugging tool that could present them with traces representing a deadlock in their program, searched for while they run the program.

We have given no consideration in this paper to the CSP notions of event hiding and renaming. Events have been taken to be globally visible, and cannot be renamed. It would be possible to augment the CHP API and trace recording mechanism to allow hiding and renaming of events. For example, in CHP one could write something like the following:

```
(p <||> q) <\\> ["c"]
```

This would represent the parallel composition of processes p and q, hiding event c in the resulting trace. This would be especially applicable to structural traces, because the hiding mechanism would be bound into the structure of the program.

6. Conclusions

We have explained how to implement the recording of CSP, VCR and structural traces, and have shown examples of each. We are not aware of any previous work on recording such traces from CSP implementations, besides the work of Barnes on compiling CSP [28]. We believe that being able to record traces is a useful tool for debugging.

One problem with recording traces that is especially apparent in our dining philosophers and token-cell examples is that traces can be large and difficult to understand. Tools to visualise and analyse traces will definitely be required for large and long-running programs. Process-oriented programming has always supported visual representations of its program layouts, and we hope that this could be integrated with replaying traces.

The work described here has been on the new Communicating Haskell Processes library, but the techniques should apply to any other CSP implementation or language. All that is really required is process-local storage and global shared data protected by a mutex – two easily available features in most settings. It would be interesting to compare the traces from different implementations of the same program.

The Communicating Haskell Processes library is publicly available under a BSD-like licence. Details on obtaining and using the library can be found at its homepage: `http://www.cs.kent.ac.uk/projects/ofa/chp/`. The tracing facilities are contained in the `Control.Concurrent.CHP.Traces` module.

Acknowledgements

We would like to thank our anonymous reviewers, and also Peter Welch, for their incredibly helpful and detailed comments on this paper.

References

[1] C. A. R. Hoare. *Communicating Sequential Processes*. Prentice-Hall, 1985.
[2] A.W. Roscoe. *The Theory and Practice of Concurrency*. Prentice-Hall, 1997.
[3] Peter H. Welch and Fred R. M. Barnes. Communicating mobile processes: introducing occam-pi. In *25 Years of CSP*, volume 3525 of *Lecture Notes in Computer Science*, pages 175–210. Springer Verlag, 2005.
[4] N.C.C. Brown. Communicating Haskell Processes: Composable explicit concurrency using monads. In *Communicating Process Architectures 2008*, September 2008.
[5] Formal Systems (Europe) Ltd. *Failures-Divergence Refinement: FDR2 Manual*. 1997.
[6] Marc L. Smith, Rebecca J. Parsons, and Charles E. Hughes. View-Centric Reasoning for Linda and Tuple Space computation. *IEE Proceedings–Software*, 150(2):71–84, April 2003.
[7] Marc L. Smith. Focusing on traces to link VCR and CSP. In East, Martin, Welch, Duce, and Green, editors, *Communicating Process Architectures 2004*, pages 353–360. IOS Press, September 2004.
[8] Tim Harris, Simon Marlow, Simon Peyton-Jones, and Maurice Herlihy. Composable memory transactions. In *PPoPP '05*, pages 48–60. ACM, 2005.
[9] P.H. Welch, F.R.M. Barnes, and F.A.C. Polack. Communicating complex systems. In Michael G Hinchey, editor, *Proceedings of the 11th IEEE International Conference on Engineering of Complex Computer Systems (ICECCS-2006)*, pages 107–117, Stanford, California, August 2006. IEEE. ISBN: 0-7695-2530-X.
[10] P.H. Welch. Emulating Digital Logic using Transputer Networks (Very High Parallelism = Simplicity = Performance). *International Journal of Parallel Computing*, 9, January 1989. North-Holland.
[11] P.H. Welch, G.R.R. Justo, and C.J. Willcock. Higher-Level Paradigms for Deadlock-Free High-Performance Systems. In R. Grebe, J. Hektor, S.C. Hilton, M.R. Jane, and P.H. Welch, editors, *Transputer Applications and Systems '93, Proceedings of the 1993 World Transputer Congress*, volume 2, pages 981–1004, Aachen, Germany, September 1993. IOS Press, Netherlands. ISBN 90-5199-140-1.
[12] J.M.R. Martin, I. East, and S. Jassim. Design Rules for Deadlock Freedom. *Transputer Communications*, 3(2):121–133, September 1994. John Wiley and Sons. 1070-454X.
[13] J.M.R. Martin and P.H. Welch. A Design Strategy for Deadlock-Free Concurrent Systems. *Transputer Communications*, 3(4):215–232, October 1996. John Wiley and Sons. 1070-454X.
[14] A. W. Roscoe and Naiem Dathi. The pursuit of deadlock freedom. *Information and Computation*, 75(3):289–327, December 1987.
[15] Mark Burgin and Marc L. Smith. Compositions of concurrent processes. In F.R.M. Barnes, J.M. Kerridge, and P.H. Welch, editors, *Communicating Process Architectures 2006*, pages 281–296. IOS Press, September 2006.
[16] Rance Cleaveland and Scott A. Smolka. Strategic directions in concurrency research. *ACM Computing Surveys*, 28(4), January 1996.
[17] C. A. Petri. Kommunikation mit automaten. Technical report, Schriften des IIm 2, Institut fur Instrumentelle Mathematik, Bonn, 1962.
[18] W. Reisig. *Petri Nets—An Introduction*, volume 4 of *EATCS Monographs on Theoretical Computer Science*. Springer-Verlag, Berlin, New York, 1985.
[19] G. Kahn. The semantics of a simple language for parallel programming. In J. L. Rosenfeld, editor, *Information Processing 74*, North-Holland, Amsterdam, 1974.
[20] A. Mazurkiewicz. Trace theory. In W. Brauer, W. Reisig, and G. Rozenberg, editors, *Petri Nets: Applications and Relationships to Other Models of Concurrency, Advances in Petri Nets, 1986, Part II; Pro-*

ceedings of an Advanced Course (Bad Honnef, Sept.), volume 255 of *Lecture Notes in Computer Science*, pages 279–324, Berlin, 1987.
[21] V. R. Pratt. Modeling concurrency with partial orders. *Int. J. Parallel Program.*, 15(1):33–71, 1986.
[22] G. Winskel. An introduction to event structures. In J. W. de Bakker, W. P. de Roever, and G. Rozenberg, editors, *REX School and Workshop on Linear Time, Branching Time and Partial Order in Logics and Models for Concurrency*, volume 354, pages 364–397, New York, 1989. Springer-Verlag.
[23] Leslie Lamport. Time, clocks, and the ordering of events in a distributed system. *Commun. ACM*, 21(7):558–565, 1978.
[24] C. J. Fidge. Timestamps in message-passing systems that preserve the partial ordering. In *Proceedings of the 11th Australian Computer Science Conference (ACSC'88)*, pages 56–66, February 1988.
[25] D.A. Khotimsky and I.A. Zhuklinets. Hierarchical vector clock: Scalable plausible clock for detecting causality in large distributed systems. In *Proc. 2nd Int. Conf. on ATM, ICATM'99*, pages 156–163, 1999.
[26] Thomas Böttcher and Frank Huch. A debugger for concurrent haskell. In *Implementation of Functional Languages 2002*, pages 129–141, 2002. Draft Proceedings.
[27] Jan Christiansen and Frank Huch. Searching for deadlocks while debugging concurrent haskell programs. *SIGPLAN Not.*, 39(9):28–39, 2004.
[28] F.R.M. Barnes. Compiling CSP. In P.H. Welch, J. Kerridge, and F.R.M. Barnes, editors, *Communicating Process Architectures 2006*, pages 377–388. IOS Press, September 2006. ISBN: 1-58603-671-8.

CSPBuilder – CSP based Scientific Workflow Modelling

Rune Møllegård FRIBORG and Brian VINTER

Department of Computer Science, University of Copenhagen,
DK-2100 Copenhagen, Denmark

{runef,vinter}@diku.dk

Abstract. This paper introduces a framework for building CSP based applications, targeted for clusters and next generation CPU designs. CPUs are produced with several cores today and every future CPU generation will feature increasingly more cores, resulting in a requirement for concurrency that has not previously been called for. The framework is CSP presented as a scientific workflow model, specialized for scientific computing applications. The purpose of the framework is to enable scientists to exploit large parallel computation resources, which has previously been hard due of the difficulty of concurrent programming using threads and locks.

Keywords. CSP, Python, eScience, computational science, workflow, parallel, concurrency, SMP.

Introduction

This paper presents a software development framework targeted for clusters and tomorrow's CPU designs. CPUs are produced with multiple cores today and every future CPU generation will feature increasingly more cores. To fully exploit this increasingly parallel hardware, more concurrency is required in developed applications.

The framework is presented as a scientific workflow model, specialized for scientific computing. The purpose of the framework is to enable scientists to gain access to large computation resources, which have previously been off limits, because of the difficulty of concurrent programming — the *threads-and-locks* approach does not scale well.

The major challenges faced in this work include creating a graphical user interface to create and edit CSP [1] networks, design a component system that works well with CSP and Python, create an execution model of the designed CSP networks and run experiments on the framework to find the possibilities and limitations. CSPBuilder can be downloaded from [2].

1. Background

Over the past few decades, companies producing CPUs have consistently increased processor speeds in each new edition by decreasing the size of transistors and increasing the complexity of the processor. The number of transistors on a chip have doubled every 2 years over the last 40 years, as declared by Moore's Law [3]. However, doubling the number of transistors does not automatically lead to faster CPU speeds, and requires additional control logic to manage these. Speed and throughput have typically been increased by adding more control logic and memory logic, in addition to increasing the length of the processor pipeline. Unfortunately more pipelines mean more branch-prediction logic, with the effect that it becomes very expensive to flush the pipeline when a branch is wrongly predicted. Many other extensions and complexities, e.g. SIMD pipelines, have been added to the CPU design during the past 40 years to increase CPU performance.

Today, numerous *walls* have been hit. The amount of transistors is still doubled every two years, so Moore's Law still applies. However, three problems have been raised: the *power wall*, the *frequency wall* and the *memory wall*. According to Intel [4], heat dissipation and power consumption increase by 3 percent for every 1 percent increase in processor performance. Intel also explain that because of bigger relative difference between memory access and CPU speeds, memory also becomes a bottleneck. Furthermore, the pipeline has become too long, so the cost of flushing outweighs the performance gained by increasing the pipeline length. All of these mean that we can go no further with current designs, and Intel suggest in [4] that the next step is parallel computation.

With several processing units, the *power wall*, *frequency wall* and *memory wall* are avoided, since there is no longer a need to increase the processor performance for a single unit. Instead you must be aware of communication and synchronisation between threads, which can cause overhead, deadlocks, livelocks and starvation if used wrongly.

Computers of tomorrow are getting more and more processing units, which can be utilized by creating concurrent applications that will scale towards many processors. We are already at 128+ cores in graphic processors, 9 cores in the CELL-BE processor from IBM, SONY and TOSHIBA and recently Intel announced that they are experimenting with an 80-core CPU [5].

1.1. Motivation

Many scientists (chemists, physicists, etc.) are not experienced programmers, but are able to do scientific computing by programming sequential applications. So far they have been relying on the hardware manufactures to produce hardware which has improved the performance of their applications — allowing for more sophisticated and computationally intensive science.

Due to the limitations of sequential computing already discussed, scientists must now develop *concurrent* applications, in order to take advantage of parallel hardware and to advance the science. The amount of difficulty involved in creating concurrent applications, depends on the programming language and methodology. Traditional concurrent programming, with *threads* and *locks*, makes it difficult to program even simple applications — adding more parallelism to an already threaded program tends to result in problems, not solutions. As a direct result, concurrent programming is seen as *hard*, and is generally avoided by the majority of programmers.

We want to encourage scientists to develop concurrent programs using a CSP [6] based approach, where applications are built as layered networks of communicating processes. Such an approach is *reliable*, no unexpected surprises; *scalable*, to different numbers of processes and processors; and *compositional*, enabling processes to be 'glued' together to build increasingly complex functionality.

A feature of CSP based designs is that every process can be completely isolated from the global namespace, only interacting with other processes through well-defined mechanisms such as channel inputs and outputs — processes are *not* context sensitive. This in turn permits a high level of code reuse within scientific communities, as previously built components can be connected in different ways, corresponding to the data-flow of a particular computation.

Recent reports of using the GPU[1] and CELL-BE for scientific computing, have reported performance increases of up to 100-fold for some scientific algorithms. However, the diffi-

---

[1]*Graphics Processing Unit* – general-purpose graphics hardware found in high-end workstations, e.g. the NVidia GeForce2.

culty of programming on a GPU or the CELL-BE is evident, and we desire a high level of code reuse — i.e. algorithms written should be able to run on a number of different architectures, without a significant porting effort. This includes within a single-processor system, heterogeneous multi-core systems, and distributed over networks of machines. A CSP based design, of communicating processes, allows us to mix and match processing architectures — selecting the best performing implementations of processes for particular architectures.

While architectures have differing performance characteristics, programming in different languages can also affect performance. Development in a high-level language such as Python is usually faster, but produces code that runs slower than a similar implementation in a low-level language, such as C. By programming the computation intensive parts in C, and using Python as the 'glue', we optimize the execution time and avoid having to program the entire application in C, saving development time.

When doing scientific work, which often relies on particular mathematics libraries to do the "number crunching", the functions provided are not necessarily all implemented in the same language. By using tools such as SWIG [7] and F2PY [8] we hope to address this issue, making it possible to use code from C, C++ and Fortran in a single scientific application.

Our solution is to provide a framework, written in Python, that assists scientists in creating concurrent applications based on a CSP design. The framework uses a graphical user interface similar to other *flow-based* programming environments already available, and as such, we hope that scientists will find our framework useful and accessible.

1.2. PyCSP

PyCSP [9] is the CSP [1] library for Python used in this paper. It is a new implementation and is currently evolving into a stable library. At the moment it supports four different channel types, that can be used for connecting parallel processes: *one-to-one*, *one-to-any*, *any-to-one* and *any-to-any*. Similar to occam, support for guarded choices is only available on the reading ends of *one-to-one* and *any-to-one* channels. When more than one process is attached to the *any* end of a channel, only one process at that end is involved in the communication, and queue in a FIFO. Communication on channels is synchronous — a channel output will not complete until the inputting process has accepted the data. In the future, we hope to support all types of guards for channel communication, as well as having full support for networked channels, and the easy distribution of CSPBuilder applications across computer networks.

The syntax of PyCSP is fairly simple and works well in Python. When executing a CSP network using PyCSP, all processes are created as kernel threads, though performance on *shared-memory* architectures is limited by the *Global Interpreter Lock* (see section 3.1.4).

1.3. Scientific Workflow Modelling and CSP

The purpose of a scientific application is usually to calculate a result based on input data. This data flows through the application and is the basis of sub-problems and sub-solutions until eventually a result, or several results, are found. With this in mind we use the term "workflow" for the data-flow of a scientific application. We use the term "scientific workflow" for the workflow of eScience applications, where "eScience" is used to describe computationally intensive science applications, normally run on shared-memory multi-processor hardware or in distributed network environments.

A typical eScience application might be anything from complex climate modelling to a simple n-body simulation. Generally, any application that does a large number of computations to produce a result within a particular scientific field.

Only a few [10,11] have previously looked at CSP and thought that this might be a good description for scientific workflows. In this paper we will produce an application that uses some of the ideas from CSP algebra and the projects mentioned above, combined in a framework that allows CSP based applications to be designed in a visual tool, and executed in a variety of ways (depending on the hardware available). We stipulate that CSP is ideal for reasoning about the dataflow of eScience applications, particularly when the target environment is concurrent execution. The compositional structure of a CSP network enables application developers to reuse networks of components as top-level components themselves.

In section 5 we cover some of the other frameworks available. Some of these are very popular today, and at the PARA '08 event there was an entire day of workshops devoted to scientific workflow modelling. The scientists there argued that they are able to understand flow-based programming environments, and use them to develop scientific applications. The future users of CSPBuilder are the same as for other frameworks, and by making CSPBuilder operate in a similar fashion, we expect that those users will be able to use the CSPBuilder framework to construct applications.

One of the reasons for working with scientific workflows is to enable access to large computation resources. The model presented in this paper, in addition to support for remote channels, will make it possible to divide scientific workflow applications from a small number of CPU cores, to hundreds of nodes on different LANs — provided that the application is designed in a way that supports this; a design method that is promoted by the CSPBuilder framework.

1.4. Summary of Contributions

A new framework is implemented, tested and benchmarked in this paper. This framework consists of a visual tool to build applications and a tool to execute the constructed applications. The framework is implemented in Python and supports to use C, C++ and Fortran code by providing 'wizards' to access these languages. The framework is called CSPBuilder and incorporates extensive use of the CSP algebra.

The visual tool provides an "easy to use" graphical user interface, enabling users to construct applications using the ideas of flow-based programming [12] to produce a CSP [1] network. In our experiments we show that the visual tool is capable of handling large and complex applications.

Applications that are constructed with CSPBuilder can be executed successfully on a single computer, combining routines from a number of different programming languages. With the future introduction of remote channels in PyCSP it will be possible to execute the applications on any number of hosts.

The framework encourages code reuse by constructing applications from reusable components. This has proven very useful during the experimentation phase.

The primary advantages of this framework lie in code reuse and constructing complex scientific applications focusing on the workflow. CSP ideas underpin the concurrency mechanisms employed in constructed applications, enabling the automatic deconstruction of whole systems into individual concurrent components.

2. The Visual Tool

This section describes a user-friendly application that can model a CSP network using a layout similar to flow-based programming [12]. This layout is required to resemble the CSP network for a scientific workflow model. Figure 1 shows an application modelled using our visual tool.

In CSPBuilder every application starts with a blank canvas, where processes and channels can be inserted. Processes appear as named boxes, with their external connections labelled. Channels are shown as lines connecting the processes. To simplify things, any in-

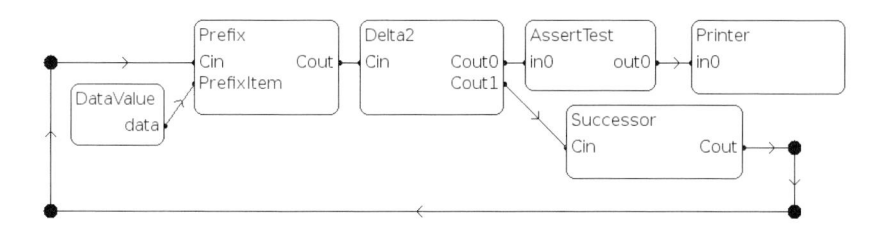

Figure 1. A CSPBuilder application that generates incrementing natural numbers.

bound or outbound connection will only accept one channel going in or out, depending on the connection type.

A number of connected processes are known as a process network, as shown in figure 1. This network could be used as a component in another application, described in section 2.1.

The remainder of this section describes the component system, connecting components with channels and connection points. Saving and loading CSP applications to and from files are then described, followed by details on component configuration and replication. These parts are necessary to construct an application, and are parts of the framework that make it possible to build CSP networks that can be run efficiently in a distributed environment.

2.1. Component System

The design of the component system is based on the following requirements:

- We need to be able to link the Python code of each process in an easy to understand framework, to make it simple to add or remove components.
- The organisation of the process network needs to be scalable, which means that the user should be able to handle large and complex applications, without losing control or an overview of the whole system.
- The user should quickly and easily be able to group parts of the process network into components, that appears and function like other processes.
- Components should be stored in a library for reuse.
- An application built with CSPBuilder must be targetable to different hardware, and have a performance better than or equal to an equivalent application written entirely in Python.

These requirements are examined in more detail in the following sections.

2.1.1. Scalable Organisation

Consider a network of 2000 processes. To handle this many processes, and even more channels, it is necessary to group parts of the network into smaller compositional processes. This can be done by allowing the user to select a group of connected processes and condense them into a single component. If this new component has unconnected inbound or outbound connections, these are added to its interface, in addition to channels that already cross the group boundary. From an external perspective, this new component looks like any other component in the system.

Collecting together components in groups, and using these to form other components, leads naturally to a tree structure, whose leaves are component implementations. Each level of the tree is assigned an increasing *rank* number, with leaf processes having a rank of 1. This is used to prevent cyclic structures.

2.1.2. Components

Components are the most important part of CSPBuilder. A component is a CSPBuilder application that has been stored in the component library. These stored components are available for use in other applications, and come in two different forms:

1. The component is a process network consisting entirely of process instances of other components and includes no actual code implementations.
2. The component includes at least one process that contains a process implementation. This process implementation has a link to a Python function that implements the process. A simple example of a process in CSPBuilder is "IDProcess", shown in listing 1, that simply forwards data received on its input channel to its output channel.

```
1  from common import *
2
3  def CSP_IdProcessFunc(cin, cout):
4      while 1:
5          t = cin()
6          cout(t)
```

Listing 1. Example CSP process implementation – the IDProcess

To make it as easy as possible for the user to create components, we specify that to create a component, you just have to copy or move your CSPBuilder application to a "Components" directory. When the CSPBuilder application reloads the library, it discovers this new component and makes it available for use in new applications.

Functions specific to building components are also incorporated. These include naming unconnected channel-ends and naming the main application. When creating components, the application name is used for the new component. Unconnected channel-ends for the component's input and output are named in similar ways.

2.1.3. Component Library

To aid in component management, each component requires a package name. This is to make it easier to find the desired component, for example, a "statistics" package containing relevant statistical components. For CSPBuilder to be an effective tool, it will need a wide variety of components, offering a range of different functionalities.

2.1.4. A Wizard for Building Components

A developer should be able to reuse code made by others, or reude code made earlier in another application. Reusing older code is made easier with components and the component library, so to increase the ease of creating new components a 'wizard' has been implemented that guides the developer through the process of creating a component.

A quick search on the Internet will show that large online archives of scientific code are available for free use. It is desirable to be able to easily use a function written in any language, and currently it could be argued that it is possible just by having the components implemented in Python. The developer can use SWIG [7] to import code from C or C++, and most programming languages are able to build libraries that can be used from C or C++. This therefore makes it possible to extend Python with code written in all kinds of languages. A project named F2PY [8] can import Fortran 77 and Fortran 90 code into Python.

The wizard guides the user through the process of creating components written in Python, C, C++, Fortran 77 and Fortran 90. These languages were chosen because of the numerous scientific libraries that use these. As mentioned earlier, most languages can build a library that is accessible from C or C++.

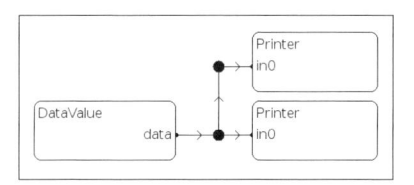

Figure 2. One2AnyChannel formed by connecting three processes to a single connection point, single outputting process, multiple inputters.

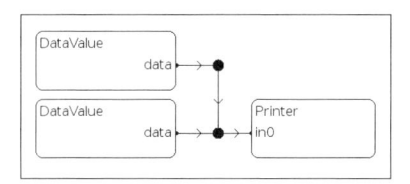

Figure 3. Any2OneChannel formed by connecting three processes to a single connection point, single inputting process, multiple outputters.

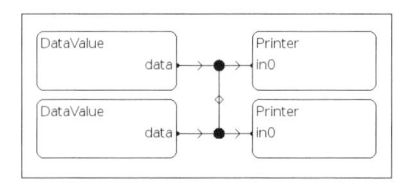

Figure 4. Any2AnyChannel formed by connecting four processes to a single connection point, multiple inputting and outputting processes.

The inclusion of other programming languages is expected to have a positive effect on application performance in CSPBuilder. Python uses the *Global Interpreter Lock* (see section 3.1.4) to access Python objects. This means that only one Python thread is allowed to access Python objects at any one time, limiting any advantage of running threads that are not dependent on each other in parallel. This lock can be freed when executing external code imported into Python, making it efficient to have certain parts written in other languages. Also, compiled languages are typically faster than interpreted languages, which further improves performance.

2.2. Channels and Connection Points

Processes connected by channels form a process network. The different types of channels available and how they work in PyCSP were introduced earlier. The types of channels are One2OneChannel, One2AnyChannel, Any2OneChannel and Any2AnyChannel.

The One2OneChannel is simple, because it can be represented by a single line going from one process to another. Representing the other types of channel is more complex. To address this issue, we introduce connection points. These can have any number of inbound and outbound connections, to processes or other connection points, enabling visualisation of all channel types and for the 'bending' of channels. Examples of these can be seen in figures 1, 2, 3 and 4.

Before any code can be generated, or process networks constructed, the connection graph for each channel is reduced to contain at most one connection point. Starting with each connection point, or node, that node's neighbours are examined. If that neighbour is another node, as opposed to a process, the connections there are moved to the current node. This is done recursively, until only single connection points remain, and runs in $O(n)$ time, where n is the number of connection points.

The visual tool does not currently indicate the type of data carried on a channel, but the channels are typed (in Python). When trying to execute a mis-connected network, the tool will generate an error.

2.3. Configuring a Component

When working with the visual tool some components will need to be configured. These components should have their individual configuration functionality specialised for their specific purpose. A method is provided for the user to configure the component and save this setting in the .csp file, for later execution. A typical example of component configuration is something that allows the user to specify the name of a data file. To handle this, a structure is defined that a component has to implement in order to provide a configuration functionality.

We will now focus on the three issues of configuring a component:

1. Activate the configuration process.
2. Save the new configuration.
3. Load saved or default configuration on execution.

As mentioned in section 2.1.2, the Python implementation of a component is a file that we import, with its own name-space. If this name-space has a function named setup(), we call this function when the user configures the component. If the function does not exist, the user will not be able to configure the component. To save the configuration, any structure returned by this setup() function is serialized and saved in the component's .csp file. When executed, the component's top-level function is provided with the previously saved unserialized data structure. An example of a small configurable component is shown in listing 2.

It is left to the individual component programmer to decide what user interface will be used to configure the component. In the example shown in listing 2, a wxWindows file dialog is used to acquire input from the user.

The configuration data may be saved on several levels. When working with CSPBuilder a configuration can be saved on the working level or on any lower level, down to the rank where the process implementation is located. As standard all saved information from setting up components is saved in the working process and not in the process with the implementation. This gives the possibility for different setups for every application, and necessary to create components that are as general as possible. Saved configurations are attached to the process instance.

Configuration data with a higher rank will override any configuration data with a lower rank. This has the desired effect: that any configured process instance of a component will use the most recent configuration, as long as it is activated in the main application, and not as part of any other component.

2.4. Process Replication

When building applications for concurrent scientific computing, a common way to organize the calculations, if the algorithms allow it, is to divide the calculation into different jobs and process these concurrently with workers. An application that use 50 workers would quickly become cumbersome in CSPBuilder because of the 50 process instances in the visual tool. To address this issue, a process multiplier is created. When enabling the process multiplier on a process instance, the user must enter the desired number of replications.

```
 1  configurable = True
 2  from common import *
 3  import pylab
 4
 5  default_data = None
 6
 7  # Configuration (called from builder.py)
 8  def setup(data = default_data):
 9      import wx
10      import os
11      wildcard = "PNG (*.png)|*.png|" \
12                 "All files (*.*)|*.*"
13
14      saveDir = os.getcwd()
15
16      dlg = wx.FileDialog(
17          None, message="Choose an image file, containing the data",
18          defaultDir=os.getcwd(),
19          defaultFile="",
20          wildcard=wildcard,
21          style=wx.OPEN | wx.CHANGE_DIR
22          )
23
24      if dlg.ShowModal() == wx.ID_OK:
25          paths = dlg.GetPaths()
26          data = paths[0].replace(saveDir + '/', '')
27
28      os.chdir(saveDir)
29      dlg.Destroy()
30      return data
31
32  # CSP Process (called from execute.py)
33  def ReadFileFunc(out0 , data = default_data):
34      img = pylab.imread(str(data))
35      out0(img)
```

Listing 2. An example of a component that has configuration enabled

Any channels connected to a process instance where a multiplier has been set, can be thought of as being multiplied by the corresponding amount. The addition of extra channels and processes is handled in the execution step.

On execution, a multiplier x will cause the specified process instance to be created in x exact copies. If the process instance is an instance of a process network this network will be multiplied in x exact copies, creating x times the number of processes and channels in the process network. When a process is multiplied, all connections are multiplied as well and will be turned into One2AnyChannels, Any2OneChannels or Any2AnyChannels.

3. Concurrent Execution

In this section we describe how a data structure, constructed by the visual tool and saved to .csp files, is executed successfully. This is done by converting the data structure into a structure resembling a CSP process network. The PyCSP library is used to construct processes and their connections, and finally to execute those processes.

All functionality presented by the visual tool in section 2 must be handled in the execution step. Here we will focus on the requirements relevant when executing on a single system. The non-trivial functionalities required include: channel poisoning; multiplication of components and their connections; importing external code; and releasing the *Global Interpreter Lock*.

3.1. Building and Executing a Process Network

The overall goal is to build a network that will have a performance similar to a network implemented entirely in Python using PyCSP. This means that all parsing and network building needs to be done before execution and cannot be done on demand. To improve performance, the tree data-structure describing processes is first flattened, as shown in figure 5.

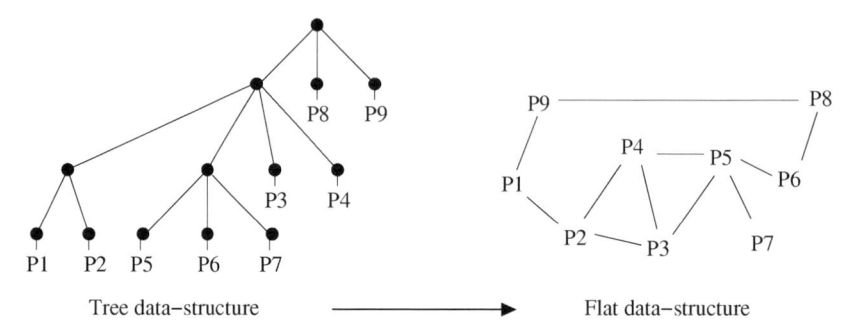

Figure 5. Data structures. In the left figure the tree data-structure is illustrated, which represents the structure of the CSP network when the .csp files are parsed. The black dots are a process structure and the lines represent any number of connection structures. This data-structure is converted into the flat data-structure illustrated in the right figure. This is a one-way conversion and can not be reversed.

An important feature in the construction of CSPBuilder has been to resemble the CSP algebra in the visual tool. During execution it is equally important to execute the CSPBuilder application exactly as it was built, and to ensure that everything is executed correctly. Here we focus on guards, channel poisoning, importing external code and releasing the *Global Interpreter Lock*, which comprise the difficult parts of executing a CSPBuilder application.

3.1.1. Multiplying Processes

Multiplying a process only makes sense in cases where a computation is embarrassingly parallel, meaning that the problem state can be sent to a process and the process can compute a result using this state data, with no dependencies, and send the partial result to a process that collects all partial results into a final result. This design is usually called a producer-worker or a producer-worker-collector setup and works best with embarrassingly parallel problems. A dynamic orchestration of processes is used where the amount of workers can be varied easily and you can have many more jobs than workers, making it easier to utilize all processes. If a computation can not be done in a dynamic orchestration design, then it does not make sense to use this multiplier flag. Instead a static design can be built with specialized components for doing a parallel computation with $2, 4, 8, ...$ processes.

Another design where multiplying processes will be applicable is in process networks handling streams. Imagine 4 processes connected in serial, doing different actions on a stream. If one of these steps is more time-consuming than any of the others, it will slow down the entire process. Multiplying this process is simple and if hardware is available for the extra process, it improves the overall performance of the process network.

3.1.2. Channel Poisoning

In CSP, without channel poisoning, a process can only terminate once it has fulfilled its task. This creates a problem when a process does not know when it has fulfilled its task. When constructing a network of communicating processes most of the processes will be the kind that will never know when they have fulfilled their task. They will read from their input channels, compute and send the resulting data to their output channels. These processes combined will compute advanced problems and loop forever. One might add a limit saying that a process will do 500 loops and it can consider its task fulfilled. In some applications this is possible, but most applications can not define the needed loops prior to execution. Also one might construct an extra set of channels that will communicate a signal to the processes letting them know that their task is fulfilled, and initiate a shut-down. Channel poisoning is a clever method to do just that, but uses communication the channels that already exist. PyCSP has support for channel poisoning, which is based on channel poisoning in JCSP [13,14].

Channel poisoning is implemented in PyCSP by raising an exception in process execution, when a channel connected to this process is poisoned. The exception is caught by the PyCSP library and poisons all other channels connected to this process. After poisoning all channels connected to the process, the process terminates. This will eventually terminate all processes and cause the entire application to exit as desired.

If a process is currently waiting on a non-poisoned channel, then nothing will happen in the process until it reads or writes from one of its poisoned channels. This might happen if a process is waiting for an action and it is another process that has poisoned the network and desires that the application terminates. The application will stall until the action happens and the process writes or reads to the poisoned network.

For this reason when constructing CSPBuilder applications it is important to consider how an application is poisoned if the user wants the application to terminate at some point.

3.1.3. Importing External Code

The wizard for CSPBuilder described in section 2.1.4 provides an easy method for building a component that calls into C, C++ or Fortran code. In this section the framework for using external code in CSPBuilder is described.

Using the import statement in Python it is possible to import modules. A module can be a Python script, package or it can be a binary shared library, as in this case where we want to use code from other programming languages.

For importing Fortran code the F2PY [8] project is used, which is capable of compiling Fortran 77/90/95 code to a binary shared library, making it accessible for Python. To import C or C++ code the SWIG [7] project is used to compile to binary shared libraries, similar to F2PY. Both projects are wrappers that make it relatively easy to handle data conversion between Python and other languages.

All external code will reside in the `External` folder in the CSPBuilder directory. A module name specifies a sub-directory in `External`, where all source and interface files are located. When compiled, the generated module will be saved as a '`.so`' file with the module name as its file name in the `External` directory. A *Makefile* is created for every component and for the entire `External` directory, so that all modules can be compiled by executing `make` in the `External` directory. This is necessary when applications are moved to different machines, where the architecture and shared library dependencies may vary.

3.1.4. Releasing the GIL

PyCSP [9] uses the Python *treading.Thread* class to handle the execution of processes in a CSP network. This class uses kernel threads to implement multi-threading which should enable PyCSP to run concurrently on SMP systems. Unfortunately concurrent execution of threads is prohibited by the GIL. The GIL (Global Interpreter Lock) is a lock that protects access to Python objects. It is described in the documentation of Python threads [15]. Accessing Python objects is not thread-safe and as such cannot be done concurrently.

To be able to utilize the processors in an SMP system we will release the GIL while doing computations outside the domain of Python. In section 3.1.3 it was explained how external code can be imported into Python. When calling into Fortran code using F2PY the

GIL is released automatically and acquired again when returning to Python. With C and C++ the situation is different, because here it is possible to access Python objects by using the API declared in python.h. It is the responsibility of the component developer to not access Python objects while the GIL is released. Releasing and acquiring is done with the following macros defined in python.h:

```
// Release GIL
Py_BEGIN_ALLOW_THREADS

// Acquire GIL
Py_END_ALLOW_THREADS
```

The effects of releasing the GIL can be seen in section 4.1 where experiments are carried out on an SMP system. We have now covered relevant issues in the building and execution of a process network and can construct a CSP network from the .csp files created in the visual tool.

3.2. Performance Evaluation

A classic performance test for CSP implementations includes the Commstime [16] test, which is commonly used for benchmarking CSP frameworks. This computes the time spent on a single channel communication. In this test we will compare the performance of the Commstime test written in "Python with PyCSP", with the CSPBuilder created "Commstime" application shown in figure 6. The CSPBuilder Commstime creates a CSP network in PyCSP and should perform the same, with perhaps only a slight overhead of having to create the extra *DataValue* process. In table 1 the result of the tests are shown. When comparing, there is a slight difference where the *DataValue* process is concerned, but this process is necessary to initialise the network and cannot be removed from the application. In "Python with PyCSP" this data-value is a simple integer.

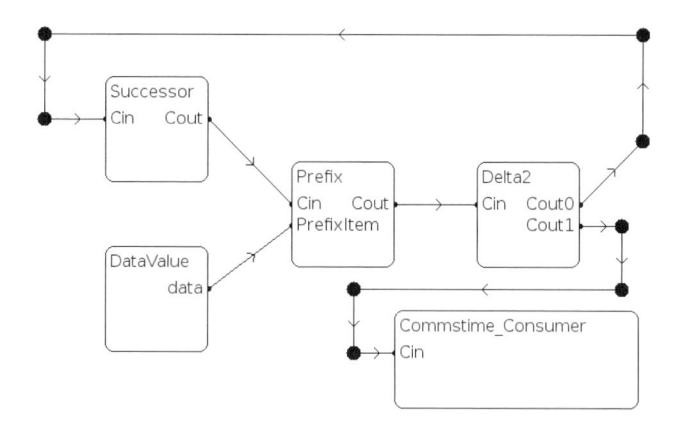

Figure 6. Commstime. A CSPBuilder application that resembles the Commstime performance test.

Table 1. CSPBuilder Commstime. A comparison of the channel communication time when using CSPBuilder vs. only Python and PyCSP. The Commstime tests were executed on a Pentium 4 2Ghz CPU.

Test	Avg. time per. chan (μs)
Python and PyCSP	91.43
CSPBuilder	96.30

The results of CSPBuilder are as expected. The performance of Python and PyCSP are not competitive to many other CSP implementations, especially compilable languages. However, Python has many other advantages that in our case outweigh the poor performance:

- Easy to use and very flexible.
- Can interact with most languages.
- Many scientists already know Python.
- Faster development cycle.
- Encourages programmers to write readable code.
- Compute intensive parts can be written in compilable languages.

4. Experiments

In this section we test the performance of CSPBuilder using a simple *Prime Factorisation* experiment. The tests will be performed with a varied amount of workers in the application. Workers are the processes that, because of the design of the process network, are meant to be identical, run concurrently and compute sub-problems of a larger problem.

The experiments show that CSPBuilder is capable of executing applications on an 8 core SMP system. On the 8 core SMP system the GIL is released to be able to utilize all cores successfully.

4.1. Prime Factorisation

As a test case for executing applications in CSPBuilder, *Prime Factorisation* was chosen. It is simple and the computation problem can easily be changed to run for varying times. In the book by Donald Knuth [17], 5 different algorithms for doing prime factorisation are explained. The simple one is the least effective and is based on doing *trial division*[2]. *Trial division* is used in the *direct search factorisation*[3] algorithm. The simple prime factorisation algorithm was chosen for the following reasons:

- Parts of the algorithm can to be written in both C and Python. The simplicity of the algorithm is an advantage here.
- The nature of the algorithm makes it possible to use the *multiplier* functionality in CSPBuilder. The algorithm is easy to divide into jobs that can be computed by workers.
- With a simple algorithm it will be easier to identify the aspects that do not perform well.
- The algorithm has limited communication, but still enough to test various cases, e.g. distributed vs. one machine.

A serialized Python implementation of the *direct search factorisation* algorithm can be found at PLEAC[4] (the Programming Language Examples Alike Cookbook). This implementation is extended and adapted to a parallel version that we implement in the CSPBuilder framework.

4.1.1. Implementation Details

The *prime factorisation* problem is built as a component reading a number as input and outputting a result. Since *direct search factorisation* is an embarrassingly parallel problem, the processing can be divided into jobs and handed over to a set of workers as illustrated in figure 7.

---

[2]Trial division: http://mathworld.wolfram.com/TrialDivision.html
[3]Direct search factorisation: http://mathworld.wolfram.com/DirectSearchFactorization.html
[4]PLEAC: http://pleac.sourceforge.net/pleac_python/numbers.html

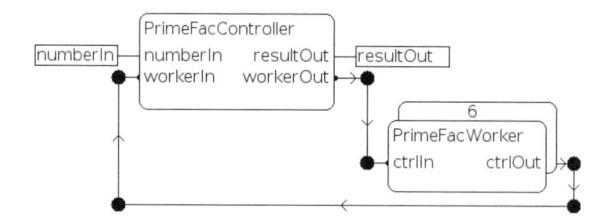

Figure 7. PrimeFac Component, consisting of a controller and a worker multiplied 6 times.

On initialisation, the worker process sends an empty result to the controller, to indicate that it is ready for more work. The controller loops until all primes have been found, sending jobs to and collecting results from workers. If a non-empty result is received, the controller waits for all workers to finish and, if any other workers also had a non-empty result, the best result is picked and the computation resumes.

If n is the number we are factorizing into primes, then all primes have been found when $d >= \sqrt{n}$, where $[2 \ldots d]$ are the divisors tested. All the prime factorisations of n can be found in $[2 \ldots \sqrt{n}]$.

Numbers that are particularly interesting to factorize into primes are those larger than the representation available generally in compilers (e.g. 32-bit and 64-bit). To work with unsigned integers larger than 18446744073709551615, which is the limit for $64bit$ registers, some special operations are needed. Numbers larger than this need software routines for doing basic operations such as addition, subtraction, multiplication and division.

Python has internal support for large numbers which makes the task of implementing prime factorisation in Python much simpler. Creating the C version is a bit more tricky. An external component is created using the wizard described in section 2.1.4. To test the implementation, a version working with numbers less than $64bits$ is created. All basic mathematical operations are then replaced with function calls to the library "LibTomMath"[5], which handles large numbers. For transferring large numbers between Python and C a decimal string format is used.

Finally we add a release for the GIL as described in section 3.1.4, which enables us to maximize concurrent execution in the application.

4.1.2. Performance Evaluation

For our experiments the Mersenne[6] number $2^{222} - 1$ is used. This number was picked by trial and error, with the purpose to find a number where the prime factorisations could be computed within 30 minutes for the least effective run. All tests have solved the problem:

$$n = 2^{222} - 1$$
$$= 6739986666787659948666753771754907668409286105635143120275902562303$$
$$= 3^2 * 7 * 223 * 1777 * 3331 * 17539 * 321679 * 25781083$$
$$* 26295457 * 319020217 * 616318177 * 107775231312019$$

In the performance test we compare the two implementations, one with the worker written in Python and one with the worker as an external component written in C which also releases the GIL. In the C implementation we use the large number library *LibTomMath*. This large number implementation is actually slower than the large number implementation

---

[5]LibTomMath: http://math.libtomcrypt.com/
[6]Mersenne number: http://mathworld.wolfram.com/MersenneNumber.html

in Python, shown in the tests where the "Python only" version outperforms the "Python and C" version for the case with only one worker. We base this conclusion on the fact that the sequential test for "Python and C" finishes in 1547 minutes, while the "Python only" version finishes in 1005 minutes. Both implementations spend all of the execution time in the worker loop with very little communication between processes.

To compare the effects of adding more workers we examine tests with 1, 2, 4, 6 and 8 workers, shown in figure 8. The "Python and C" version performs well, and by looking at the speedup in figure 9, we see that performance scales almost linearly. This means that adding double the amount of workers on a system with double the capacity doubles the performance and halves the run-time. The speedup shown in figure 9 is not quite linear. The drop in performance is caused by having to flush the workers every time a result is found. Time is then spent sending new jobs to workers. This overhead increases with the number of workers, but is largely acceptable given the advantages and benefits of this approach. All benchmarks were run on an 8 core SMP system.

The increase in run-time, when adding workers to the "Python only" version in figure 8, is caused by the unnecessary context-switching and communication, since the added workers will only steal CPU time from the first worker. The reason that the run-time only increases by a little even though many workers are added, is that the other workers are starved and therefore will never ask for a job to compute.

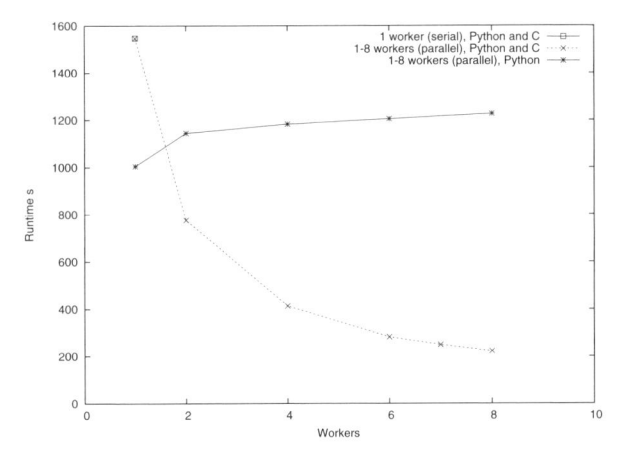

Figure 8. Prime factorisation of the Mersenne number $2^{222} - 1$.

The sequential benchmark is based on single worker execution. This is arranged by setting the job size to 10^{16} iterations, which causes only one job to be sent to the single worker waiting. This benchmark provides a baseline reference for sequential execution speed in CSPBuilder, and is used as the basis when calculating the speedup of the parallel benchmark shown in figure 9.

These results show us that when constructing a scientific workflow in CSPBuilder, it is possible to get a reasonable performance and avoid the GIL, by programming the computationally intensive components in compilable languages. CSPBuilder is usable for both coarse-grained and fine-grained construction of whole systems. With a coarse-grained process network, we require the computation intensive components to execute concurrently internally, if a reasonable performance is desired. With a fine-grained process network, internal concurrency in the components is not necessary. The *prime factorisation* implementation is somewhere in between a coarse-grained and fine-grained network.

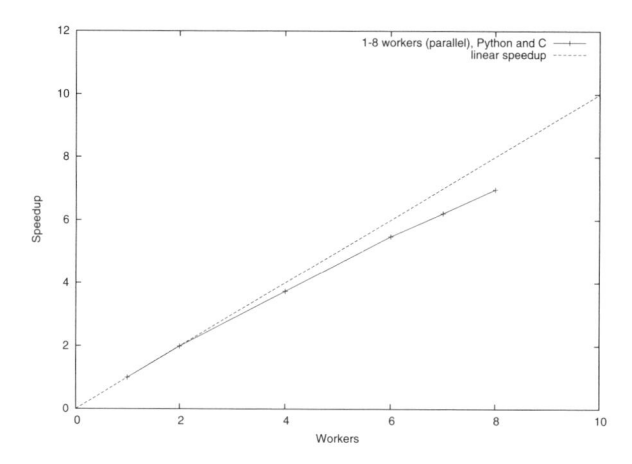

Figure 9. Speedup of prime factorisation of the Mersenne number $2^{222} - 1$.

5. Related Work

Several different frameworks exist that can handle scientific workflows in different ways. To mention some of the more common, there are *The Kepler Project*[7] [18], *Knime*[8], *LabVIEW*[9], *FlowDesigner*[10] and *Taverna*[11]. The graphical tool of CSPBuilder is a quite similar to these frameworks, though currently less functionality is available in CSPBuilder. CSPBuilder differs by having a basic graphical tool, that assists in constructing a CSP network and manages a component library. The power of the CSPBuilder framework lies in the communication model based on CSP.

On the CSP side, Hilderink [19] has created a graphical modelling language, GML, in which CSP networks can be defined.

6. Conclusions and Future Work

In this paper we have presented a graphical framework for designing and building concurrent applications based on CSP. Ideally suited to current and future multi-processor and multi-core machines, CSPBuilder provides a simple and intuitive means for designing concurrent applications. The graphical tool compiles directly to Python using PyCSP, and supports transparent integration of C, C++ and Fortran functions. Experiments have shown that near linear speedup can be obtained on embarrassingly parallel applications, which demonstrates that the CSPBuilder tool dos not impose any significant overheads.

This paper has hinted at the distribution of CSPBuilder applications on networks of workstations and other distributed memory architectures. Although PyCSP does support networked channels, some modifications to the basic channel code in PyCSP have been made as part of the work presented here. Similar changes will need to be made to the network channel code in PyCSP before CSPBuilder is able to target these architectures.

It might also be interesting and useful to add more descriptive visual representations of channels, inspired by Hilderink, such as identifying guarded choice on channel inputs to a process.

---

[7]The Kepler Project: http://www.kepler-project.org/
[8]Knime: http://www.knime.org/
[9]LabVIEW: http://www.ni.com/labview/
[10]FlowDesigner: http://flowdesigner.sourceforge.net/
[11]Taverna: http://taverna.sourceforge.net/

Although CSPBuilder is at a relatively early stage of development, we hope that it will grow and flourish, eventually becoming a useful tool to aid scientists in constructing scientific workflows, as well as for the programming of CSP based concurrent applications generally.

References

[1] C. A. R. Hoare. *Communicating Sequential Processes*. Prentice Hall International, june 21, 2004 edition, 2004.
[2] The CSPBuilder Framework. http://www.migrid.org/vgrid/CSPBuilder/.
[3] Description of Moores Law. http://www.intel.com/technology/mooreslaw/. Viewed Online January 2008.
[4] S. Borkar, P. Dubey, K. Kahn, D. Kuck, H. Mulder, S. Pawlowski, and J. Rattner. Platform 2015: Intel Processor and Platform Evolution for the Next Decade. *Intel White Paper*, 2005.
[5] Annoncement: 80 core CPU. http://www.intel.com/pressroom/archive/releases/20070204comp.htm. Viewed online september 2007.
[6] C. A. R. Hoare. Communicating sequential processes. *Commun. ACM*, 21(8):666–677, 1978.
[7] Simplified Wrapper and Interface Generator (SWIG). http://www.swig.org. Viewed online january 2007.
[8] F2PY - Fortran to Python interface generator. http://www.scipy.org/F2py. Viewed online January 2008.
[9] Otto J. Anshus, John Markus Bjørndalen, and Brian Vinter. PyCSP - Communicating Sequential Processes for Python. In Alistair A. McEwan, Wilson Ifill, and Peter H. Welch, editors, *Communicating Process Architectures 2007*, pages 229–248, jul 2007.
[10] Peter Y. H. Wong and Jeremy Gibbons. A Process-Algebraic Approach to Workflow Specification and Refinement. In *Proceedings of 6th International Symposium on Software Composition*, March 2007.
[11] Peter Y. H. Wong. Towards A Unified Model for Workflow Processes. In *1st Service-Oriented Software Research Network (SOSoRNet) Workshop*, Manchester, United Kingdom, June 2006.
[12] Flow-Based Programming. http://en.wikipedia.org/wiki/Flow-based_programming. Viewed online september 2007.
[13] Communicating Sequential Processes for Java. http://www.cs.kent.ac.uk/projects/ofa/jcsp/. Viewed online january 2008.
[14] Berhnard H.C Sputh and Alastair R. Allan. JCSP-Poison: Safe Termination of CSP Process Networks. *Communicating Process Architectures 2005*, pages 71–107, 2005.
[15] Thread State and the Global Interpreter Lock. http://docs.python.org/api/threads.html. Viewed online january 2008.
[16] Neil C. Brown and Peter H. Welch. An Introduction to the Kent C++CSP Library. In Jan F. Broenink and Gerald H. Hilderink, editors, *Communicating Process Architectures 2003*, pages 139–156, sep 2003.
[17] Donald E. Knuth. *The Art of Computer Programming - Volume 2 - Seminumerical Algorithms*. Addison-Wesley, third edition, 1998.
[18] Bertram Ludäscher, Ilkay Altintas, Chad Berkley, Dan Higgins, Efrat Jaeger, Matthew Jones, Edward A. Lee, Jing Tao, and Yang Zhao. Scientific workflow management and the Kepler system: Research Articles. *Concurr. Comput. : Pract. Exper.*, 18(10):1039–1065, 2006.
[19] G.H. Hilderink. Graphical Modelling Language for Specifying Concurrency Based on CSP. *IEE Proceedings - Software*, 150(2):108–120, 2003.

Visual Process-Oriented Programming for Robotics

Jonathan SIMPSON and Christian L. JACOBSEN

Computing Laboratory, University of Kent, Canterbury, Kent, CT2 7NF, England.
{jon,christian}@transterpreter.org

Abstract. When teaching concurrency, using a *process-oriented* language, it is often introduced through a visual representation of programs in the form of *process network diagrams*. These diagrams allow the design of and abstract reasoning about programs, consisting of concurrently executing communicating processes, without needing any syntactic knowledge of the eventual implementation language. Process network diagrams are usually drawn on paper or with general-purpose diagramming software, meaning the program must be implemented as syntactically correct program code before it can be run. This paper presents *POPed*, an introductory parallel programming tool leveraging process network diagrams as a visual language for the creation of process-oriented programs. Using only visual layout and connection of pre-created components, the user can explore process orientation without knowledge of the underlying programming language, enabling a "processes first" approach to parallel programming. POPed has been targeted specifically at basic robotic control, to provide a context in which introductory parallel programming can be naturally motivated.

Introduction

At the University of Kent, parallel programming is introduced using the occam-π programming language [1]. This is a relatively small language, based on the formalisms of Milner's π-calculus [2] and Hoare's Communicating Sequential Process (CSP) algebra [3]. One of the strengths of occam-π is its simplicity in expressing parallel programs, and that knowledge of the π-calculus or CSP algebra is not required to make effective use of the language. The result is that parallel programming can be taught without referring to the underlying formalisms, whilst maintaining these as subjects for later study within the context of the programming language.

Given the nature of occam-π programs, made up of concurrently executing processes that communicate over well-defined channel interfaces, it is often useful to visualise these as networks of processes. Indeed, when teaching the occam-π language to students, as the focus is on the parallel aspects of the language, they are often asked to draw such diagrams as part of their first programming exercises. As the students complete these first exercises, they are shown the steps that one can perform in order to turn a process diagram into a running occam-π program. They must however, also be taught the syntax of the language, how to operate the compiler, and how to write the sequential parts of a program[1]. It is our view that this early introduction to syntax and tools distracts from learning programming [4]. The initial need to learn sequential programming distracts from what should be the goal of the students' first encounter with occam-π: illustrating the power of process-oriented parallel programming.

---
[1]While it is possible to provide a toolbox of occam-π processes, and thus require no sequential programming, the syntax and tools must be taught in all cases.

This involves them gaining hands on experience in designing programs composed of multiple, concurrently operating processes which communicate using synchronous message passing. By familiarising themselves with the composition of processes and channels to construct programs, it is our hope that introductory parallel programmers will have a reduced number of concepts to understand when introduced to the occam-π programming language.

We will start by detailing in Section 1 the specific motivation behind the need for another visual programming tool for occam-π, and will then, in Section 2, explore previous work both specifically in relation to the occam family of programming languages as well as popular visual programming tools for other languages. Section 3 will describe the current prototype of the visual programming environment, followed by examples of the kinds of programs we envision using for an introduction to parallel programming using the tool (Section 4). Finally, Section 5 will present the conclusions and the future work required to make the tool described in this paper accessible to a wider audience.

1. Motivation

Programs written in process-oriented languages such as occam-π often have their state represented visually through the use of process network diagrams. When teaching concurrency at the University of Kent these diagrams are used from the very first lecture, allowing students to begin to understand the concepts involved with networks of communicating processes without having any knowledge of occam-π. The use of process network diagrams allows parallel programs to be designed on paper, given a number of well-defined processes, simply by drawing networks of those processes like the diagram shown in Figure 1.

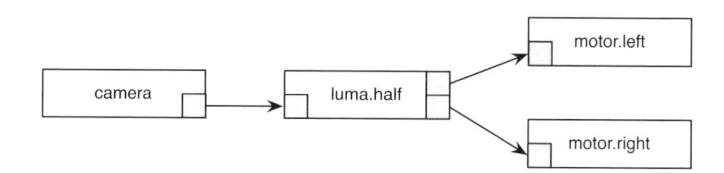

Figure 1. A process network for a simple robotics program

Given the `camera`, `luma.half`, `motor.left` and `motor.right` processes and a knowledge of their interfaces, a student can construct the process network above in occam-π program code, as shown in Listing 1. Nearly all of the requisite information is encapsulated within the process network diagram itself with the exception of the channel types, which are part of the interface to each process.

```
PROC light.seek ()
  CHAN FRAME frame:
  CHAN INT left, right:
  PAR
    camera(frame!)
    luma.half(frame?, left!, right!)
    motor.left(left?)
    motor.right(right?)
:
```

Listing 1. Construction of the process network for the example simple robotics program in occam-π

Robotics lends a useful context in which to frame the visual design of parallel programs. Robotic control can be reduced to a problem of transforming sensory input (the view the robot has of the world around it) into action, allowing the robot to react to the environment around it. The data-flow model of communicating processes fits naturally into this simplified model of robotic control, with hardware interface processes providing streams of sensory information and outputs for commands to the effectors of the robotics platform. The choice of hardware interface processes used in a program can capture the hardware configuration of a re-configurable robotics platform where different input and output devices can be connected to a selection of ports on the robot.

The visual programming environment presented in this paper has been limited in scope. Specifically, the tool will be limited to creating introductory-level parallel programs for use on small-scale robots, a target platform for introductory programming which the authors have been exploring for some time [5]. The application of simple design patterns on a small robotics platform, for introducing users to parallel programming, has influenced our approach to designing a visual programming tool [6].

Limiting the scope to introductory parallel programming with robotics makes the tasks of creating and distributing a visual editor manageable and provides an environment (a robot) which we feel provides an authentic setting. However there are further applications for the concepts behind our visual editor in the graphical construction of large scale process-oriented systems such as those currently being explored by the RMoX [7] and CoSMoS [8] projects.

1.1. Surveyor SRV-1

We have chosen to focus our prototype environment to support process-oriented robotic control on the Surveyor SRV-1, a tracked mobile robotics platform equipped with two laser pointers and a 1.3-megapixel camera [9]. This platform uses a 500MHz Analog Devices Blackfin processor along with 32MB of SDRAM and equipped with Wi-Fi connectivity, provided as a serial IO interface to the control board. This network capability allows the creation of programming environments for the SRV-1 which communicate with the robot directly to upload new programs "over the air" and offers many opportunities for streamlining the process of working with a remote host separate to the development machine.

The SRV-1 provides significant motivation for our work as we have recently completed a port of the Transterpreter virtual machine to the platform [10]. The Transterpreter is a highly portable runtime environment for occam-π with a cross-platform interpreter core and platform-specific wrappers interfacing the core to the hardware platform to provide a given port. Our SRV-1 port is unique in that it includes a replacement for the standard firmware for the platform written in occam-π which provides a hardware interface composed of parallel processes. This hardware interface makes the SRV-1 a good platform for occam-π robotics and also makes it a compelling environment in which to introduce new users to parallel robotics.

2. Related Work

In designing a graphical editor for use in parallel robotics, a depth of previous work with visual languages, both in robotics and for parallel programming is available. We have separated the two areas and explored each, allowing identification of the best features of visual languages and environments designed for robotics independently of those best suited to representing process-oriented systems.

2.1. Visual Programming Languages in Robotics

The applicability of visual languages to robotic control was given significant attention in academia during the 1996 and 1997 competitions at the Symposium on Visual Languages, presenting tasks promoting the use of visual languages in robotic control and leading to the development of languages such as VBBL. Commercial visual programming environments such as robotics-focused variants of LabVIEW and the recent Microsoft Robotics Studio have a strong presence within educational robotics due to their widespread availability. A number of these tools are examined below, presented in chronological order.

2.1.1. LEGOsheets

An early, rule-based environment called LEGOsheets allowed users to build a representation of a LEGO robot and configure its behaviour based on simulated "cables" connected to virtual sensors with user-supplied values [11]. The use of external connections within the graphical environment to receive values from sensors and provide values to actuators at development time is an interesting feature, and one that would be useful for a tool accompanied by a simulation environment.

2.1.2. Visual Behaviour-based Language (VBBL)

Cox et al. [12] made use of a visual object-oriented dataflow language called ProGraph to develop Visual Behaviour-based Language (VBBL), a rule-based visual language making use of finite state machines (FSM's). These FSM's define various sets of behaviours that are switched between based on conditions and message flows. ProGraph allows the use of a message passing model between components, but all language operations must be represented graphically. Improving on VBBL, Cox et al. proposed a visual programming environment based on representations of the actual robot itself, similar to LEGOsheets, noting that VBBL could be improved by leveraging "the obvious visual representations" of objects instead of focusing on the visualisation of abstract control concepts [13].

A second, improved environment allowed for the creation of robotic control programs through direct manipulation of a user-specified simulated robot within a defined environment, prompting for the specification of behaviour when unknown combinations of sensor input were encountered at runtime. This approach reduced complexity and avoided the manual user creation of finite state machines. Having separate behaviours allowed for the concurrent execution of each within a subsumption architecture, albeit one missing the ability to prioritise behaviours over one another. Requiring the full specification of the robot's environment limits the utility of the model for problems outside simulation, as the physical world represents an inherently unknown environment. The hardware definition module (HDM) model defines a robotics platform as a programming target by composing classes of objects into a graphical and functional representation of the robot and its abilities.

The concept of hardware definition modules maps directly into our approaches with occam-π robotics, using processes to represent hardware on the robotics platform which can be connected into process networks as required, providing a coherent model for specifying interfaces to and configuration of hardware from within the program. This model is particularly useful on re-configurable robotics platforms such as the LEGO Mindstorms RCX or NXT and its utility and application has been previously explored [14].

2.1.3. LabVIEW and LEGO Mindstorms

There are a commercially available visual programming environments used for robotics, a number of which are derived from National Instruments' LabVIEW product. The LEGO Mindstorms RCX and NXT series of robots has had much influence in promoting visual programming languages in education, having included several visual languages with both platforms. The RCX was supplied to educators with RoboLab, a flowchart language based on LabVIEW [15]. RoboLab presents a palette of possible actions for the robot to perform, which are connected together by the user on a canvas to indicate their execution sequence. A number of additional components allow the modification of execution flow, including looping structures and conditionals. A simplified and unrelated language called *RCX Code* was present in the consumer version of the RCX, which used components which slotted together as puzzle pieces. RCX Code was developed by LEGO from a prototype environment developed by the MIT Learning and Epistemology group called LogoBlocks [16] from which the visual model has been carried forward into Scratch [17], a visual language focused on allowing children eight and up to learn mathematical and computational skills.

The LEGO Group continue to bundle a visual language with the newer Mindstorms NXT, another LabVIEW variant called NXT-G. The NXT-G language borrows metaphors from LEGO blocks in terms of visual layout, and is changed from RoboLab in significantly emphasising data-flow over wires between components. The full LabVIEW product is not specifically a robotics environment but can it be used for robotics, and sets of components are supplied to allow its use with the Mindstorms NXT. LabVIEW's graphical control language, *G* has shown potential for the design of robotic control systems outside of small, educational robotics platforms, having been used by the Virginia Tech Team in the 2007 DARPA Urban Challenge to claim third prize [18].

2.1.4. Microsoft Robotics Studio

Another recent development in commercial visual programming environments for robotics is the Microsoft Visual Programming Language (MSVPL), included only with Microsoft's Robotics Studio (MSRS) [19]. This language is much like NXT-G in being a robotics focused dataflow language, and it generates code which runs on top of the Concurrency and Coordination Runtime (CCR), used to execute robotics programs designed with MSRS. The CCR uses asynchronous communication between components to allow the design of parallel systems. The toolbox and component canvas model used in the MSVPL is very similar to that which we propose for POPed. MSVPL can be used to program the Surveyor SRV-1, our target robotics platform. It should be noted that there is no underlying textual representation for programs written using this visual environment, a constraint which MSVPL shares with LabVIEW. Lacking the ability to write sequential code textually, the visual paradigm in these tools must contain all primitive actions. As a result of this constraint, MVPL has 'Data' blocks which input values specified by the programmer to named 'Variable' blocks, a clunky workaround for assignment, as shown in Figure 2.

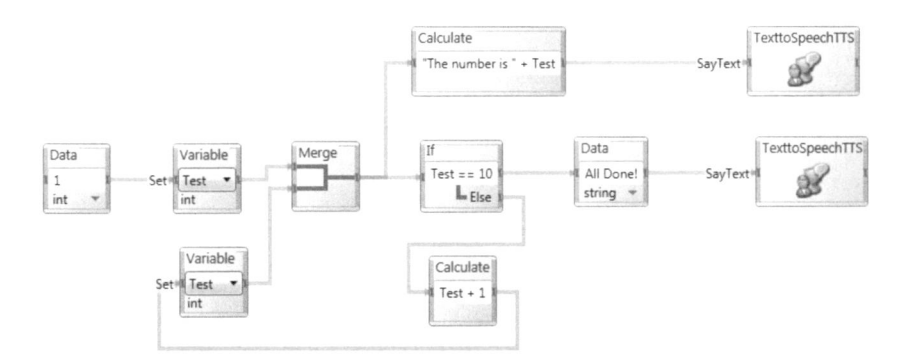

Figure 2. A sample program in the Microsoft Visual Programming language, showing its visual representation for variable assignment, conditional and hardware interface, from [19]

2.2. Visual occam and CSP-based Tools

Historically a number of visual occam programming tools, as well as program visualisation tools have been created and used; where the former is used for creating new programs visually, and the latter for visualising existing programs. It is also possible to find a number of other, CSP-based tools that are not specific to occam. Notably, of the occam tools which can be found in the literature, very few of them are in a state in which they can be used today.

Whilst examining occam tools designed for use with networks of Transputers, it is clear that there are a number of additional features commonly present to support the implementation and optimisation of occam programs on such parallel hardware. Visual layout of software processes across processors in a Transputer network and the instrumentation of programs to provide processor and link utilisation data for optimisation are two development processes that currently have no equivalent in the development of modern occam-π applications, and these may be opportunities for future expansion.

2.2.1. GRAIL

Stepney's GRAIL offered a visual representation of the parallel and sequential structure of occam programs, using the hierarchical structure of the language to manage large programs by "folding" (hiding) sections of the program [20,21]. GRAIL allowed for the additional display of channel information, but the representation used bears little resemblance to those used for process networks and is tied to the hierarchical display of parallel and sequential occam code. No editing capability was provided within GRAIL, but its representation of process internals could be developed into a visual language for the creation of sequential occam code, as discussed in Section 5.1.

2.2.2. Visputer and Millipede

Other graphical tools have traditionally focused on the design of programs for execution on networks of Transputers, a once widespread hardware processor designed for use with the occam programming language. Visputer by Zhang and Marwaha provided a complete set of graphical tools for program composition, processor allocation, performance monitoring and debugging [22]. Visputer provided the option for nested processes to be expressed in a visual language, with low-level logic provided textually. Millipede by Aspnas *et al.* provided visual process layout integrated along with performance monitoring of processes and communication links [23]. Millipede used processes and channel ends as its primitives in a 'palette' of graphical components, providing a way to place channels between processes, configuring the network and specifying process interfaces. As with Visputer, Millipede made use of a text-based editor for specifying process logic and allowed the graphical expression of 'compound' process networks containing nested sub-networks of processes.

2.2.3. TRAPPER

TRAPPER by Scheidler *et al.* offered a graphical programming environment comprising four separate component tools: Designtool, Configtool, Vistool and Perftool [24]. The entire TRAPPER environment covers a similar scope to Visputer, but its separation allows us to focus on Designtool, the most relevant component of the four to visual programming. TRAPPER allowed for the reduction of large process networks into a hierarchy of sub-networks which can be collapsed, an essential feature for managing complexity. The use of connection points at the edge of process network diagrams to represent connections to the outside world in Designtool is novel and serves well to highlight the interface between program and hardware interfaces, as shown in Figure 3 on the next page. The remaining three components of the TRAPPER environment relate to management of tasks involving the configuration, instrumentation and optimisation of programs running on actual Transputers, and as such are of limited relevance to our objectives at this point.

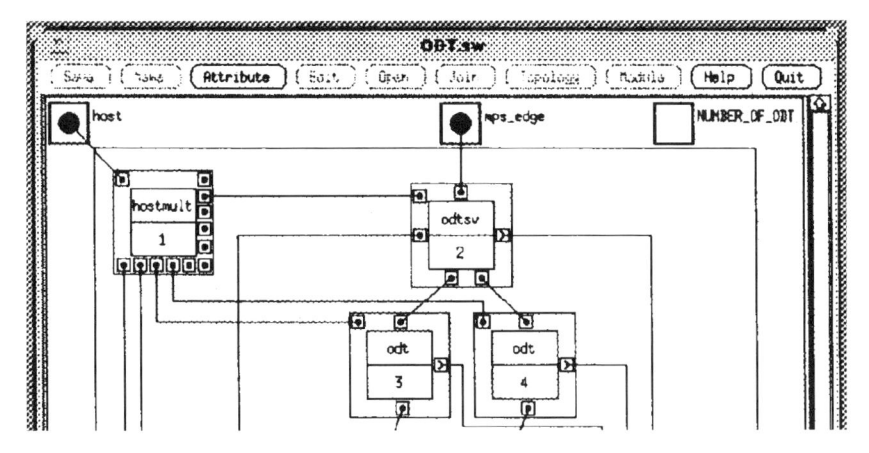

Figure 3. TRAPPER's Designtool showing its connection points outside of the canvas for interfacing to external components, from [24]

2.2.4. The occam Design Tool)

The occam Design Tool (ODT) by Beckett and Welch allowed for process networks to be designed graphically using common paradigms of occam programming [25]. The tool was specifically aimed at the creation of deadlock-free systems and made use of 'interfaces', which defined roles for processes which specified their communication behaviour. These interfaces were used in ODT to enforce the Client/Server, IO-SEQ and IO-PAR parallel design patterns [26], ensuring that networks designed with ODT were deadlock-free and that components were composed in these patterns which have been proven correct. Visual representations for various complex capabilities of the occam programming language were described as future expansions to ODT. These expansions are of interest when designing a visual environment capable of manipulating more complex process architectures, especially those using features such as replication to create pipelines, rings and grids of processes.

2.2.5. gCSP

The gCSP tool by Hilderink and Broenink et al. [27] offers a visual environment for designing concurrent programs using a graphical modelling language based on CSP [28]. Both dataflow and the concurrency of the program along with a hierarchical view of the program's structure are presented to the user, more thoroughly discussed for the design of user programs in [29]. The ability to provide an information-rich outlined structure of a process-oriented program, whilst being beyond the scale of our current aims for an introductory tool, is a potentially desirable feature for developing more complex programs.

gCSP has multiple code generators which allow it to output C++, occam or CSPm from the user-facing graphical representation of the program, essentially making the graphical representation an intermediary between many different process-oriented languages [30]. The model of code generation from connected components provides benefits we also seek in our solution, such as the removal of syntax. The visual language used in gCSP is designed for the fully express programs graphically and has stronger ties to CSP than occam-π. The generality of gCSP is problematic for its use as an introductory tool for occam-π programmers.

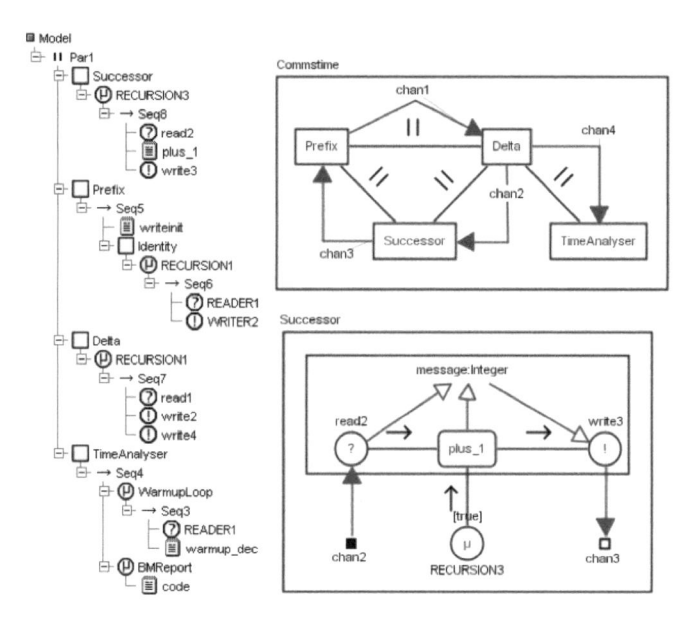

Figure 4. A gCSP session showing sequential and parallel composition of processes, along with the hierarchical browser, from [30]

2.2.6. GATOR

The Graphical Analysis Tool for occam Resources (GATOR) by Slowe and Tanner aimed to provide a debugging aid for occam programs and to progress towards a graphical environment for the program creation [31]. GATOR parsed already written textual occam programs to generate visual representations and offered no way to edit create new or edit existing processes in a loaded network, instead focusing on providing basic interrogation of an executing occam program.

2.2.7. LOVE

The Live occam Visual Environment or LOVE presents a graphical framework for audio processing through the creation of networks constructed from a palette of synthesiser components [32]. Popular synthesis tools have used a dataflow approach, beginning with the MAX graphical language by Puckette et al. [33] and its contemporary, the open source Pure-Data [34] which builds upon four primary "atoms" to construct more complex objects which are constructed in dataflow networks.

LOVE shares scope with our intended tool, in presenting a predefined set of components for the user to connect together on a canvas on which they can be arranged. It has no structuring tools for larger networks, allowing the users to see the entire network at once, but restricting the size of networks that can be created. LOVE also has graphical selection of channel ends to aid in making connections, offering visual cues for correct input connections from a selected output. Enforcing type rules, through the control of connections made, is a desirable feature in a visual editor as type errors can be eliminated.

2.2.8. POPExplorer

POPExplorer by Jacobsen takes a very different approach to manipulating process networks, interfacing with the Transterpreter virtual machine runtime for occam-π [35]. POPExplorer contains a toolbox of processes whose bytecode is loaded into the virtual machine when they are dropped onto the canvas, and subsequently executed when the user chooses for them to be 'live'. Channels can be connected and disconnected whilst processes are running, and the communication state and type of a given channel end is displayed. Problems are caused by the capabilities for run-time connection and disconnection of channels, caused mainly by the design of the occam language to assume a reliable and fault-free environment. Solving these problems and providing an interactive process canvas is an area for further work.

3. Prototyping POPed

To explore the notion of a visual process editing tool for parallel robotics, we have created POPed, a prototype with which to further develop and evaluate our ideas. We have written our prototype in the Python programming language using the wxPython user interface library, as Python has a good range of inbuilt libraries and the use of wxWidgets for the GUI allows our software to run across multiple platforms. Python also streamlines the process of generating distributions for platforms like Windows and Mac OS X.

The POPed tool contains a set of processes designed for use in creating simple robotic control programs to run on the Surveyor SRV-1 mobile robot. To write programs built-in processes are dropped onto a canvas and connected together with channels, using connection points representing the channel ends of the processes. When the user wishes to execute the program, the environment checks that the process network is fully connected and then creates an executable occam-π program by combining the code for each of the processes used and generating a top level process containing all of the processes and channel definitions specified graphically by the user. This automation removes an entire category of potential errors relating to the wiring together of process networks. The set of processes supplied with POPed could be customised to allow the use of the tool for a number of applications and the underlying code generation technique could equally use another process-oriented language such as PyCSP [36] or JCSP [37].

The POPed user interface has three main components: a Toolbox, Canvas and Information Panel. These are shown in the interface diagram in Figure 5.

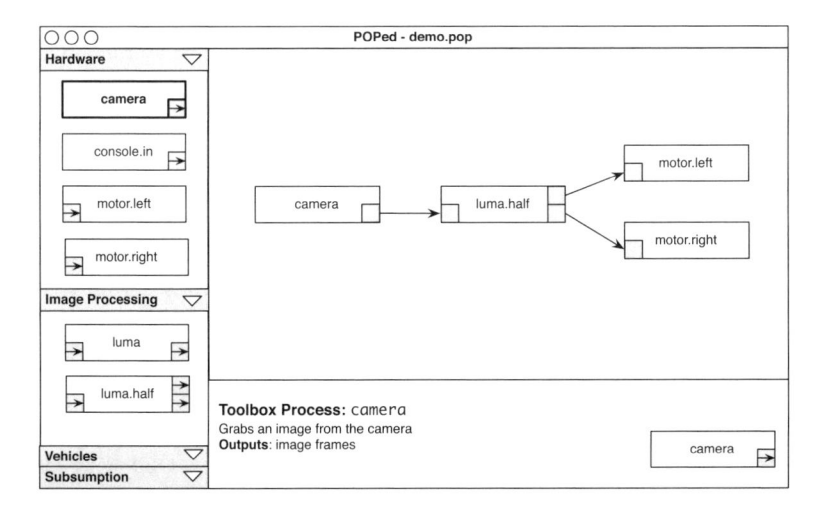

Figure 5. An Illustration of the POPed User Interface

3.1. Toolbox

The 'toolbox' is located on the left-hand side of the window and contains a graphical list of the processes available to the user with small diagrams of each process. The diagrams have connection points showing the direction of the channel ends specified in their interfaces to help clarify the connectivity of the process. The location of the arrows follows the general principle of data-flow from left to right, with inputs on the left and outputs on the right. Using a visual representation of the process in the toolbox provides cues for the data-flow behaviour of processes. When a process is selected in the toolbox, information about its inputs, outputs and parameters is displayed in the information panel at the bottom of the screen along with a description to aid the user in understanding the purpose of the process. The information panel is fully discussed in Section 3.3 on the next page. Processes within the toolbox are logically grouped together. To make use of components to build a program, users drag them from the toolbox onto the canvas, at which point a instance of the process in the toolbox appears and can be freely positioned by the end user.

Toolbox processes can be specified in generic terms, despite the lack of generics in occam-π, whereby a process can be parameterised by a type. POPEd makes use of templating to generate valid code and substitute real types for the generic placeholder types, ensuring that occam-π's strict type rules are met. The ability to use generics means that we can specify processes such as `delta` generally, and allow the tool to customise the delta for basic types such as `INT`, or `BOOL` depending on what is connected to it. It is necessary to place a constraint such that once one of the channel ends on a process using generics is connected to, that the generic type is set for all channel ends. For example, if a `delta` process were placed on the canvas and had its input connected to a channel of `INT` from another process, its outputs would at that point be become of type `CHAN INT`. This approach to generating code instead of real generics is limited in terms of its use as channels using `PROTOCOL`s cannot be substituted into these false generics. As our interfaces to hardware on the Surveyor SRV-1 make extensive use of `PROTOCOL`s, this will need to be improved.

3.2. Canvas

The canvas is the central focus of POPed, being the area in which the user builds their program. Channel ends are represented by *connection points* on the process, allowing the user to easily connect the processes together. The user selects a connection point and the potential points to which a connection can be made are highlighted, giving a visual representation of both type-compatibility and unconnected nodes. Only the points to which connections can be made to are made active reducing the potential for erroneous clicks on other channel ends and attempting to reduce the possibility for mistakes. This selection mechanism is shown in the figure below. Once a connection has been made, a directed arrow is placed between the two connection points. On attempting to compile a program which is not fully connected with channels between all connection points on the canvas, an error is displayed in the information panel informing the user of the processes which are not properly connected, and the offending channel connection points are highlighted.

Layout of the process network is managed entirely by the user, capturing the way paper or diagramming tools are typically used in the first few weeks of parallel program design. Processes on the canvas are able to have their program code inspected, allowing the user to gain an insight into the code behind the diagram. This functionality could be extended to include editing of existing processes, but that is outside the scope of our prototype. Ensuring the underlying code is not hidden is important as we aim for POPed to be a first tool, with its users moving on to write programs textually in the occam-π language.

(a) Possible connection points, no selection made

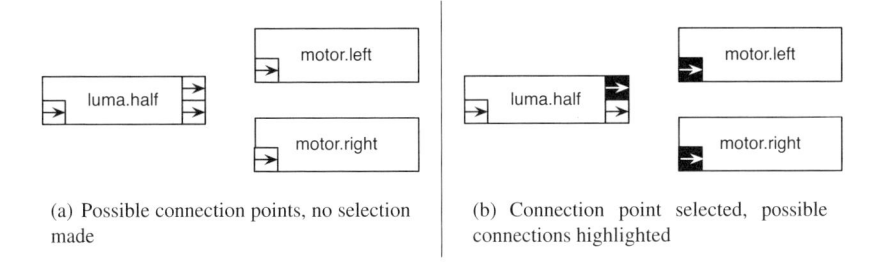

(b) Connection point selected, possible connections highlighted

Figure 6. Connection points and their type highlighting mechanism

3.3. Information Panel

The information panel is located at the bottom of the screen and allows contextual information to be provided to the student about processes selected from the toolbox and canvas. An example of the information displayed when a toolbox process is selected is shown in Figure 5 on page 373, while an example of a selected canvas process is shown in Figure 7.

Process Instance: `luma.half`
Calculates luminance values for the left and right halves of an image frame.
Inputs: Image frames from `camera`
Outputs: Left half luminance value to `motor.left`,
Right half luminance value to `motor.right`
Parameters: none

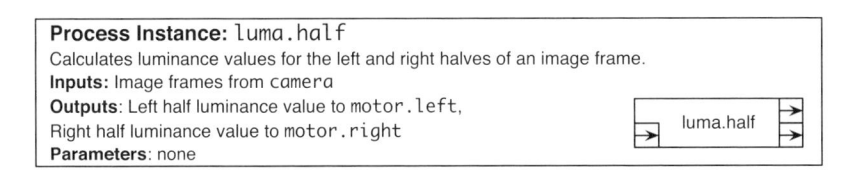

Figure 7. The information panel with a process instance selected on the canvas

This presentation of additional textual information is intended to allow the student to fully understand the components and connections that make up their program, along with the relationship between the connected processes. We can infer information from the connections between processes to present the user with textual descriptions of the inputs and outputs from a process instance. These descriptions provide a simplified explanation of the component's operation within the system, given well named processes and described types. An example of these explanations for a `luma.half` process is shown in Figure 7, where a `camera` process has been connected to input and two motor control processes (`motor.left, motor.right`) are connected to the outputs.

4. Example Scenarios

Two specific robotics control applications are being considered in the construction of POPed's default process set. A number of hardware interface processes and general purpose components have been provided to allow the creation of simple pipelines, but specific components have been included to facilitate the creation of programs based on Braitenberg's *Vehicles* and the subsumption architecture.

4.1. Behavioural Control: Vehicles

When targeting first explorations in concurrent robotics, Braitenberg's *Vehicles* offers a useful introduction to what can be accomplished with small numbers of processes connected together [38]. By connecting sources of sensory input directly (or almost directly) to outputs, very simple programs can be created.

The program shown in Figure 1 on page 366 shows a simple program for the Surveyor SRV-1 which uses very few processes to achieve a useful result. Image data from the camera is averaged to provide a light level reading for the left and right of the image. These light level values are subsequently sent to the `motor.left` and `motor.right` processes, which change the speed that the motors on each side of the robot run at proportionally to the value received. By connecting the light levels and motor speeds together using a direct relationship, the robot will turn away from light sources, as a stronger light reading on the left-hand side will cause the left track to speed up, and the right track to be slowed down. The inverse is also true: by slowing the motors as the light level increases, we can program a robot that seems to 'like' light and heads towards it. In the Vehicles set of processes, an `invert` process is included, to make switching between positive and negative relations of inputs and outputs easier. Also worth noting for the implementation of Vehicles is that processes which generate values within the network have had their values scaled to the range 0-100, such that the brightness reading from `luma.half` can be used directly as a motor speed.

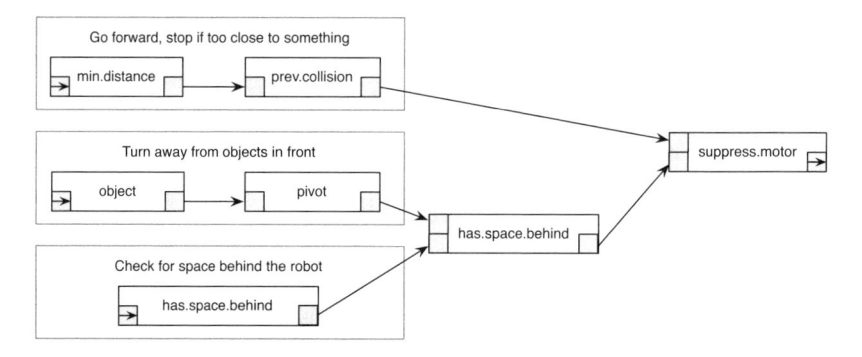

Figure 8. A small robot control program built with the subsumption architecture which uses two types of sensor input and three behaviours to manoeuvre around a space

4.2. Subsumption Architecture

Brooks' *subsumption architecture* involves building robot control systems with increasing levels of competence composed of concurrently operating modules [39]. It offers a design paradigm for robotic control that emphasises re-use, decentralisation and concurrent, communicating processes. The application of the subsumption architecture for designing parallel robotics programs has been previously explored in [40], and provides substantial motivation for the use of a graphical process network configuration tool.

We have found the use of diagrams representing subsumption architectures particularly useful due to the additional complexity of the interactions between their levels of behaviour. A first step towards supporting their development of within POPed has been to include processes for suppression and inhibition, the two primary operations carried out between different levels of behaviour. To fully support the development of large subsumption-based control programs, it would be necessary to allow sub-networks of processes to be collapsed into a single top-level behaviour and allow the expansion of that behaviour on demand, hiding complexity when it is not required. The ability to expand and contract sub-networks of processes is discussed further as an area of future work in Section 5.1 on page 378.

To illustrate the advantages of a visual approach to the design of subsumption architectures, we can look at a process network diagram for a simple program, and the occam-π code required to set up the network. The diagram in Figure 8 shows a graphical representation of a simple program which moves around a space and backs away from objects if the robot gets too close to them. Each behaviour is broken out separately in the diagram, with an outer box enclosing the sets of processes which are composed to create the behaviour. The occam-π code to set up the process network as shown in the diagram (assuming that all components and hardware interfaces exist and are available) is shown in Listing 2. The complexity inherent to this code, having to write code to connect the network with named channels, can be completely eliminated with a visual approach. The removal of this complexity is desirable as subsumption architectures grow beyond two or three behaviours to have ten or fifteen levels of behaviour based on many different input sensors.

```
PROC explore.space ()
  CHAN MOTORS motor.control:
  CHAN INT minimum.distance:
  CHAN INT motor.command.in, motor.command.out, motor.command.suppress:
  CHAN INT pivot.motor.command.in, pivot.motor.command.out:
  SHARED ? CHAN LASER laser.data:
  CHAN SONAR sonar.data:
  CHAN BOOL object, inhibit:
  PAR
    motor(motor.command.out?, motor.control!)
    brain.stem(motor.control?, laser.data!, sonar.data!,
               default.player.host, default.player.port)
    min.distance(laser.data?, minimum.distance!)
    prevent.collision(minimum.distance?, motor.command.in!)
    object.in.front(laser.data?, object!)
    pivot.if.object(object?, pivot.motor.command.in!)
    motor.command.suppressor(suppress.time, pivot.motor.command.out?,
                             motor.command.in?, motor.command.out!)
    pivot.inhibitor(inhibit.time, inhibit?, pivot.motor.command.in?,
                    pivot.motor.command.out!)
    check.has.space(sonar.data?, inhibit!)
:
```

Listing 2. Construction of the process network for the example simple robotics program in occam-π

5. Conclusions

In this paper we have discussed the design features of an environment for introducing parallel programming in a "processes first" methodology, allowing the demonstration of parallel design patterns and manipulation of process networks without having to learn the syntax of a parallel language. These design features are the product of studying other tools for robotics programming and process network design in occam and related languages. We continue to develop our prototype tool, with the aim of including it along with our port of the Transterpreter runtime to the Surveyor SRV-1 as an illustration of process-oriented design for robotic control systems. It is our intention to improve upon this prototype to the stage where it can be used in the classroom, allowing the use of feedback from students to shape further development.

By focusing on robotics applications we have been able to begin designing a graphical parallel program development environment for the occam-π program language and create a tool which can be used to illustrate and explore simple process-oriented robotics with those unfamiliar to occam-π. We feel there is much left to do in this area, and a large number of questions remain in the development of effective visual tools for pedagogic process-oriented programming.

5.1. Future Work

The occam-π programming language contains features for mobility of channels and processes at run-time which are heavily used in larger applications. At the University of Kent there are currently two research projects which could see potential benefit from graphical system construction tools.

One is RMoX, an experimental occam-π operating system which could allow the generation of operating systems for a given embedded hardware platform by the visual composition of the correct hardware drivers and system components, an approach previously illustrated by gCSP [30]. The second is CoSMoS, a complex systems modelling and analysis project which aims to allow non-programmers to be able to build their own systems. The CoSMoS project could see benefit from an environment to allow these users to express systems visually.

Both of these systems share a heavy use of the dynamic language features in occam-π, with large amounts of process creation and mobility meaning that static process network diagrams are significantly reduced in utility, useful only as 'snapshots' of an executing system. Developing a visual language and model for representing and designing these dynamic systems is an open area of research. A pedagogic tool which encompasses the entire feature-set of occam-π would have significant impact in introducing new users to building complex applications.

Work is currently ongoing into addition of debugging facilities to the Transterpreter virtual machine to provide high-quality debugging of occam-π programs, allowing the user to single-step, slow down execution and monitor the activity of programs as they execute. Exposing these facilities to users in ways that allow them gain a better understanding of how their code executes will be an exercise in user interface design. We consider the problems of representation in debugging and performance monitoring similar to those involved in creating a visual programming language. Integrating the program design and implementation stages with a live runtime would provide similar facilities to the POPExplorer tool discussed in Section 2 on page 367 to be built upon futher.

Having worked with the BlueJ environment for Java where objects can be instantiated individually on an "object bench" and interacted with, we see potential in allowing users to interact with individual processes to explore their run-time behaviour [41,42]. We envisage that the additional control in the runtime could be used to provide a live test environment with provided values or data sets for channel inputs, giving a "process bench" for vivisection of processes. The integration of testing abilities and live experimentation with processes or networks of processes has much potential for use with introductory parallel programmers.

Stepney's GRAIL [20] included a visual representation of programs written in occam based on the structure of the program code itself, and this representation bears a resemblance to Scratch [17], a visual programming language developed at MIT and used to encourage school-age children to learn programming. This resemblance serves to highlight the potential for a fully-visual programming language inspired by occam-π, and exploring the possibilities for graphical representations to remove syntax and provide an augmented editing environment for the textual occam-π language may yield improvements for those new to occam-π.

Acknowledgements

Many thanks to all who continue to work on and support the Transterpreter project. Howard Gordon of Surveyor Corporation has provided support for our work with the SRV-1 by supplying robotics platforms to work with. Carl Ritson has both made improvements to the Transterpreter and provided an excellent, fully-featured port to the Surveyor, facilitating a wide range of further robotics work. Additional thanks to Adam Sampson, Fred Barnes and

Peter Welch who have provided both formative input and extremely helpful feedback on this work.

References

[1] P.H. Welch and F.R.M. Barnes. Communicating Mobile Processes: Introducing occam-π. In A.E. Abdallah, C.B. Jones, and J.W. Sanders, editors, *25 Years of CSP*, volume 3525 of *Lecture Notes in Computer Science*, pages 175–210. Springer Verlag, April 2005.
[2] Robin Milner. *Communicating and mobile systems: the π-calculus*. Cambridge University Press, New York, NY, USA, 1999.
[3] C. A. R. Hoare. *Communicating sequential processes*. Prentice-Hall, Inc., Upper Saddle River, NJ, USA, 1985.
[4] Matthew C. Jadud. Methods and tools for exploring novice compilation behaviour. In *ICER '06: Proceedings of the 2006 international workshop on Computing education research*, pages 73–84, New York, NY, USA, 2006. ACM Press.
[5] Christian L. Jacobsen and Matthew C. Jadud. Towards concrete concurrency: occam-π on the LEGO Mindstorms. In *SIGCSE '05: Proceedings of the 36th SIGCSE technical symposium on Computer science education*, pages 431–435, New York, NY, USA, 2005. ACM Press.
[6] Matthew C. Jadud, Jonathan Simpson, and Christian L. Jacobsen. Patterns for programming in parallel, pedagogically. *SIGCSE Bulletin*, 40(1):231–235, 2008.
[7] F.R.M. Barnes, C.L. Jacobsen, and B. Vinter. RMoX: a Raw Metal occam Experiment. In J.F. Broenink and G.H. Hilderink, editors, *Communicating Process Architectures 2003*, WoTUG-26, Concurrent Systems Engineering, ISSN 1383-7575, pages 269–288, IOS Press, Amsterdam, The Netherlands, September 2003. ISBN: 1-58603-381-6.
[8] Susan Stepney and Peter H. Welch. The CoSMoS Research Project. http://www.cosmos-research.org/caseforsupport.html, October 2007.
[9] Surveyor Corporation. Surveyor SRV-1 Blackfin Robot. http://www.surveyor.com/, July 2008.
[10] Christian L. Jacobsen and Matthew C. Jadud. The Transterpreter: A Transputer Interpreter. In *Communicating Process Architectures 2004*, pages 99–107, 2004.
[11] J. Gindling, A. Ioannidou, J. Loh, O. Lokkebo, and A. Repenning. Legosheets: a rule-based programming, simulation and manipulation environment for the lego programmable brick. In *VL '95: Proceedings of the 11th International IEEE Symposium on Visual Languages*, page 172, Washington, DC, USA, 1995. IEEE Computer Society.
[12] Philip T. Cox, Christopher C. Risley, and Trevor J. Smedley. Toward concrete representation in visual languages for robot control. *Journal of Visual Languages & Computing*, 9(2):211–239, 1998.
[13] Philip T. Cox and Trevor J. Smedley. Visual programming for robot control. In *VL '98: Proceedings of the IEEE Symposium on Visual Languages*, page 217, Washington, DC, USA, 1998. IEEE Computer Society.
[14] Jonathan Simpson, Christian L. Jacobsen, and Matthew C. Jadud. A Native Transterpreter for the LEGO Mindstorms RCX. In Alistair A. McEwan, Steve Schneider, Wilson Ifill, and Peter H. Welch, editors, *Communicating Process Architectures 2007*, volume 65 of *Concurrent Systems Engineering*, Amsterdam, The Netherlands, July 2007. IOS Press.
[15] Ben Erwin, Martha Cyr, and Chris Rogers. LEGO engineer and ROBOLAB: Teaching engineering with LabVIEW from kindergarten to graduate school. *International Journal of Engineering Education*, (16):2000, 2000.
[16] Andrew Begel. LogoBlocks: A Graphical Programming Language for Interacting with the World. Technical report, MIT Media Laboratory, Cambridge, MA, USA, May 1996.
[17] John Maloney, Leo Burd, Yasmin Kafai, Natalie Rusk, Brian Silverman, and Mitchel Resnick. Scratch: A sneak preview. In *C5 '04: Proceedings of the Second International Conference on Creating, Connecting and Collaborating through Computing*, pages 104–109, Washington, DC, USA, 2004. IEEE Computer Society.
[18] Team VictorTango - 2007 DARPA Urban Challenge. http://www.me.vt.edu/urbanchallenge/, November 2007.
[19] Microsoft Corp. Microsoft Robotics Studio: VPL Introduction. http://msdn.microsoft.com/en-us/library/bb483088.aspx, 2008.
[20] Susan Stepney. GRAIL: Graphical representation of activity, interconnection and loading. In Traian Muntean, editor, *7th Technical meeting of the occam User Group,Grenoble, France*. IOS Amsterdam, 1987.

[21] Susan Stepney. Pictorial representation of parallel programs. In Alistair Kilgour and Rae A. Earnshaw, editors, *Graphical Tools for Software Engineering*, BCS conference proceedings. CUP, 1989.
[22] K. Zhang and G. Marwaha. Visputer–A Graphical Visualization Tool for Parallel Programming. *The Computer Journal*, 38(8):658–669, 1995.
[23] M. Aspnas, R. Back, and T. Langbacka. Millipede: A programming environment providing visual support for parallel programming, 1992.
[24] C. Scheidler, L. Schafers, and O. Kramer-Fuhrmann. Software engineering for parallel systems: the TRAPPER approach. In *HICSS '95: Proceedings of the 28th Hawaii International Conference on System Sciences (HICSS'95)*, page 349, Washington, DC, USA, 1995. IEEE Computer Society.
[25] D.J. Beckett and P.H. Welch. A Strict occam Design Tool. In C.R. Jesshope and A. Shafarenko, editors, *Proceedings of UK Parallel '96*, pages 53–69, Guildford, UK, July 1996. Springer-Verlag, London.
[26] P.H. Welch, G.R.R. Justo, and C.J. Willcock. Higher-Level Paradigms for Deadlock-Free High-Performance Systems. In R. Grebe, J. Hektor, S.C. Hilton, M.R. Jane, and P.H. Welch, editors, *Transputer Applications and Systems '93, Proceedings of the 1993 World Transputer Congress*, volume 2, pages 981–1004, Aachen, Germany, September 1993. IOS Press, Netherlands.
[27] Jan F. Broenink and Dusko S. Jovanovic. Graphical Tool for Designing CSP Systems. In Ian R. East, David Duce, Mark Green, Jeremy M. R. Martin, and Peter H. Welch, editors, *Communicating Process Architectures 2004*, pages 233–252, September 2004.
[28] Gerald H. Hilderink. A Graphical Modeling Language for Specifying Concurrency based on CSP. In James Pascoe, Roger Loader, and Vaidy Sunderam, editors, *Communicating Process Architectures 2002*, pages 255–284, September 2002.
[29] H. J. Volkerink, Gerald H. Hilderink, Jan F. Broenink, W.A. Veroort, and André W. P. Bakkers. CSP Design Model and Tool Support. In Peter H. Welch and André W. P. Bakkers, editors, *Communicating Process Architectures 2000*, pages 33–48, September 2000.
[30] Jan F. Broenink, Marcel A. Groothuis, and Geert K. Liet. gCSP occam Code Generation for RMoX. In *Communicating Process Architectures 2005*, pages 375–383, sep 2005.
[31] Matthew Slowe and Ben Tanner. Graphical Analysis Tool for Occam Resources. Final year B.Sc project report, University of Kent, 2004.
[32] Adam Sampson. What is LOVE? Available at: https://www.cs.kent.ac.uk/research/groups/sys/wiki/LOVE, September 2006.
[33] Miller S. Puckette. Combining Event and Signal Processing in the MAX Graphical Programming Environment. *Computer Music Journal*, 15(3):68–77, Fall 1991.
[34] Miller S. Puckette. Pure Data: another integrated computer music environment. In *Proceedings of the Second Intercollege Computer Music Concerts*, pages 37–41, 1996.
[35] Christian L. Jacobsen. *A Portable Runtime for Concurrency Research and Application*. PhD thesis, University of Kent, Canterbury, Kent, England, December 2006.
[36] Otto J. Anshus, John Markus Bjørndalen, and Brian Vinter. PyCSP - Communicating Sequential Processes for Python. In Alistair A. McEwan, Wilson Ifill, and Peter H. Welch, editors, *Communicating Process Architectures 2007*, volume 65 of *Concurrent Systems Engineering*, pages 229–248, Amsterdam, The Netherlands, jul 2007. IOS Press.
[37] Peter H. Welch and Neil Brown. The JCSP Home Page: Communicating Sequential Processes for Java. http://www.cs.kent.ac.uk/projects/ofa/jcsp/, March 2008.
[38] Valentino Braitenberg. *Vehicles: Experiments in Synthetic Psychology*. MIT Press, Cambridge, MA, USA, 1986.
[39] Rodney A. Brooks. A robust layered control system for a mobile robot. Technical report, MIT, Cambridge, MA, USA, 1985.
[40] Jonathan Simpson, Christian L. Jacobsen, and Matthew C. Jadud. Mobile Robot Control: The Subsumption Architecture and occam-π. In Frederick R. M. Barnes, Jon M. Kerridge, and Peter H. Welch, editors, *Communicating Process Architectures 2006*, pages 225–236, Amsterdam, The Netherlands, September 2006. IOS Press.
[41] Michael Kölling and John Rosenberg. Tools and techniques for teaching objects first in a Java course. *SIGCSE Bulletin*, 31(1):368, 1999.
[42] Michael Kölling and John Rosenberg. Objects first with Java and BlueJ (seminar session). In *SIGCSE '00: Proceedings of the thirty-first SIGCSE technical symposium on Computer science education*, page 429, New York, NY, USA, 2000. ACM Press.

Solving the Santa Claus Problem: a Comparison of Various Concurrent Programming Techniques

Jason HURT and Jan B. PEDERSEN

School of Computer Science, University of Nevada

`jleehurt@gmail.com, matt@cs.unlv.edu`

Abstract. The Santa Claus problem provides an excellent exercise in concurrent programming and can be used to show the simplicity or complexity of solving problems using a particular set of concurrency mechanisms and offers a comparison of these mechanisms. Shared-memory constructs, message passing constructs, and process oriented constructs will be used in various programming languages to solve the Santa Claus Problem. Various concurrency mechanisms available will be examined and analyzed as to their respective strengths and weaknesses.

Keywords. concurrency, distributed memory, shared memory, process-oriented programming.

Introduction

Concurrent or parallel computing has always been an area of interest in the computer science community. Historically computer clusters and multiprocessor computers that provide hardware environments for parallel applications were expensive and used only in highly specialized development environments such as weather prediction, environmental modeling, and nuclear simulation. With the popularity of the Internet, distributed applications have become more widely used. *PlanetLab* [1] and *BOINC* [2] which powers *SETI* [3] are two examples of this. There are an increasing number of developer APIs that have a distributed architecture and applications that are implemented as a set of disparate, functional components connected via web services [4,5]. In addition, due to the physical limitations of silicon, chip makers such as Intel and AMD have begun moving towards multi-processor/multi-core architectures as opposed to increasing the clock speed of a single CPU, and consequently parallel computing is moving further into mainstream application development [6]. In the past, the increase in clock speed for single CPU systems meant that programmers got "free" speedup in their applications with every new generation of CPU. Now, programmers have to find ways to utilize multiple-CPU architectures in order to improve application performance. Simply moving an old sequential program to a multi-core architecture will not utilize the full potential of the processor. A number of ways to write parallel programs will be examined here, as well as the advantages and disadvantages of each method.

First, we will look at the popular shared memory model (threads). Next, a message passing architecture known as MPI will be explored, and lastly, a process oriented approach using the JCSP library for Java is examined. The metrics for comparing these models will be the readability, writability, and reliability of programs written using each model. In order to compare the metric properties of the models, a simple description of a problem is explored that lends itself naturally to a concurrent solution, the Santa Claus Problem, introduced by

John Trono in 1994 [7]. In addition to the Polyphonic C# [8] and Ada [9] versions, solutions to the problem have been written in Java using Java threads, Groovy using Java threads, C# using the .NET threading mechanism, C using pthreads [10], C using MPI, and Java using JCSP.

1. Problem Definition

The Santa Claus problem is stated in [7] as follows: "Santa Claus sleeps at the North Pole until awakened by either all of the nine reindeer, or by a group of three out of ten elves. He then performs one of two indivisible actions. If awakened by the group of reindeer, Santa harnesses them to a sleigh, delivers toys, and finally unharnesses the reindeer who then go on vacation. If awakened by a group of elves, Santa shows them into his office, consults with them on toy R&D, and finally shows them out so they can return to work constructing toys. A waiting group of reindeer must be served by Santa before a waiting group of elves. Since Santa's time is extremely valuable, marshaling the reindeer or elves into a group must not be done by Santa."

1.1. Correctness

As noted by Ben-ari [9], John Trono's solution using semaphores is incorrect because it incorrectly "assumes that a process released from waiting on a semaphore will necessarily be scheduled for execution." In a concurrent context, correctness refers to a system that is free from the possibility of deadlocks, livelocks, and race conditions. The system in question may have bugs, but they will not be due to the concurrency of the system. Note that the original problem description is up for interpretation, so in order to test the correctness of our solutions we will add a set of messages that can be reported by Santa, the Elves, and the Reindeer. Each message is an indication of the state of a particular entity: Santa, an Elf, or a Reindeer. In this way a visual representation of the states of the entities at runtime is available for debugging and testing purposes. We will impose a partial ordering on the union of three sets of messages: Santa, Elf, and Reindeer messages. The union set of these messages, which we will call S, is the set of all messages that the program can possibly report. In the real world, complex systems may be controlled as opposed to printing messages to a console. Keep in mind that the ordering of these messages may be important in order to not cause chaos to the system under control, and therefore the correctness of a solution would be necessary in order to trust a system is free from bugs due to concurrency. Let SR be the subset of S that includes the set of messages that the Reindeer can print and that Santa can print while dealing with the Reindeer. We define here a message ordering on SR, and by the transitivity property we get a complete ordering on SR:

1. Reindeer $<id>$: on holiday ... wish you were here, :)
2. Reindeer $<id>$: back from holiday ... ready for work, :(
3. Santa: Ho-ho-ho ... the reindeer are back!
4. Santa: harnessing reindeer $<id> ...$
5. Santa: mush mush ...
6. Reindeer $<id>$: delivering toys
7. Santa: woah ... we're back home!
8. Reindeer: $<id>$: all toys delivered
9. Santa: un-harnessing reindeer $<id> ...$

In addition to the above, all Reindeer must report 2 before Santa can report 3, Santa must report 4 to all Reindeer before reporting 5, and all Reindeer must report 6 before Santa can report 7.

Let SL be the subset of S that includes the set of messages that the Elves can print and that Santa can print while dealing with the Elves. We define here a message ordering on SL, and by the transitivity property we get a complete ordering on SL:

1. Elf $<id>$: need to consult santa, :(
2. Santa: Ho-ho-ho ... some elves are here!
3. Santa: hello elf $<id>$ ...
4. Elf $<id>$: about these toys ... ???
5. Santa: consulting with elves ...
6. Santa: OK, all done - thanks!
7. Elf $<id>$: OK ... we'll build it
8. Santa: goodbye elf $<id>$ ...
9. Elf $<id>$: working, :)

In addition, three Elves must report 2 before Santa can report 3 and the same three Elves must report 5 before Santa can report 6.

Note that the intersection of SR and SL is the empty set. In addition to the message ordering, the Reindeer have priority over the elves, and only three Elves at a time may consult with Santa. Moreover, freedom from deadlock and livelock are necessary; no process may halt its execution indefinitely and the states of the entities must proceed as per the problem description.

2. Shared Memory

Shared memory solutions are often implemented in the form of threads. The thread model is a shared memory model that provides a way for a program to split itself into two or more tasks of execution. Since the threads operate on shared data, the programmer must be careful not to modify a piece of data that another thread may be reading or modifying at the same time. Constructs such as barriers, locks, semaphores, mutexes, and monitors can be used for this purpose and for two or more threads to synchronize at a certai n point.

Threads on a single processor system will be swapped in and out very fast by the underlying operating system and scheduled for execution in an interleaved manner, giving the appearance of parallelism, while those in multi-processor or multi-core systems will actually run in parallel (i.e., true parallelism). In order to ensure message ordering, threads must have a way to synchronize (i.e., pause at specific points of execution).

Shared memory models do not offer an easy way to derive a correctness proof for the solutions; heavy testing is usually done on the system to build confidence, but often possible paths of execution are missed in testing environments. Formally, to prove correctness, all possible interleavings must be considered. To make matters worse, the scheduling mechanisms of the Java Virtual Machine (JVM) [11], Common Language Runtime (CLR) [12], Linux, Solaris, and Windows are complex and do not offer much guarantee in terms of when a thread is interrupted and when it will get its next burst of CPU time. Thread libraries are available in many popular programming languages. The focus here is on C, Java, and C#.

2.1. Java

In Java the threading mechanism is tightly coupled to the language. Every object has an implicit monitor that can be used both as a locking mechanism and a wait/notify mechanism. In addition, method signatures may contain the *synchronized* keyword, which tell the JVM to use the current object's implicit monitor for locking to facilitate transactional method calls, or methods that will ensure the atomicity of their instruction sets with respect to other methods of the same object that contain the *synchronized* keyword.

A thread library is available for Java that is based on the threading mechanism of the Java Virtual Machine. Sun [13] added this library in Java 1.5 to help ease the pains of writing multi-threaded code. Of particular interest with respect to thread synchronization is the addition of a *CyclicBarrier* [14] type whose job it is to provide a re-entrant barrier for multiple threads to synchronize on. Here is code from the Santa thread that shows the use of two *CyclicBarrier* objects to synchronize with Elf threads:

```
// notify elves of "OK" message
try {
    m_elfBarrierTwo.await();
}
catch (InterruptedException e) {
    e.printStackTrace();
}
catch (BrokenBarrierException e) {
    e.printStackTrace();
}

// wait for elves to say "ok we'll build it"
try {
    m_elfBarrierThree.await();
}
catch ...  // exception handling logic
```

And the corresponding Elf code:

```
// wait for santa to say "ok all done"
try {
    m_elfBarrierTwo.await();
}
catch ...  // exception handling logic

System.out.println("Elf " + this + ": OK ... we'll build it\n");

// notify santa of "ok" message
try {
    m_elfBarrierThree.await();
}
catch ...  // exception handling logic
```

Note the two checked exceptions that the *await()* method of a $CyclicBarrier$ can throw, $InterruptedException$ and $BrokenBarrierException$. Proper handling of these exceptions requires additional effort if the program needs to recover from errors due to single points of failure in a single thread. Handling exceptions from failed calls to the thread library can become increasingly difficult as an application grows in size. In our Java solution if either of the CyclicBarrier exceptions are thrown in a single thread, a stack trace will be printed but the other threads will not be made aware of this failure, they will continue operating. However, if the system depended on the thread that failed it will eventually come to a state of deadlock. In this case, if the Santa thread fails, the program eventually comes to a halt, but if a single Elf thread fails and it is not in the waiting queue at the time, the program will continue operating minus the failed Elf. If the desired behavior is to restore threads to an acceptable state then additional thread logic would have been needed.

Although $CyclicBarrier$ has eliminated the need for shared state when synchronizing, a wait and notify mechanism is needed when either a group of Elves or a group of Reindeer is ready to see Santa. Due to the asynchronous nature of the $Object.notify()$ method, in the

Java version Santa must query the Elves and Reindeer in case of missed notifications. This leads to code that is more coupled among multiple threads than we would prefer:

```
if (elfQueue.size() % 3 == 0) {
    synchronized (m_santaLock) {
        m_santaLock.notify();
        notifiedCount++;
    }
}
```

In addition, a JVM implementation is permitted to perform spurious wake-ups in order to remove threads from wait sets and thus enable resumption without explicit instructions to do so [11]. This requires all calls to $Object.wait()$ to have extra logic to check if the condition the thread was waiting on is true or it was a spurious wakeup by the JVM. This hurts the readability of the wait/notify mechanism. It also requires both the notifying thread and waiting thread to share some state so that the notifier can set a condition before notifying, and the waiting thread can check the condition in the case of a wakeup.

2.2. Pthreads

Pthreads are a POSIX standard for threads. Pthread libraries are available for a number of operating systems including Windows, Linux, and Solaris. Pthreads provide developers access to mutexes, semaphores, and monitors for thread synchronization. For the Santa Claus problem a custom partial barrier implementation similar to $CyclicBarrier$ in Java was implemented in order to improve the readability of the solution and because the library does not have a built-in partial barrier implementation. For the synchronization between Santa and the Reindeer, the pthread libraries' $join$ function is sufficient, but the $join$ function cannot be used for Santa to wait for a group of elves because Santa cannot foresee which three elves will need to consult with him. A simple implementation of a barrier can be defined with the following:

```
void AwaitBarrier(barrier_t *barrier) {
    /* ensure that only one thread can execute this method at a time */
    pthread_mutex_lock(&barrier->mutex);
    /* if this barrier has reached its total count */
    if (++barrier->currentCount == barrier->count) {
        /* notify all threads waiting on the barrier's condition */
        pthread_cond_broadcast(&barrier->condition);
    }
    else { /* at least one thread has not entered this barrier */
        /* wait on the barrier's condition */
        pthread_cond_wait(&barrier->condition, &barrier->mutex);
    }
    /* allow other threads to enter the body of AwaitBarrier */
    pthread_mutex_unlock(&barrier->mutex);
}
```

A mutex and a condition variable from the library are used in order to synchronize a group of threads. Each thread will wait on a condition variable, and the last thread to synchronize will notify each of the threads waiting on the condition variable. A mutex is used to ensure no two threads will ever be in the body of $AwaitBarrier$ at a time. This will ensure that exactly $N-1$ threads in the group will wait for the condition, and only the Nth thread will notify the group of waiting threads. Here is a portion of the Santa thread code that uses the barriers to ensure message ordering:

```
/* notify elves of "OK" message */
AwaitBarrier(&elfBarrierTwo);

/* wait for elves to say "ok we'll build it" */
AwaitBarrier(&elfBarrierThree);
```

Here is the corresponding code for the Elf threads:

```
/* wait for santa to say "ok all done" */
AwaitBarrier(&elfBarrierTwo);

printf("Elf %d: OK ... we'll build it\n", elfId);

/* notify santa of "ok" message */
AwaitBarrier(&elfBarrierThree);
```

In addition to thread synchronization to ensure message ordering, the program must also satisfy the "three elves at a time" constraint. The Elf threads share state in the form of a counter so that every third Elf in the waiting room will go and wake Santa. When one thread signals another the signal will be lost if the thread on the receiving end of the signal was not currently waiting on the condition. This can be handled a number of ways. Here, as in the Java version, Santa queries the Elves to see if they are ready in case Santa was with the Reindeer and at the time an Elf notification was sent. A different implementation could remove the query and instead have a waiting group of Elves send notifications to Santa until he responds to one. An Elf thread uses the following code to wake Santa:

```
pthread_mutex_lock(&santaMutex);
pthread_cond_signal(&santaCondition);
pthread_mutex_unlock(&santaMutex);
```

Missed notifications are recorded via use of a shared memory counter. Both the Santa and the Elf code must be aware of the fact that Santa needs a way to query for a group of three elves, which leads to tightly coupled code between the Santa and Elf threads.

2.3. C#

.NET provides a threading library that is tied to the memory model of the Common Language Runtime (CLR). The memory model for the CLR [12] provides an additional layer of abstraction over the underlying operating system's memory model, including its own thread scheduling mechanism. Unlike Java, C# objects do not have implicit monitor associations. In the C# version we use the provided threading library to write a barrier for thread synchronization. A monitor object is used for locking and unlocking of code blocks:

```
public void Await() {
    // ensure only one thread is in this method at a time
    Monitor.Enter(_awaitLock);

    // if all threads have arrived at the barrier
    if (++_lockValue == _count) {
        // notify the earlier threads waiting on this barrier
        Monitor.PulseAll(awaitLock);
    } else { // at least one thread has not entered the barrier
        // wait for a notification
        Monitor.Wait(awaitLock);
    }
```

```
        // allow another thread to call Await()
        Monitor.Exit(awaitLock);
}
```

Here is a sample usage of the barriers in the Santa code:

```
// harness reindeer
deerBarrierOne.Await();

// wait for all deer to say "delivering toys"
deerBarrierTwo.Await();
```

And the corresponding reindeer code:

```
// wait for santa to harness us
Santa.deerBarrierOne.Await();

m_Form.WriteMessage("Reindeer " + this.ToString()
                    + ": delivering toys\n");

// notify santa of "delivering toys"
Santa.deerBarrierTwo.Await();
```

Note that in both the C and the C# barrier implementation a wait/notify mechanism is used along with a shared variable to keep track of state. The wait/notify behaves like an asynchronous messaging system, in that a thread may notify even if there is a thread that is not waiting yet, so lost notifications are possible. Here we use a monitor to avoid this in our barrier implementation, in the C version a mutex and a condition variable are used for the same purpose. The monitor ensures that only the Nth thread will call $notify$ after the other $N-1$ threads are waiting for notification.

2.4. Groovy

Groovy [15] is touted as "an agile dynamic language for the Java Platform". Groovy compiles to Java byte code and adds a number of features not present in Java, most notably dynamic typing and closures. Closures are anonymous chunks of code that may take arguments, return a value, and reference and use variables declared in its surrounding scope. Although Groovy does not expand upon Java's threading mechanism, the Groovy version is more readable than the Java version due to the use of closures to remove the replication of exception handling code. Thread related calls are wrapped in closures, allowing them to be passed to methods or other closures which use a different closure for the exception handling logic. Here, a method named $performOperation$ is used which takes a closure as an argument that is the operation to be executed and uses a member variable closure which handles the exception handling logic. In this case we are using a closure named $simpleExceptionHandler$ which simply prints the reason for the exception to the console. More complicated exception handlers can be defined in a similiar manner. Here is the Groovy version of the above Java Elf code:

```
// wait for santa to say "ok all done"
barrierAwait(m_elfBarrierTwo)

println("Elf " + id + ": OK ... we'll build it\n")

// notify santa of "ok" message
barrierAwait(m_elfBarrierThree)
```

2.5. Differences

Due to possible spurious wake-ups by a JVM, whenever a call to $Object.wait()$ is made in the Java and Groovy versions, condition checking code, typically in the form of a $while$ loop, must be added in order to differentiate between a call to $Object.notify()$ and an unexpected spurious wakeup. Spurious wake-ups are cited in the Java Language Specification [11], this may be due to a bug in the JVM or a flaw in operating systems such as older versions of Linux that used LinuxThreads [16]. In the description of the Native POSIX Thread Library for Linux [17] spurious wake-ups and "the misuse of signals to implement synchronization primitives" add to the problems of the Linux Threads library, concluding that "delivering signals is a very heavy-handed approach to ensure synchronization."

With regards to error handling, the Java and Groovy versions must do something with the checked exceptions that a $CyclicBarrier$ can throw. These problems make the Java code less readable and even more so if the program must attempt to recover from these types of errors. Often the simplest, and therefore easiest to read, solution is to restart all threads in the program. Thanks to closures in Groovy some of the readability can be recovered. In the C and C# versions, the respective library call errors are ignored, as these languages do not require the handling of error conditions at compile time.

In Java, threads are more tightly coupled to the language than C#. Aside from this, Java's spurious wake-ups, C#'s lack of a built-in partial barrier, and the lack of checked exceptions in C#, the Java and C# differences are syntactic. One example is Java's use of the $synchronized$ keyword which is comparable to C#'s explicit $Monitor$ type. An advantage of the pthread library is the ability to distinguish between the condition and mutex primitives. This decouples the locking mechanism from the notification mechanism.

2.6. Ada

A solution to the problem was proposed in Ada [9] that used two constructs that were added to Ada 95 [18]. One was a construct that enables a group of processes to be collected and then released together, similar to the barriers that were presented above. The second is a rendezvous that enables a process to wait simultaneously for more than one possible event. This can be done in Java with $Object.wait()$, C# with $Monitor.Wait()$, and C with a mutex and condition variable.

2.7. Polyphonic C#

A solution which claims to be easier to reason about than the Ada solution has also been written using Polyphonic C# [8]. Here a concept known as a chord is used, which associates a code body with a set of method headers, and the body of a chord can only run once all of the methods in the set have been called. This solution shows how a wait/notify mechanism that can prioritize notifications can be implemented with shared memory if chords are available to the programmer.

2.8. Haskell STM

A solution to the problem has been written in Haskell using Software Transactional Memory, or STM [19]. STM provides atomic transactions to protect against race hazards and choice operators for prioritizing actions. The concept of a gate is used for marshalling the Reindeer and Elves. The choice operator allows for a simple way to give the Reindeer priority over a group of Elves. This solution shows how these constructs can help to de-couple and modularize multi-threaded code.

2.9. Summary

The readability, writability, and reliability of shared memory solutions is crippled by the amount of time it can take in order to search a large codebase for all threads that may possibly cause race conditions, deadlock, livelock and starvation. Preventing race conditions means ensuring the atomicity of a set of one or more instructions and usually involves the use of mutexes, semaphores, and monitors. In each of the shared memory solutions, locks are used to ensure data integrity between each thread. These locks ensure the atomicity of one or more instructions in a thread. This can slow down performance when many threads are waiting on the same lock, in addition to adding complexity to the code. When a Reindeer or Elf is added or removed from a queue, it must happen as an atomic transaction, and in cases where the logic that follows depends on the size of a queue, this logic must be included in the atomic transaction. Preventing deadlock and livelock involves examining all possible instruction interleavings that a program can generate, a daunting task. In addition, the more state that threads share the harder refactoring becomes. Constraints such as "three elves at a time" or "Reindeer have priority over Elves" require the use of wait/notify and shared counter variables. For example, a shared counter variable is used between the Reindeer threads because Santa must have the ability to query the Reindeer to see if they are ready to deliver toys after consulting with a group of elves. A wait/notify mechanism is also used by the Reindeer and Elves to knock on Santas door. Thus refactoring the Elf or Reindeer threads requires knowledge of the Santa thread. Worse, refactoring the Santa thread requires knowledge of the Elf and Reindeer threads. In small programs such as the Santa Claus solution this is manageable, but mutli-threaded code can become entangled in a larger application. The modification of this code is time consuming and error prone, and there is no quick and simple way to check if a code change has introduced concurrency bugs into the code.

Verifying the correctness of a multi-threaded application is also complicated. Attempts at examining all possible instruction interleavings have been made [20] and shown to be feasible under certain conditions. "Given N threads, each with a local store of size L, and the threads communicate via a shared global store of size G, if each thread is finite-state (without a stack), the naive model checking algorithm requires $O(G * L_n)$ space, but if each thread has a stack, the general model checking problem is undecidable." Successful thread verification has been shown feasible when threads are largely independent and loosely coupled [21], requiring only $O(n * G * (G + L))$ space.

In addition to the complexity of using shared state and the various problems with wait/notify, the thread scheduling mechanisms introduce non-determinism into various measures of thread execution. These include when and for how long a thread will get CPU time, and when and for how long it will get interrupted and wait for CPU time.

3. Distributed Memory

MPI [22] is one of the more popular libraries designed for taking advantage of a distributed memory architecture, such as a cluster, or a shared memory multiprocessor architecture with a large amount of processors, such as those found in supercomputers. Here, an application's N separate tasks are separated into N processes and appropriate code for each process is written. In the case of MPI, the runtime will ensure the application behaves like a distributed memory model regardless of the underlying hardware. In MPI, groups of processes can be formed at runtime, and a $MPI\_Barrier$ method can be used to synchronize a particular group of processes. Synchronous receives can be used in place of wait/notify thus removing the need to handle the more complicated asynchronous logic. These techniques are used in the MPI solution to the Santa Claus problem in order for the Santa, Elf, and Reindeer processes to synchronize with each other, pausing at various points of execution. Notice how the same barrier is used for two different synchronization points without an explicit method call to reset the barrier. The barrier will implicitly reset itself after the required number of processes has reached the barrier. Here is part of Santa process code that uses a $MPI\_Barrier()$ to synchronize with the Reindeer:

```
// wait for all reindeer to say "delivering toys" message
mpiReturnValue = MPI_Barrier(commSantaReindeer);
CHECK_MPI_ERROR(globalRank, mpiReturnValue);
printf("Santa: woah ... we're back home!\n");
```

Here is part of the Reindeer process code that uses a $MPI\_Barrier()$ to synchronize with Santa:

```
// wait for santa to harness us
mpiReturnValue = MPI_Barrier(commSantaReindeer);
CHECK_MPI_ERROR(globalRank, mpiReturnValue);
printf("Reindeer %d: delivering toys\n", id);
```

Due to the lack of shared memory, solutions to problems in MPI typically have more processes running than a C, C# or Java program would have threads. In the MPI solution to the Santa Claus Problem, there is a separate process for a Reindeer queue for when the Reindeer come back from vacation, and a separate process for an Elf queue for when the Elves get confused and join the waiting room. Instead of using shared memory synchronized data, each process gets data, performs operations on that data, and passes data on to one or more other processes. The same piece of data is never operated on by more than one process at a time, thus eliminating the need for data protection.

3.1. Erlang

There are other solutions to the problem written in Erlang [23,24]. Erlang is a functional language with built-in message passing semantics. Although Erlang processes can share memory via use of the *ets* [25] module, the solutions here use message passing semantics. Similiar to the MPI version, these solutions show the lack of shared state can simplify concurrent programming.

3.2. Summary

The lack of shared state improves the readability of MPI beyond that of shared memory models because it will not allow two or more processes to modify the same piece of data at a time, allowing programmers to focus on one process. Understanding the implications of the use of asynchronous sends and/or receives, however, can be even more challenging than the wait/notify problem for shared memory. With a wait/notify, only the notifying thread behaves asynchronously, while a waiting thread will block until it receives a notify. With the use of $MPI\_IRecv()$, a receive will not block, but can be probed later in the code using $MPI\_Probe()$. In our solution we chose to use only synchronous messages due to the constraints of the problem.

Writing distributed memory code is also made simpler because of the lack of shared state. As long as the corresponding receives and sends in a process are handled properly, MPI code is easier to refactor than thread code. An example in the Santa Claus problem is the way that the three elves at a time constraint is handled. A separate Elf queue process handles this constraint, decoupling the Santa and Elf code further than the shared memory solutions.

For distributed systems, reliability can be hard to measure. The time to send data between nodes on the network can vary greatly, making it difficult to determine whether or not a node has gone down or is just taking a long time to respond. Peer to peer systems are more reliable because of the lack of specialized nodes, but for many applications the idea of every node executing the same code does not fit the problem. In this case, there are requirements that dictate one Santa process, nine Reindeer processes, and ten Elf processes. On Shared Memory Multiprocessor (SMP) machines, reliability is greatly enhanced by eliminating the network. One difficulty that exists in both SMP and distributed environments is the inability to verify asynchronous message behavior. Another problem with asynchronous messaging is that the channel buffers can become full and then start behaving like synchronous channels, causing errors that can only be seen in unfortunate runtime scenarios. Tools such as SPIN [26] are available for model checking the correctness of MPI programs. Traditionally, SPIN has been used for checking only deterministic MPI primitives, but recently work has been done to use the tool for checking non-deterministic primitives [27].

4. Process Oriented

Yet another way to write parallel applications is by using a process oriented architecture, based on the ideas of CSP [28]. Communicating Sequential Processes, or CSP, is a process algebra used to describe interactions in concurrent systems [29]. Relationships between processes and the way they communicate with their environment can be described using CSPs process algebraic operators. Parallelism, synchronization, deterministic and nondeterministic choices, timeouts, interrupts, and other operators are used to express complex process descriptions in CSP. Languages such as occam [30] are based on CSP and there are libraries that provide support for the CSP model for many other languages. Here we focus on one such library for Java, JCSP [31]. The JCSP library is open source and is implemented using Java threads, so it is portable across JVM implementations. Similar to MPI, the system is viewed as a set of independent processes that communicate via channels. There is no concept of shared memory, even if the underlying library is implemented in this way. Channels can be shared by many processes. In the JCSP version, synchronous messages and the library's built-in barrier type are used to ensure message ordering. In addition, we implemented a barrier that will synchronize Santa and a group of three elves using two many-to-one channels to enforce the "three Elves at a time" constraint. The $MyBarrier$ class is implemented using two shared channels, and will read, in succession, two sets of messages for each process on the writing end of the barrier:

```
class MyBarrier implements CSProcess {
    private int count;
    private ChannelInput inA, inB;

    public MyBarrier(int n, ChannelInput inA, ChannelInput inB) {
        this.count = n;
        this.inA = inA;
        this.inB = inB;
    }

    public void run() {
        while (true) {
            for (int i = 0; i < count; i++) {
                inA.read();
            }
            for (int i = 0; i < count; i++) {
                inB.read();
            }
        }
```

}
    }

The corresponding *Sync* class represents only the writing end of the barrier. Note that only when all members of the barrier have sent their first message will a process start to send its second message to the reading end of the barrier:

```
class Sync implements CSProcess {
    private SharedChannelOutput outA, outB;

    public Sync(SharedChannelOutput outA, SharedChannelOutput outB) {
        this.outA = outA;
        this.outB = outB;
    }

    public void run() {
        outA.write(Boolean.TRUE);
        outB.write(Boolean.TRUE);
    }
}
```

Here is part of the Santa code that uses the barrier:

```
// wait for Elves to say "about these toys"
new Sync(outSantaElvesA, outSantaElvesB).run();
outReport.write("Santa: consulting with Elves ...\n");
```

And the corresponding Elf code:

```
outReport.write("\t\t\tElf: " + id + ": about these toys ... ???\n");
// notify Santa of "about these toys"
new Sync(outSantaElvesA, outSantaElvesB).run();
```

One part of occam that is missing from JCSP is extended rendezvous, an addition to standard occam available in KRoC [32]. Although long standing, these extensions are to some extent experimental. This is a way to force a process A to wait for another process B to do some work before returning from a blocking send or receive. This allows two processes to synchronize without having to explicitly send/receive another message, which is what is done in the JCSP version in replace of an extended rendezvous. The Reindeer waits to be unharnessed with a blocking receive:

```
// Reindeer waits to be unharnessed
inFromSanta.read();
```

For each Reindeer, Santa outputs the unharness message and does a blocking send to the Reindeer to unharness the Reindeer:

```
outReport.write("Santa: un-harnessing reindeer " + id + " ...\n");
// unharness this Reindeer
channelsSantaReindeer[id - 1].out().write(0);
```

In order to ensure priority of the Reindeer over the Elves a construct known as an alternation is used. An alternation enables a process to wait passively for and choose between a number of guarded events, and in this case the guarded events are a notification to Santa from a Reindeer or an Elf. Priority can be given to a guarded event an alternation will choose, and in this case priority is given to the Reindeer notification. In this way, if both the Reindeer and a group of Elves are ready at the same time, Santa will handle the Reindeer first.

Table 1. Program line counts.

	SM				DM	PO
	C#	C	Java	Groovy	MPI	JCSP
Total	642	420	564	315	352	315
Synchronization/Communication	48	49	46	46	34	27
Prevent Race Condition	14	8	8	8	N/A	N/A
Exception/Error Handling	35	0	177	18	41	0
Custom Barrier Implementation	42	35	N/A	N/A	N/A	55
GUI	145	N/A	N/A	N/A	N/A	N/A

SM = Shared Memory, DM = Distributed Memory, PO = Process Oriented

4.1. Summary

Assuming familiarity with the JCSP library, the readability of the JCSP version is better than the shared memory Java solution due to the lack of shared state. Similar to MPI, the code for a process can be looked at without worrying about other processes which might possibly access the same instance data. One advantage that JCSP has over MPI is the ability to specify one-to-one and many-to-one channels on which to send and receive messages. This ensures that a data read will only happen when the sending process has the writing end of a channel, and that a data write will only happen when the receiving process has a reading end of the channel. With MPI, the transportation mechanism for sending and receiving data is abstracted away from the programmer, allowing any process that wishes to communicate with the system to do so. Thus, there is no easy way to ensure message integrity in MPI.

A process oriented application can be refactored easily. The way a process interfaces with the rest of the system is readily available by looking at the input and output channels associated with a process. This allows the programmer to focus on a processes input and output channels, that is, the data a process sends and receives, and not on any of the data it reads or modifies. Again, JCSP's use of channels improves refactorability over the MPI version. A channel's reading and writing ends must be explicitly referenced in order for communication to happen over the channel. This makes it clearer to the programmer which processes are communicating.

In addition, the mapping between JCSP and CSP allows JCSP code to be formally reasoned about using the process algebra. A tool called FDR [33] exists which is a model checker for CSP. The CSP that corresponds to a piece of occam or JCSP code can be run through the tool to check for possible concurrency errors, such as existence of deadlock, livelock, starvation, and fairness. Therefore, JCSP and programs written using different implementations of the CSP model can be formally verified for correct multi-parallel behavior before or after development, and the mapping is almost a one-to-one mapping in both directions, reducing the risk of introducing errors in the model checking/verification code.

5. Results

5.1. Line Count Comparisons

Comparing the line count for each solution can give an indication of the readability and writability of each solution. To be fair, the line counts for the Ada, Polyphonic C#, Haskell STM, and Erlang solutions are omitted because they did not include the message ordering constraints that we imposed on the problem. Table 1 gives an overview of the line count for each version:

5.1.1. Synchronization and Communication

The shared memory solutions all have roughly the same number of lines of synchronization and communication code. The line counts are higher than the distributed memory and process oriented counterparts due to the fact that wait/notify must be wrapped in a lock and unlock statement and extra logic is required in case of lost notifications. The line count here for all of the solutions would be much higher without the use of barriers, and would require duplicated synchronization logic each time a group of threads or processes needed to synchronize with each other.

5.1.2. Preventing Race Conditions

Each of the shared memory solutions includes locking and unlocking code to prevent race conditions whenever inserting and removing a Reindeer or Elf into a queue. Locks must also be in place anytime there is logic that depends on the size of the Elf and Reindeer queue, in this case when the last Reindeer or every third Elf must notify Santa. For the distributed memory and process oriented solutions there is no shared memory so there is no code needed to prevent a race condition.

5.1.3. Exception and Error Handling

In each solution the error handling code will simply print an error message to the console. For a real world problem, the error handling code would include logic that attempts to recover from errors. Gracefully recovering from errors in a concurrent system often requires additional communication and synchronization so that a particular thread or process can determine the state of the system as a whole and then bring itself to an appropriate state. Due to checked exceptions and the two exceptions that a *CyclicBarrier.await* can throw the Java error handling line count is much higher than the other solutions. For the Groovy solution, closures have been used to wrap *CyclicBarrier.await* and *Object.wait*, consolidating the error handling into the closure and reducing the error handling line count. For the C# version the unchecked exceptions in the .NET threading library were ignored. The C version ignores the errors that the pthread library calls can generate. The MPI version includes a macro for error handling which is used to check for errors every time a call to a method in the library takes place. The parts of JCSP that are used in this solution do not throw checked exceptions so there is no exception handling code in the JCSP version.

5.1.4. Custom Barrier Implementation

For C, C# custom barriers were implemented in order to reduce the amount of code required for thread synchronization and also prevent errors. If a barrier implementation is broken in can be fixed once without having to change any of the Santa, Elf, or Reindeer code. Custom barriers were implemented in the JCSP solution both to simplify process synchronization and to place the logic for the "three elves at a time" constraint into a special barrier type, eliminating the need for a separate Elf queue process like the one used in the MPI solution. The Java thread library and MPI version have built-in barrier implementations that are robust enough to solve the Santa Claus Problem and so no custom barriers were implemented in the Java, Groovy, and MPI solutions.

5.1.5. GUI

The additional line count for the C# version can be attributed to two things. The first is the additional lines added for a Windows Forms GUI. The second is that Visual Studio formats code with beginning braces on their own line, while in the other solutions they are not.

6. Conclusions

We have implemented solutions to the Santa Claus Problem using various concurrent programming models as a basis for comparing the three models, shared memory, distributed memory, and process oriented.

For the shared memory solutions two difficulties were identified. The first issue was the overly complicated use of mutexes, condition variables, and monitors to ensure mutual exclusion between various sets of instructions when using multiple threads. The second was the difficulty of using the wait/notify mechanism due to the asynchronous nature of the notify and the possibility of lost notifications by the receiver. Comparisons between the shared memory models themselves were discussed to show that the minor implementation differences do not alleviate the issues with the model.

We have shown how a distributed memory model such as MPI can simplify process ordering with the use of synchronous messages and how a a distributed memory model also gives built-in data integrity and allows for much simpler implementations of various program constraints. Furthermore, we have introduced JCSP and shown how the use of libraries and languages based on CSP further simplify the development of a concurrent application in a distributed memory model. The additional benefit of a one-to-one mapping with CSP, an algebra designed to describe process interaction, allows for the ability to formally verify various correctness measures of the application. A final benefit of JCSP over MPI is the use of channels which allows for increased trust between communicating processes.

The code for all of the solutions we implemented is available at:

http://www.santaclausproblem.net

7. Future Work

The Santa Claus Problem requires heavy synchronization, but future work should include the comparison of all three models for a wider range of problem types. A more rigorous comparison should be done between the three models, including the performance of the models in various settings. In addition to comparing the models themselves, it may be useful to compare the model checking tools available for each model, such as MPI-Spin and FDR. Ease of use and what can and cannot be proved are two interesting criteria for comparison here.

Acknowledgements

We acknowledge the comments of one of the referees who suggested that a solution that exploits Groovy Parallel [34] and which also employs the JCSP synchronisation primitives; Alting Barrier [35] and Bucket might be even shorter. In fact such a solution is about 75 lines shorter than the JCSP solution described previously.

References

[1] PlanetLab home page. http://www.planet-lab.org.
[2] BOINC home page. http://boinc.berkeley.edu.
[3] SETI Institute home page. http://www.seti.org.
[4] Google APIs. http://code.google.com/more.
[5] Yahoo! Developer Network. http://developer.yahoo.com.
[6] C. Ajluni. Multicore Trends Continue to Drive the Embedded Market in 2007. *Chip Design*, Jan/Feb 2008.

[7] J. A. Trono. A new exercise in concurrency. *SIGCSE Bull.*, 26(3):8–10, 1994.
[8] N. Benton. Jingle Bells: Solving the Santa Claus Problem in Polyphonic C#. Technical report, Microsoft Research, Cambdridge UK, 2003.
[9] M. Ben-Ari. How to solve the Santa Claus problem. *Concurrency: Practice and Experience*, 10(6):485–496, 1998.
[10] B. Nichols, D. Buttlar, and J. P. Farrell. *Pthreads Programming, First Edition*. O'Reilly Media, 1996.
[11] J. Gosling, B. Joy, G. Steele, and G. Bracha. *The Java Language Specification, Third Edition*. Addison-Wesley Professional, 2005.
[12] R. F. Stark and E. Borger. An ASM specification of C# threads and the .NET memory model. In *ASM 2004*, pages 38–60, 2004.
[13] Sun Microsystems home page. http://www.planet-lab.org.
[14] CyclicBarrier API.
http://java.sun.com/j2se/1.5.0/docs/api/java/util/concurrent/CyclicBarrier.html.
[15] D. Koenig, A. Glover, P. King, G. Laforge, and J.Skeet. *Groovy in Action*. Manning Publications Co., 2007.
[16] LinuxThreads library. http://pauillac.inria.fr/~xleroy/linuxthreads/.
[17] U. Drepper and I. Molnar. The Native POSIX Thread Library for Linux. 2005.
[18] ANSI/ISO/IEC. *Ada 95 Language Reference Manual*, 1995.
[19] S. P. Jones. *Beautiful concurrency*. O'Reilly, 2007.
[20] C. Flanagana and S. N. Freund and S. Qadeerc and S. A. Seshia. Modular verification of multithreaded programs. In *Theoretical Computer Science*, 2005.
[21] C. Flanagana and S. Qadeerc. Thread-modular model checking. In *Model Checking Software*, 2003.
[22] J. Dongarra. MPI: A message passing interface standard. *The International Journal of Supercomputers and High Performance Computing*, 8:165–184, 1994.
[23] The Santa Claus Problem (Erlang). http://www.crsr.net/Notes/SantaClausProblem.html.
[24] Solving the Santa Claus Problem in Erlang.
http://www.cs.otago.ac.nz/staffpriv/ok/santa/index.htm.
[25] ets. http://www.erlang.org/doc/man/ets.html.
[26] G. J. Holzmann. *The SPIN Model Checker: Primer and Reference Manual*. Addison-Wesley Professional, 2003.
[27] S. R. Siegel. Model Checking Nonblocking MPI Programs. *Verification, Model Checking, and Abstract Interpretation*, 4349:44–58, 2007.
[28] P. H. Welch. Process Oriented Design for Java: Concurrency for All. In *ICCS '02: Proceedings of the International Conference on Computational Science-Part II*, page 687, London, UK, 2002. Springer-Verlag.
[29] C. A. R. Hoare. *Communicating Sequential Processes*. Prentice Hall International, 1985.
[30] occam 2.1 reference manual.
http://www.wotug.org/occam/documentation/oc21refman.pdf, 1995.
[31] P. H. Welch. Communicating Sequential Processes for Java.
http://www.cs.kent.ac.uk/projects/ofa/jcsp.
[32] P. H. Welch and F. R. M. Barnes. occam-π: blending the best of CSP and the π-calculus.
http://www.cs.kent.ac.uk/projects/ofa/kroc/.
[33] Formal Systems (Europe) Ltd. *Failures-Divergence Refinement: FDR2 manual*, 1998.
[34] J. M. Kerridge, K. Barclay, and J. Savage. Groovy Parallel! A Return to the Spirit of occam? In *Communicating Process Architectures*, pages 13–28, 2005.
[35] P. H. Welch, N. C. C. Brown, J. Moores, K. Chalmers, and B. Sputh. Integrating and Extending JCSP. In *Communicating Process Architectures*, pages 349–370, 2007.

Mobile Agents and Processes using Communicating Process Architectures

Jon KERRIDGE, Jens-Oliver HASCHKE and Kevin CHALMERS

School of Computing, Napier University, Edinburgh UK, EH10 5DT
{j.kerridge, k.chalmers}@napier.ac.uk, jens.haschke@gmx.de

Abstract. The mobile agent concept has been developed over a number of years and is widely accepted as one way of solving problems that require the achievement of a goal that cannot be serviced at a specific node in a network. The concept of a mobile process is less well developed because implicitly it requires a parallel environment within which to operate. In such a parallel environment a mobile agent can be seen as a specialization of a mobile process and both concepts can be incorporated into a single application environment, where both have well defined requirements, implementation and functionality. These concepts are explored using a simple application in which a node in a network of processors is required to undertake some processing of a piece of data for which it does not have the required process. It is known that the required process is available somewhere in the network. The means by which the required process is accessed and utilized is described. As a final demonstration of the capability we show how a mobile meeting organizer could be built that allows friends in a social network to create meetings using their mobile devices given that they each have access to the others' on-line diaries.

Keywords. mobile agents, mobile processes, mobile devices, social networking, ubiquitous computing, pervasive adaptation.

Introduction

The advent of small form factor computing devices that are inherently mobile and the widespread use of various wireless networking technologies mean that techniques have to be developed which permit the easy but correct use of these technologies. The goal of ubiquitous computing is to provide an environment in which the dynamic aspects of the environment become irrelevant and as the users of mobile devices move around their devices seamlessly integrate with both their immediate surroundings and those which are fixed in some way to predetermined locations [1, 2].

The π-calculus [3] has provided a means of reasoning about such capabilities and some of its concepts have been implemented in environments such as occam-π. In these environments the emphasis is to provide a means whereby mobility is achieved by the movement of communication channel ends from one process to another regardless of the processing node upon which the process resides. The occam-π environment has mainly been used in large computing cluster based experiments using the *pony* environment [4] and the creation of highly parallel models of complex systems [5]. The nature of the occam-π environment tends to promote channel mobility due to the static nature of the process workspace at a node.

The agent based community has developed a number of frameworks and design patterns that promote the use of agents. An agent, based upon the actor model, is a component that is able to move around a network interacting with each individual node's environment to achieve some pre-defined goal. Once the goal has been achieved the agent is destroyed. The goal of an agent is usually to obtain some service or outcome that either changes the originating node's capabilities, the visited nodes' capabilities or some combination of these. The agent community tends to use the object oriented paradigm and thus a number of design patterns have been created that provide a basis for designing such systems [6] describe a concurrent environment that exploits multi-agent systems and agent frameworks to build systems that employ a behavioural and goal oriented approach to system design that are able to evolve as a result of co-operation between components within the resulting implementation.

The JCSP community [7] has taken a different approach to mobility in which the mobility of processes and communication channels is seen as being equally important [8]. The primary advantage of using a Java based technology is its widespread use in a large number of mobile devices and hence it provides some degree of portability. Further, it has already been shown that transferring processes from one node to another is feasible [8, 9] as a result of specific changes made to the dynamic Java class loading system. The impact of ubiquitous computing and mobility is discussed in [9], where the emphasis was using wi-fi access to enable the mobility of processes. In that case, only the mobility of processes was considered, whereas, in this paper we demonstrate how the concept of a mobile agent can be implemented in the JCSP context.

The development of both occam-π and JCSP has been undertaken by the same group and thus many of the concepts are shared and build upon the same theoretical frameworks (CSP and the π-calculus). The JCSP developments have always been more widely based in terms of their use of network capability. In particular, the ability to achieve process mobility has always been present in the networked version of JCSP, though few people have exploited the capability.

The approach taken in this paper is to exploit the JCSP approach and to extend the dynamic process loading capability so that agents can be sent around a network to obtain processes that can be returned to the agent's originating node where they are installed and can execute as if they had been running there from the outset.

In the next section we describe the structure of an agent in our implementation, which is followed by a description of corresponding interaction between an agent and a node that it is visiting. In the third section a description of an initial test environment is described. We then present a case study concerning a mobile social networking application. Finally, conclusions are drawn and further work is identified.

1. Agent Formulation

The goal of an agent is to find and retrieve, on behalf of its originating node, a copy of a process that is currently unavailable at the node. The node has been asked to process some data for which the required process is not available. The node initializes the agent with the identity of the required process. The agent then travels around the network containing the node until it finds a node with the required process. The process is then transferred to the agent, which then returns to its originating node where the process is added to the processing capability of the node. This means that the node is now able to process this new type of data as if it had been able to do so from the outset.

In the experiments reported in this paper an agent has the following requirements:

- At the originating node it needs to be initialized with its goal, which in this case is the identity of the required process,

- At a visited node it needs to determine whether the required process is available. If it is available then the process should be copied into the agent, which then returns to its originating node. If the required process is not available, the agent should cause itself to move to another node in the network.
- An agent returning to its originating node should transfer the required process to the node, which will then cause it to be incorporated into its execution environment.
- Additionally, an agent requires the ability to connect itself to a node's internal communication structure so that it can interact with the node. In order to leave a node, an agent also requires the ability to disconnect itself from the node's internal communication structure. This latter requirement arises because the agent is a `serializable` object and any local channel communication ends are not serializable.

In its simplest form an agent only requires one input and one output channel end, which it then connects with a node in a manner depending on the context. Once the connection(s) have been made, the agent can interact with the node, which has to provide the complimentary set of channel ends. The channel structure for each of the interconnections is shown in Figure 1. Figure 1a shows the connection between the Agent and its originating node. The Agent simply needs to receive data from its Node which comprises the addresses of any Nodes the Agent can visit until such time as the Node determines it needs another process. The Node sends the identity of the required process to the Agent, at which point the Agent disconnects itself from its originating Node and using the list of other Nodes travels around the network until it finds the process it requires. The Agent is thus implementing an Itinerary pattern [12]. It is assumed at this stage that the required process will be found. When the Agent visits another Node (Figure 1b) it requires two channel connections; the first is used to send the identity of the required process to the Node and the other receives the response from the Node, which is either a copy of the process or an indication that the Node does not have the required process. Finally, the Agent returns to its originating Node (Figure 1c), where only one channel is required, which is used to send the required process to the Node. The Node is then able to incorporate the process into its internal infrastructure and is thus able to process data that was previously not possible.

(a) An agent waiting to be initialised

(b) An agent visiting another node

(c) An agent returned to its original node

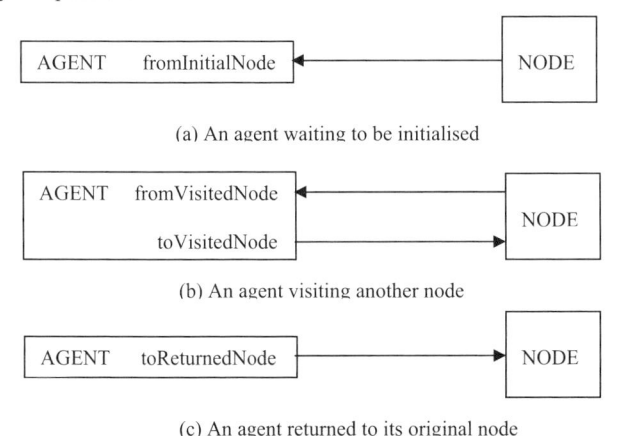

Figure 1: agent–node channel connections.

Each aspect of agent behaviour is now described in turn. The coding is presented using the Groovy Parallel formulation [11].

1.1 Agent Properties and Basic Methods

The interface `MobileAgent` {1}[1] defines two methods `connect` and `disconnect` and also implements the interface `Serializable`. In this formulation each channel that is required is specifically named {2-5}, rather than reusing channel ends for different purposes in different situations. The channel variables have the same name as used in Figure 1. An agent can be in one of three states represented by three boolean variables {6-8} of which only one can be `true` at any one time. (A single state variable could have been used but, for ease of explanation, three are used). The remaining variables {9-13} are either given values during the initialization phase or assigned values as the agent visits other nodes.

```
01    class Agent implements MobileAgent {

02      def ChannelInput fromInitialNode
03      def ChannelInput fromVisitedNode
04      def ChannelOutput toVisitedNode
05      def ChannelOutput toReturnedNode

06      def initial = true
07      def visiting = false
08      def returned = false

09      def availableNodes = [ ]
10      def requiredProcess = null
11      def returnLocation
12      def processDefinition = null
13      def homeNode

14      def connect (List c) {
15        if (initial) {
16          fromInitialNode = c[0]
17          returnLocation = c[1]
18          homeNode = c[2]
19        }
20        if (visiting) {
21          fromVisitedNode = c[0]
22          toVisitedNode = c[1]
23        }
24        if (returned) {
25          toReturnedNode = c[0]
26        }
27      }

28      def disconnect() {
29        fromInitialNode = null
30        fromVisitedNode = null
31        toVisitedNode = null
32        toReturnedNode = null
33      }
```

The `connect` method {14-27} is passed a `List` of values, the number and contents of which vary depending on the state of the agent. The `connect` method is always called by the node at which the agent is located because the values passed to the agent are local to the node. In the initial state {15-19}, the list contains a local channel input end, the net channel input location to which the agent will return when the goal has been achieved and the name of the originating node. In the case where the agent is visiting another node {20-23} the list

---
[1] The notation {n} refers to a line number in a code listing.

comprises a local channel input end and a local channel output end which is uses to create channels to communicate with the local node. Finally, in the case where the agent has returned to its originating node {24-26} the list simply comprises a local channel output end which the agent uses to transfer the code of the process that has been obtained. The `disconnect` method {28-33} simply sets all the channel end variables to `null`, which is a value that can be serialized.

The coding associated with each state of the agent is such that it is guaranteed to terminate and the agent will only ever be in one of its possible states.

1.2 Agent Initialisation

An agent is, by definition, in the initial state when it is first constructed by an originating node. In the initial state, an agent can receive two types of input on its `fromInitialNode` channel {37}. In the case of a `List` {38-40}, the communication comprises one or more net channel input ends from nodes that have been connected to the network. Nodes can be created dynamically in the system being described. In this case the net channel locations are appended (<<) to the `List` of `availableNodes` {39}. The `List` `availableNodes` therefore holds the itinerary around which the Agent will travel until it finds the required process.

```
34      if (initial) {
35        def awaitingTypeName = true
36        while (awaitingTypeName) {
37          def d = fromInitialNode.read()
38          if ( d instanceof List) {
39            for ( i in 0 ..< d.size) { availableNodes << d[i] }
40          }
41          if ( d instanceof String) {
42            requiredProcess = d
43            awaitingTypeName = false
44            initial = false
45            visiting = true
46            disconnect()
47            def nextNodeLocation = availableNodes.pop()
48            def nextNodeChannel = NetChannelEnd.createOne2Net(nextNodeLocation)
49            nextNodeChannel.write(this)
50          }
51        }
52      }
```

If the input is a String {41-50} then the agent has received the name of the process it is to locate from a node elsewhere in the network. It is presumed that the required process is always available. The name of the process is assigned to `requiredProcess` {42} and the loop control variable is set `false` {43}. The state of the agent is changed from `initial` to `visiting` {44-45}. The agent then `disconnect`s itself from its originating node {46}. The first net channel location is then `pop`ped from the list of `availableNodes` and assigned to `nextNodeLocation` {47}. This is then used to create a net channel output end as `nextNodeChannel` {48}. The agent then writes itself to this net channel, thereby transferring itself to the first node in the list of available nodes. This simple formulation essentially provides the itinerary agent design pattern.

1.3 Agent Visiting Another Node

The agent writes the value of `requiredProcess` on its `toVisitedNode` local channel {54} and then reads a response from the visited node on its `fromVisitedNode` channel {55}.

If the returned value is not null then the required process has been located {56} and the agent writes the identity of the agent's originating node to the visited node {57}. The state of the agent is changed from `visiting` to `returned` {58-59}. The agent then creates a net channel output end {60-61} using the value in `returnLocation`, `disconnect`s itself from the visited node {60} and writes itself to its originating node {63}. If the returned value is the `null` value then the agent simply disconnects itself and writes itself to the next node in the list of `availableNodes` {66-70}.

```
53      if (visiting) {
54        toVisitedNode.write(requiredProcess)
55        processDefinition = fromVisitedNode.read()
56        if ( processDefinition != null ) {
57          toVisitedNode.write(homeNode)
58          visiting = false
59          returned = true
60          disconnect()
61          def nextNodeLocation = returnLocation
62          def nextNodeChannel = NetChannelEnd.createOne2Net(nextNodeLocation)
63          nextNodeChannel.write(this)
64        }
65        else {
66          disconnect()
67          def nextNodeLocation = availableNodes.pop()
68          def nextNodeChannel = NetChannelEnd.createOne2Net(nextNodeLocation)
69          nextNodeChannel.write(this)
70        }
71      }
```

1.4 Returned Agent

An agent that has returned to its originating node simply writes a `List` comprising the process definition and the name of the process for which the agent was searching to the local channel `toReturnedNode` {73}. A node can create more than one agent, each of which is searching for a different process. Each of these agents can be active in the network at the same time. Once an agent has written the required process to the local node it terminates

```
72      if (returned) {
73        toReturnedNode.write([processDefinition, requiredProcess])
74      }
```

2. Node Processing Functionality

A node operates in an environment whereby it has to register with a specific authority node thereby indicating that it is willing to accept agents. Additionally it is initialized with zero or more of the required processes that agents will be sent to retrieve. Nodes can be created dynamically. A node registers itself with the authority by sending it the net channel input location of a channel upon which the node is willing to receive agents. The authority then sends this new node location to all previously registered nodes. Additionally, a node registers a net channel input location on which it receives data in the form of data objects. These data objects require a specific process to undertake manipulation of the data. It is this process that may have to be located and returned to a node by an agent if the required process is not already available at the node.

Once a node is registered with the authority it alternates over a small set of net input channels. It can receive:

- an input from the authority indicating that a new node has been created comprising the new node's agent visit net channel input location.
- a data object which needs to be processed. It may be able to process this data immediately because the required process is already available. If the process is not available, it initiates an agent with the name of the required process and sends the agent to find it.
- a visiting agent from which it reads the name of the required process. It can either write to the agent if the node has an instance of that process or null otherwise.
- a returned agent from which it reads the process that has been located, which it then installs in its own processing environment.

Each of these alternatives has a slightly different formulation specific to each case. We describe only the case of a returned agent. The agent is read from the net channel input `agentReturnChannel` {75} as the variable `returnAgent`. The address of the `agentReturnChannel` was passed to the agent when it was created as one of the parameters of the `connect` method {17}.

```
75    def returnAgent = agentReturnChannel.read()
76    returnAgent.connect([NodeFromReturningAgentOutEnd])
77    def returnPM = new ProcessManager (returnAgent)
78    returnPM.start()
79    def returnList = NodeFromReturningAgent.in().read()
80    returnPM.join()
81    def returnedType = returnList[1]
82    currentSearches.remove([returnedType])
83    typeOrder << returnList[1]
84    connectChannels[cp] = Channel.createOne2One()
85    processList << returnList[0]
86    def pList = [connectChannels[cp].in(), nodeId, toGathererName]
87    processList[cp].connect(pList)
88    def pm = new ProcessManager(processList[cp])
89    pm.start()
90    cp = cp + 1
```

The `returnAgent` is then `connected` {76} to a local node and passed the output end of a local channel in `NodeFromReturningAgentOutEnd`. A new `ProcessManager` is then created {77} for `returnAgent` which is then `started` {78} to enable the agent to run in parallel with the node process. The node then reads a `returnList` from the agent {79} using the `in()` end of the channel that connects the agent to the node. The data in the `List` is that written by the agent {73} and comprises the process definition and the name of the process. The node process then waits for the agent to terminate {80}. The name of the returned process is assigned to `returnedType` {81}. This name is then removed from the list of current searches {82}. Recall that a node can initiate a search for a number of processes at the same time. The list `currentSearches` is used to ensure that an agent is not initiated for a search that has been previously started. The name of the returned process is then appended to this list of processes available to this node {83} as the `cp`'th element. Each of these returned processes has a single input channel by which data that is input on the node's data input channel, connected to the Data Generator process, can be written to the process. This channel has to be dynamically created {84}. The returned process, `returnList[0]` is then appended to the list of available processes as the `cp`'th element of `processList` {85}.

Returned processes implement an interface that is very similar to the `MobileAgent` interface in that it has a `connect` method but no `disconnect` method. In this case a process requires a list, pList, of values comprising its local input channel end, the identity of the node on which it is executing and the name of net channel of a net any to one channel upon which the process can write the resulting effect of the process {86}. The cp'th element of processList is then connected to the node {87}. A new ProcessManager for this element can now be constructed {88} and started {89}. Finally the value of cp is incremented {90} ready to accept the next process that may be returned.

A client-server style design has been adopted for all interactions within the system, yielding deadlock and livelock freedom. This applies to both the interactions between the primary nodes of the system and also between nodes and agents.

3. Initial Evaluation

Figure 1 shows the basic architecture in which the Authority node and the node which creates instances of data objects are combined into a single node. Each arrow represents a networked channel. The arrows with the heavier lines are named network channels, managed by the JCSP Channel Name Server (CNS). Each of these channels is implemented as an any-to-one channel so that any number of Nodes can write to the Authority or Gatherer nodes. The Gatherer is simply a node that records the effect of the processing on any of the data objects. Each Node creates three net channel inputs, shown by the dashed lines.

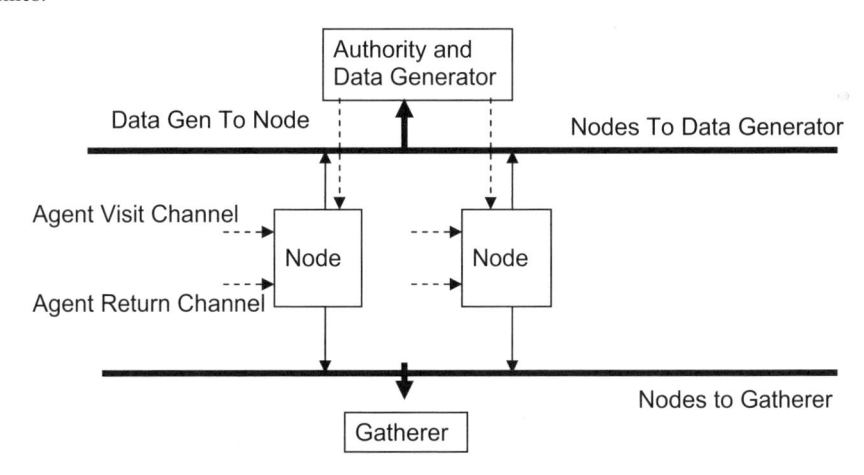

Figure 2: network architecture of the basic mobile process and agent system.

The Agent Visit Channel and Data Gen To Node net channel locations are sent to the Authority as part of the node creation mechanism. The Authority then sends the location of the Agent Visit Channel to each registered Node. The Data Gen To Node channel is used to send either Agent Visit Channel locations to a Node or to send data objects to a Node. The type of the data object and the Node it is sent to are determined randomly. The Agent Return Channel location is used to initialize any agent the node might create so that the agent knows the address to which it should write itself when it returns with a process. It is presumed that any returned process can be integrated into a Node simply by creating an internal channel which is incorporated into a data distribution function within the Node. Data objects are read from the Data Generator and their type is determined so that the data can be sent to a data object process using the internal channel. Output from the data object process is always written to the named net channel Nodes to Gatherer by means of an any-to-one net channel.

3.1 Basic Version

The network shown in Figure 2 was implemented with a proviso that all the required data processes were available at one or other of the initial Nodes. The system was tested with three different types of data object. The basic version read a data object and if able, processed it, because the required data object process was already present in the Node. If the data object process was not available then an agent was initialised and launched into the network and the data object was not processed. While the Node was waiting for the Agent to return with the required data object process, other data objects of the same type were also not processed. This enabled easier interpretation of the output from the Gatherer process because gaps in the list of processed data objects were immediately visible as each data object has a unique identifier regardless of its type.

3.2 First Revision

In this version the Authority node and Data Generator were divided into two separate processes. Each had their own named net channel. Appropriate modifications were made to the Node process but the Agent needed no modification. The aim in doing this was to separate processing functionality within the Node process so that updates to the List of Nodes an Agent could visit were received on a different input channel from that upon which data objects were received.

3.3 Second Revision

The restriction on all data object processing processes being present in the network was removed. This meant that an agent could return to its originating node with its `processDefinition` property {12, 66} `null`. In this case the originating Node recorded this fact and did not send a further agent for this type of data until a predetermined time period had elapsed. Yet again no modification of the Agent coding was required. This revision had the effect of creating an Agent that could not achieve its goal and which did not cause the system to fail. This mimics a typical situation in a real network where an Agent may fail.

3.4 Final Revision

In the final revision to the basic system, a system of two authorities was created. One authority kept a record of which node had which data object processes. This therefore represented an Authority that was more trusted in that it held private information about a node. A node could choose the authority with which it registered. The agent did require modification because in this case the Agent was sent to the Trusted Authority first to see if the required process was known to it. If it was then the agent was sent directly to the required node where it could obtain the required process and return to its originating node. If the Trusted Authority had no knowledge of required process then the agent behaved like the original Agent in that it obtained a list of node visiting addresses from the Other Authority and then went on a trip around the nodes until, it found the process if it was available. The Nodes could choose whether they placed information in the Trusted or Other Authority. The aim of this revision was to explore whether a system could be built that used more than one authority as often happens in networked environments. Various versions could have been built that captured behaviours depending on whether the process was obtained from a Node that used the Trusted Authority or not and whether the originating Node used the Trusted Authority. Some processes might never appear in the Trusted Authority in some applications. We chose not to explore all the possible combinations but simply wanted to demonstrate that the use of multiple authorities was possible.

3.5 Results and Evaluation

The above systems were all found to work as anticipated. The main result demonstrated was that it is feasible to create a system in which Nodes in a network do not have all the processes they require in order to function correctly. Nodes have the ability to obtain processes for data that they have never processed before, provided they can find a source of the required process. Systems could be built with more than one authority so that Agents could travel to several authorities in order to determine the Nodes they should visit in order to obtain a required process.

The Agent Visit Channel and the Agent Return Channel are essentially IP addresses if the underlying network is based upon TCP/IP technology. Thus the system could be implemented on top of any TCP/IP based network. World Wide Web requests for access to a server are received on a specific port of the IP address upon which the server resides. This allows the external access to pass through any firewall that might be present. If this Agent system were to be implemented in the same environment then the Agent Visit Channel and the Agent Return Channel would have to be placed at known port locations on a Node that was connected directly to the internet. This would then require a Node process that could interact with these channels in the manner described to permit access by the Agents. Obviously such access would need to be carefully controlled and monitored to ensure that unwanted access to a node does not occur. Inherently there is some security because the Node is only expecting communications conforming to a specific protocol in terms of the data it can read and write to an Agent.

4. Case Study: A Mobile Social Networking Application

The model was then expanded to deal with its application to a social networking service in which people can specify their friends to organize ad-hoc meetings. The person's list of friends is recorded in their mobile device. The aim of the service is to determine whether any of their friends are currently registered with any of the network(s) where the mobile device's owner is currently located. If this is the case then the diaries of the person and their friends are compared to see if they both have free time at the same time and if so to inform both of them that this is an opportunity for them to meet face-to-face. In this case a network refers typically to a wi-fi network which people can join and leave dynamically. In this experiment we were not concerned with managing the underlying network connection required in Mobile-IP but simply the ability of a person to join a network and send an agent into the system to see if any of their friends was already registered with the network and if so determine whether there was a possibility of a face-to-face meeting in the immediate future because they both had free times in their diaries for forthcoming period.

When a person enters a new wireless environment (Node G in Figure 3) their mobile device creates a new agent that has list of their friends together with a list of their free times for that day, or whatever period they have chosen. The agent is also initialized with the type of diary system the person's mobile device uses so that other friends will be able to determine the diary system used by this person's mobile device. The Agent is transferred to the network's Authority node, where it first registers this person as being present in the network. It then determines whether or not any of the person's friends are already registered. If this is the case the Authority creates an Agent initialized with the diary information of the person in a form suitable for the type of the friend's diary system. The Agent then transfers to the friends mobile device, for example Nodes A and E, invokes the diary access process of the friends' mobile device and determines whether or not there are times when the two people can meet. The Agent then returns to the originating person's node with any possible meetings. The Agent is thus able to visit all the friends' nodes to determine whether or not a meeting is possible and to suggest possible meeting times without the need for multiple interactions between each of the nodes.

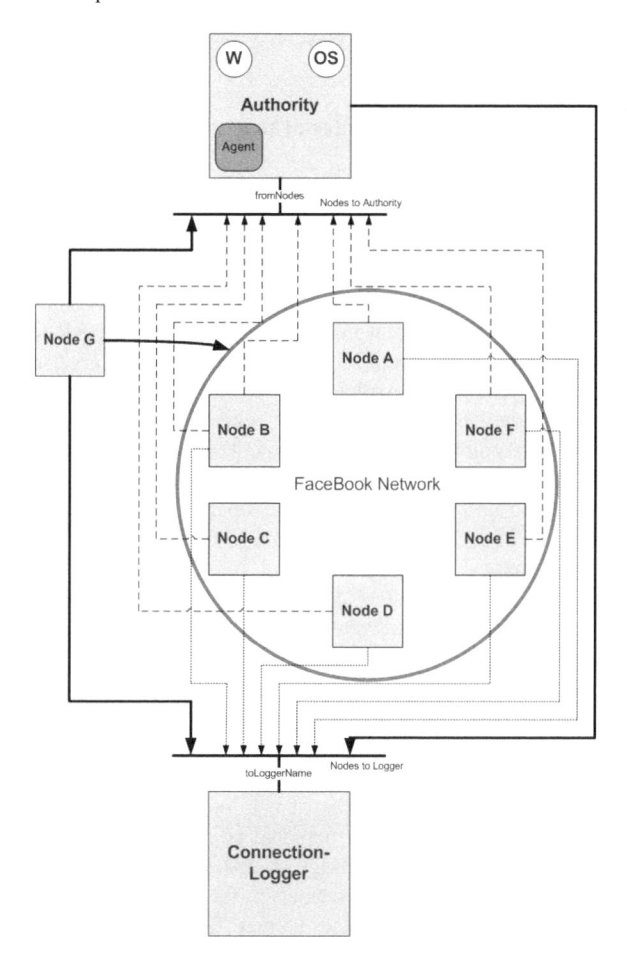

Figure 3: a mobile social networking application.

This approach means that the Authority does need to know the format required by different diary systems. Users of the service do not need to have copies of all the possible diary access mechanisms and in particular the various releases of software that might be in use at any one time within such a diverse mobile environment. Each user registers dynamically with the Authority, identifying the particular version of the diary mechanism they are using. If they happen to be the first person registering with the Authority that is using a new software release then the Authority could be enabled to send an Agent to the manufacturer of the diary system to obtain the required format information.

The system does not have to be symmetric in that a person can be a friend of someone else but the other person does not also need to have identified the other as a friend. The order in which people register does therefore have a bearing on the meetings that might be arranged. A version of the system was implemented that enabled an agent to visit a number of friends who were all registered with the authority such that a multi-person meeting could be arranged if several people were all free at the same time.

The goal of the case study was to investigate the feasibility of the approach rather than produce a fully working system. Thus aspects such as fault tolerance and optimizations that could be used, such as refining the Agent itinerary to improve system performance, were not considered.

5. Comparison with other Agent Frameworks

One of the most commonly used agent framework in the Java community is JADE [13]. This framework uses the technology of multi-agent applications based on peer-to-peer connection forwarding messages between hosts. This framework can be created on different hosts and each framework requires just one JVM (Java Virtual Machine) for execution. JADE machines can adapt themselves in respect to different protocols which will be used for the data transfer between different hosts. An agent which expects to communicate with another agent need not know about the exact name of the other agent or the agent that receives the message must not be available or executed at the same time the sending agent is available, for example the sending agent can send a message to a destination (all agents interested in baseball) and each executing agent which is interested can receive this message. JADE has also a security mechanism whereby JADE can verify and authenticate the sender of a message and can decline some operations as related to the rights of the sender for example an agent can receive a message from a sending host but cannot send a message to the same host.

Comparing JADE with the system described in this paper, there are some similarities. Both systems can transfer an agent from one host to another host and execute them at the stage the agent was stopped. But a difference of both is that the system described in this project has the opportunity that an agent can write itself from one host to another, which JADE is unable to do. This is a big advantage because an agent can take a process from a host and transfer it to another host to execute it locally. However it is not possible to communicate with another agent as in JADE which allows communication between two agents. Another advantage of JADE is that it is already possible to run this framework on different devices like mobile devices or personal computers and also on wired or wireless networks. Thereby JADE provides the usage of different environments like J2EE, J2SE or J2ME. In contrast to that the system described in this paper has not yet been tested in the manner of its execution on different devices using different environments.

6. Conclusions and Future Work

This paper has shown that it is possible to create a parallel environment that exhibits aspects of agent based technology, thereby enabling a node to adapt to the processing requirements imposed upon in it as a response to external requirements. For example, say a rendering node was sent a new type of data file of which it had no previous knowledge; it could send an agent to find the required process. Thus the capability of the rendering node has been increased by enabling it to adapt to its changing environment. This is one aspect of pervasive adaptation [10], a new action recently proposed by the Framework 7 Programme.

The main disadvantage of using Java is that objects that are communicated over a network have to be serialized using the underlying Java mechanisms. This means that it is more difficult to incorporate non-Java platforms. To this end a platform independent protocol needs to be developed that enables processes to communicate with each other regardless of the underlying software systems [14]. This would allow the communication of any data structure regardless of its underlying data types. In particular a process can be communicated as a byte array but of course cannot be platform independent because the byte array is interpreted by its virtual machine.

The current JCSP networked implementation does not always recognize when a node has failed or disconnected from the network, in a manner that is easily accessible to a system developer building process networks. The platform independent protocol referred to previously could be extended, quite simply, so that it could enable communication of node failures in a consistent manner. This would mean that as mobile devices move in and out of a particular network it would be possible to deal with some of the failure conditions at the applications level. For example the ability to determine that a node is no longer present in a network could be used by a returning agent to ensure that it did not try to reconnect with a node that was no longer accessible. However, if the underlying network was able to manage the movement of a mobile device from one network to another then this functionality could be incorporated into the application [15].

Currently, we are exploring how this technology could be used to implement a distributed version of a Learning Environment. In such an environment we would exploit the fact that much of the material that is held in a Learning Environment, such as webCT [16] is also available on the individual lecturer's office computer. Thus when a student logs into the Institute's network and places say a USB memory stick into a PC then an agent could be transferred from the stick which holds the modules for which this student is enrolled. This registration status could be checked with an authority, which also knows the IP location of the computers used by the lecturers that maintain material for the student's modules. The agent could then travel round the network finding out whether any new material had been available for the modules and this could then be transferred automatically to the agent and returned to the USB memory stick. Furthermore if the student had any special needs, which could also be recorded by the agent, then any files that require modification before they can be used by the student could be sent for transformation at a special node in the Institute's network.

Acknowledgments

Jens Haschke acknowledges the support of the Student Awards Agency Scotland which supported him, in part, during the course of his Bachelor's programme of study that contained a project element upon which parts of this paper are based.

References

[1] R. Milner, "Ubiquitous Computing: Shall we Understand It?," *The Computer Journal,* 49(4), pp. 383-389, 2006.
[2] M. Weiser, "The Computer for the 21st Century," *Scientific American,* September, 1991.
[3] R. Milner, J. Parrow, and D. Walker, "A Calculus of Mobile Processes, I," *Information and Computation,* 100(1), pp. 1-40, 1992.

[4] M. Schweigler and A. T. Sampson, "pony - The occam-π Network Environment," in P. H. Welch, J. Kerridge, and F. R. M. Barnes (Eds.), *Communicating Process Architectures 2006*, pp. 77-108, IOS Press, Amsterdam, 2006.

[5] C. G. Ritson and P. H. Welch, "A Process-Oriented Architecture for Complex System Modelling," in A. McEwan, S. Schneider, W. Ifill, and P. H. Welch (Eds.), *Communicating Process Architectures 2007*, pp. 249-266, IOS Press, Amsterdam, 2007.

[6] E. Gonzalez, C. Bustacara, and J. Avila, "Agents for Concurrent Programming," in J. F. Broenink and G. H. Hilderink (Eds.), *Communicating Process Architectures 2003*, pp. 157-166, IOS Press, Amsterdam, 2003.

[7] P. H. Welch, "Process Oriented Design for Java: Concurrency for All," in H. R. Arabnia (Ed.), *International Conference on Parallel and Distributed Processing Techniques and Applications (PDPTA '2000) Volume 1*, pp. 51-57, CSREA Press, 2000.

[8] K. Chalmers, J. Kerridge, and I. Romdhani, "Mobility in JCSP: New Mobile Channel and Mobile Process Models," in A. McEwan, S. Schneider, W. Ifill, and P. H. Welch (Eds.), *Communicating Process Architectures 2007*, pp. 163-182, IOS Press, Amsterdam, 2007.

[9] J. Kerridge and K. Chalmers, "Ubiquitous Access to Site Specific Services," in P. H. Welch, J. Kerridge, and F. R. M. Barnes (Eds.), *Communicating Process Architectures 2006*, pp. 41-58, IOS Press, Amsterdam, 2006.

[10] European Union Framework Programme 7, "Pervasive Adaptation: Background Document," 2008. Available from: ftp://ftp.cordis.europa.eu/pub/ist/docs/fet/ie-jan07-peradapt-01.pdf

[11] J Kerridge, K Barclay and J Savage, "Groovy Parallel! A Return to the Spirit of occam?", in JJ Broenink et al (Eds.), *Communicating Process Architectures 2005*, pp. 13-28, IOS Press, Amsterdam, 2005.

[12] DB Lange and M Oshima, "*Programming and Deploying Java Mobile Agents with Aglets*", Addison-Wesley, ISBN 0-201-32582-9, 1998.

[13] Bellifemine, F., Caire, G., Poggi, A., and Rimassa, G. "*JADE - Java Agent DEvelopment Framework*", retrieved March 2008, from http://jade.tilab.com/papers/2003/WhitePaperJADEEXP.pdf, 2003

[14] K Chalmers, J Kerridge and I Romdhani, "Critique of JCSP Networking", in P.H. Welch et al. (eds), *Communicating Process Architectures 2008,* ibid, IOS Press, Amsterdam, 2008.

[15] K Chalmers, J Kerridge and I Romdhani, "Mobility in JCSP: New Mobile Channel and Mobile Process Models", in AA McEwan et al (eds), *Communicating Process Architectures 2007*, pp. 163-182, IOS Press, Amsterdam, 2007.

[16] Wikipedia, " WebCT", http://en.wikipedia.org/wiki/WebCT, retrieved 18-6-2008.

YASS: a Scaleable Sensornet Simulator for Large Scale Experimentation

Jonathan TATE and Iain BATE

Department of Computer Science, University of York

{jt,iain.bate} @cs.york.ac.uk

Abstract. Sensornets have been proposed consisting of thousands or tens of thousands of nodes. Economic and logistical considerations imply predeployment evaluation must take place through simulation rather than field trials. However, most current simulators are inadequate for networks with more than a few hundred nodes. In this paper we demonstrate some properties of sensornet application and protocols that only emerge when considered at scale, and cannot be effectively addressed by representative small-scale simulation. We propose a novel multi-phase approach to radio propagation modelling which substantially reduces computational overhead without significant loss in accuracy.

Keywords. sensornets, networks, simulation, scalability.

Introduction

Wireless Sensor Networks, or *sensornets*, are an emerging discipline of embedded system and network design. Large numbers of minimally resourced nodes are equipped with sensors to monitor their physical environment. Nodes cooperate to manage ad-hoc wireless networks, within which distributed applications distill voluminous raw data about sensed physical phenomena into meaningful information with utility to end users. Real-time interaction with the real world is not merely a factor to consider in sensornet design; it is the fundamental purpose of the sensornet.

Wireless sensor networks are currently at an interesting point in their evolution. Some real-world deployments have been implemented but at relatively small scale. These small trials have validated the concept as workable and useful. However, despite considerable interest, wireless sensor networks have not yet made the transition from the laboratory to commonplace real-world usage. What is holding back these real-world deployments?

One contributing factor is the lack of confidence that a given network design will function adequately in a given environmental scenario. Few experimental studies employ large numbers of sensornet nodes, and consequently there is relatively little experimental measurement of sensornet protocol scalability [12]. Until such time as there exists a large number of previous sensornet installations from which to draw experience, simulation offers a low-risk and low-cost environment in which to assess the viability of proposed solutions, and to improve the quality and relevance of any putative solutions. Answers offered by simulation can be only as good as the model from which simulation results derive. Nevertheless, simulation is a valuable first stage in weeding out unworkable solutions, and in identifying which options are worthy of further study.

Good practice generally suggests the reuse of existing tools wherever possible to avoid wasteful duplication of effort. However, where existing tools are inappropriate, unusable, or simply unavailable, it is often necessary to develop new tools to address pressing needs. Most existing simulators focus on low-level simulation of small networks [34]. However, this is infeasible where many simulation instances are required to rigorously explore parameter landscapes, for example in protocol evolution or multi-objective optimisation experiments.

Poor performance also precludes the simulation of large sensornets. This latter point is of key importance where interesting behaviours are evident only in large sensornets, and in assessing the capability of candidate protocols at scale. Most existing simulators can handle at most a few hundred nodes before runtime becomes intolerable, as poor time complexity implies lengthy wall time for network problem instances of moderate but realistic scale [16]. Real sensornets may contain thousands of nodes, and proposals exist for sensornets containing millions of nodes. It is clear that simulation scalability cannot be ignored in this context.

The decision was therefore taken to develop *yass*, "Yet Another Sensornet Simulator". *yass* is a high-level sensornet simulator which prioritises speed over accuracy to render feasible the exploration of large sets of scenarios in acceptable time. Our current work uses *yass* to model sensornets for a wide range of purposes including the tuning of existing protocols, the design of new protocols, and multi-objective optimisation of sensornet applications.

The structure of the remainder of this document is as follows. Section 1 discusses related work on sensornet simulation. Section 2 enumerates the research objectives addressed by this paper. Section 3 describes the *yass* simulator and the novel optimisations it implements. Section 4 considers the performance profile of these optimisations, and section 5 assesses the impact on accuracy. Section 6 describes experiments to validate *yass*. Finally, section 8 presents conclusions against the research objectives.

1. Related Work

Sensornet design and evaluation frequently requires *simulation* and *emulation*; large-scale *testbeds* or *field trials* are infeasible and costly [4]. Formal analysis of sensornets may [18] or may not [30] be feasible; regardless of feasibility, it is rarely attempted.

Investigators must determine acceptable accuracy-scalability tradeoffs [31]; simulation-derived results are meaningless if simulated behaviour does not sufficiently match real behaviour [29] and are particularly sensitive to timing discrepancies [23]. Wireless communication models are usually the component with highest computational cost [27] but represent the greatest source of inaccuracy [20].

Discrete event simulators are well suited to computer network simulation [3]. *Simulation models* [4] are constructed, similar to those used in model checking [33], and executed in *simulation engines* [13]. Incorporating real application code, execution environments, hardware, network connections, or other real entities, into *simulation models* yields *emulation models* [13], improving accuracy but harming scalability [35]. Real and simulated entities interact directly in *hybrid simulations* [24].

Numerous sensornet-relevant simulators and emulators exist. Unfortunately, no current examples offer total accuracy or reach desired scalability. *TOSSIM* offers cycle-accurate low-level emulation of Berkeley motes running TinyOS but very simplistic network modelling [25]. More detailed modelling might be required where observable phenomena are very sensitive to minor variation in network conditions [20].

ns-2, the predominant network simulator in sensornet research [25,5], uses highly-detailed network models [5] but is single-threaded [16] and scales only to around 100 simulated nodes [27]. *ns-2* was not originally designed for wireless network simulation, support for which must be added through extensions [5]. Much of the popularity of *ns-2* can be attributed to the breadth of reusable libraries and protocol models developed by numerous researchers. However, the complexity stemming from this lack of focus and the underlying architecture can make working with or extending *ns-2* time-consuming and difficult [16].

J-Sim offers similar facilities to *ns-2*, also providing a component model which can be scripted and customised through the *Tcl* language. *J-Sim* is less widely used than *ns-2* but better suited to sensornet simulation as it was designed for this purpose, and has more detailed support for modelling physical environments and network-environment interaction. It also offers limited multithreading support within a single processing host [2].

The high computation cost of network simulation might be addressed by task parallelisation. Interentity communication across parallel simulation hosts [32] may negate some benefit of additional processors by Amdahl's Law [1,34]. However, this is not to say that parallel processing has no role to play, merely that care must be taken to ensure that simulator designs work harmoniously with the characteristics of a given target parallel processing environment.

Simulating sensornet-scale networks of millions of nodes requires entity concatenation [3] and layer concatenation [5] to reduce memory footprint [6], further sacrificing simulation accuracy. *GloMoSim* exploits parallel execution by multithreaded simulation, scaling to 10000 simulated nodes across 10 processors [38]. However, to achieve this scale *GloMoSim* consolidates many independent entities of the simulated system into single compound entities, necessarily sacrificing low-level accuracy for performance.

2. Research Objectives

Given a set of typical and broadly comparable sensornet configurations, and a typical sensornet application implementing a tuneable networking protocol, we define the following objectives that form the principal contributions of this paper:

Objective 1: *Identify techniques through which to improve upon the performance of existing sensornet simulators*

Objective 2: *Measure the extent to which these optimising techniques improve performance and enable larger-scale simulation*

Objective 3: *Improve the range of simulated network measures made available to the investigator*

Objective 4: *Validate optimised simulation to ensure that performance gains do not sacrifice solution quality*

3. Yet Another Sensornet Simulator

yass (Yet Another Sensornet Simulator) is distributed under the GNU General Public License. Source code and API documentation can be downloaded from the project web pages: http://www.cs.york.ac.uk/rts/yass/

3.1. Motivation and Concept

Simulation is an important and accepted tool for network researchers, offering reduced cost, time and risk in comparison to real-world experiments. They represent a sensible first step in development, and are of particular importance where real-world experiments are infeasible.

It could be argued that researchers would be well advised to reuse one of the numerous extant network simulators in their investigative work. However, common practice in the network research community does not follow this strategy. One study of 287 peer-to-peer network research papers [26] finds that approximately 49% obtain experimental results through simulation. Of these 50% do not mention which simulator was used, and 30% use a custom simulator for which there has usually been no attempt to perform empirical validation. Although some cases may be explained by "*Not Invented Here syndrome*", these figures suggest that experimenters are often forced to develop their own tools where no existing tool is adequate for the task in hand.

As the name suggests, *yass* is simulation software for evaluation of sensornets and sensornet applications *in silico*. The main principle underpinning the design of *yass* is that significant performance gains are possible without sacrificing accuracy, simply by avoiding unnecessary work. If the results of a calculation will never be used, then that calculation should not be performed. *yass* addresses two aspects of simulation in which this approach yields useful performance gains; the production of statistical measures, and radio propagation modelling.

3.2. Statistical Measures and Event Traces

Statistics offer a powerful tool with which to understand the behaviour of networks and networking protocols at an appropriate level, without becoming swamped with the great quantities of unnecessary detail pertaining to the mechanisms underpinning the higher-level behaviour. Unfortunately, many existing simulators are weak at revealing statistical information required by investigators [26] such as delivery latency or throughput.

This is particularly true where investigators do not know in advance which statistical measures will provide the answers they require, and hence these required measures are not taken; the experiments must be repeated. Conversely, valuable resources are often squandered in the production of a myriad of metrics in which the investigator has no interest and are not required by the simulator itself.

yass takes a different approach. As the simulation progresses noteworthy network events are recorded in an *event trace* [17]. This event trace records the behaviour of the simulated network elements rather than that of the simulator, and is backed by a relational DBMS to enable large datasets to be handled efficiently and to allow the usage of common data analysis tools. Each event is timestamped by simulation time rather than wall time to allow events to be logged out of their natural ordering, for example when two or more event-generating processes are scheduled unpredictably.

Standard trace operations can be applied during trace analysis. The trace sequence consists of the *interleaving* of all subsequences deriving from individual nodes, such that *restriction* might be applied to filter those events relating to network elements of interest to an investigator, or specific traffic flows. If the desired behaviour of a given network protocol is well-defined it is possible to determine whether a given trace records correct or incorrect network behaviour by determining whether that trace is in the set of all possible correct traces. This offers a powerful non-statistical tool for network analysis by simulation.

Statistical measures are not taken online during simulation but are instead obtained *post hoc* from the event trace offline, allowing data mining across multiple scenarios and extraction of metrics not originally considered by investigators. We demonstrate this principle in section 6 by running sets of simulations, then upon completion analysing multiple recorded simulation traces to derive metrics. These observations address Objective 3's requirements.

The tradeoff for these performance gains is that metrics are not available during simulation. Should investigators require online metric availability for online scenario adaptation it is possible to calculate these from a partial trace at the point of usage by the same method as applied to the complete trace at the end of the simulation.

3.3. Radio Propagation Model

Scalability is a weakness of many existing simulators [4]. Proposed sensornets may involve thousands or tens of thousands of nodes, but most existing simulators struggle with more than a few hundred simulated nodes [16]. Scalability problems generally stem from $O(n^2)$

growth in possible node pair interactions, depending ultimately on interacting broadcasts in the shared wireless medium.

yass implements a three-phase radio propagation model to calculate damage sustained to messages being received at sensornet nodes inflicted by other concurrent transmissions that cannot be prevented by the CSMA mechanism. A corollary is that nodes which are not receiving messages need not be tested at all. This considerably reduces the computation overhead for lightly-loaded networks.

The phases are ordered by increasing cost such that expensive tests are only applied when strictly necessary. As soon as the simulator has determined that a given packet reception has already been damaged beyond the capability for error detection and correction processes to recover, there is no benefit in applying further checks. This effectively implements a *lazy evaluation* approach explicitly in the simulation model, rather than implicitly through a language which supports lazy evaluation.

Phase one considers random environmental noise not influenced by network activity. Phases two and three apply a clipping strategy to determine nodes posing an interaction risk due to proximity. Phase two considers nearby nodes which are very likely to cause reception corruption, obtaining a fast first approximation. Phase three obtains a better approximation using a more expensive calculation. This multi-phase approach, outlined in Algorithm 1, addresses the requirements of Objective 1.

3.3.1. Three-Phase Radio Algorithm

Consider a sensornet composed of similar nodes distributed in a plane. Assume some node, N, is currently receiving a message being transmitted in the wireless medium by some other node, T. Background *1/f noise* is present at all times but can be rejected at N provided it is sufficiently weak. Inevitably, however, bursts of noise above the rejection threshold will be observed at a predictable rate but at unpredictable times [22]. Lines 1 to 1 of Algorithm 1 model this effect, phase 1 of the algorithm.

Within a circle of radius r, the typical communication range of the node hardware, exist other nodes with which N can reliably detect, receive and send messages. Nodes enclosed by r can reliably communicate with N through the wireless medium, or can refrain from transmitting if the local wireless medium is determined busy by CSMA.

However, it is feasible that N could lie between two other nodes O and P, such that $||\overrightarrow{NO}|| \leq r$ and $||\overrightarrow{NP}|| \leq r$ but $||\overrightarrow{OP}|| > r$. If N is receiving from O at some time, P cannot detect this and may start to broadcast simultaneously. This broadcast by P is very likely to corrupt the unrelated reception at N, as $||\overrightarrow{NP}||$ is within broadcast range. This *hidden terminal* problem [11] is addressed by lines 1 to 1 of Algorithm 1 which describe this second phase.

This offers a good first approximation and has been successfully employed in wireless ad-hoc network research [9] but may not in isolation capture all relevant behaviour. It is known that nodes can occasionally exert influence at a surprisingly long distance [12], as signals are merely attenuated with distance in the wireless medium rather than abruptly disappearing. On the other hand, if two nodes are sufficiently distant the probability of their interaction is vanishingly small, and the impact on network behaviour is negligible.

We address this by considering nodes falling within an annulus defined by radii r and s, where $r < s$ and s is beyond the communication range of nodes. Consider a node Q falling within this annulus. Reliable pairwise communication between N and Q is impossible as $||\overrightarrow{NQ}|| > r$. However, as $||\overrightarrow{NQ}|| \leq s$, N and Q are sufficiently close that some interaction between N and Q is possible due to random fluctuations in the wireless medium, transmission gain of Q, and reception gain of N.

In other words, should Q broadcast at full power the effective power received at N is

Algorithm 1 : Three-phase radio algorithm

1: **for** each node, n **do**
2:   determine if n is actively receiving data
3:   **if** n is currently receiving **then**
4:     determine if environmental noise corrupts reception at n
5:     **if** reception corrupted at n by noise **then**
6:       reception at n fails
7:       **jump** back to line 1 for next n
8:     **end if**
9:     find set of nodes R with $distance < r$
10:     **for** each node, m, in R **do**
11:       **if** m is transmitting **then**
12:         reception at n fails
13:         **jump** back to line 1 for next n
14:       **end if**
15:     **end for**
16:     find set of nodes S with $r < distance < s$
17:     **for** each node, m, in S **do**
18:       **if** m is transmitting **then**
19:         apply expensive radio model to find effective received power, p, from m at n
20:         **if** $p > sensitivity(n)$ **then**
21:           determine if error detection + correction at n can nullify influence of p
22:           **if** error correction fails at n **then**
23:             reception at n fails
24:             **jump** back to line 1 for next n
25:           **end if**
26:         **end if**
27:       **end if**
28:     **end for**
29:   **end if**
30: **end for**

below the sensitivity threshold but at times might interfere with an unrelated signal being received at N. For nodes like Q we must consider the distribution function for effective received power at N; sometimes the received power will be above the threshold and at other times below. It is for nodes like Q that the higher cost of sophisticated but expensive radio interference models can be justified. Lines 1 to 1 of Algorithm 1 describe this third phase.

Finally, consider a node X located such that $||\overrightarrow{NX}|| > s$. X is sufficiently distant from N that, should X transmit at full power, the effective received power at N is below the sensitivity threshold even when random fluctuations are taken into account. Transmissions from X cannot be distinguished from background noise at N, and hence need not be considered at all in Algorithm 1. In large networks there may be many such distant nodes, and hence a significant saving can be obtained by this optimisation.

In non-planar networks radii r and s define spheres rather than circles and annuli but the algorithm remains unchanged. For a given point isotropic source the enclosing surface defined by a given radius is a hollow sphere of zero thickness provided that transmission occurs in an empty void. An infinite number of such surfaces can be defined for the continuous range of possible attenuation, with zero radius representing zero attenuation and infinite radius representing full attenuation. If transmission does not occur within an empty void it is appropriate to instead interpret radii r and s as nonspherical surfaces of equivalent attenuation, with equivalent results. Surfaces of equivalent attenuation for complex radio environments may be defined by surveying deployment regions or by specialised propagation models, but only spherical surfaces are considered within the scope of this paper.

3.3.2. Cost Analysis

From the definition of Algorithm 1 it is evident that the number of computational steps is dependent on the size of various sets of nodes defined for each message-receiving node. Assume all network nodes are found in the set N, and that all nodes are alike. For each node $x \in N$, the following sets are defined:

- A_x: all other nodes within radius r with which reliable communication is possible
- B_x: all other nodes within radius s with which interaction is possible, where $A_x \subseteq B_x$
- $C_x = B_x \setminus A_x$: all other nodes with which interaction is possible but reliable communication is impossible
- $D_x = N \setminus B_x$: all other nodes with which no interaction is possible

A_x, B_x and C_x can either be defined implicitly by deciding set membership at the point of use, or precalculated and cached for improved efficiency under a *staged simulation* approach [36]. Note that the cost of maintaining these cached sets is non-trivial for networks containing mobile nodes, being $O(n^2)$ in total node count to rebuild. Membership of D_x need not be defined explicitly as members cannot interact with x and play no role in Algorithm 1.

The cardinalities of A_x, B_x and C_x are independent of total network size, and are dependent only on spatial density in the region around x. Clipping spheres of radius r and s centred on a given node, x, enclose the volumes v_r and v_s respectively. Multiplying these volumes by the local node density (measured in *node m^{-3}*) yields the number of nodes m_{xr} and m_{xs} falling within clipping spheres of radii r and s respectively.

From these node counts we see that $|A_x| = m_{xr}$, $|B_x| = m_{xs}$, and $|C_x| = |B_x| - |A_x|$. Each is $O(d)$ in local node density d and $O(1)$ in overall network size. Assuming uniform spatial node distribution and homogenous nodes throughout the network we can ignore the identity of node x, yielding $|A|$, $|B|$ and $|C|$. These set cardinalities, and the corresponding computational cost per individual node, are independent of total network size, a highly desirable property for scalable simulation of large networks.

3.3.3. Complexity Analysis

Consider the total cost for all nodes passively monitoring the wireless medium, waiting to begin receiving data. For a given node $x \in N$ we need check only members of A_x for starting transmissions as other nodes cannot establish communications with x. Assuming each pairwise check completes in constant time, uniform $|A|$ for all nodes, and i idly listening nodes, the total cost is $i|A|$. If i is $O(n)$ in network size and $|A|$ is $O(d)$ in spatial density d, total cost is $O(nd)$ and hence linear in total network size $|N|$. A similar cost is observed for nodes establishing that the local wireless medium is free prior to packet transmission as this entails similar local passive monitoring.

Consider the total cost for all nodes actively transmitting to the wireless medium. Trivially this is zero, and hence $O(1)$ in network size $|N|$, because after transmission of the packet body begins the transmitting node makes no further checks on the local wireless medium.

Now consider the total cost for all nodes actively receiving from the wireless medium. As described in Algorithm 1 in the worst case this requires for each node $x \in N$ to perform pairwise checks between x and all potentially interfering nodes $y \in B_x$. In the worst case the radii r and s are such that for each node $x \in N$ a significant proportion of N is represented in A_x and B_x, and every ongoing message reception must be tested against every ongoing message transmission.

Assume at some time t that some proportion α_t of nodes are actively transmitting and some proportion β_t of nodes are actively receiving, with $\alpha_t, \beta_t \in [0,1]$ and $\alpha_t + \beta_t \leq 1$. The set of transmitters, F, has $|F| = \alpha_t |N|$, and the set of receivers, G, has $|G| = \beta_t |N|$. Under the optimised model the number of pairwise interaction tests required is no greater than $|F||G| = \alpha_t \beta_t |N|^2$, as opposed to the $|N|^2$ tests required under the unoptimised model. Optimised cost remains $O(n^2)$ in network size $|N|$ if total network activity is dependent on network size as described above. However, as $\alpha_t, \beta_t \in [0,1]$ optimised cost is generally lower, and never higher, than unoptimised cost.

As $\alpha_t, \beta_t \to 0$, network activity decreases and cost savings under optimisation increase. Total cost under the optimised model only reaches that of the unoptimised model when every node is actively participating in network activity, and every node potentially interacts with every other node. This is of particular relevance to the simulation of sensornets, in which network traffic is minimised in order that energy consumption resulting from network activity be minimised. Energy-efficient network applications accordingly require less real time for simulation to complete.

3.4. Parallelism

yass is a time-driven rather than event-driven simulator, in which any simulated entity can interact with any other and simulated time progresses in small increments. There is no requirement that consistency must be maintained during the calculations which progress from one consistent state to another, but each simulated time increment transforms the simulation model from one consistent state to another consistent state. A dependency exists from any given state to the preceding state from which it was derived, implicitly requiring that the sequence of simulated periods must be addressed sequentially. However, there is no requirement that calculation within a simulated period be conducted serially.

The radio model described in section 3.3 enables the most costly element of the simulation to be addressed using parallelism. Within each simulated period the three phases of the radio model must be executed serially. Within each phase the work can readily be divided between any number of processing units, and the results combined prior to commencing the next phase. This is feasible because each calculation considers only a single pairing of nodes which can be performed independently of all other pairings, and can be performed without large volumes of shared state. However, this implies the existence of a single *master* which delegates work to numerous *slaves*, regardless of whether this delegation uses multiple CPU cores, multiple clustered hosts, or any other similar architecture. The delegation of node pairings need only be performed once if the set of simulated nodes and the set of processing units does not change, though fresh data would need to be distributed for each simulation quantum.

Although there is no fundamental impediment to concurrent execution of sections of the simulation model, the current implementation of *yass* uses a purely sequential approach as the bottleneck is I/O throughput rather than CPU speed. Recording the event trace from which statistical measures and network metrics are derived (see section 3.2) can generate data at a rate orders of magnitude greater than typical hard disks can store it. It follows that no real performance gain would be observed by splitting the processing work between multiple processing units unless atypically high-performance storage resources are available. This is a consequence of prioritising offline processing over online processing in the simulator design.

In the absence of performance gain in a typical execution environment the decision was taken to favour a simple serial implementation of the radio algorithm. However, if I/O volume could be reduced by capturing fewer simulated events, or by storing only a representative sample of all simulated events, CPU capacity could become the primary bottleneck. Under this condition it would be highly desirable to re-implement the radio model in *yass* to exploit fine-grained parallelism as described above. In the first stages of experimental work the key challenge is often to establish which of the measurable values yield useful insight. Once this is known it is possible to reduce the volume of data captured to the bare minimum necessary to derive the required measures and discard the remainder, increasing the feasibility of obtaining useful performance improvements through exploitation of parallelism.

The current implementation of *yass* does exploit coarse-grained parallelism. One of the main use cases of the simulator is to efficiently explore parameter landscapes through large numbers of simulations, each test case evaluating some fixed point of the landscape. Several such test case simulations can be executed in parallel in separate threads as simulations do not interact, allowing simple test suite coordination with no interthread coordination overhead other than managing access to disks constituting the database backing storage.

yass has built-in support for scripting large numbers of simulation scenarios and scheduling n such simulations to be running at all times until the set of test cases is exhausted. Usually n would be defined to be equal to the number of CPU cores available on the host; setting it lower allows these cores to be devoted to other work, whereas setting it higher yields no further performance gains. Again, I/O throughput is the limiting factor on overall simulation throughput, but more efficient use of I/O is possible as other threads can continue to make progress while any given thread performs I/O activity.

3.5. Implementation Details

yass is a time-driven rather than event-driven simulator. The times at which simulated events occur are quantised to the boundaries of discrete periods during which the simulation model transitions from one consistent state to another. Shorter quanta yield greater accuracy but slower progress. The length of all such quanta need not be equal, such that a smaller simulation quantum could be applied for network periods of particular interest to the researcher or during which network behaviour is particularly sensitive to inaccuracy.

The simulator is written in the Java language and as such can be executed without recompilation on any host with a Java Virtual Machine capable of supporting at least version 5.0 of the Java language. The event trace data is managed by a DBMS accessed through JDBC. The Apache Derby DBMS is used by default to minimise end-user configuration, but any DBMS which supports at least SQL-92 and is accessible through JDBC should be compatible.

All simulated entities are represented as objects which are responsible for managing their own state. As a consequence there is very little global state to be managed. Nodes are composed of hardware modules, each of which can adopt a number of states to balance performance and energy efficiency. Hardware module state can be changed by other modules or by the application, either immediately or through a built-in scheduler. Applications are composed of a set of packet producers, a chain of packet consumers, and a packet dispatcher.

4. Evaluating Performance

A set of timing experiments was run to assess the performance of *yass*. Each experiment was repeated nine times on the same machine to minimise influence of outliers and to smooth the effects of the stochastic simulated application. The GNU *time* command was used to measure CPU time allocated by the scheduler, with timing data given to $\pm 1 \times 10^{-2} s$ precision. Each problem instance reused the same simulated network, to block effects of uncontrolled factors, for each of 21 distinct rebroadcast probabilities distributed evenly in the interval $[0, 1]$ running for a fixed number of simulated seconds. Packet source and destination node was randomly selected for each packet.

The first two experiments considered performance for simulated networks of size ranging between 50 and 1000 nodes, varying in 50 node increments, with constant spatial density of 4.0×10^{-8} *node* m^{-3}. Simulated nodes were based on the MICA2 mote [8] with an IEEE

802.11 MAC layer and radio range of around 150m, although this detail is irrelevant to simulation runtime. The first experiment was conducted with the optimisations of Algorithm 1 enabled, and the second with these optimisations disabled. The results are shown in figure 1.

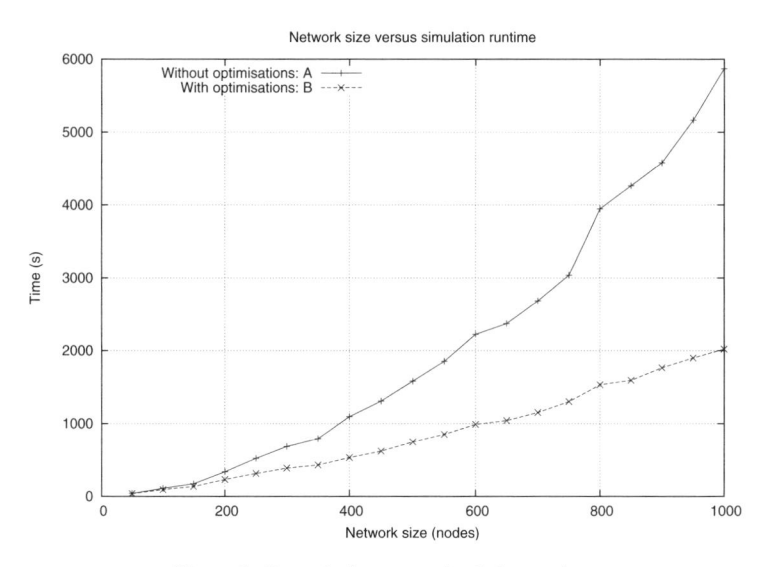

Figure 1. Network size versus simulation runtime

Trace A shows the runtime with simulation optimisations disabled, and trace B with simulation optimisations enabled. It can be seen that in all cases the runtimes illustrated in trace B are less than those of A, indicating that in all problem instances considered here the optimisations offer measurable time savings.

Trace A approximates a quadratic curve, which is as expected from the $O(n^2)$ behaviour predicted by section 3.3. The plotted curve is not perfectly smooth, an expected consequence of experimental noise induced by the simulated network application's stochastic behaviour.

Trace B also approximates a quadratic curve but with much shallower curvature, again as predicted by section 3.3. The curve of trace B is sufficiently shallow when considered against trace A that a linear approximation is reasonable within this range of network sizes.

Very large networks of the order of tens of thousands of nodes may require adoption of further abstractions in order for simulations to complete in acceptable time [38]. Most such abstractions would adversely affect simulation accuracy, so care must be taken to minimise this effect. Ideally these abstractions would follow the natural structure of the network.

For example, with clusters of multiple cooperating nodes, each cluster could be represented as a single entity when modelling intercluster traffic. However, experiments using simulations of the order of thousands of nodes are entirely feasible with commodity hardware without these additional abstractions and sources of inaccuracy.

The latter two experiments considered performance for networks of density ranging between 1.0×10^{-8} and 1.2×10^{-7} $node\ m^{-3}$, with constant network size of 500 nodes. The third experiment was conducted with simulation optimisations enabled, and the fourth with optimisations disabled. Nodes were based on the MICA2 mote [8] with an IEEE 802.11 MAC layer and communication range of around 150m, although again this detail is irrelevant to runtime. Results are shown in figure 2.

Figure 2 shows two traces for simulations of the same problem instances. Trace C shows runtime with simulation optimisations disabled, and trace D shows runtime with simulation optimisations enabled. It can be seen that in all cases the runtimes illustrated in trace D are

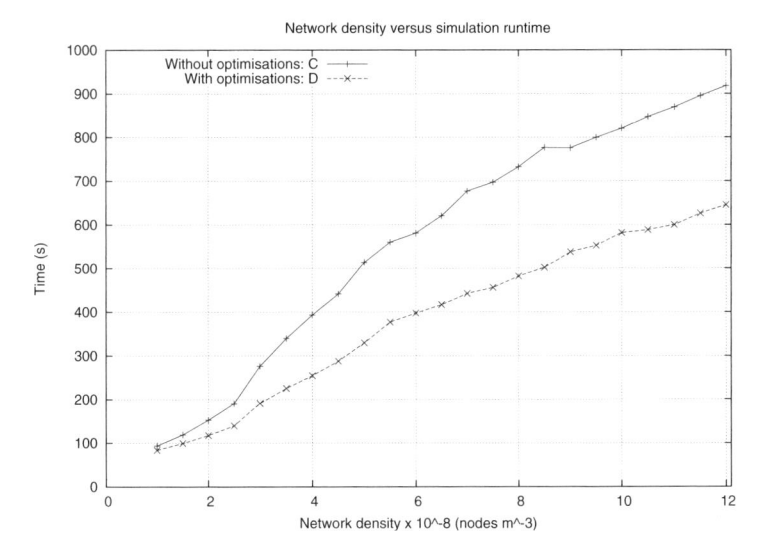

Figure 2. Network density versus simulation runtime

less than those of C, indicating that in problem instances considered here the optimisations offer measurable time savings.

Simulation runtime increases in network density, but traces C and D show runtime asymptotically approaching maximum values as density increases. Potential communication partner count per node grows with increasing density, with commensurate growth in pairwise communication as packets are flooded to more neighbours.

However, as each message transmission occupies the wireless medium for non-zero time there is also growth in network congestion. Simulated packets have length randomly selected in the interval [128, 1024] bits, including header. With the MICA2 radio having transmit speed of $3.84 \times 10^4 bits^{-1}$ [8] this gives per-packet transmit times in the interval $[3.33 \times 10^{-3}, 2.67 \times 10^{-2}]$ seconds. When packet transmission begins the local wireless medium is occupied for some duration in this interval.

The tendency for network traffic to increase as more pairwise message exchanges become possible is countered by the tendency for increased congestion to restrict pairwise message exchange, as illustrated in figure 3. As density increases the level of attempted network activity increases, but there is decreasing spare capacity in which to accommodate the additional potential traffic.

Small periods in which the wireless medium remains unused are an expected consequence of the simulated stochastic application, becoming sparser and smaller in the time domain as the network becomes busier. CSMA with exponential backoff defers packet transmission until the wireless medium is free, when it is implicitly claimed for the full transmission period. There is no local or global coordination of traffic beyond this mechanism.

As network utilisation increases there are decreasingly many unoccupied periods fitted between occupied periods because increasingly many nodes attempt to claim the wireless medium per unit volume. Unoccupied periods tend to become shorter as many simultaneously-running backoff procedures attempt to claim the medium.

Smaller packets occupy the wireless medium for a smaller period, accommodating more packets within a given duration. Packets are forbidden from becoming smaller than the minimal header size, an integral number of bits, and hence it is not possible for the network to fill all unoccupied timeslots with sufficiently small transmissions. Eventually the network becomes saturated with the wireless medium occupied in places at all usable times.

This pattern of behaviour in which each subsequent increase in network density yields a smaller increase in simulation runtime suggests that a reasonable approximation to the observed behaviour could be obtained by fitting a logarithmic or harmonic series to the data points over the range of densities considered in the experiment.

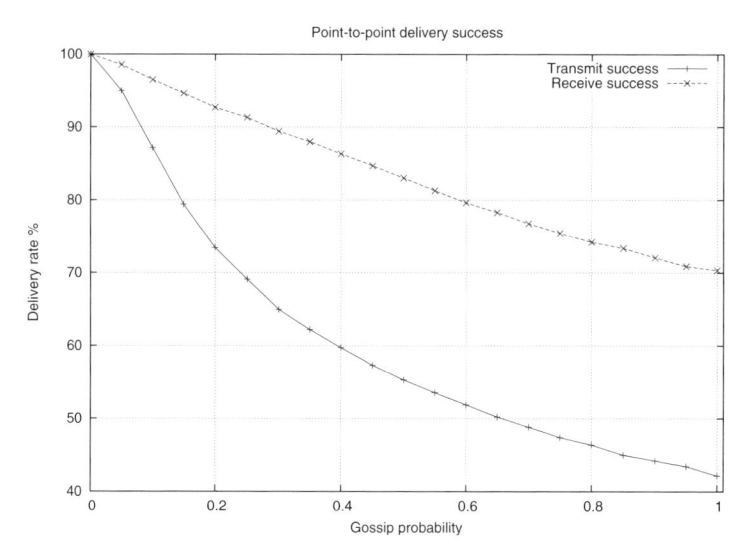

Figure 3. Point-to-point success vs rebroadcast probability

5. Evaluating Accuracy

A set of experiments were run to assess the influence of the proposed simulation optimisations on the accuracy of measured network metrics. Note that the scope of this section is restricted to comparison within the *yass* context, and does not address other simulation environments. General simulation validation is considered in section 6.

Each simulation reused the same 200-node network and simulated application, thus providing blocking of uncontrolled factors. Simulated nodes were based on the MICA2 mote [8] with an IEEE 802.11 MAC layer and radio range of around 150m. A single controlled factor, rebroadcast probability, was varied within the interval [0,1] with 10 intermediate steps. Each simulation was repeated 9 times and the arithmetic mean taken for each of three network performance metrics. The simulation set was repeated both with and without optimisations enabled, and the results plotted on the same axes for comparison.

Figure 4 illustrates the success rates for point-to-point packet transmission, point-to-point packet reception, and end-to-end packet delivery. For each metric there are a pair of traces; one obtained with, and one without, simulation optimisations enabled. It can be seen in each case that the paired traces are very similar across the range of gossip probability values considered in these experiments.

Experimental errors introduced by simulation optimisations are indicated by any divergence between the plots for a given metric. Trace pairs for the point-to-point metrics are so close as to be indistinguishable. The trace pair for the end-to-end metric shows slight divergence, resulting from compounding of multiple smaller point-to-point errors and borderline cases along delivery paths.

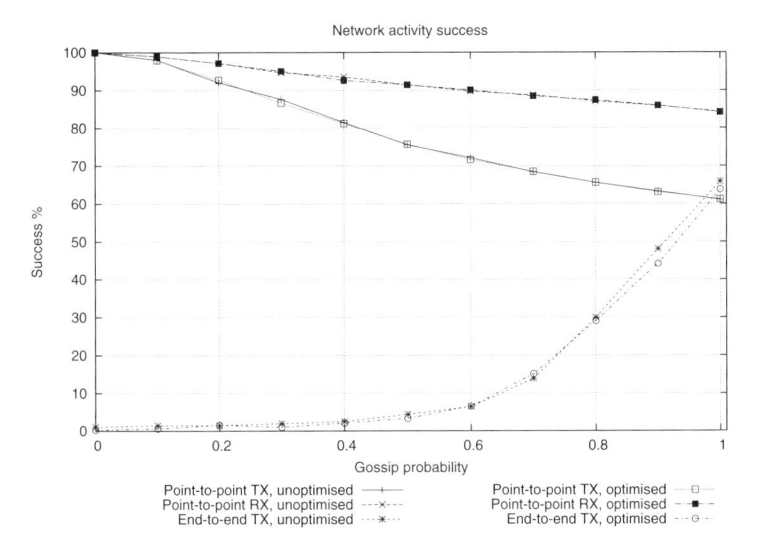

Figure 4. Network metrics with/without optimisation

With each point-to-point packet exchange the abstraction of the simulation model is a source of inaccuracy. A multi-hop delivery path has numerous such point-to-point exchanges, and hence compounds this inherent error for each hop. However, we see from Figure 4 that cumulative experimental error magnitude remains very small in comparison with the magnitude of the measured values.

More importantly, the overall trends and general shapes of response curves are retained when comparing the paired unoptimised and optimised traces. This is significant for experimenters seeking to identify emergent effects in large-scale network behaviour, or to establish the relative merit of two or more candidate protocol configurations. Under these experimental goals it is more important that relative trends and orderings across the parameters are demonstrated than to obtain accurate absolute values for any solution quality metric.

Where accuracy is prioritised, experimenters can employ a two-phase approach. *yass* can be used to quickly establish which input configuration gives the most desirable simulated behaviour. A more detailed but slower simulator can then be used to obtain accurate absolute response measures yielded by this optimal input configuration.

Combining the findings of sections 3.3, 4 and 5 it can be seen that the proposed optimisations allow simulations to be completed in significantly reduced time with only a minimal impact on solution accuracy. This addresses Objective 2 as defined in section 2.

6. Validating Simulation Results

We have demonstrated that the novel optimisations implemented by the *yass* simulator offer real and substantial performance improvements. However, these improvements are of little consequence if the accuracy of simulation results is unduly compromised. We examine the quality of solutions obtained under optimised simulation by validating *yass* against theoretical findings and experimental results published elsewhere, and examining observed trends against expected behaviour. We also consider the advantage conferred by simulating larger sensornets by comparing against results derived from smaller-scale simulations. The experiment designs and results described in this section address the requirements of Objective 4.

All experiments simulate networks employing the *GOSSIP1(p)* protocol defined by Haas et. el. [14] with gossip probability, *p*, taking 21 values with equidistant spacing in the interval $[0, 1]$. This simple protocol was selected as it has been thoroughly examined in the literature [14,37], which predicts a readily-identifiable and theoretically explicable sigmoid curve when gossip probability is plotted versus end-to-end delivery success rates. Presence (or absence) of this characteristic curve is deemed indicative of the accuracy (or inaccuracy) of the *yass* simulator. Additionally, flooding and gossiping protocols are essential components of many more sophisticated protocols. For example, AODV [28] and DSR [19] use simple flooding for route discovery, but may observe improved performance by substituting probabilistic gossiping [14].

Each value of p is evaluated within an otherwise-identical simulated problem instance for blocking of uncontrolled factors. Each simulation instance is repeated 10 times and the arithmetic mean of the analytical results plotted. Each plotted curve therefore represents the outcome of 210 independent simulation runs. Each simulation continues until 60 seconds of simulated time elapse, allowing performance metrics to settle on stable values.

Simulated nodes are modelled on the MICA2 mote [8] with IEEE 802.11 MAC layer and radio range of around 150m, and always transmit at maximum available power and maximum available rate. Two networks are considered of 100 nodes and 1000 nodes respectively, with the former being a subset of the latter. Spatial distribution of nodes within a bounding volume is random and even, with constant spatial density of $4.0 \times 10^{-8}\ node\ m^{-3}$.

The simulated application assumes a flat network hierarchy. Each node acts as a packet destination and periodic source, with any source-destination pair being equally likely in a unicast communication paradigm. Offset, period, and per-period production probability differ between nodes, but remain constant throughout and between simulations. Simulated packets have length randomly selected in the interval [128, 1024] bits, including header.

It is assumed that delivery of a given packet fails if it does not arrive at the intended destination node prior to the end of the simulation. Point-to-point packet exchange may fail due to corruption wherever a nearby node is transmitting with sufficient power, for example due to the *hidden terminal* problem [11]. Stochastic failure may also result from background noise with probability of 0.05 over the duration of each transmission.

6.1. Bimodal Delivery Behaviour

Previous results published in the literature predict bimodal behaviour for probabilistic rebroadcast algorithms such as *gossiping* [14,37]. As rebroadcast probability is increased it passes through a critical value. Below this critical value it is expected that packet distribution dies out quickly, and few nodes receive the packet. Above this critical value it is expected that most packets will reach most nodes.

Figure 5 demonstrates the end-to-end delivery ratio versus rebroadcast probability for a typical network and networked application employing the *gossiping* protocol. Graphing experimental results for the 1000-node network yields the sigmoid shape predicted in the literature [14,37], in which the steepest curve section contains the critical probability predicted by theoretical analysis. This demonstrates qualitatively that the relationship between rebroadcast probability and end-to-end delivery ratio is as expected. Quantitatively, rebroadcast probabilities in the interval [0.6, 0.8] are expected to deliver most packets to most nodes [14]. In figure 5 we see that for $p > 0.65$ delivery ratio remains within 90-92%, near-constant within experimental error, matching expected behaviour.

The 100-node network shows similar but less acutely defined behaviour, as discussed in the next section. Note also that end-to-end delivery success is higher for the 100-node network than the 1000-node network for small rebroadcast probabilities. This is because a larger proportion of the network falls within the range of a transmitting node, and hence a greater proportion of the network is directly reachable without any packet routing mechanism.

The proportion of the network covered by packet flooding grows exponentially in continuous time, represented by a number of discrete *rounds* of the flooding algorithm. In each round, every node which received a given packet in the previous round is a rebroadcast candidate. If each node has on average i nodes with which pairwise communication is possible, the initial packet broadcast at the source node will reach up to i rebroadcast candidates in the first round. In the second round, if each candidate is able to rebroadcast, then each can reach up to another i nodes, and so on for each subsequent round.

Assume packet transmission takes negligible time such that network congestion is negligible. Ignoring corrupted broadcasts, and assuming no other network activity, after r rounds of flooding the number of nodes rebroadcasting the packet in the nth round is i^r having received the packet in the $r-1$th round. If the rebroadcast probability is now changed from 1.0 to p we find that the average number of concurrent rebroadcasts is $(pi)^r$. This gives the level of network activity, and hence potential network congestion, due to this packet at this time. It is entirely possible that none of the neighbours of a given node will choose to rebroadcast as there is no local coordination.

At the completion of the rth round, all neighbours of each broadcasting node have been exposed to the packet, as have the outermost nodes reached in previous rounds. As the rth round completes, all neighbours of broadcasting nodes have been exposed to the packet in addition to those exposed in previous rounds. Flooding terminates when there are no further rebroadcast candidates. Disallowing multiple rebroadcasts of any given packet ensures flooding terminates in finite time, preventing packets endlessly circulating within the network.

Take the number of nodes covered by a packet as n_r at the end of the rth round, and $n_0 = 1$ as the source node has already got the packet before flooding begins. Assume that a node which has already been considered as a rebroadcast candidate for a given packet in a previous round is still able to receive that same packet again, even though it will not be considered as a rebroadcast candidate again. In particular, a node will be able to re-receive during the next flooding round after it has itself rebroadcast.

If each node has on average i neighbours, and during the rth round there are $(pi)^r$ concurrent rebroadcasts, up to $i(pi)^r$ nodes can receive a given packet during the rth round. Assume that q rounds complete before flooding terminates, q being $O(x)$ in network diameter x [21]. The total number of possible packet receptions completed at the end of the rth round is given by $\sum_{r=1}^{q} i(pi)^r$.

Each packet reception represents a delivery of a packet to *some* node. It is entirely feasible that a given node will receive a given packet multiple times, but will only be a rebroadcast candidate on the first reception. Provided that at least one of these successful packet receptions occurs at the intended destination, the packet has been delivered.

Any given node is likely to be a potential communication partner for a number of other nodes. The communication range sphere for any given node is likely to enclose a number of other nodes, such that the neighbour-node sets corresponding to these spheres are unlikely to be identical in the majority of cases. Put another way, it is likely that any given node pairing can be joined by multiple paths through the network graph defined by communication range, and may therefore receive a given packet multiple times along different routes.

In an otherwise quiet network it is therefore very likely that most nodes will receive at least one copy of a given packet even if not all nodes rebroadcast this packet on reception. However, as the expanding ring of network coverage radiates from the source node, it is not guaranteed that all nodes will have received this packet. For example, nodes may have received a corrupted signal, or may have been occupied with unrelated wireless communications, or may simply have not been near enough to a broadcast.

As p approaches 1.0 the delivery success improvement achieved by increasing it further becomes smaller. This is because each node can receive a packet from multiple neighbours but only the first occurrence is necessary for delivery or rebroadcast candidacy. As p grows larger the proportion of nodes which have received at least one copy of a given packet increases, but most additional network activity induced by this growth simply increases repeated reception. The average number of receipts of a given successfully delivered packet per node grows from exactly 1 to just over 3 as p grows from 0 to 1 in the experiments from which figures 5 and 6 derive.

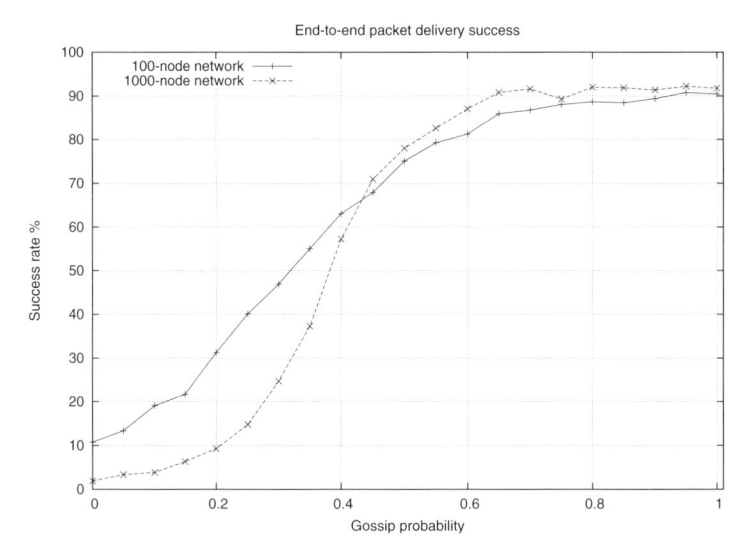

Figure 5. End-to-end packet delivery ratio versus rebroadcast probability for 100-node and 1000-node networks

6.2. Emergent Effects at Scale

Further evidence of the merit of larger-scale simulation is found in comparing network performance metrics obtained by simulation of otherwise-similar networks differing in node count. To observe the behavioural differences between small and large networks we generate network performance statistics for two networks. The networks differ in size by an order of magnitude; a 100-node network and a 1000-node network, in which the former is a subset of the latter. We measure *packet delivery success ratio* and *end-to-end delivery latency*, as measured by Das *et al.* in earlier comparative work [10], but omit *routing overhead* as it is always zero for gossip-based networks.

Consider figure 6 in which end-to-end delivery latency is plotted versus rebroadcast probability, normalised against maximum latency observed for the corresponding network size. In both cases we see average end-to-end latency increases until rebroadcast probability reaches around 0.4. For the 1000-node network we see that latency peaks around this rebroadcast probability, and then falls as rebroadcast probability increases further. However, in the 100-node network we do not see a similar fall in latency as rebroadcast probability increases further.

Now consider figure 5 in which end-to-end delivery success rate is plotted versus rebroadcast probability. Both plots follow similarly-shaped sigmoid curves in which delivery success transitions from *subcritical* to *supercritical* state as rebroadcast probability passes a

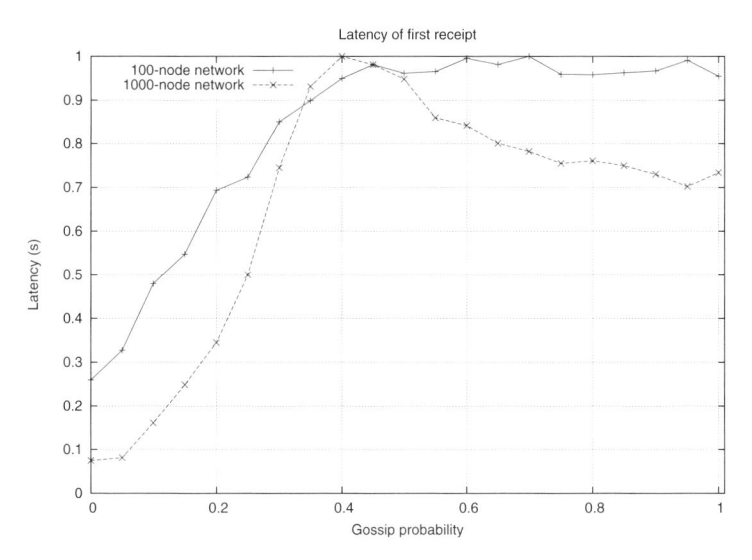

Figure 6. Normalised end-to-end latency versus rebroadcast probability for 100-node and 1000-node networks

critical value. However, the expected sigmoid shape is much more clearly defined in the 1000-node network plot than the 100-node network plot. Increasing network size further yields ever more sharply defined sigmoid curves. This is consistent with *percolation theory* which states that the transition between *subcritical* to *supercritical* states becomes more abrupt as network size increases [37], because the message blocking influence of individual nodes on the greater network becomes lesser [14].

We observe a qualitative difference in simulated behaviour between the small and large networks. This is of key importance in the design of real-time networks carrying time-sensitive data. Although the 100-node network is a subset of the 1000-node network, simulating the smaller network is not sufficient to capture all behaviour demonstated in the larger network. Whereas other behavioural effects might be evident in simulations of varying scale, the extent to which these effects are expressed may also vary. This latter observation is of critical importance where we hope to exploit simulation to discover new effects, rather than to confirm already-known effects.

Sensornet designers utilising simulation techniques must ensure that the scale of the simulated network is comparable to that of the proposed real network. This precludes use of simulators with poor scalability characteristics.

Consider a 1000-node network in which there exists a requirement for average end-to-end latency to be no greater than 0.8s. A designer with access only to data for 100-node networks may determine that this requirement can be met only where gossip probability $p < 0.25$ (figure 6). However, a designer with access to data for 1000-node networks would determine that this requirement can be met with $0.00 \leq p < 0.30$, or $0.65 < p \leq 1.00$. If critical probability p_c at which bimodal transition [14] is observed is around 0.4 (figure 5) it may be greatly preferable to select some value of $p > 0.65$ to achieve acceptable delivery.

7. Future work

The three-phase radio algorithm described in section 3.3.1 is inherently parallelisable because each phase consists of a set of independent pairwise node-node tests. The model requires the preservation of the phase execution ordering only, and does not require any specific ordering of calculations within phases or prohibit these being performed concurrently. However, the current implementation in *yass* performs these tests serially, with coarse parallelism achieved by running several independent simulations in parallel. Plans exist to reimplement the algorithm in a parallelised form, allowing larger individual simulations to be executed in a given time, in which each set of tests is divided between a set of threads which may or may not execute on the same machine. Plans also exist for a GPGPU implementation, exploiting the cheap power offered by commodity many-core GPUs. The algorithm is an excellent candidate for this type of acceleration as it consists of many independent but similar calculations and small shared datasets, mapping neatly onto GPU *kernels* and *textures* respectively [15].

The *yass* tool serves both as a simulation environment and as a testbed for experimental simulator components. It would be feasible to extract the three-phase radio algorithm described in section 3.3.1 and reimplement this for other network simulators such as *ns-2* [5] or *J-Sim* [2]. If successful this would permit similar performance improvement in widely-used simulation tools as in *yass*. However, the work may be non-negligible and require modification of underlying component architecture and interfaces if the data used within the algorithm are not currently exposed to the simulation components responsible for radio modelling.

Section 6 showed that simulation results obtained using *yass* are consistent with theoretical results and simulation results published by other researchers. It would be useful to make further comparisons, across a representative cross-section of typical sensornet scenarios, against other simulators and real testbed networks. Showing the results produced by *yass* to be equivalent, within some acceptable margin of error, would be sufficient validation to accept usage of *yass* in any situation where otherwise a real testbed network or some other less efficient simulator would otherwise have been required. Unfortunately, within the sensornet research community there does not currently exist any broadly accepted validation test suite, the development and acceptance of which is a necessary prerequisite for this work.

Section 3.2 discusses the post-hoc analysis of simulation trace data captured during execution of test cases. Recording this trace data is generally very expensive due to the number of distinct elements and overall volume, such that overall simulation performance is I/O-bound. Reducing I/O overhead would permit more simulated behaviour to be evaluated per unit time, enabling evaluation of longer simulated periods, evaluation of larger simulated systems, or simply to obtain results more quickly.

One approach is to record only a partial trace containing some fraction of simulated events and discard the remainder, with the assumption that the recorded subset is sufficiently representative of the whole. Sampling approaches such as *Cisco NetFlow* [7] record every nth packet, hence reducing the volume of recorded data to $\frac{1}{n}$ of the unexpurgated dataset. It would be possible to mandate that simulated time progressed as some multiple of wall time, dropping any trace records which cannot be recorded within deadline, dynamically balancing quality and performance. It remains an open question whether these partial sampling approaches are appropriate in low-traffic wireless sensor networks which are fundamentally unlike the conventional high-traffic wired LANs for which they were developed.

8. Conclusions

In section 2 a set of desired research objectives was defined, against which we now state our findings.

Objective 1: *Identify techniques through which to improve upon the performance of existing sensornet simulators*

A novel multi-phase model is presented by which computational costs of the most expensive element of sensornet simulation, modelling effects of radio propagation on low-level communications activity, is reduced substantially.

Objective 2: *Measure the extent to which these optimising techniques improve performance and enable larger-scale simulation*

The proposed optimisations transform the quadratic relationship between simulated network size and simulation runtime to near-linear for networks of the order of 1000 nodes, and that increased time savings are obtained as network size and network density increase.

Objective 3: *Improve the range of simulated network measures made available to the investigator*

Offline analysis performed once at the end of simulation activity can produce any measure that is producable by online analysis performed continually during simulation but with reduced overhead. Offline analysis also enables production of analytical results not foreseen prior to commencement of simulation activity.

Objective 4: *Validate optimised simulation to ensure that performance gains do not sacrifice solution quality*

Simulated results are as predicted by theoretical analysis, and as demonstrated by other simulators. Simulation results obtained under optimised simulation match closely those obtained for unoptimised simulation.

Acknowledgements

The authors would like to thank the anonymous reviewers for the helpful and constructive criticism which was invaluable in the preparation of this document.

References

[1] G. Amdahl. Validity of the single processor approach to achieving large scale computing capabilities. In *Proceedings of the AFIPS Spring Joint Computer Conference*, pages 483–485, Atlantic City, April 1967.
[2] A. Sobeih, W. Chen, J. Hou, L. Kung, N. Li, H. Lim, H. Tyan, and H. Zhang. J-Sim: A simulation environment for wireless sensor networks. In *Annual Simulation Symposium*, pages 175–187, 2005.
[3] R. Bagrodia and R. Meyer. PARSEC: A Parallel Simulation Environment for Complex Systems. *IEEE Computer*, pages 77–85, October 1998.
[4] S. Bajaj, L. Breslau, D. Estrin, K. Fall, S. Floyd, P. Haldar, M. Handley, A. Helmy, J. Heidemann, P. Huang, S. Kumar, S. McCanne, R. Rejaie, P. Sharma, K. Varadhan, Y. Xu, H. Yu, and D. Zappala. Improving simulation for network research. Technical Report 99-702, USC Computer Science Dept., March 1999.
[5] L. Breslau, D. Estrin, K. Fall, S. Floyd, J. Heidemann, A. Helmy, P. Huang, S. McCanne, K. Varadhan, Y. Xu, and H. Yu. Advances in network simulation. *Computer*, 33(5):59–67, 2000.
[6] D. Cavin, Y. Sasson, and A. Schiper. On the accuracy of MANET simulators. In *Proceedings of the Second ACM International Workshop on Principles of Mobile Computing*, pages 38–43, 2002.
[7] B. Claise. RFC 3954: Cisco Systems NetFlow Services Export Version 9. Downloaded from http://www.ietf.org/rfc/rfc3954.txt (checked 12/06/2008), 2004.
[8] Crossbow Technology Inc. MICA2 datasheet, part number 6020-0042-08 Rev A.
[9] B. Das and V. Bharghavan. Routing in ad-hoc networks using minimum connected dominating sets. In *IEEE International Conference on Communications (ICC)*, pages 376–380, 1997.
[10] S. Das, C. Perkins, and E. Royer. Performance comparison of two on-demand routing protocols for ad hoc networks. In *INFOCOM (1)*, pages 3–12, 2000.
[11] C. Fullmer and J. Garcia-Luna-Aceves. Solutions to hidden terminal problems in wireless networks. In *SIGCOMM'97*, pages 39–49, New York, NY, USA, 1997. ACM.
[12] D. Ganesan, B. Krishnamachari, A. Woo, D. Culler, D. Estrin, and S. Wicker. Complex behavior at scale: An experimental study of low-power wireless sensor networks. Technical Report CSD-TR 02-0013, UCLA, February 2002.

[13] E. Goturk. Emulating ad hoc networks: Differences from simulations and emulation specific problems. In *New Trends in Computer Networks*, volume 1. Imperial College Press, October 2005.
[14] Z. Haas, J. Halpern, and L. Li. Gossip-based ad hoc routing. *IEEE/ACM Transactions on Networking*, 14(3):479–491, 2006.
[15] M. Harris. *GPU Gems 2*, chapter 31: Mapping Computational Concepts to GPUs, pages 493–508. Addison Wesley, 2005.
[16] T. Henderson, S. Roy, S. Floyd, and G. Riley. ns-3 project goals. In *WNS2 '06: Proceedings of the 2006 workshop on ns-2: the IP network simulator*, pages 13–20, New York, NY, USA, 2006. ACM Press.
[17] C. Hoare. *Communicating Sequential Processes*. Prentice Hall International, 1985.
[18] P. Huang, D. Estrin, and J. Heidemann. Enabling large-scale simulations: Selective abstraction approach to the study of multicast protocols. In *MASCOTS*, pages 241–248, 1998.
[19] D. Johnson, D. Maltz, and J. Broch. *DSR: The Dynamic Source Routing Protocol for Multihop Wireless Ad Hoc Networks*, chapter 5, pages 139–172. Addison-Wesley, 2001.
[20] D. Kotz, C. Newport, and C. Elliott. The mistaken axioms of wireless-network research. Technical Report TR2003-467, Dept. of Computer Science, Dartmouth College, July 2003.
[21] J. Kulik, W. Heinzelman, and H. Balakrishnan. Negotiation-based protocols for disseminating information in wireless sensor networks. *Wireless Networks*, 8(2-3):169–185, 2002.
[22] K. Kundert. Introduction to RF simulation and its application. *IEEE Journal of Solid-State Circuits*, 34(9):1298–1319, Sep 1999.
[23] O. Landsiedel, K. Wehrle, B. Titzer, and J. Palsberg. Enabling detailed modeling and analysis of sensor networks. *Praxis der Informationsverarbeitung und Kommunikation*, 28(2):101–106, 2005.
[24] J. Lehnert, D. Gsrgen, H. Frey, and P. Sturm. A scalable workbench for implementing and evaluating distributed applications in mobile ad hoc networks. In *Proceedings of Mobile Ad Hoc Networks, Western Simulation MultiConference WMC'04*, 2004.
[25] P. Levis, N. Lee, M. Welsh, and D. Culler. TOSSIM: Accurate and scalable simulation of entire TinyOS applications. In *SenSys '03: Embedded Network Sensor Systems*, pages 126–137, 2003.
[26] S. Naicken, A. Basu, B. Livingston, and S. Rodhetbhai. Towards yet another peer-to-peer simulator. In *Performance Modelling and Evaluation of Heterogeneous Networks (HET-NETs) 2006*, 2006.
[27] V. Naoumov and T. Gross. Simulation of large ad hoc networks. In *MSWIM '03: Modeling analysis and simulation of wireless and mobile systems*, pages 50–57, New York, NY, USA, 2003. ACM Press.
[28] C. Perkins and E. Royer. Ad-hoc On-demand Distance Vector routing. In *Proceedings of Second IEEE Workshop on Mobile Computer Systems and Applications*, pages 90–100, New Orleans, LA, Feb 1999.
[29] L. F. Perrone and D. M. Nicol. A scalable simulator for TinyOS applications. In *Proceedings of the 2002 Winter Simulation Conference (WSC'02)*, volume 1, pages 679–687, 2002.
[30] D. Rao and P. Wilsey. Modeling and simulation of active networks. In *Proceedings of the 34th Annual Simulation Symposium (SS01)*, pages 177–184, Washington, DC, USA, 2001. IEEE Computer Society.
[31] D. Rao and P. Wilsey. Multi-resolution network simulations using dynamic component substitution. In *Proceedings of the Ninth International Symposium in Modeling, Analysis and Simulation of Computer and Telecommunication Systems*, pages 142–149, Washington, DC, USA, 2001. IEEE Computer Society.
[32] G. Riley, R. Fujimoto, and M. Ammar. A generic framework for parallelization of network simulations. In *MASCOTS'99*, pages 128–144, 1999.
[33] A. Sobeih, M. Viswanathan, and J. Hou. Check and simulate: A case for incorporating model checking in network simulation. In *Proc. of ACM-IEEE MEMOCODE'04*, pages 27–36, June 2004.
[34] B. Titzer, D. Lee, and J. Palsberg. Avrora: scalable sensor network simulation with precise timing. In *Proceedings of Information Processing in Sensor Networks (IPSN'05)*, pages 67–72. IEEE Press, 2005.
[35] A. Vahdat, K. Yocum, K. Walsh, P. Mahadevan, D. Kostic, J. Chase, and D. Becker. Scalability and accuracy in a large-scale network emulator. In *Proc. of the 5th symposium on Operating Systems Design and Implementation*, pages 271–284, New York, NY, USA, December 2002. ACM.
[36] K. Walsh and E. Sirer. Staged simulation: A general technique for improving simulation scale and performance. *ACM Transactions on Modeling and Computer Simulation*, 14(2):170–195, 2004.
[37] A. Schiper Y. Sasson, D. Cavin. Probabilistic broadcast for flooding in wireless mobile ad hoc networks. In *Proceedings of IEEE Wireless Communications and Networking (WCNC03)*, volume 2, pages 1124–1130, March 2003.
[38] X. Zeng, R. Bagrodia, and M. Gerla. GloMoSim: a library for parallel simulation of large-scale wireless networks. In *Proceedings of the 12th workshop on Parallel and Distributed Simulation*, pages 154–161, Washington, DC, USA, 1998. IEEE Computer Society.

Modelling a Multi-Core Media Processor Using JCSP

Anna KOSEK [a], Jon KERRIDGE [a] and Aly SYED [b]

[a] *School of Computing, Napier University, Edinburgh, EH10 5DT, UK*
[b] *NXP Semiconductors Research, Eindhoven, The Netherlands*

Abstract. Manufacturers are creating multi-core processors to solve specialized problems. This kind of processor can process tasks faster by running them in parallel. This paper explores the usability of the Communicating Sequential Processes model to create a simulation of a multi-core processor aimed at media processing in hand-held mobile devices. Every core in such systems can have different capabilities and can generate different amounts of heat depending on the task being performed. Heat generated reduces the performance of the core. We have used mobile processes in JCSP to implement the allocation of tasks to cores based upon the work the core has done previously.

Keywords. JCSP, multi-core processor, simulation, task allocation.

Introduction

Many manufacturers of semiconductor computer processors are designing multi-core systems these days [1]. In multi-core processing systems, allocation of work to processors can be seen as similar to the task of allocating work to people in a human society. A person responsible for controlling this process has to know the abilities of their employees and estimate the time in which a task can be finished. Tasks can often be finished faster if more workers are assigned to work on them. Generally, tasks can also be finished faster if they can be divided into smaller sub-tasks and sub-tasks can be processed concurrently. One very important condition that has to be met is that these sub-tasks have to be allocated wisely so that co-workers working on different sub-tasks can not hinder each other's progress. The manager has to allocate the task to the worker that is the best for the assignment in current circumstances. Using this idea many contemporary scientists and engineers are building multi-core processing systems.

Multi-core processor technology is one of the fastest developing hardware domains [2]. Modern personal computers already have multiple computing cores to increase a computer's performance. Multi-core systems for consumer electronics however have different challenges than those in personal computers.

Targeted media processors have been a goal of research of many scientists. In paper [3] the authors are presenting a heterogeneous multiprocessor architecture designed for media processing. The multi-core architecture presented in [4] consist of three programmable cores specialized for frequently occurring media processing operations of higher complexity. Cores are fully programmable so they can be adapted to new algorithm developments in this field [4]. More advanced research was shown in [5] presenting heterogeneous multi-core processor capable of self reconfiguring to fit new requirements.

1. Heterogeneous Multi-core Systems in Consumer Electronics

In a heterogeneous multi-core system for a consumer electronics device, it is often not possible to predict what functions the user would like to perform at what time and what data the user needs to process. A system manager has to deal with incoming requests, divide them into tasks and allocate them to one or more cores. In a heterogeneous multi-core system, every core has different capabilities and also generates different amounts of heat depending on the task being performed. The heat generated not only reduces the performance of the core and if unchecked could result in a thermal runaway leading to incorrect behaviour. Heat generation also makes a hand-held device less comfortable to use. Power consumption reduction, associated with heat generation, is very important issues when considering handheld mobile devices, only the latter is considered in this project. Heat reduction can be mitigated by an appropriate allocation of tasks to cores. In paper [6] authors show that a choice of appropriate core and switching between cores in heterogeneous multi-core architecture can optimize functions of performance and energy.

Our premise is that the amount of processing required depends on the task content, which is not always known during allocation, Therefore the system management should dynamically allocate tasks to cores based upon their previous use. If a task can not be finished in a given time, allocate it to a different core. Envisaging how such a system can work with a real application is very difficult.

A hardware simulation is usually used to shorten the hardware design and development process [7]. Hardware simulations are undertaken to prove some assumptions or new ideas about hardware architecture without actually creating the hardware itself. Computer systems used in environments where software execution should meet timing limitations are defined as real-time systems [7]. The real-time behaviour of devices is very important for their correct functioning, especially in systems designed to render various forms of media such as audio and video.

The aim of simulation presented in this paper is to show that an appropriate allocation can be done and if, for some reason, tasks can not be accomplished a different fallback plan can be adopted. The simulation uses the capabilities of the Communicating Sequential Processes (CSP) model [8] to create parallel systems. CSP is a formal model describing concurrent systems that consist of processes working simultaneously and communicating with each other [8]. The CSP model enables real-time aspects by scheduling and priority handling [9]. The concept of a system in CSP comprises a set of processes running concurrently connected through channels to exchange data and control. It has been shown that if the principles of CSP model are preserved, a correct implementation of a system can be built [9]. We have used JCSP (Java-CSP) that is an implementation of CSP and provides libraries allowing the development of Java applications.

We have built a simulation framework to investigate processing and scheduling requirements on heterogeneous multi-core processors that will be used in future devices. We can simulate the performance of a chosen architecture running a real application.

2. Function of the Simulated System

The operation of the system is captured in Figure 1. The diagram in Figure 1 was drafted as part of a project specification [10] with NXP Semiconductors. The system receives a data stream comprising audio and video compressed broadcast data. A selected channel is decoded (*Channel decode*) to extract the audio-video multiplex of streams. The audio and video components are then de-multiplexed (*Demux*) and sent on to the audio and video decoders, respectively.

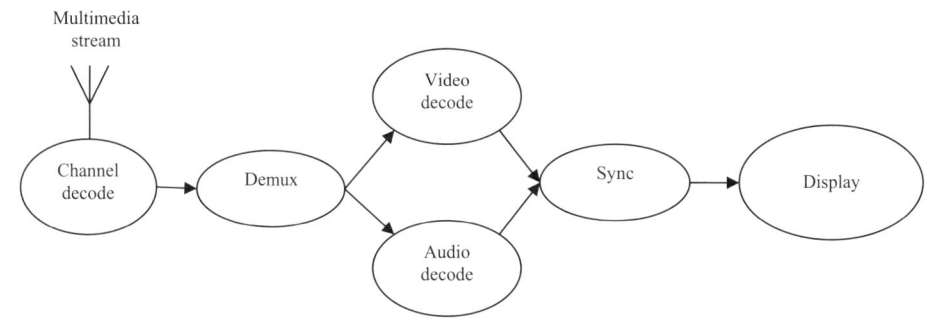

Figure1: de-multiplexing, decoding and re-synchronising broadcast multimedia data.

These decoders send their result streams to be re-synchronized for correct presentation. After this step, the display of the audio-video stream can take place, usually employing some form of rendering process. The modelled system assumes that cores of different capabilities are available.

The *Demux* process has the task of deciding which core is most appropriate to deal with the incoming task based upon: task requirements, core capability and the current state of the chosen core. The state of the core depends upon the amount of heat yet to be dissipated. The greater the residual heat, the slower the core will operate.

3. Subdividing Audio and Video Streams

As described in the section above, the system receives blocks of audio-video stream which are demultiplexed. We assign the demultiplexer also the function of dividing these streams into tasks and subtasks. These subtasks are then allocated to the different processing cores in the system. Figure 2 shows that multimedia data stream consists of blocks of data containing audio and video data. The demultiplexer then divides this block of data into different number of tasks depending on the amount of data in a block which in turn depends on the audio/video content. Blocks and tasks are numbered beginning from 0. Each task is then also divided into separate subtasks for audio and video. These subtasks can now be assigned to different cores.

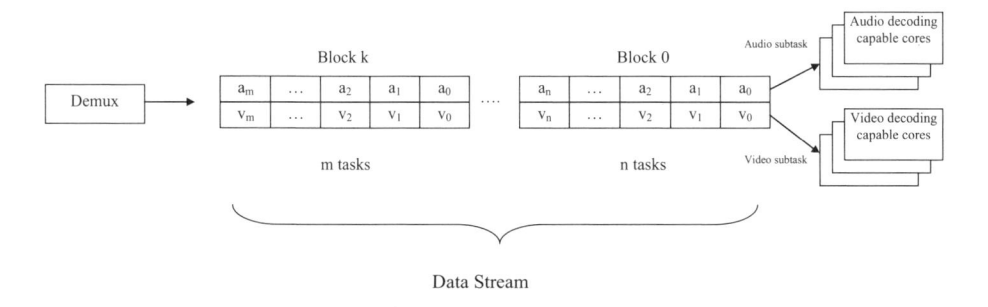

Figure 2: multimedia data stream.

In a multiplexed multimedia stream, it is unlikely that the audio-video subtasks will line up as regularly as indicated in Figure 2. An important aim of this project was to investigate the feasibility of allocating units of work to different cores. For the purpose of the designed architecture audio and video blocks always appear together.

A subtask contains the requested task type, the amount of work that has to be done and the data necessary for synchronization. Blocks are sent in sequence and the order of task synchronizing and displaying is vital. The order for synchronization is important simply because the video and audio tasks processed by the system have to match. Corresponding audio and video tasks can take a different amount of time to be processed, but have to be displayed at the same time.

4. Architecture of the Simulated System

Figure 3 shows the process architecture that has been implemented to test the feasibility of modelling a multi-core processor using JCSP and mobile processes. The system is designed to receive an audio-video data stream in a given structure (Fig. 2) consisting of blocks of data divided into audio and video subtasks. The system allocates subtasks to cores. The implemented system can run concurrently using any number of cores with different capabilities. In the present implementation, the number of cores is chosen to be nine; however, the same architecture can be applied to an arbitrary number of cores. The capabilities of these cores are defined in Table 2. The spatial arrangement of cores on Figure 3 is not relevant.

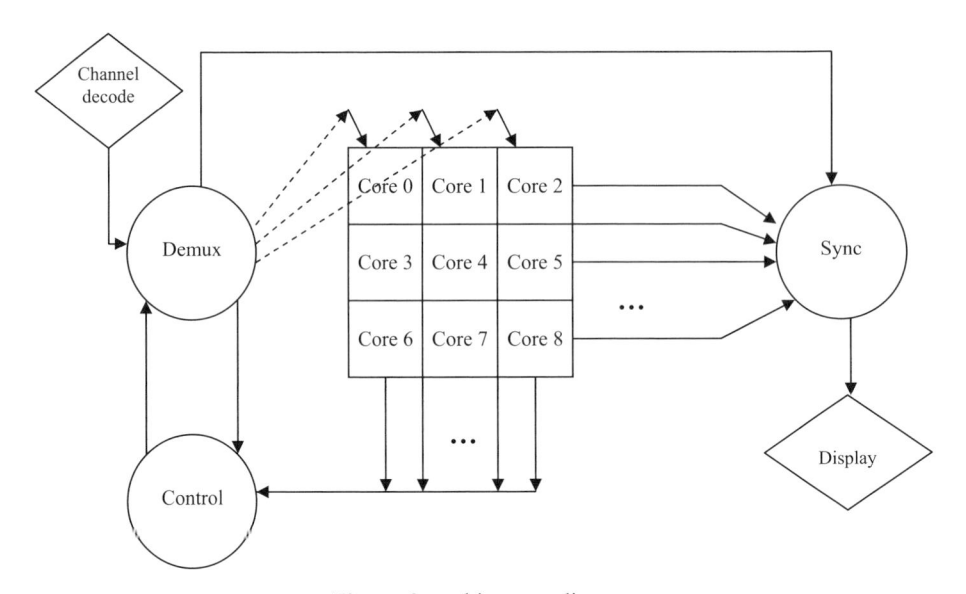

Figure 3: architecture diagram.

The system consists of processes running in parallel and connected with network channels (Fig. 3). This architecture better mimics the final design goal in that each process in the system is executed in its own JVM and hence all processes are running in parallel using a communication network implemented using network channels. This meant that we could more easily model network latency in future experiments. Every network channel is assigned a name which is unique and recognized by a CNS Server process, from the JCSP package [11], running in the background.

In addition to the basic concepts of CSP as discussed in section 1, we base the architecture in particular on the concept of mobile processes in CSP to achieve correct scheduling of subtasks on the different cores in the system. We consider every subtask as defined in section 3 to be a mobile process that can be sent to any core in the system for execution. A scheduling of such a system is considered correct if it makes optimal use of the many cores in the system with differing capabilities.

The Channel Decode process is responsible for receiving the audio-video stream, forming it into separate blocks of audio and video data representing subtasks and sending them to the Demux process which is also responsible for scheduling them to run on different cores in the system. The Demux has a variety of cores with differing capabilities available to choose from. The decision of where to send subtask data depends on capabilities of the cores and their availability. In order to make this decision, the Demux process sends a message to the Control process with a request to identify a core where a subtask should be allocated.

The capabilities of a core are dynamic and can change when a core heats up. Therefore every time a core changes its capabilities it informs the Control process which in turn updates its knowledge about the capabilities of each core. Thus when a Control process receives a message from the Demux process that also identifies the type of subtask, the Control process makes a decision about which core to assign the data for processing based on the capabilities and availability of the cores in the system at that moment in time. The Control process then sends a message to the Demux process identifying the suitable core.

When the Demux process receives the identity of the suitable core from the Control process, it creates a mobile process containing the data structure, shown in Table 1 and sends the mobile process to the designated core. To do this, a network channel is created dynamically using the core's location and CNS Server. The connection is established only once and when the core is selected for another subtask, the channel will be used again. Only the necessary connections between Demux and core processes are created

The Demux also sends a message to the Sync process informing it of the order in which it should receive data from cores in order to re-synchronise the video and audio subtasks contained within a block.

All of the cores in the system are connected to the Sync process. This process is responsible for synchronizing data in the right order, first audio then video subtasks from blocks. The subtasks are processed on cores in parallel so the Sync process might not get them in the right order. Therefore the Sync process is informed by the Demux about a sequence of cores used for processing data. The Sync waits for audio and video parts of one task and waits for a message only from those two cores that were processing corresponding subtasks. When the connection takes place the merged task is sent to the Display process.

In the simulation, a core processor is responsible for the required processing of a subtask but the time taken, is dependent upon the core's specific capabilities and also its present heat content which can lower its processing capabilities. The core then determines the processing required for the subtask and using its own look-up table determines the factor that will be applied to a subtask to determine the actual time taken to process a subtask's data. The core then waits for that number of time units (e.g. milliseconds) before sending the updated mobile process to the Sync process.

To make the multi-core processor simulation more realistic the core that executes a subtask gets heated. Heat gained changes a core's capability, increasing the time to process a task. If a core doesn't process any data it loses heat until it reaches its initial state. Decreasing temperature also affects a core's capability and makes it work more quickly.

The system uses the properties of CSP channels and avoids including a process that will be a buffer holding subtasks that have been processed, therefore finished subtasks wait on cores for the Sync process to be ready to read them.

4.1 Data Structure and Mobility

As described in section 3, the incoming stream of data is divided into blocks, tasks and audio and video subtasks. A data structure described in this section is designed to describe these entities and to track their progress through the system.

The data structure used by a mobile process is shown in Table 1; all values are integer type. The input stream is a sequence of blocks (**b**). Each block is divided into tasks (**t**) each of which comprises a subtask that can be processed by a core. Subtasks are separated into audio and video categories (**c**). Each subtask is given a required processing type (**r**) representing the nature of the processing required. The amount of work required to process the subtask is specified as the time (**w**) taken to process on an ideal core. An ideal core is one that can undertake the requested subtask in the minimum possible time and that has not been previously used and is thus at minimum possible heat content. The output variable (**o**) is the actual time to process the subtask on the core to which it was allocated. The data structure is designed to describe the subtask and to keep information about it and update it in the system.

Table 1: structure of data carried in the system.

Name	Description	Special values
b	Block number	Starts from 0
t	Task number	Starts from 0
c	Subtask category	Audio = 0 Video = 1
r	Requested subtask type	Starts from 0
w	Amount of work of requested subtask type (units of work)	Starts from 1
o	Variable reserved for actual core performance (outcome) time needed to execute requested subtask	By default equals 0, but changes after subtask processing

The value of **o** is determined as follows:

$$o = F[r] \cdot w$$

Each core process has a table **F** such that each element of **F** contains a factor (greater than 1) which is used to determine the time it takes to process a subtask of a particular requested subtask type **r**. A core's capabilities change dynamically so entries in the **F** table will not always be the same for the same core. Therefore the output variable (**o**) holds the actual time that the core needed to process a subtask and it is updated after a subtask's execution.

Heat management is essential to avoid thermal runaways and can be done by proper dynamic scheduling of tasks so that cores do not heat up beyond a certain point. The mechanism helps overheated cores to cool down by allocating tasks to other cores. However, this has to be done in such a manner that real-time constraints are maintained. This aspect was not considered in the work represented in this paper. If a task is sent to a core that is not ideal, because the ideal core is already hot, then the time to process the task may be longer and hence the chosen core may heat up more than would have occurred if the ideal core had been available. In a real system heat can be transferred between cores due to their adjacency. This aspect was not considered in this project.

4.2 Mobile Processes

A mobile process can visit a processing node and perform operations using processing nodes resources [12]. The mobile process interface defines two methods *connect* and *disconnect* that allow the mobile process to connect to a node's channels and disconnect when it has completed its task. Mobile processes exist in one of two states: active and passive. In the active state it can carry out some instructions and write and read from the node it is connected to. In the passive state a mobile process can be moved or activated. When a mobile process becomes passive it can be suspended or terminated. After suspension a mobile process saves its state and when it is reactivated starts from the same state. When a mobile process is terminated it may not be reactivated. When a mobile process is activated by some other process it starts working in parallel with it, channels between those two processes can be created to achieve communication.

The data processed by the simulation is carried by the mobile processes. The system uses mobility of processes to track information about every subtask. One mobile process is responsible for carrying only one subtask and storing information about it. The mobile processes are created in the Demux process and initialised with values representing a subtask (Table 1). The mobile process is disconnected from the Demux process and sent to the appropriate core through a network channel. The mobile process goes into a passive state and after it arrives at the core it has to be reactivated. The mobile process can be connected and communicates with the core process exchanging data and capturing information necessary to evaluate the core performance. The mobile process is next sent to the Sync process and connected to it to retrieve data about the task and the time in which it was processed on the core.

One objective of the project was to determine the feasibility of using the mobile process concept as a means of distributing work in such a heterogeneous multi-core environment. The idea being that a complete package comprising process and data might be easier to move around the system, especially if there is a threat that the real-time constraints may be broken and the task has to be moved to a faster core that has now become available.

Tracking information about a subtask can be performed using mobile processes as well: a process can be sent over a network that can connect to other processes, exchange data, and update its resources. This function of the system can be performed using mechanisms to run, reactivate, suspend and send processes over the network included in JCSP package.

5. Results

The simulation system as described above was built using a standard Laptop computer. In this section, we provide results.

5.1 System Performance

We have verified the simulation model by doing some experiments that confirm the correctness of its behaviour.

These experiments are:

- We have used different data streams to explore system performance to see if the stream content makes a difference to the system behaviour.
- Two different instances of the system should show the same functional behaviour if the input data streams are identical but the number and characteristics of processors in the system is changed.
- We have verified that the system output remains unchanged, although the order of communication between processes can vary.

Table 2: capabilities of the cores.

Core number:	Requested type:	Time to process the requested type:
0	0	1
	3	1
	10	1
1	3	1
	4	2
	8	3
	12	2
	16	1
	19	1
2	1	10
	7	10
	13	5
3	5	1
	6	5
	11	3
	14	1
	15	1
	17	2
4	0	10
	1	10
	…	…
	19	10
5	0	3
	9	2
	10	1
	12	1
	13	1
	14	1
6	6	1
	7	2
	8	10
7	0	2
	3	4
	5	1
	6	1
	10	7
	11	9
8	15	1

Some results are shown in tables with a short description of the system state and different data streams used to explore the system's performance. All data sets are represented as follows:

- Number – number of subtask for testing
- Block – data block number
- Task – task number

- A/V – type of subtask, audio or video
- Type – type of requested subtask
- Unit – units of work of requested subtask type

Data used to show the system's performance:

- Expected value
- Actual value

All cores have different capabilities, therefore the system will choose a core with the best suited capabilities for a given subtask type. *Expected value* is evaluated by multiplication of a core's capabilities and units of work needed for the requested subtask type. For evaluating the *Expected value* the core capabilities are listed (Table 2), both *Expected* and *Actual values* are in milliseconds. When a core starts to process the subtask it cannot be interrupted. Both expected and actual values do not take into account the total time that subtasks have been in the system. The time to send a subtask to a different core is excluded.

There are 20 different possible request types. Core 4 can perform all types but at a slow speed and was created to simulate a slow general-purpose processor. Core 8 on the other hand can perform only one task type but is very fast.

Three scenarios are presented to evaluate the system's performance using different sets of data and various numbers of cores running in parallel. The input data is randomly chosen before simulation execution, therefore the system can be run many times with the same blocks of experimental data; requested types are drawn from $\langle 0,19 \rangle$. We define that cores can process only 20 types of tasks. This number is defined for this simulation, but it can be easily modified. The scenarios show how the system works with different sets of cores and various data streams.

5.1.1 Scenario 1

The first Scenario explores the system's performance with a number of data subtasks. The data stream consists of many types of tasks. There are two blocks in the stream: the first consists of 6 subtasks and the second has 8 subtasks. All of the main system processes and all of the cores are running.

Table 3: results of scenario 1.

Number	Block	Task	A/V	Type	Unit	Expected value:	Actual value:
0	0	0	A	0	50	50	50
1	0	0	V	10	100	100	100
2	0	1	A	6	100	100	100
3	0	1	V	19	200	200	200
4	0	2	A	4	70	140	140
5	0	2	V	11	200	600	600
6	1	0	A	9	50	100	100
7	1	0	V	14	100	100	100
8	1	1	A	7	100	100	100
9	1	1	V	19	200	200	200
10	1	2	A	6	70	70	70
11	1	2	V	10	200	200	200
12	1	3	A	1	70	350	350
13	1	3	V	12	200	200	200
				Total:	**1910**	**2510**	**2510**

The results of this simulation are shown in table 3. Table 3 shows that the system reacts as expected; subtasks are allocated to cores with the best capabilities for a particular subtask. It shows that each subtask was processed in the minimum possible time because they were sent to different cores and previously used cores had sufficient time to cool down between uses.

5.1.2 Scenario 2

The second Scenario explores system's performance with the same data subtasks as those of Scenario 1. Only three cores are used in the system. Therefore all of the main system processes are running and only cores 1, 4 and 7 are running.

Table 4: results of scenario 2.

Number	Block	Task	A/V	Type	Unit	Expected value:	Actual value:
0	0	0	A	0	50	100	100
1	0	0	V	10	100	700	700
2	0	1	A	6	100	100	100
3	0	1	V	19	200	200	200
4	0	2	A	4	70	140	140
5	0	2	V	11	200	200	200
6	1	0	A	9	50	500	500
7	1	0	V	14	100	1000	1000
8	1	1	A	7	100	1000	1000
9	1	1	V	19	200	200	200
10	1	2	A	6	70	70	70
11	1	2	V	10	200	1400	1400
12	1	3	A	1	70	700	700
13	1	3	V	12	200	400	400
				Total:	1910	6710	6710

The results of this scenario are shown in Table 4. We observe that the expected values also equal actual values as one would expect for this scenario. The processing time in this case has increased as compared to the scenario 1 in accordance with expectation as lesser number of cores is used to perform the same task. The functional result of the system was verified to be correct.

5.1.3 Scenario 3

This scenario is designed to show how the system will perform when the heating effect of cores is taken into account. In this scenario the data stream is designed in a way that only two cores are used, although all of the available cores are running. This is because only core 2 and 4 can run task type 1 and core 4 can only handle task type 18.

The table of results (Table 5) shows that for subtasks with numbers 0-3 *Expected value* equals *Actual value*. From subtask number 4 the *Actual value* increases. This happens because only three cores are used and they heat up every time a subtask is processed. The heat influences the core's capabilities and the actual value rises. It should be noted that while calculating the *Expected values* in the table, the heating up effect is not taken into account.

This scenario demonstrates how the system reacts on allocating tasks always to the same cores. Allocating tasks only to one core can decrease the system's performance.

Those three test cases were chosen to show how the system works with different sets of cores. In Scenario 1 and 2 data blocks are the same, but the number of cores has decreased. In Scenario 1 total of actual values is 2510 milliseconds where in Scenario 2 equals 6710 milliseconds. This difference is caused by number of cores used. The third Scenario shows how overheated cores decrease the system's performance.

Table 5: results of scenario 3.

Number	Block	Task	A/V	Type	Unit	Expected value:	Actual value:
0	0	0	A	1	20	200	200
1	0	0	V	18	40	400	400
2	0	1	A	1	100	1000	1000
3	0	1	V	18	130	1300	1300
4	0	2	A	1	10	100	110
5	0	2	V	18	50	500	700
6	1	0	A	1	110	1100	2090
7	1	0	V	18	300	3000	8700
8	1	1	A	1	10	100	5100
9	1	1	V	18	20	200	1740
				Total:	790	7900	21340

6. Conclusions

We have presented an architecture to simulate a heterogeneous multi-core media processor. In a multi-core processor the most important aspect is to allocate tasks depending on a core's capability and to run tasks simultaneously. The software was designed to simulate work of many cores in a single processor. Task scheduling and allocation is targeted to efficient use of available heterogeneous cores. The process of allocating tasks was designed and implemented with most of the required functionalities. The system assignment is to receive a data stream, divide it into audio and video streams, process and synchronise both data streams to enable display.

The designed processor consists of cores designed to be separate processes and have different capabilities. Tasks are allocated to cores and run simultaneously achieving faster overall performance. Because of the variety of capabilities some cores are used more frequently to process some tasks. The simulation can model heat generation, its influence on cores capabilities and dissipation of heat. Task allocation is designed to reduce heat generation.

The system is built using CSP principles and consists of processes running in parallel responsible for dealing with data streams, allocating tasks, synchronising data and simulating task execution. A parallel architecture can better reflect the dynamics of the real world, and can be used to model real-time systems. The simulation captures the desired operational characteristics and requirements in terms of the utility of JCSP to model the internal dynamics of a heterogeneous multimedia processor. The JCSP mobile processes were very useful when building concurrent real-time systems. The mobile processes are used to distribute work to cores. Mobility of processes and the ability to run in parallel with other processes are the main capabilities used to build a simulation of the multi-core media processor. The CSP model can also be used to build simulations of other equipment that can help test new ideas and define problems before creating the hardware itself. In particular, the architecture did not involve the use of any buffers between system components. Modern processors often use buffers between system components to improve parallelism. In this design we used the underlying non-buffered CSP communication concept whereby data is held at the output end of a channel until the input end is ready.

7. Future Work

To meet further requirements the final version of the system needs to be extended with additional functionality. The most important system function, that was not included in prototype, is dealing with core failure. Cores can fail in two ways: either stop working or be unable to complete a subtask in time.

Cores can stop working for many reasons. When this happens a core can inform the Controller about its failure. In this case the Controller can stop allocating tasks to this core. The core can be reactivated at any time and send the Controller a signal that it is ready to process tasks. There can be unexpected core failures, where the core will not have time to inform the Controller. To avoid a complete system crash the cores could be scanned periodically. A simple signal may be sent to the core and, if it is still working, in some previously defined time it will respond. This operation can be done by the Controller. For example, every 10 seconds, the Controller can request a response from all of the cores and erase the table responsible for storing cores' capabilities. To ensure the system deals with this kind of problem functions should be added to both Controller and core processes.

In the case where a core cannot complete the task in time, the core should send appropriate information to the Controller. This would also require an additional element in the subtask data structure (Table 1) to include a maximum allowable delay.

In both core failure cases the subtask should be sent back to the Demux to be allocated again. Of course if the subtask allocation is changed the Sync process has to be informed. The Demux process has to send revised information about the sequence in which the Sync process should receive subtasks from cores. To deal with core failure all processes in the system would need to be rewritten.

The system is designed to receive subtasks of a type that can be processed in at least one core. If cores with capabilities suitable for a particular subtask stop working the Controller will not allocate the subtask. This behaviour is recognized by the current version of the system, but functions to deal with this problem are not included. The best way to solve this problem is to make the Demux repeat the request to the Controller about task allocation until some of the cores become available. Of course this loop of requests cannot last forever; there should be other conditions preventing the system from livelock.

In the current version of the system the Channel Decode process initializes the data stream directly with subtasks. This causes confusion and makes testing the system more difficult. In the final system, data should be read from an external file or a channel input. An external file can be for example an XML file with information about blocks of tasks.

One of the system functions that should be also added is to include interconnect latency into the model based upon the size of the subtask. This function would increase simulation accuracy. To make the simulation easier to use there needs to be a user interface added to the system instead of displaying results and process states in a console window.

References

[1] May, D., Processes Architecture for Multicores. Communicating Process Architectures, 2007.
[2] Ramanathan, R.M., Intel® Multi-Core Processors, Making the Move to Quad-Core and Beyond.
[3] Rutten, M.J., et al., Eclipse: Heterogeneous Multiprocessor Architecture for Flexible Media Processing. International Parallel and Distributed Processing Symposium: IPDPS 2002 Workshops, 2002.
[4] Stolberg, H.-J., et al., HiBRID-SoC: A Multi-Core System-on-Chip Architecture for Multimedia Signal Processing Applications. Design, Automation and Test in Europe Conference and Exhibition, 2003.
[5] Pericas, M., et al., A Flexible Heterogeneous Multi-Core Architecture. Proceedings of the 16th International Conference on Parallel Architecture and Compilation Techniques, 2007: pp. 13-24.

[6] Kumar, R., et al. Single-ISA Heterogeneous Multi-Core Architectures: The Potential for Processor Power Reduction. in 36th Annual IEEE/ACM International Symposium on Microarchitecture (MICRO'03) 2003.
[7] Shobaki, M.E., Verification of Embedded Real-Time Systems Using Hardware/Software Co-simulation. IEEE Computer Society, 1998.
[8] Hoare, C.A.R., Communicating Sequential Processes. 1985: Prentice Hall International Series in Computer Science.
[9] Bakkers, A., G. Hilderink, and J. Broenink, A Distributed Real-Time Java System Based on CSP. Architectures, Languages and Techniques, 1999.
[10] NXP, Private communication concerning the project specification. 2007.
[11] Welch, P.H., J.R. Aldous, and J. Foster. CSP Networking for Java (JCSP.net). in International Conference Computational Science - ICCS 2002. 2002. Amsterdam, The Netherlands: Springer Berlin / Heilderberg.
[12] Chalmers, K., J. Kerridge, and I. Romdhani, Mobility in JCSP: New Mobile Channel and Mobile Process Models. Communicating Process Architectures, 2005.

How to Make a Process Invisible

Neil C.C. BROWN

Computing Laboratory, University of Kent

`neil@twistedsquare.com`

Abstract. Sometimes it is useful to be able to invisibly splice a process into a channel, allowing it to observe (log or present to a GUI) communications on the channel without breaking the synchronous communication semantics. occam-*pi*'s extended rendezvous when reading from a channel made this possible; the invisible process could keep the writer waiting until the real reader had accepted the forwarded communication. This breaks down when it is possible to have choice on outputs (also known as output guards). An extended rendezvous for writing to a channel fixes this aspect but in turn does not support choice on the input. It becomes impossible to keep your process invisible in all circumstances. This talk explains the problem, and proposes a radical new feature that would solve it.

Outline

An occam-π extended rendezvous on a synchronous channel allows a reader to delay a writer (after the data has been communicated) until the reader has also performed an extended action. This can be used to create an invisible process (α in the diagram below), that reads from one channel, c, and writes to another, d, as part of an extended action. The processes writing to channel c (process P) and reading from channel d (process Q) have synchronous communication semantics, as if c and d were the same channel, despite the process α in between (Frigure 1).

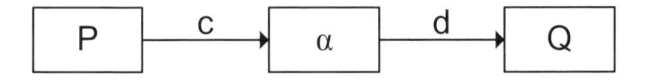

Figure 1. The middle process is not there.

This only holds if choice is available on inputs, but outputs are committed. If P offers an output on c as part of a larger choice, the supposedly-invisible process α will accept the communication immediately and wait for a reader on d during the extended action. Thus P will engage in a communication on c even if Q never reads from d, breaking the synchronous semantics and revealing the presence of the intermediate process α.

An extended output (where a writer performs an extended action after the reader has arrived but before the data is communicated) solves this part of the problem; the invisible process α waits to write on channel d, but then performs an extended action to read from c. But if Q is also involved in a choice, this again breaks the synchronous channel semantics; process P may never write to the channel c, but Q will have chosen to read from d anyway.

We believe that this problem cannot be solved without the introduction of a new feature: the ability to wait for multiple actions. The invisible process α must wait for *both* the reading end of c *and* the writing end of d to be ready, and must perform an extended action on *both* channels to forward the data across. Introducing this ability requires a different implementation and may have far-reaching consequences for how we can construct our process-oriented programs.

Designing Animation Facilities for gCSP

Hans T.J. VAN DER STEEN, Marcel A. GROOTHUIS and Jan F. BROENINK

*Control Engineering, Faculty EEMCS, University of Twente,
P.O. Box 217 7500 AE Enschede, The Netherlands.*

{T.T.J.vanderSteen,M.A.Groothuis,J.F.Broenink}@utwente.nl

Abstract. To improve feedback on how concurrent CSP-based programs run, the graphical CSP design tool (gCSP [3,2]) has been extended with animation facilities. The state of processes, constructs, and channel ends are indicated with colours both in the gCSP diagrams and in the composition tree (hierarchical tree showing the structure of the total program). Furthermore, the contents of the channels are also shown. In the Fringe session, we will present and demonstrate this prototype animation facility, being the result of the MSc project of Hans van der Steen [5], and ask for feedback.

Keywords. graphical CSP tools, IDE, code generation.

Outline

The CTC++ run time library [1,4] has been augmented, such that it generates the status information gCSP needs. The content of the ready queue of the scheduler is also made available.

The animation is controlled from within the gCSP design tool. Breakpoints and the animation speed can be set. Choosing the animation speed between about 0.2 and 1 s (i.e. time duration between two state changes of processes, constructs or channel ends) allows the user to follow the behaviour of the program. The execution history (the stream of status events coming from the running CSP program) is shown in a log window. The stream of status events can be filtered, to focus on those parts of the program one is interested in. The contents of the channels and of the ready queue are shown in a separate log window.

Tests were performed on the practical use of animation, the execution behavior of gCSP models and the C++ code generator of gCSP. Using a prototype version in our MSc class on real-time software development showed that this animation helps the students' understanding of concurrency. At least, significantly fewer questions were asked during the lab exercises.

References

[1] G.H. Hilderink, A.W.P. Bakkers, and J.F. Broenink. A Distributed Real-Time Java System Based on CSP. In *The third IEEE International Symposium on Object-Oriented Real-Time Distributed Computing ISORC 2000*, pages 400–407. IEEE, Newport Beach, CA, 2000.
[2] D.S. Jovanovic. *Designing dependable process-oriented software, a CSP approach*. PhD thesis, University of Twente, Enschede, NL, 2006.
[3] Dusko S. Jovanovic, Bojan Orlic, Geert K. Liet, and Jan F. Broenink. gCSP: A Graphical Tool for Designing CSP systems. In Ian East, Jeremy Martin, Peter H. Welch, David Duce, and Mark Green, editors, *Communicating Process Architectures 2004*, pages 233–251. IOS press, Oxford, UK, 2004.
[4] Bojan Orlic and Jan F. Broenink. Redesign of the C++ Communicating Threads library for embedded control systems. In Frank Karelse, editor, *5th PROGRESS Symposium on Embedded Systems*, pages 141–156. STW, Nieuwegein, NL, 2004.
[5] T.T.J. van der Steen. Design of animation and debug facilities for gCSP. MSc Thesis 020CE2008, University of Twente, 2008.

Tock: One Year On

Adam T. SAMPSON and Neil C.C. BROWN

Computing Laboratory, University of Kent

A.T.Sampson@kent.ac.uk, neil@twistedsquare.com

Abstract. Tock is a compiler for concurrent programming languages under development at the University of Kent. It translates occam-π and Rain into portable, high-performance C or C++. It is implemented in Haskell using the nanopass approach, and aims to make it easy to experiment with new language and compiler features. Since our initial presentation of Tock at CPA 2007, we have added new frontends and backends, implemented a parallel usage checker based on the Omega test, improved the effectiveness of Tock's test suite, developed more efficient tree traversals using generic programming – and more besides! In this fringe session, we will describe our recent work on Tock, discuss our plans for the project, and show how it can be of use to other process-oriented programming researchers.

Keywords. concurrency, compilation, generics, Haskell, nanopass, occam-π, Rain.

Introducing JCSP Networking 2.0

Kevin CHALMERS

School of Computing, Napier University, Edinburgh, EH10 5DT
`k.chalmers@napier.ac.uk`

Abstract. The original implementation of JCSP Networking is based on the T9000 model of virtual channels across a communications mechanism, and can be considered sufficiently adequate for applications which are not resource constrained or liable to connection failure. However, work undertaken has revealed a number of limitations due to excessive resource usage, lack of sufficient error handling, reliance on Java serialization, and reliance on now deprecated features of JCSP. These problems reflect badly when considering JCSP Networking in a broader sense beyond the normal desktop. In this talk, a brief overview on how these problems have been overcome is presented. This will be followed by some tutorial examples on how to use JCSP Networking 2.0. This should be familiar to current JCSP Networking users, but new additions to the library should make it easier for novices to get started. The new underlying protocol is also presented, which should enable interoperability between various platforms beyond the Java desktop environment. The new version of JCSP Networking is currently available from the JCSP Subversion repository, under the Networking-2 branch. Details are available at `http://www.cs.kent.ac.uk/projects/ofa/jcsp/`.

Keywords. JCSP, JCSP Networking, distributed systems, CSP.

Mobile Processes in an Ant Simulation

Eric BONNICI

*Computing Laboratory, University of Kent,
Canterbury, Kent, CT2 7NF, England.*
eb708@kent.ac.uk

Abstract. The term self-organisation, or emergent behaviour, may be used to describe behaviour structures that emerge at the global level of a system due to the interactions between lower level components. Components of the system have no knowledge about global state; each component has only private internal data and data that it can observe from its immediate locality (such as environmental factors and the presence of other components). Resulting global phenomina are, therefore, an emergent property of the system as a whole. An implication of this when creating artificial systems is that we should not attempt to program such kinds of complex behaviour *explicitly* into the system. It may also help if the programmer approaches the design from a radically different perspective than that found in traditional methods of software engineering. This talk outlines a process-oriented approach, using massive fine-grained concurrency, and explores the use of occam-π's mobile processes in the simulation of a classical ant colony.

Keywords. process orientation, mobile processes, complex systems, emergence.

Santa Claus – with Mobile Reindeer and Elves

Peter H. WELCH [a] and Jan B. PEDERSEN [b]

[a] *Computing Laboratory, University of Kent*
[b] *School of Computer Science, University of Nevada, Las Vegas*
p.h.welch@kent.ac.uk, matt@cs.unlv.edu

Abstract. Mobile processes, along with mobile channels, enable process networks to be dynamic: they may change their size (number of processes, channels, barriers) and shape (connection topology) as they run – much like living organisms. One of the benefits is that all connections do not have to be established statically, in advance of when they are needed and open to abuse. In classical occam, care had to be taken by processes not to use channels when they were not in the right state to use them. With occam-π mobiles, we can arrange that processes simply do not have those channels until they get into the right state – and not having such channels means that their misuse cannot even be expressed! Of course, it is a natural consequence of mobile system design that the arrivals of channels (or barriers or processes) are the very events triggering their exploitation. In our explorations so far with occam-π, we have taken advantage of the mobility of data, channels and barriers and seen very good results. Very little work has been done with mobile processes: the ability to send and receive processes through channels, plug them into local networks, fire them up, stand them down and move them on again. This talk illustrates mobile process design through a solution to Trønø's classical *Santa Claus Problem*. The reindeer and elves are modeled as mobile processes that move through holiday resorts, stables, work benches, waiting rooms, Santa's Grotto and back again. All those destinations are also processes – though static ones. As the reindeer and elves arrive at each stage, they plug in and do business. We will show the occam-π mechanisms supporting mobile processes, confess to one weakness and consider remedies. The occam-π solution did, of course, run correctly the first time it passed the stringent safety checks of the compiler and is available as open source (http://www.santaclausproblem.net).

Keywords. mobile processes, mobile design, safety, occam-π, Santa Claus.

Subject Index

20-sim	135	hardware process networks	163
active objects	237	hardware software co-design	219
asynchronous active objects	237	Haskell	67, 449
Bluetooth	255	IDE	447
channels	17	interaction	1
client	1	Java	237
client-server	85	JCSP	35, 255, 271, 431, 451
client-server protocol	99	JCSP Networking	451
clusters	309	JCSP.net	271
code generation	135, 447	LEGO NXT	255
communicating sequential processes	67, 329	LeJOS	255
		mobile agents	397
compilation	449	mobile design	455
complex systems	453	mobile devices	397
computational science	347	mobile processes	397, 453, 455
concurrency	35, 85, 293, 309, 347, 381, 449	mobility	17
		modeling	17
co-operative	149	monads	67
correctness-by-design	99	motion control	135
CSP	17, 67, 135, 163, 293, 309, 329, 347, 451	MPI	309
		MPSoC	149
CSP‖B	115	multi-core	149, 219, 431
deadlock-freedom	99	multiprocessor	149
debugging	293	nanopass	449
distributed computing	237	network on chip	219
distributed memory	381	networks	411
distributed systems	271, 451	occam	55
earliest deadline first	55	occam model	35
EDF	55	occam-π	17, 85, 293, 309, 449, 455
embedded systems	135, 219	orthogonality	1
emergence	453	parallel	255, 309, 347
eScience	347	pervasive adaptation	397
explicit concurrency	67	PID	135
FIFO queues	179	point-to-point communication	203
formal development	115	process orientation	453
FPGA	135	processes	17
gCSP	135	process-oriented programming	381
genericity	1	programming language	99
generics	449	protocols	85
graphical CSP tools	447	PVS	179
guard resolution	163	PyCSP	309
Handel-C	115, 135	Python	309, 347
hardware channel	219	Rain	449

real-time programming	55	SMP	347
relation matrix	163	social networking	397
reliability	203	system on chip	219
robot controllers	255	task allocation	431
safety	455	traces	329
Santa Claus	455	translation	115
Scala	35	type	1
scalability	411	ubiquitous computing	271, 397
scheduling	55	unstable sender	203
sensornets	411	VCR	329
serial link	219	View-Centric Reasoning	329
server	1	virtual machine	293
session types	85	visual design	99
shared-clock	149	visual programming	99
shared memory	381	workflow	347
simulation	411, 431	μCRL	179
sliding window protocols	179	π-calculus	17

Author Index

Abramsky, S.	1	McEwan, A.	v, 115
Allen, A.R.	163, 203, 219	O'Halloran, C.	15
Athaide, K.F.	149	Oprean, G.	237
Ayavoo, D.	149	Panayotopoulos, A.	255
Badban, B.	179	Pedersen, J.B.	237, 381, 455
Barnes, F.R.M.	v, 17	Polack, F.	v
Bate, I.	411	Pont, M.J.	149
Bjørndalen, J.M.	309	Ritson, C.G.	293
Bonnici, E.	453	Romdhani, I.	271
Broenink, J.F.	v, 135, 447	Sampson, A.T.	v, 85, 309, 449
Brown, N.C.C.	67, 329, 445, 449	Schneider, S.	115
Chalmers, K.	271, 397, 451	Simpson, J.	293, 365
East, I.R.	99	Smith, M.L.	329
Faust, O.	163, 203, 219	Sputh, B.H.C.	163, 203, 219
Fokkink, W.	179	Stepney, S.	v
Friborg, R.M.	347	Stiles, D.	v
Groothuis, M.A.	135, 447	Sufrin, B.	35
Haschke, J.-O.	397	Syed, A.	431
Hendseth, S.	55	Tate, J.	411
Hurt, J.	381	Treharne, H.	115
Ifill, W.	115	van de Pol, J.	179
Jacobsen, C.L.	365	van der Steen, H.T.J.	447
Kerridge, J.	255, 271, 397, 431	van Zuijlen, J.J.P.	135
Korsgaard, M.	55	Vinter, B.	347
Kosek, A.	431	Welch, P.H.	v, 17, 455
Lismore, P.	255		